Privacy-Preserving Data Mining

Models and Algorithms

ADVANCES IN DATABASE SYSTEMS

Volume 34

Series Editors

Ahmed K. Elmagarmid
Purdue University
West Lafayette, IN 47907

Amit P. Sheth
Wright State University
Dayton, Ohio 45435

Other books in the Series:

SEQUENCE DATA MINING, Guozhu Dong, Jian Pei; ISBN: 978-0-387-69936-3
DATA STREAMS: *Models and Algorithms*, edited by Charu C. Aggarwal; ISBN: 978-0-387-28759-1
SIMILARITY SEARCH: ***The Metric Space Approach***, *P. Zezula, G. Amato, V. Dohnal, M. Batko*; ISBN: 0-387-29146-6
STREAM DATA MANAGEMENT, *Nauman Chaudhry, Kevin Shaw, Mahdi Abdelguerfi*; ISBN: 0-387-24393-3
FUZZY DATABASE MODELING WITH XML, *Zongmin Ma*; ISBN: 0-387- 24248-1
MINING SEQUENTIAL PATTERNS FROM LARGE DATA SETS, *Wei Wang and Jiong Yang*; ISBN: 0-387-24246-5
ADVANCED SIGNATURE INDEXING FOR MULTIMEDIA AND WEB APPLICATIONS, *Yannis Manolopoulos, Alexandros Nanopoulos, Eleni Tousidou*; ISBN: 1-4020-7425-5
ADVANCES IN DIGITAL GOVERNMENT: *Technology, Human Factors, and Policy*, *edited by William J. McIver, Jr. and Ahmed K. Elmagarmid*; ISBN: 1-4020-7067-5
INFORMATION AND DATABASE QUALITY, *Mario Piattini, Coral Calero and Marcela Genero*; ISBN: 0-7923-7599-8
DATA QUALITY, *Richard Y. Wang, Mostapha Ziad, Yang W. Lee:* ISBN: 0-7923-7215-8
THE FRACTAL STRUCTURE OF DATA REFERENCE: *Applications to the Memory Hierarchy*, *Bruce McNutt*; ISBN: 0-7923-7945-4
SEMANTIC MODELS FOR MULTIMEDIA DATABASE SEARCHING AND BROWSING, *Shu-Ching Chen, R.L. Kashyap, and Arif Ghafoor*; ISBN: 0-7923-7888-1
INFORMATION BROKERING ACROSS HETEROGENEOUS DIGITAL DATA: A Metadata-based Approach, *Vipul Kashyap, Amit Sheth*; ISBN: 0-7923-7883-0
DATA DISSEMINATION IN WIRELESS COMPUTING ENVIRONMENTS, *Kian-Lee Tan and Beng Chin Ooi*; ISBN: 0-7923-7866-0
MIDDLEWARE NETWORKS: Concept, Design and Deployment of Internet Infrastructure, *Michah Lerner, George Vanecek, Nino Vidovic, Dad Vrsalovic*; ISBN: 0-7923-7840-7
ADVANCED DATABASE INDEXING, *Yannis Manolopoulos, Yannis Theodoridis, Vassilis J. Tsotras*; ISBN: 0-7923-7716-8
MULTILEVEL SECURE TRANSACTION PROCESSING, *Vijay Atluri, Sushil Jajodia, Binto George* ISBN: 0-7923-7702-8
FUZZY LOGIC IN DATA MODELING, *Guoqing Chen* ISBN: 0-7923-8253-6
PRIVACY-PRESERVING DATA MINING: Models and Algorithms, *edited by Charu C. Aggarwal and Philip S. Yu*; ISBN: 0-387-70991-8

Privacy-Preserving Data Mining

Models and Algorithms

Edited by

Charu C. Aggarwal
IBM T.J. Watson Research Center, USA

and

Philip S. Yu
University of Illinois at Chicago, USA

Springer

Editors:

Charu C. Aggarwal
IBM Thomas J. Watson Research Center
19 Skyline Drive
Hawthorne NY 10532
charu@us.ibm.com

Philip S. Yu
Department of Computer Science
University of Illinois at Chicago
854 South Morgan Street
Chicago, IL 60607-7053
psyu@cs.uic.edu

Series Editors
Ahmed K. Elmagarmid
Purdue University
West Lafayette, IN 47907

Amit P. Sheth
Wright State University
Dayton, Ohio 45435

ISBN 978-1-4419-4371-2 e-ISBN 978-0-387-70992-5
DOI 10.1007/978-0-387-70992-5

Printed on acid-free paper

9 8 7 6 5 4 3 2 1

springer.com

Preface

In recent years, advances in hardware technology have lead to an increase in the capability to store and record personal data about consumers and individuals. This has lead to concerns that the personal data may be misused for a variety of purposes. In order to alleviate these concerns, a number of techniques have recently been proposed in order to perform the data mining tasks in a privacy-preserving way. These techniques for performing privacy-preserving data mining are drawn from a wide array of related topics such as data mining, cryptography and information hiding. The material in this book is designed to be drawn from the different topics so as to provide a good overview of the important topics in the field.

While a large number of research papers are now available in this field, many of the topics have been studied by different communities with different styles. At this stage, it becomes important to organize the topics in such a way that the relative importance of different research areas is recognized. Furthermore, the field of privacy-preserving data mining has been explored independently by the cryptography, database and statistical disclosure control communities. In some cases, the parallel lines of work are quite similar, but the communities are not sufficiently integrated for the provision of a broader perspective. This book will contain chapters from researchers of all three communities and will therefore try to provide a balanced perspective of the work done in this field.

This book will be structured as an edited book from prominent researchers in the field. Each chapter will contain a survey which contains the key research content on the topic, and the future directions of research in the field. Emphasis will be placed on making each chapter self-sufficient. While the chapters will be written by different researchers, the topics and content is organized in such a way so as to present the most important models, algorithms, and applications in the privacy field in a structured and concise way. In addition, attention is paid in drawing chapters from researchers working in different areas in order to provide different points of view. Given the lack of structurally organized information on the topic of privacy, the book will provide insights which are not easily accessible otherwise. A few chapters in the book are not surveys, since the corresponding topics fall in the emerging category, and enough material is not available to create a survey. In such cases, the individual results have been included to give a flavor of the emerging research in the field. It is expected that the book will be a great help to researchers and graduate students interested in the topic. While the privacy field clearly falls in the emerging category because of its recency, it is now beginning to reach a maturation and popularity point, where the development of an overview book on the topic becomes both possible and necessary. It is hoped that this book will provide a reference to students, researchers and practitioners in both introducing the topic of privacy-preserving data mining and understanding the practical and algorithmic aspects of the area.

Contents

Preface v

List of Figures xvii

List of Tables xxi

1
An Introduction to Privacy-Preserving Data Mining 1
Charu C. Aggarwal, Philip S. Yu

1.1. Introduction 1
1.2. Privacy-Preserving Data Mining Algorithms 3
1.3. Conclusions and Summary 7
References 8

2
A General Survey of Privacy-Preserving Data Mining Models and Algorithms 11
Charu C. Aggarwal, Philip S. Yu

2.1. Introduction 11
2.2. The Randomization Method 13
2.2.1 Privacy Quantification 15
2.2.2 Adversarial Attacks on Randomization 18
2.2.3 Randomization Methods for Data Streams 18
2.2.4 Multiplicative Perturbations 19
2.2.5 Data Swapping 19
2.3. Group Based Anonymization 20
2.3.1 The k-Anonymity Framework 20
2.3.2 Personalized Privacy-Preservation 24
2.3.3 Utility Based Privacy Preservation 24
2.3.4 Sequential Releases 25
2.3.5 The l-diversity Method 26
2.3.6 The t-closeness Model 27
2.3.7 Models for Text, Binary and String Data 27
2.4. Distributed Privacy-Preserving Data Mining 28
2.4.1 Distributed Algorithms over Horizontally Partitioned Data Sets 30
2.4.2 Distributed Algorithms over Vertically Partitioned Data 31
2.4.3 Distributed Algorithms for k-Anonymity 32

2.5. Privacy-Preservation of Application Results 32
2.5.1 Association Rule Hiding 33
2.5.2 Downgrading Classifier Effectiveness 34
2.5.3 Query Auditing and Inference Control 34
2.6. Limitations of Privacy: The Curse of Dimensionality 37
2.7. Applications of Privacy-Preserving Data Mining 38
2.7.1 Medical Databases: The Scrub and Datafly Systems 39
2.7.2 Bioterrorism Applications 40
2.7.3 Homeland Security Applications 40
2.7.4 Genomic Privacy 42
2.8. Summary 43
References 43

3
A Survey of Inference Control Methods for Privacy-Preserving Data Mining 53
Josep Domingo-Ferrer
3.1. Introduction 54
3.2. A classification of Microdata Protection Methods 55
3.3. Perturbative Masking Methods 58
3.3.1 Additive Noise 58
3.3.2 Microaggregation 59
3.3.3 Data Wapping and Rank Swapping 61
3.3.4 Rounding 62
3.3.5 Resampling 62
3.3.6 PRAM 62
3.3.7 MASSC 63
3.4. Non-perturbative Masking Methods 63
3.4.1 Sampling 64
3.4.2 Global Recoding 64
3.4.3 Top and Bottom Coding 65
3.4.4 Local Suppression 65
3.5. Synthetic Microdata Generation 65
3.5.1 Synthetic Data by Multiple Imputation 65
3.5.2 Synthetic Data by Bootstrap 66
3.5.3 Synthetic Data by Latin Hypercube Sampling 66
3.5.4 Partially Synthetic Data by Cholesky Decomposition 67
3.5.5 Other Partially Synthetic and Hybrid Microdata Approaches 67
3.5.6 Pros and Cons of Synthetic Microdata 68
3.6. Trading off Information Loss and Disclosure Risk 69
3.6.1 Score Construction 69
3.6.2 R-U Maps 71
3.6.3 k-anonymity 71
3.7. Conclusions and Research Directions 72
References 73

4
Measures of Anonymity 81
Suresh Venkatasubramanian
4.1. Introduction 81
4.1.1 What is Privacy? 82
4.1.2 Data Anonymization Methods 83
4.1.3 A Classification of Methods 84
4.2. Statistical Measures of Anonymity 85
4.2.1 Query Restriction 85
4.2.2 Anonymity via Variance 85
4.2.3 Anonymity via Multiplicity 86
4.3. Probabilistic Measures of Anonymity 87
4.3.1 Measures Based on Random Perturbation 87
4.3.2 Measures Based on Generalization 90
4.3.3 Utility vs Privacy 94
4.4. Computational Measures of Anonymity 94
4.4.1 Anonymity via Isolation 97
4.5. Conclusions and New Directions 97
4.5.1 New Directions 98
References 99

5
k-Anonymous Data Mining: A Survey 105
V. Ciriani, S. De Capitani di Vimercati, S. Foresti, and P. Samarati
5.1. Introduction 105
5.2. k-Anonymity 107
5.3. Algorithms for Enforcing k-Anonymity 110
5.4. k-Anonymity Threats from Data Mining 117
5.4.1 Association Rules 118
5.4.2 Classification Mining 118
5.5. k-Anonymity in Data Mining 120
5.6. Anonymize-and-Mine 123
5.7. Mine-and-Anonymize 126
5.7.1 Enforcing k-Anonymity on Association Rules 126
5.7.2 Enforcing k-Anonymity on Decision Trees 130
5.8. Conclusions 133
Acknowledgments 133
References 134

6
A Survey of Randomization Methods for Privacy-Preserving Data Mining 137
Charu C. Aggarwal, Philip S. Yu
6.1. Introduction 137
6.2. Reconstruction Methods for Randomization 139
6.2.1 The Bayes Reconstruction Method 139
6.2.2 The EM Reconstruction Method 141
6.2.3 Utility and Optimality of Randomization Models 143

6.3. Applications of Randomization 144
6.3.1 Privacy-Preserving Classification with Randomization 144
6.3.2 Privacy-Preserving OLAP 145
6.3.3 Collaborative Filtering 145
6.4. The Privacy-Information Loss Tradeoff 146
6.5. Vulnerabilities of the Randomization Method 149
6.6. Randomization of Time Series Data Streams 151
6.7. Multiplicative Noise for Randomization 152
6.7.1 Vulnerabilities of Multiplicative Randomization 153
6.7.2 Sketch Based Randomization 153
6.8. Conclusions and Summary 154
References 154

7
A Survey of Multiplicative Perturbation for Privacy-Preserving Data Mining 157
Keke Chen and Ling Liu
7.1. Introduction 158
7.1.1 Data Privacy vs. Data Utility 159
7.1.2 Outline 160
7.2. Definition of Multiplicative Perturbation 161
7.2.1 Notations 161
7.2.2 Rotation Perturbation 161
7.2.3 Projection Perturbation 162
7.2.4 Sketch-based Approach 164
7.2.5 Geometric Perturbation 164
7.3. Transformation Invariant Data Mining Models 165
7.3.1 Definition of Transformation Invariant Models 166
7.3.2 Transformation-Invariant Classification Models 166
7.3.3 Transformation-Invariant Clustering Models 167
7.4. Privacy Evaluation for Multiplicative Perturbation 168
7.4.1 A Conceptual Multidimensional Privacy Evaluation Model 168
7.4.2 Variance of Difference as Column Privacy Metric 169
7.4.3 Incorporating Attack Evaluation 170
7.4.4 Other Metrics 171
7.5. Attack Resilient Multiplicative Perturbations 171
7.5.1 Naive Estimation to Rotation Perturbation 171
7.5.2 ICA-Based Attacks 173
7.5.3 Distance-Inference Attacks 174
7.5.4 Attacks with More Prior Knowledge 176
7.5.5 Finding Attack-Resilient Perturbations 177
7.6. Conclusion 177
Acknowledgment 178
References 179

8
A Survey of Quantification of Privacy Preserving Data Mining Algorithms 183
Elisa Bertino, Dan Lin and Wei Jiang
8.1. Introduction 184
8.2. Metrics for Quantifying Privacy Level 186
8.2.1 Data Privacy 186

8.2.2 Result Privacy 191
8.3. Metrics for Quantifying Hiding Failure 192
8.4. Metrics for Quantifying Data Quality 193
8.4.1 Quality of the Data Resulting from the PPDM Process 193
8.4.2 Quality of the Data Mining Results 198
8.5. Complexity Metrics 200
8.6. How to Select a Proper Metric 201
8.7. Conclusion and Research Directions 202
References 202

9
A Survey of Utility-based Privacy-Preserving Data Transformation Methods 207
Ming Hua and Jian Pei
9.1. Introduction 208
9.1.1 What is Utility-based Privacy Preservation? 209
9.2. Types of Utility-based Privacy Preservation Methods 210
9.2.1 Privacy Models 210
9.2.2 Utility Measures 212
9.2.3 Summary of the Utility-Based Privacy Preserving Methods 214
9.3. Utility-Based Anonymization Using Local Recoding 214
9.3.1 Global Recoding and Local Recoding 215
9.3.2 Utility Measure 216
9.3.3 Anonymization Methods 217
9.3.4 Summary and Discussion 219
9.4. The Utility-based Privacy Preserving Methods in Classification Problems 219
9.4.1 The Top-Down Specialization Method 220
9.4.2 The Progressive Disclosure Algorithm 224
9.4.3 Summary and Discussion 228
9.5. Anonymized Marginal: Injecting Utility into Anonymized Data Sets 228
9.5.1 Anonymized Marginal 229
9.5.2 Utility Measure 230
9.5.3 Injecting Utility Using Anonymized Marginals 231
9.5.4 Summary and Discussion 233
9.6. Summary 234
Acknowledgments 234
References 234

10
Mining Association Rules under Privacy Constraints 239
Jayant R. Haritsa
10.1. Introduction 239
10.2. Problem Framework 240
10.2.1 Database Model 240
10.2.2 Mining Objective 241
10.2.3 Privacy Mechanisms 241
10.2.4 Privacy Metric 243
10.2.5 Accuracy Metric 245

10.3. Evolution of the Literature 246
10.4. The FRAPP Framework 251
10.4.1 Reconstruction Model 252
10.4.2 Estimation Error 253
10.4.3 Randomizing the Perturbation Matrix 256
10.4.4 Efficient Perturbation 256
10.4.5 Integration with Association Rule Mining 258
10.5. Sample Results 259
10.6. Closing Remarks 263
Acknowledgments 263
References 263

11
A Survey of Association Rule Hiding Methods for Privacy 267
Vassilios S. Verykios and Aris Gkoulalas-Divanis
11.1. Introduction 267
11.2. Terminology and Preliminaries 269
11.3. Taxonomy of Association Rule Hiding Algorithms 270
11.4. Classes of Association Rule Algorithms 271
11.4.1 Heuristic Approaches 272
11.4.2 Border-based Approaches 277
11.4.3 Exact Approaches 278
11.5. Other Hiding Approaches 279
11.6. Metrics and Performance Analysis 281
11.7. Discussion and Future Trends 284
11.8. Conclusions 285
References 286

12
A Survey of Statistical Approaches to Preserving Confidentiality of Contingency Table Entries 291
Stephen E. Fienberg and Aleksandra B. Slavkovic
12.1. Introduction 291
12.2. The Statistical Approach Privacy Protection 292
12.3. Datamining Algorithms, Association Rules, and Disclosure Limitation 294
12.4. Estimation and Disclosure Limitation for Multi-way Contingency Tables 295
12.5. Two Illustrative Examples 301
12.5.1 Example 1: Data from a Randomized Clinical Trial 301
12.5.2 Example 2: Data from the 1993 U.S. Current Population Survey 305
12.6. Conclusions 308
Acknowledgments 309
References 309

13
A Survey of Privacy-Preserving Methods Across Horizontally Partitioned Data 313
Murat Kantarcioglu
13.1. Introduction 313

13.2. Basic Cryptographic Techniques for Privacy-Preserving Distributed Data Mining 315
13.3. Common Secure Sub-protocols Used in Privacy-Preserving Distributed Data Mining 318
13.4. Privacy-preserving Distributed Data Mining on Horizontally Partitioned Data 323
13.5. Comparison to Vertically Partitioned Data Model 326
13.6. Extension to Malicious Parties 327
13.7. Limitations of the Cryptographic Techniques Used in Privacy-Preserving Distributed Data Mining 329
13.8. Privacy Issues Related to Data Mining Results 330
13.9. Conclusion 332
References 332

14
A Survey of Privacy-Preserving Methods Across Vertically Partitioned Data 337
Jaideep Vaidya
14.1. Introduction 337
14.2. Classification 341
14.2.1 Naïve Bayes Classification 342
14.2.2 Bayesian Network Structure Learning 343
14.2.3 Decision Tree Classification 344
14.3. Clustering 346
14.4. Association Rule Mining 347
14.5. Outlier detection 349
14.5.1 Algorithm 351
14.5.2 Security Analysis 352
14.5.3 Computation and Communication Analysis 354
14.6. Challenges and Research Directions 355
References 356

15
A Survey of Attack Techniques on Privacy-Preserving Data Perturbation Methods 359
Kun Liu, Chris Giannella, and Hillol Kargupta
15.1. Introduction 360
15.2. Definitions and Notation 360
15.3. Attacking Additive Data Perturbation 361
15.3.1 Eigen-Analysis and PCA Preliminaries 362
15.3.2 Spectral Filtering 363
15.3.3 SVD Filtering 364
15.3.4 PCA Filtering 365
15.3.5 MAP Estimation Attack 366
15.3.6 Distribution Analysis Attack 367
15.3.7 Summary 367
15.4. Attacking Matrix Multiplicative Data Perturbation 369
15.4.1 Known I/O Attacks 370
15.4.2 Known Sample Attack 373
15.4.3 Other Attacks Based on ICA 374

15.4.4 Summary 375
15.5. Attacking k-Anonymization 376
15.6. Conclusion 376
Acknowledgments 377
References 377

16
Private Data Analysis via Output Perturbation 383
Kobbi Nissim
16.1. Introduction 383
16.2. The Abstract Model – Statistical Databases, Queries, and Sanitizers 385
16.3. Privacy 388
16.3.1 Interpreting the Privacy Definition 390
16.4. The Basic Technique: Calibrating Noise to Sensitivity 394
16.4.1 Applications: Functions with Low Global Sensitivity 396
16.5. Constructing Sanitizers for Complex Functionalities 400
16.5.1 k-Means Clustering 401
16.5.2 SVD and PCA 403
16.5.3 Learning in the Statistical Queries Model 404
16.6. Beyond the Basics 405
16.6.1 Instance Based Noise and Smooth Sensitivity 406
16.6.2 The Sample-Aggregate Framework 408
16.6.3 A General Sanitization Mechanism 409
16.7. Related Work and Bibliographic Notes 409
Acknowledgments 411
References 411

17
A Survey of Query Auditing Techniques for Data Privacy 415
Shubha U. Nabar, Krishnaram Kenthapadi, Nina Mishra and Rajeev Motwani
17.1. Introduction 415
17.2. Auditing Aggregate Queries 416
17.2.1 Offline Auditing 417
17.2.2 Online Auditing 418
17.3. Auditing Select-Project-Join Queries 426
17.4. Challenges in Auditing 427
17.5. Reading 429
References 430

18
Privacy and the Dimensionality Curse 433
Charu C. Aggarwal
18.1. Introduction 433
18.2. The Dimensionality Curse and the k-anonymity Method 435
18.3. The Dimensionality Curse and Condensation 441
18.4. The Dimensionality Curse and the Randomization Method 446
18.4.1 Effects of Public Information 446
18.4.2 Effects of High Dimensionality 450
18.4.3 Gaussian Perturbing Distribution 450
18.4.4 Uniform Perturbing Distribution 455

18.5. The Dimensionality Curse and l-diversity 458
18.6. Conclusions and Research Directions 459
References 460

19
Personalized Privacy Preservation 461
Yufei Tao and Xiaokui Xiao
19.1. Introduction 461
19.2. Formalization of Personalized Anonymity 463
19.2.1 Personal Privacy Requirements 464
19.2.2 Generalization 465
19.3. Combinatorial Process of Privacy Attack 467
19.3.1 Primary Case 468
19.3.2 Non-primary Case 469
19.4. Theoretical Foundation 470
19.4.1 Notations and Basic Properties 471
19.4.2 Derivation of the Breach Probability 472
19.5. Generalization Algorithm 473
19.5.1 The Greedy Framework 474
19.5.2 Optimal SA-generalization 476
19.6. Alternative Forms of Personalized Privacy Preservation 478
19.6.1 Extension of k-anonymity 479
19.6.2 Personalization in Location Privacy Protection 480
19.7. Summary and Future Work 482
References 485

20
Privacy-Preserving Data Stream Classification 487
Yabo Xu, Ke Wang, Ada Wai-Chee Fu, Rong She, and Jian Pei
20.1. Introduction 487
20.1.1 Motivating Example 488
20.1.2 Contributions and Paper Outline 490
20.2. Related Works 491
20.3. Problem Statement 493
20.3.1 Secure Join Stream Classification 493
20.3.2 Naive Bayesian Classifiers 494
20.4. Our Approach 495
20.4.1 Initialization 495
20.4.2 Bottom-Up Propagation 496
20.4.3 Top-Down Propagation 497
20.4.4 Using NBC 499
20.4.5 Algorithm Analysis 500
20.5. Empirical Studies 501
20.5.1 Real-life Datasets 502
20.5.2 Synthetic Datasets 504
20.5.3 Discussion 506
20.6. Conclusions 507
References 508

Index 511

List of Figures

5.1	Simplified representation of a private table	108
5.2	An example of domain and value generalization hierarchies	109
5.3	Classification of k-anonymity techniques [11]	110
5.4	Generalization hierarchy for QI=$\{$Marital_status, Sex$\}$	111
5.5	Index assignment to attributes Marital_status and Sex	112
5.6	An example of set enumeration tree over set $\mathcal{I} = \{1, 2, 3\}$ of indexes	113
5.7	Sub-hierarchies computed by Incognito for the table in Figure 5.1	114
5.8	Spatial representation (a) and possible partitioning (b)-(d) of the table in Figure 5.1	116
5.9	An example of decision tree	119
5.10	Different approaches for combining k-anonymity and data mining	120
5.11	An example of top-down anonymization for the private table in Figure 5.1	124
5.12	Frequent itemsets extracted from the table in Figure 5.1	127
5.13	An example of binary table	128
5.14	Itemsets extracted from the table in Figure 5.13(b)	128
5.15	Itemsets with support at least equal to 40 (a) and corresponding anonymized itemsets (b)	129
5.16	3-anonymous version of the tree of Figure 5.9	131
5.17	Suppression of occurrences in non-leaf nodes in the tree in Figure 5.9	132
5.18	Table inferred from the decision tree in Figure 5.17	132
5.19	11-anonymous version of the tree in Figure 5.17	132
5.20	Table inferred from the decision tree in Figure 5.19	133
6.1	Illustration of the Information Loss Metric	149
7.1	Using known points and distance relationship to infer the rotation matrix	175

9.1 A taxonomy tree on categorical attribute *Education* 221
9.2 A taxonomy tree on continuous attribute *Age* 221
9.3 Interactive graph 232
9.4 A decomposition 232
10.1 CENSUS ($\gamma = 19$) 261
10.2 Perturbation Matrix Condition Numbers ($\gamma = 19$) 262

13.1 Relationship between Secure Sub-protocols and Privacy Preserving Distributed Data Mining on Horizontally Partitioned Data 323
14.1 Two dimensional problem that cannot be decomposed into two one-dimensional problems 340
15.1 Wigner's semi-circle law: a histogram of the eigenvalues of $\frac{A+A'}{2\sqrt{2p}}$ for a large, randomly generated A 363
17.1 Skeleton of a simulatable private randomized auditor 423
18.1 Some Examples of Generalization for 2-Anonymity 435
18.2 Upper Bound of 2-anonymity Probability in an Non-Empty Grid Cell 439
18.3 Fraction of Data Points Preserving 2-Anonymity with Data Dimensionality (Gaussian Clusters) 440
18.4 Minimum Information Loss for 2-Anonymity (Gaussian Clusters) 445
18.5 Randomization Level with Increasing Dimensionality, Perturbation level $= 8 \cdot \sigma^o$ $(UniDis)$ 457
19.1 Microdata and generalization 462
19.2 The taxonomy of attribute *Disease* 463
19.3 A possible result of our generalization scheme 466
19.4 The voter registration list 468
19.5 Algorithm for computing personalized generalization 474
19.6 Algorithm for finding the optimal SA-generalization 478
19.7 Personalized k-anonymous generalization 480
20.1 Related streams / tables 489
20.2 The join stream 489
20.3 Example with 3 streams at initialization 496
20.4 After bottom-up propagations 498
20.5 After top-down propagations 499
20.6 UK road accident data (2001) 502
20.7 Classifier accuracy 503
20.8 Time per input tuple 503

20.9 Classifier accuracy vs. window size 505

20.10 Classifier accuracy vs. concept drifting interval 505

20.11 Time per input tuple vs. window size 506

20.12 Time per input tuple vs. blow-up ratio 506

20.13 Time per input tuple vs. number of streams 507

List of Tables

3.1	Perturbative methods vs data types. "X" denotes applicable and "(X)" denotes applicable with some adaptation	58
3.2	Example of rank swapping. Left, original file; right, rankswapped file	62
3.3	Non-perturbative methods vs data types	64
9.1a	The original table	209
9.2b	A 2-anonymized table with better utility	209
9.3c	A 2-anonymized table with poorer utility	209
9.4	Summary of utility-based privacy preserving methods	214
9.5a	3-anonymous table by global recoding	215
9.6b	3-anonymous table by local recoding	215
9.7a	The original table	223
9.8b	The anonymized table	223
9.9a	The original table	225
9.10b	The suppressed table	225
9.11a	The original table	229
9.12b	The anonymized table	229
9.13a	*Age* Marginal	229
9.14b	(*Education, AnnualIncome*) Marginal	229
10.1	CENSUS Dataset	260
10.2	Frequent Itemsets for $sup_{min} = 0.02$	260
12.1	Results of clinical trial for the effectiveness of an analgesic drug	302
12.2	Second panel has LP relaxation bounds, and third panel has sharp IP bounds for cell entries in Table 1.1 given $[R\|CST]$ conditional probability values	303
12.3	Sharp upper and lower bounds for cell entries in Table 12.1 given the $[CSR]$ margin, and LP relaxation bounds given $[R\|CS]$ conditional probability values	304
12.4	Description of variables in CPS data extract	305

12.5 Marginal table $[ACDGH]$ from 8-way CPS table 306
12.6 Summary of difference between upper and lower bounds for small cell counts in the full 8-way CPS table under Model 1 and under Model 2 307
14.1 The Weather Dataset 338
14.2 Arbitrary partitioning of data between 2 sites 339
14.3 Vertical partitioning of data between 2 sites 340
15.1 Summarization of Attacks on Additive Perturbation 368
15.2 Summarization of Attacks on Matrix Multiplicative Perturbation 375
18.1 Notations and Definitions 441

Chapter 1

An Introduction to Privacy-Preserving Data Mining

Charu C. Aggarwal
IBM T. J. Watson Research Center
Hawthorne, NY 10532
charu@us.ibm.com

Philip S. Yu
University of Illinois at Chicago
Chicago, IL 60607
psyu@cs.uic.edu

Abstract The field of privacy has seen rapid advances in recent years because of the increases in the ability to store data. In particular, recent advances in the data mining field have lead to increased concerns about privacy. While the topic of privacy has been traditionally studied in the context of cryptography and information-hiding, recent emphasis on data mining has lead to renewed interest in the field. In this chapter, we will introduce the topic of privacy-preserving data mining and provide an overview of the different topics covered in this book.

Keywords: Privacy-preserving data mining, privacy, randomization, k-anonymity.

1.1 Introduction

The problem of privacy-preserving data mining has become more important in recent years because of the increasing ability to store personal data about users, and the increasing sophistication of data mining algorithms to leverage this information. A number of techniques such as randomization and k-anonymity [1, 4, 16] have been suggested in recent years in order to perform privacy-preserving data mining. Furthermore, the problem has been discussed in multiple communities such as the database community, the statistical disclosure control community and the cryptography community. In some cases, the different communities have explored parallel lines of work which are quite similar. This book will try to explore different topics from the perspective of different communities, and will try to give a fused idea of the work in different communities.

The key directions in the field of privacy-preserving data mining are as follows:

- **Privacy-Preserving Data Publishing:** These techniques tend to study different transformation methods associated with privacy. These techniques include methods such as randomization [1], k-anonymity [16, 7], and l-diversity [11]. Another related issue is how the perturbed data can be used in conjunction with classical data mining methods such as association rule mining [15]. Other related problems include that of determining privacy-preserving methods to keep the underlying data useful (utility-based methods), or the problem of studying the different definitions of privacy, and how they compare in terms of effectiveness in different scenarios.

- **Changing the results of Data Mining Applications to preserve privacy:** In many cases, the results of data mining applications such as association rule or classification rule mining can compromise the privacy of the data. This has spawned a field of privacy in which the results of data mining algorithms such as association rule mining are modified in order to preserve the privacy of the data. A classic example of such techniques are association rule hiding methods, in which some of the association rules are suppressed in order to preserve privacy.

- **Query Auditing:** Such methods are akin to the previous case of modifying the results of data mining algorithms. Here, we are either modifying or restricting the results of queries. Methods for perturbing the output of queries are discussed in [8], whereas techniques for restricting queries are discussed in [9, 13].

- **Cryptographic Methods for Distributed Privacy:** In many cases, the data may be distributed across multiple sites, and the owners of the data across these different sites may wish to compute a common function. In such cases, a variety of cryptographic protocols may be used in order to communicate among the different sites, so that secure function computation is possible without revealing sensitive information. A survey of such methods may be found in [14].

- **Theoretical Challenges in High Dimensionality:** Real data sets are usually extremely high dimensional, and this makes the process of privacy-preservation extremely difficult both from a computational and effectiveness point of view. In [12], it has been shown that optimal k-anonymization is NP-hard. Furthermore, the technique is not even effective with increasing dimensionality, since the data can typically be combined with either public or background information to reveal the identity of the underlying record owners. A variety of methods for adversarial attacks in the high dimensional case are discussed in [5, 6].

This book will attempt to cover the different topics from the point of view of different communities in the field. This chapter will provide an overview of the different privacy-preserving algorithms covered in this book. We will discuss the challenges associated with each kind of problem, and discuss an overview of the material in the corresponding chapter.

1.2 Privacy-Preserving Data Mining Algorithms

In this section, we will discuss the key stream mining problems and will discuss the challenges associated with each problem. We will also discuss an overview of the material covered in each chapter of this book. The broad topics covered in this book are as follows:

General Survey. In chapter 2, we provide a broad survey of privacy-preserving data-mining methods. We provide an overview of the different techniques and how they relate to one another. The individual topics will be covered in sufficient detail to provide the reader with a good reference point. The idea is to provide an overview of the field for a new reader from the perspective of the data mining community. However, more detailed discussions are deferred to future chapters which contain descriptions of different data mining algorithms.

Statistical Methods for Disclosure Control. The topic of privacy-preserving data mining has often been studied extensively by the data mining community without sufficient attention to the work done by the conventional work done by the statistical disclosure control community. In chapter 3, detailed methods for statistical disclosure control have been presented along with some of the relationships to the parallel work done in the database and data mining community. This includes methods such as k-anonymity, swapping, randomization, micro-aggregation and synthetic data generation. The idea is to give the readers an overview of the common themes in privacy-preserving data mining by different communities.

Measures of Anonymity. There are a very large number of definitions of anonymity in the privacy-preserving data mining field. This is partially because of the varying goals of different privacy-preserving data mining algorithms. For example, methods such as k-anonymity, l-diversity and t-closeness are all designed to prevent identification, though the final goal is to preserve the underlying sensitive information. Each of these methods is designed to prevent disclosure of sensitive information in a different way. Chapter 4 is a survey of different measures of anonymity. The chapter tries to define privacy from the perspective of anonymity measures and classifies such measures. The chapter also compares and contrasts different measures, and discusses the relative advantages of different measures. This chapter thus provides an overview and perspective of the different ways in which privacy could be defined, and what the relative advantages of each method might be.

The k-anonymity Method. An important method for privacy de-identification is the method of k-anonymity [16]. The motivating factor behind the k-anonymity technique is that many attributes in the data can often be considered pseudo-identifiers which can be used in conjunction with public records in order to uniquely identify the records. For example, if the identifications from the records are removed, attributes such as the birth date and zip-code an be used in order to uniquely identify the identities of the underlying records. The idea in k-anonymity is to reduce the granularity of representation of the data in such a way that a given record cannot be distinguished from at least $(k - 1)$ other records. In chapter 5, the k-anonymity method is discussed in detail. A number of important algorithms for k-anonymity are discussed in the same chapter.

The Randomization Method. The randomization technique uses data distortion methods in order to create private representations of the records [1, 4]. In most cases, the individual records cannot be recovered, but only aggregate distributions can be recovered. These aggregate distributions can be used for data mining purposes. Two kinds of perturbation are possible with the randomization method:

- **Additive Perturbation:** In this case, randomized noise is added to the data records. The overall data distributions can be recovered from the randomized records. Data mining and management algorithms re designed to work with these data distributions. A detailed discussion of these methods is provided in chapter 6.

- **Multiplicative Perturbation:** In this case, the random projection or random rotation techniques are used in order to perturb the records. A detailed discussion of these methods is provided in chapter 7.

In addition, these chapters deal with the issue of adversarial attacks and vulnerabilities of these methods.

Quantification of Privacy. A key issue in measuring the security of different privacy-preservation methods is the way in which the underlying privacy is quantified. The idea in privacy quantification is to measure the risk of disclosure for a given level of perturbation. In chapter 8, the issue of quantification of privacy is closely examined. The chapter also examines the issue of utility, and its natural tradeoff with privacy quantification. A discussion of the relative advantages of different kinds of methods is presented.

Utility Based Privacy-Preserving Data Mining. Most privacy-preserving data mining methods apply a transformation which reduces the effectiveness of the underlying data when it is applied to data mining methods or algorithms. In fact, there is a natural tradeoff between privacy and accuracy, though this tradeoff is affected by the particular algorithm which is used for privacy-preservation. A key issue is to maintain maximum utility of the data without compromising the underlying privacy constraints. In chapter 9, a broad overview of the different utility based methods for privacy-preserving data mining is presented. The issue of designing utility based algorithms to work effectively with certain kinds of data mining problems is addressed.

Mining Association Rules under Privacy Constraints. Since association rule mining is one of the important problems in data mining, we have devoted a number of chapters to this problem. There are two aspects to the privacy-preserving association rule mining problem:

- When the input to the data is perturbed, it is a challenging problem to accurately determine the association rules on the perturbed data. Chapter 10 discusses the problem of association rule mining on the perturbed data.

- A different issue is that of output association rule privacy. In this case, we try to ensure that none of the association rules in the output result in leakage of sensitive data. This problem is referred to as *association rule hiding* [17] by the database community, and that of *contingency table privacy-preservation* by the statistical community. The problem of output association rule privacy is briefly discussed in chapter 10. A detailed survey of association rule hiding from the perspective of the database community is discussed in chapter 11, and a discussion from the perspective of the statistical community is discussed in chapter 12.

Cryptographic Methods for Information Sharing and Privacy. In many cases, multiple parties may wish to share *aggregate private data*, without leaking any sensitive information at their end [14]. For example, different superstores with sensitive sales data may wish to coordinate among themselves in knowing aggregate trends without leaking the trends of their individual stores. This requires *secure and cryptographic protocols* for sharing the information across the different parties. The data may be distributed in two ways across different sites:

- **Horizontal Partitioning:** In this case, the different sites may have different sets of records containing the same attributes.

- **Vertical Partitioning:** In this case, the different sites may have different attributes of the same sets of records.

Clearly, the challenges for the horizontal and vertical partitioning case are quite different. In chapters 13 and 14, a variety of cryptographic protocols for horizontally and vertically partitioned data are discussed. The different kinds of cryptographic methods are introduced in chapter 13. Methods for horizontally partitioned data are discussed in chapter 13, whereas methods for vertically partitioned data are discussed in chapter 14.

Privacy Attacks. It is useful to examine the different ways in which one can make adversarial attacks on privacy-transformed data. This helps in designing more effective privacy-transformation methods. Some examples of methods which can be used in order to attack the privacy of the underlying data include SVD-based methods, spectral filtering methods and background knowledge attacks. In chapter 15, a detailed description of different kinds of attacks on data perturbation methods is provided.

Query Auditing and Inference Control. Many private databases are open to querying. This can compromise the security of the results, when the adversary can use different kinds of queries in order to undermine the security of the data. For example, a combination of range queries can be used in order to narrow down the possibilities for that record. Therefore, the results over multiple queries can be combined in order to uniquely identify a record, or at least reduce the uncertainty in identifying it. There are two primary methods for preventing this kind of attack:

- **Query Output Perturbation:** In this case, we add noise to the output of the query result in order to preserve privacy [8]. A detailed description of such methods is provided in chapter 16.

- **Query Auditing:** In this case, we choose to deny a subset of the queries, so that the particular combination of queries cannot be used in order to violate the privacy [9, 13]. A detailed survey of query auditing methods have been provided in chapter 17.

Privacy and the Dimensionality Curse. In recent years, it has been observed that many privacy-preservation methods such as k-anonymity and randomization are not very effective in the high dimensional case [5, 6]. In chapter 18, we have provided a detailed description of the effects of the dimensionality curse on different kinds of privacy-preserving data mining algorithm. It is clear from the discussion in the chapter that most privacy methods are not very effective in the high dimensional case.

Personalized Privacy Preservation. In many applications, different subjects have different requirements for privacy. For example, a brokerage customer with a very large account would likely have a much higher level of privacy-protection than a customer with a lower level of privacy protection. In such case, it is necessary to *personalize* the privacy-protection algorithm. In personalized privacy-preservation, we construct anonymizations of the data such that different records have a different level of privacy. Two examples of personalized privacy-preservation methods are discussed in [3, 18]. The method in [3] uses condensation approach for personalized anonymization, while the method in [18] uses a more conventional generalization approach for anonymization. In chapter 19, a number of algorithms for personalized anonymity are examined.

Privacy-Preservation of Data Streams. A new topic in the area of privacy-preserving data mining is that of data streams, in which data grows rapidly at an unlimited rate. In such cases, the problem of privacy-preservation is quite challenging since the data is being released incrementally. In addition, the fast nature of data streams obviates the possibility of using the past history of the data. We note that both the topics of data streams and privacy-preserving data mining are relatively new, and there has not been much work on combining the two topics. Some work has been done on performing randomization of data streams [10], and other work deals with the issue of condensation based anonymization [2] of data streams. Both of these methods are discussed in Chapters 2 and 5, which are surveys on privacy and randomization respectively. Nevertheless, the literature on the stream topic remains sparse. Therefore, in chapter 20, we have added a chapter which specifically deals with the issue of privacy-preserving classification of data streams. While this chapter is unlike other chapters in the sense that it is not a survey, we have included it in order to provide a flavor of the emerging techniques in this important area of research.

1.3 Conclusions and Summary

In this chapter, we introduced the problem of privacy-preserving data mining and discussed the broad areas of research in the field. The broad areas of privacy are as follows:

- **Privacy-preserving data publishing:** This corresponds to sanitizing the data, so that its privacy remains preserved.

- **Privacy-Preserving Applications:** This corresponds to designing data management and mining algorithms in such a way that the privacy remains preserved. Some examples include association rule mining, classification, and query processing.

- **Utility Issues:** Since the perturbed data may often be used for mining and management purposes, its utility needs to be preserved. Therefore, the data mining and privacy transformation techniques need to be designed effectively, so to to preserve the utility of the results.

- **Distributed Privacy, cryptography and adversarial collaboration:** This corresponds to secure communication protocols between trusted parties, so that information can be shared effectively without revealing sensitive information about particular parties.

We also discussed a broad overview of the different topics discussed in this book. In the remaining chapters, the surveys will provide a comprehensive treatment of the topics in each category.

References

[1] Agrawal R., Srikant R. Privacy-Preserving Data Mining. *ACM SIGMOD Conference*, 2000.

[2] Aggarwal C. C., Yu P. S.: A Condensation approach to privacy preserving data mining. *EDBT Conference*, 2004.

[3] Aggarwal C. C., Yu P. S. On Variable Constraints in Privacy Preserving Data Mining. *ACM SIAM Data Mining Conference*, 2005.

[4] Agrawal D. Aggarwal C. C. On the Design and Quantification of Privacy Preserving Data Mining Algorithms. *ACM PODS Conference*, 2002.

[5] Aggarwal C. C. On k-anonymity and the curse of dimensionality. *VLDB Conference*, 2004.

[6] Aggarwal C. C. On Randomization, Public Information, and the Curse of Dimensionality. *ICDE Conference*, 2007.

[7] Bayardo R. J., Agrawal R. Data Privacy through optimal k-anonymization. *ICDE Conference*, 2005.

[8] Blum A., Dwork C., McSherry F., Nissim K.: Practical Privacy: The SuLQ Framework. *ACM PODS Conference*, 2005.

[9] Kenthapadi K.,Mishra N., Nissim K.: Simulatable Auditing, *ACM PODS Conference*, 2005.

[10] Li F., Sun J., Papadimitriou S. Mihaila G., Stanoi I.: Hiding in the Crowd: Privacy Preservation on Evolving Streams through Correlation Tracking. *ICDE Conference*, 2007.

[11] Machanavajjhala A., Gehrke J., Kifer D. ℓ-diversity: Privacy beyond k-anonymity. *IEEE ICDE Conference*, 2006.

[12] Meyerson A., Williams R. On the complexity of optimal k-anonymity. *ACM PODS Conference*, 2004.

[13] Nabar S., Marthi B., Kenthapadi K., Mishra N., Motwani R.: Towards Robustness in Query Auditing. *VLDB Conference*, 2006.

[14] Pinkas B.: Cryptographic Techniques for Privacy-Preserving Data Mining. *ACM SIGKDD Explorations*, 4(2), 2002.

[15] Rizvi S., Haritsa J. Maintaining Data Privacy in Association Rule Mining. *VLDB Conference*, 2002.

[16] Samarati P., Sweeney L. Protecting Privacy when Disclosing Information: k-Anonymity and its Enforcement Through Generalization and Suppression. *IEEE Symp. on Security and Privacy*, 1998.

[17] Verykios V. S., Elmagarmid A., Bertino E., Saygin Y.,, Dasseni E.: Association Rule Hiding. *IEEE Transactions on Knowledge and Data Engineering*, 16(4), 2004.

[18] Xiao X., Tao Y.. Personalized Privacy Preservation. *ACM SIGMOD Conference*, 2006.

Chapter 2

A General Survey of Privacy-Preserving Data Mining Models and Algorithms

Charu C. Aggarwal
IBM T. J. Watson Research Center
Hawthorne, NY 10532
charu@us.ibm.com

Philip S. Yu
University of Illinois at Chicago
Chicago, IL 60607
psyu@cs.uic.edu

Abstract In recent years, privacy-preserving data mining has been studied extensively, because of the wide proliferation of sensitive information on the internet. A number of algorithmic techniques have been designed for privacy-preserving data mining. In this paper, we provide a review of the state-of-the-art methods for privacy. We discuss methods for randomization, k-anonymization, and distributed privacy-preserving data mining. We also discuss cases in which the output of data mining applications needs to be sanitized for privacy-preservation purposes. We discuss the computational and theoretical limits associated with privacy-preservation over high dimensional data sets.

Keywords: Privacy-preserving data mining, randomization, k-anonymity.

2.1 Introduction

In recent years, data mining has been viewed as a threat to privacy because of the widespread proliferation of electronic data maintained by corporations. This has lead to increased concerns about the privacy of the underlying data. In recent years, a number of techniques have been proposed for modifying or transforming the data in such a way so as to preserve privacy. A survey on some of the techniques used for privacy-preserving data mining may be found in [123]. In this chapter, we will study an overview of the state-of-the-art in privacy-preserving data mining.

Privacy-preserving data mining finds numerous applications in surveillance which are naturally supposed to be "privacy-violating" applications. The key is to design methods [113] which continue to be effective, without compromising security. In [113], a number of techniques have been discussed for bio-surveillance, facial de-dentification, and identity theft. More detailed discussions on some of these sssues may be found in [96, 114–116].

Most methods for privacy computations use some form of transformation on the data in order to perform the privacy preservation. Typically, such methods reduce the granularity of representation in order to reduce the privacy. This reduction in granularity results in some loss of effectiveness of data management or mining algorithms. This is the natural trade-off between information loss and privacy. Some examples of such techniques are as follows:

- *The randomization method:* The randomization method is a technique for privacy-preserving data mining in which noise is added to the data in order to mask the attribute values of records [2, 5]. The noise added is sufficiently large so that individual record values cannot be recovered. Therefore, techniques are designed to derive aggregate distributions from the perturbed records. Subsequently, data mining techniques can be developed in order to work with these aggregate distributions. We will describe the randomization technique in greater detail in a later section.

- *The k-anonymity model and l-diversity:* The k-anonymity model was developed because of the possibility of indirect identification of records from public databases. This is because combinations of record attributes can be used to exactly identify individual records. In the k-anonymity method, we reduce the granularity of data representation with the use of techniques such as generalization and suppression. This granularity is reduced sufficiently that any given record maps onto at least k other records in the data. The l-diversity model was designed to handle some weaknesses in the k-anonymity model since protecting identities to the level of k-individuals is not the same as protecting the corresponding sensitive values, especially when there is homogeneity of sensitive values within a group. To do so, the concept of intra-group diversity of sensitive values is promoted within the anonymization scheme [83].

- *Distributed privacy preservation:* In many cases, individual entities may wish to derive *aggregate results* from data sets which are partitioned across these entities. Such partitioning may be horizontal (when the records are distributed across multiple entities) or vertical (when the attributes are distributed across multiple entities). While the individual entities may not desire to share their entire data sets, they may consent to limited information sharing with the use of a variety of protocols. The overall effect of such methods is to maintain privacy for each individual entity, while deriving aggregate results over the entire data.

- *Downgrading Application Effectiveness:* In many cases, even though the data may not be available, the output of applications such as association rule mining, classification or query processing may result in violations of privacy. This has lead to research in downgrading the effectiveness of applications by either data or application modifications. Some examples of such techniques include association rule hiding [124], classifier downgrading [92], and query auditing [1].

In this paper, we will provide a broad overview of the different techniques for privacy-preserving data mining. We will provide a review of the major algorithms available for each method, and the variations on the different techniques. We will also discuss a number of combinations of different concepts such as k-anonymous mining over vertically- or horizontally-partitioned data. We will also discuss a number of unique challenges associated with privacy-preserving data mining in the high dimensional case.

This paper is organized as follows. In section 2, we will introduce the randomization method for privacy preserving data mining. In section 3, we will discuss the k-anonymization method along with its different variations. In section 4, we will discuss issues in distributed privacy-preserving data mining. In section 5, we will discuss a number of techniques for privacy which arise in the context of sensitive output of a variety of data mining and data management applications. In section 6, we will discuss some unique challenges associated with privacy in the high dimensional case. A number of applications of privacy-preserving models and algorithms are discussed in Section 7. Section 8 contains the conclusions and discussions.

2.2 The Randomization Method

In this section, we will discuss the randomization method for privacy-preserving data mining. The randomization method has been traditionally used in the context of distorting data by probability distribution for methods such as surveys which have an evasive answer bias because of privacy concerns [74, 129]. This technique has also been extended to the problem of privacy-preserving data mining [2].

The method of randomization can be described as follows. Consider a set of data records denoted by $X = \{x_1 \dots x_N\}$. For record $x_i \in X$, we add a noise component which is drawn from the probability distribution $f_Y(y)$. These noise components are drawn independently, and are denoted $y_1 \dots y_N$. Thus, the new set of distorted records are denoted by $x_1 + y_1 \dots x_N + y_N$. We denote this new set of records by $z_1 \ldots z_N$. In general, it is assumed that the variance of the added noise is large enough, so that the original record values cannot be easily guessed from the distorted data. Thus, the original records cannot be recovered, but the distribution of the original records can be recovered.

Thus, if X be the random variable denoting the data distribution for the original record, Y be the random variable describing the noise distribution, and Z be the random variable denoting the final record, we have:

$$Z = X + Y$$
$$X = Z - Y$$

Now, we note that N instantiations of the probability distribution Z are known, whereas the distribution Y is known publicly. For a large enough number of values of N, the distribution Z can be approximated closely by using a variety of methods such as kernel density estimation. By subtracting Y from the approximated distribution of Z, it is possible to approximate the original probability distribution X. In practice, one can combine the process of approximation of Z with subtraction of the distribution Y from Z by using a variety of iterative methods such as those discussed in [2, 5]. Such iterative methods typically have a higher accuracy than the sequential solution of first approximating Z and then subtracting Y from it. In particular, the EM method proposed in [5] shows a number of optimal properties in approximating the distribution of X.

We note that at the end of the process, we only have a *distribution* containing the behavior of X. Individual records are not available. Furthermore, the distributions are available only along individual dimensions. Therefore, new data mining algorithms need to be designed to work with the uni-variate distributions rather than the individual records. This can sometimes be a challenge, since many data mining algorithms are inherently dependent on statistics which can only be extracted from either the individual records or the multi-variate probability distributions associated with the records. While the approach can certainly be extended to multi-variate distributions, density estimation becomes inherently more challenging [112] with increasing dimensionalities. For even modest dimensionalities such as 7 to 10, the process of density estimation becomes increasingly inaccurate, and falls prey to the curse of dimensionality.

One key advantage of the randomization method is that it is relatively simple, and does not require knowledge of the distribution of other records in the data. This is not true of other methods such as k-anonymity which require the knowledge of other records in the data. Therefore, the randomization method can be implemented at *data collection time*, and does not require the use of a trusted server containing all the original records in order to perform the anonymization process. While this is a strength of the randomization method, it also leads to some weaknesses, since it treats all records equally irrespective of their local density. Therefore, outlier records are more susceptible to adversarial attacks as compared to records in more dense regions in the data [10]. In order to guard against this, one may need to be needlessly more aggressive in adding noise to all the records in the data. This reduces the utility of the data for mining purposes.

The randomization method has been extended to a variety of data mining problems. In [2], it was discussed how to use the approach for classification. A number of other techniques [143, 145] have also been proposed which seem to work well over a variety of different classifiers. Techniques have also been proposed for privacy-preserving methods of improving the effectiveness of classifiers. For example, the work in [51] proposes methods for privacy-preserving boosting of classifiers. Methods for privacy-preserving mining of association rules have been proposed in [47, 107]. The problem of association rules is especially challenging because of the discrete nature of the attributes corresponding to presence or absence of items. In order to deal with this issue, the randomization technique needs to be modified slightly. Instead of adding quantitative noise, random items are dropped or included with a certain probability. The perturbed transactions are then used for aggregate association rule mining. This technique has shown to be extremely effective in [47]. The randomization approach has also been extended to other applications such as OLAP [3], and SVD based collaborative filtering [103].

2.2.1 Privacy Quantification

The quantity used to measure privacy should indicate how closely the original value of an attribute can be estimated. The work in [2] uses a measure that defines privacy as follows: If the original value can be estimated with $c\%$ confidence to lie in the interval $[\alpha_1, \alpha_2]$, then the interval width $(\alpha_2 - \alpha_1)$ defines the amount of privacy at $c\%$ confidence level. For example, if the perturbing additive is uniformly distributed in an interval of width 2α, then α is the amount of privacy at confidence level 50% and 2α is the amount of privacy at confidence level 100%. However, this simple method of determining privacy can be subtly incomplete in some situations. This can be best explained by the following example.

EXAMPLE 2.1 *Consider an attribute X with the density function $f_X(x)$ given by:*

$$f_X(x) = \begin{cases} 0.5 & 0 \leq x \leq 1 \\ 0.5 & 4 \leq x \leq 5 \\ 0 & \textit{otherwise} \end{cases}$$

Assume that the perturbing additive Y is distributed uniformly between $[-1, 1]$. *Then according to the measure proposed in [2], the amount of privacy is 2 at confidence level 100%.*

However, after performing the perturbation and subsequent reconstruction, the density function $f_X(x)$ will be approximately revealed. Let us assume for a moment that a large amount of data is available, so that the distribution function is revealed to a high degree of accuracy. Since the (distribution of the) perturbing additive is publically known, the two pieces of information can be combined to determine that if $Z \in [-1, 2]$, then $X \in [0, 1]$; whereas if $Z \in [3, 6]$ then $X \in [4, 5]$.

Thus, in each case, the value of X can be localized to an interval of length 1. This means that the actual amount of privacy offered by the perturbing additive Y is at most *1 at confidence level 100%. We use the qualifier 'at most' since X can often be localized to an interval of length less than one. For example, if the value of Z happens to be -0.5, then the value of X can be localized to an even smaller interval of* $[0, 0.5]$.

This example illustrates that the method suggested in [2] does not take into account the distribution of original data. In other words, the (aggregate) reconstruction of the attribute value also provides a certain level of knowledge which can be used to guess a data value to a higher level of accuracy. To accurately quantify privacy, we need a method which takes such side-information into account.

A key privacy measure [5] is based on the *differential entropy* of a random variable. The differential entropy $h(A)$ of a random variable A is defined as follows:

$$h(A) = -\int_{\Omega_A} f_A(a) \log_2 f_A(a)\, da \tag{2.1}$$

where Ω_A is the domain of A. It is well-known that $h(A)$ is a measure of uncertainty inherent in the value of A [111]. It can be easily seen that for a random variable U distributed uniformly between 0 and a, $h(U) = \log_2(a)$. For $a = 1$, $h(U) = 0$.

In [5], it was proposed that $2^{h(A)}$ is a measure of privacy inherent in the random variable A. This value is denoted by $\Pi(A)$. Thus, a random variable U distributed uniformly between 0 and a has privacy $\Pi(U) = 2^{log_2(a)} = a$. For a general random variable A, $\Pi(A)$ denote the length of the interval, over which a uniformly distributed random variable has the same uncertainty as A.

Given a random variable B, the *conditional* differential entropy of A is defined as follows:

$$h(A|B) = -\int_{\Omega_{A,B}} f_{A,B}(a, b) \log_2 f_{A|B=b}(a)\, da\, db \tag{2.2}$$

Thus, the average conditional privacy of A given B is $\Pi(A|B) = 2^{h(A|B)}$. This motivates the following metric $\mathcal{P}(A|B)$ for the conditional privacy loss of A, given B:

$$\mathcal{P}(A|B) = 1 - \Pi(A|B)/\Pi(A) = 1 - 2^{h(A|B)}/2^{h(A)} = 1 - 2^{-I(A;B)}.$$

where $I(A;B) = h(A) - h(A|B) = h(B) - h(B|A)$. $I(A;B)$ is also known as the *mutual information* between the random variables A and B. Clearly, $\mathcal{P}(A|B)$ is the fraction of privacy of A which is lost by revealing B.

As an illustration, let us reconsider Example 2.1 given above. In this case, the differential entropy of X is given by:

$$h(X) = -\int_{\Omega_X} f_X(x) \log_2 f_X(x)\, dx$$
$$= -\int_0^1 0.5 \log_2 0.5\, dx - \int_4^5 0.5 \log_2 0.5\, dx$$
$$= 1$$

Thus the privacy of X, $\Pi(X) = 2^1 = 2$. In other words, X has as much privacy as a random variable distributed uniformly in an interval of length 2. The density function of the perturbed value Z is given by $f_Z(z) = \int_{-\infty}^{\infty} f_X(\nu) f_Y(z - \nu)\, d\nu$.

Using $f_Z(z)$, we can compute the differential entropy $h(Z)$ of Z. It turns out that $h(Z) = 9/4$. Therefore, we have:

$$I(X;Z) = h(Z) - h(Z|X) = 9/4 - h(Y) = 9/4 - 1 = 5/4$$

Here, the second equality $h(Z|X) = h(Y)$ follows from the fact that X and Y are independent and $Z = X + Y$. Thus, the fraction of privacy loss in this case is $\mathcal{P}(X|Z) = 1 - 2^{-5/4} = 0.5796$. Therefore, after revealing Z, X has privacy $\Pi(X|Z) = \Pi(X) \times (1 - \mathcal{P}(X|Z)) = 2 \times (1.0 - 0.5796) = 0.8408$. This value is less than 1, since X can be localized to an interval of length less than one for many values of Z.

The problem of privacy quantification has been studied quite extensively in the literature, and a variety of metrics have been proposed to quantify privacy. A number of quantification issues in the measurement of privacy breaches has been discussed in [46, 48]. In [19], the problem of privacy-preservation has been studied from the broader context of the tradeoff between the privacy and the information loss. We note that the quantification of privacy alone is not sufficient without quantifying the utility of the data created by the randomization process. A framework has been proposed to explore this tradeoff for a variety of different privacy transformation algorithms.

2.2.2 Adversarial Attacks on Randomization

In the earlier section on privacy quantification, we illustrated an example in which the reconstructed distribution on the data can be used in order to reduce the privacy of the underlying data record. In general, a systematic approach can be used to do this in multi-dimensional data sets with the use of spectral filtering or PCA based techniques [54, 66]. The broad idea in techniques such as PCA [54] is that the correlation structure in the original data can be estimated fairly accurately (in larger data sets) even after noise addition. Once the broad correlation structure in the data has been determined, one can then try to remove the noise in the data in such a way that it fits the aggregate correlation structure of the data. It has been shown that such techniques can reduce the privacy of the perturbation process significantly since the noise removal results in values which are fairly close to their original values [54, 66]. Some other discussions on limiting breaches of privacy in the randomization method may be found in [46].

A second kind of adversarial attack is with the use of public information. Consider a record $X = (x_1 \ldots x_d)$, which is perturbed to $Z = (z_1 \ldots z_d)$. Then, since the distribution of the perturbations is known, we can try to use a maximum likelihood fit of the *potential perturbation* of Z to a public record. Consider the publicly public record $W = (w_1 \ldots w_d)$. Then, the *potential perturbation* of Z with respect to W is given by $(Z - W) = (z_1 - w_1 \ldots z_d - w_d)$. Each of these values $(z_i - w_i)$ should fit the distribution $f_Y(y)$. The corresponding log-likelihood fit is given by $-\sum_{i=1}^{d} \log(f_y(z_i - w_i))$. The higher the log-likelihood fit, the greater the probability that the record W corresponds to X. If it is known that the public data set always includes X, then the maximum likelihood fit can provide a high degree of certainty in identifying the correct record, especially in cases where d is large. We will discuss this issue in greater detail in a later section.

2.2.3 Randomization Methods for Data Streams

The randomization approach is particularly well suited to privacy-preserving data mining of streams, since the noise added to a given record is independent of the rest of the data. However, streams provide a particularly vulnerable target for adversarial attacks with the use of PCA based techniques [54] because of the large volume of the data available for analysis. In [78], an interesting technique for randomization has been proposed which uses the auto-correlations in different time series while deciding the noise to be added to any particular value. It has been shown in [78] that such an approach is more robust since the noise correlates with the stream behavior, and it is more difficult to create effective adversarial attacks with the use of correlation analysis techniques.

2.2.4 Multiplicative Perturbations

The most common method of randomization is that of additive perturbations. However, multiplicative perturbations can also be used to good effect for privacy-preserving data mining. Many of these techniques derive their roots in the work of [61] which shows how to use multi-dimensional projections in order to reduce the dimensionality of the data. This technique preserves the inter-record distances approximately, and therefore the transformed records can be used in conjunction with a variety of data mining applications. In particular, the approach is discussed in detail in [97, 98], in which it is shown how to use the method for privacy-preserving clustering. The technique can also be applied to the problem of classification as discussed in [28]. Multiplicative perturbations can also be used for distributed privacy-preserving data mining. Details can be found in [81]. A number of techniques for multiplicative perturbation in the context of masking census data may be found in [70]. A variation on this theme may be implemented with the use of distance preserving fourier transforms, which work effectively for a variety of cases [91].

As in the case of additive perturbations, multiplicative perturbations are not entirely safe from adversarial attacks. In general, if the attacker has no prior knowledge of the data, then it is relatively difficult to attack the privacy of the transformation. However, with some prior knowledge, two kinds of attacks are possible [82]:

- **Known Input-Output Attack:** In this case, the attacker knows some linearly independent collection of records, and their corresponding perturbed version. In such cases, linear algebra techniques can be used to reverse-engineer the nature of the privacy preserving transformation.

- **Known Sample Attack:** In this case, the attacker has a collection of independent data samples from the same distribution from which the original data was drawn. In such cases, principal component analysis techniques can be used in order to reconstruct the behavior of the original data.

2.2.5 Data Swapping

We note that noise addition or multiplication is not the only technique which can be used to perturb the data. A related method is that of data swapping, in which the values across different records are swapped in order to perform the privacy-preservation [49]. One advantage of this technique is that the lower order marginal totals of the data are completely preserved and are not perturbed at all. Therefore certain kinds of aggregate computations can be exactly performed without violating the privacy of the data. We note that this technique does not follow the general principle in randomization which allows the value of a record to be perturbed independent;y of the other records. Therefore, this technique can be used in combination with other frameworks such as k-anonymity, as long as the swapping process is designed to preserve the definitions of privacy for that model.

2.3 Group Based Anonymization

The randomization method is a simple technique which can be easily implemented at *data collection time*, because the noise added to a given record is independent of the behavior of other data records. This is also a weakness because outlier records can often be difficult to mask. Clearly, in cases in which the privacy-preservation does not need to be performed at data-collection time, it is desirable to have a technique in which the level of inaccuracy depends upon the behavior of the locality of that given record. Another key weakness of the randomization framework is that it does not consider the possibility that publicly available records can be used to identify the identity of the owners of that record. In [10], it has been shown that the use of publicly available records can lead to the privacy getting heavily compromised in high-dimensional cases. This is especially true of outlier records which can be easily distinguished from other records in their locality. Therefore, a broad approach to many privacy transformations is to construct groups of anonymous records which are transformed in a group-specific way.

2.3.1 The k-Anonymity Framework

In many applications, the data records are made available by simply removing key identifiers such as the name and social-security numbers from personal records. However, other kinds of attributes (known as pseudo-identifiers) can be used in order to accurately identify the records. Foe example, attributes such as age, zip-code and sex are available in public records such as census rolls. When these attributes are also available in a given data set, they can be used to infer the identity of the corresponding individual. A combination of these attributes can be very powerful, since they can be used to narrow down the possibilities to a small number of individuals.

In k-anonymity techniques [110], we reduce the granularity of representation of these pseudo-identifiers with the use of techniques such as *generalization* and *suppression*. In the method of *generalization*, the attribute values are generalized to a range in order to reduce the granularity of representation. For example, the date of birth could be generalized to a range such as year of birth, so as to reduce the risk of identification. In the method of *suppression*, the value of the attribute is removed completely. It is clear that such methods reduce the risk of identification with the use of public records, while reducing the accuracy of applications on the transformed data.

In order to reduce the risk of identification, the k-anonymity approach requires that every tuple in the table be indistinguishability related to no fewer than k respondents. This can be formalized as follows:

DEFINITION 2.2 *Each release of the data must be such that every combination of values of quasi-identifiers can be indistinguishably matched to at least k respondents.*

The first algorithm for k-anonymity was proposed in [110]. The approach uses *domain generalization hierarchies* of the quasi-identifiers in order to build k-anonymous tables. The concept of k-minimal generalization has been proposed in [110] in order to limit the level of generalization for maintaining as much data precision as possible for a given level of anonymity. Subsequently, the topic of k-anonymity has been widely researched. A good overview and survey of the corresponding algorithms may be found in [31].

We note that the problem of optimal anonymization is inherently a difficult one. In [89], it has been shown that the problem of optimal k-anonymization is NP-hard. Nevertheless, the problem can be solved quite effectively by the use of a number of heuristic methods. A method proposed by Bayardo and Agrawal [18] is the *k-Optimize* algorithm which can often obtain effective solutions.

The approach assumes an ordering among the quasi-identifier attributes. The values of the attributes are discretized into intervals (quantitative attributes) or grouped into different sets of values (categorical attributes). Each such grouping is an *item*. For a given attribute, the corresponding items are also ordered. An index is created using these attribute-interval pairs (or items) and a set enumeration tree is constructed on these attribute-interval pairs. This set enumeration tree is a systematic enumeration of all possible generalizations with the use of these groupings. The root of the node is the null node, and every successive level of the tree is constructed by appending one item which is lexicographically larger than all the items at that node of the tree. We note that the number of possible nodes in the tree increases exponentially with the data dimensionality. Therefore, it is not possible to build the entire tree even for modest values of n. However, the k-Optimize algorithm can use a number of pruning strategies to good effect. In particular, a node of the tree can be pruned when it is determined that no descendent of it could be optimal. This can be done by computing a bound on the quality of all descendents of that node, and comparing it to the quality of the current best solution obtained during the traversal process. A branch and bound technique can be used to successively improve the quality of the solution during the traversal process. Eventually, it is possible to terminate the algorithm at a maximum computational time, and use the current solution at that point, which is often quite good, but may not be optimal.

In [75], the *Incognito* method has been proposed for computing a k-minimal generalization with the use of bottom-up aggregation along domain generalization hierarchies. The Incognito method uses a bottom-up breadth-first search of the domain generalization hierarchy, in which it generates all the possible minimal k-anonymous tables for a given private table. First, it checks k-anonymity for each single attribute, and removes all those generalizations which do not satisfy k-anonymity. Then, it computes generalizations in pairs, again pruning those pairs which do not satisfy the k-anonymity constraints. In general, the Incognito algorithm computes $(i + 1)$-dimensional generalization *candidates* from the i-dimensional generalizations, and removes all those those generalizations which do not satisfy the k-anonymity constraint. This approach is continued until, no further candidates can be constructed, or all possible dimensions have been exhausted. We note that the methods in [76, 75] use a more general model for k-anonymity than that in [110]. This is because the method in [110] assumes that the value generalization hierarchy is a tree, whereas that in [76, 75] assumes that it is a graph.

Two interesting methods for top-down specialization and bottom-up generalization for k-anonymity have been proposed in [50, 125]. In [50], a top-down heuristic is designed, which starts with a general solution, and then specializes some attributes of the current solution so as to increase the information, but reduce the anonymity. The reduction in anonymity is always controlled, so that k-anonymity is never violated. At the same time each step of the specialization is controlled by a goodness metric which takes into account both the gain in information and the loss in anonymity. A complementary method to top down specialization is that of *bottom up generalization*, for which an interesting method is proposed in [125].

We note that generalization and suppression are not the only transformation techniques for implementing k-anonymity. For example in [38] it is discussed how to use micro-aggregation in which clusters of records are constructed. For each cluster, its representative value is the average value along each dimension in the cluster. A similar method for achieving anonymity via clustering is proposed in [15]. The work in [15] also provides constant factor approximation algorithms to design the clustering. In [8], a related method has been independently proposed for condensation based privacy-preserving data mining. This technique generates pseudo-data from clustered groups of k-records. The process of pseudo-data generation uses principal component analysis of the behavior of the records within a group. It has been shown in [8], that the approach can be effectively used for the problem of classification. We note that the use of pseudo-data provides an additional layer of protection, since it is difficult to perform adversarial attacks on synthetic data. At the same time, the aggregate behavior of the data is preserved, and this can be useful for a variety of data mining problems.

Since the problem of k-anonymization is essentially a search over a space of possible multi-dimensional solutions, standard heuristic search techniques such as genetic algorithms or simulated annealing can be effectively used. Such a technique has been proposed in [130] in which a simulated annealing algorithm is used in order to generate k-anonymous representations of the data. Another technique proposed in [59] uses genetic algorithms in order to construct k-anonymous representations of the data. Both of these techniques require high computational times, and provide no guarantees on the quality of the solutions found.

The only known techniques which provide guarantees on the quality of the solution are *approximation algorithms* [13, 14, 89], in which the solution found is guaranteed to be within a certain factor of the cost of the optimal solution. An approximation algorithm for k-anonymity was proposed in [89], and it provides an $O(k \cdot \log k)$ optimal solution. A number of techniques have also been proposed in [13, 14], which provide $O(k)$-approximations to the optimal cost k-anonymous solutions. In [100], a large improvement was proposed over these different methods. The technique in [100] proposes an $O(\log(k))$-approximation algorithm. This is significantly better than competing algorithms. Furthermore, the work in [100] also proposes a $O(\beta \cdot \log(k))$ approximation algorithm, where the parameter β can be gracefully adjusted based on running time constraints. Thus, this approach not only provides an approximation algorithm, but also gracefully explores the tradeoff between accuracy and running time.

In many cases, associations between pseudo-identifiers and sensitive attributes can be protected by using multiple views, such that the pseudo-identifiers and sensitive attributes occur in different views of the table. Thus, only a small subset of the selected views may be made available. It may be possible to achieve k-anonymity because of the lossy nature of the join across the two views. In the event that the join is not lossy enough, it may result in a violation of k-anonymity. In [140], the problem of violation of k-anonymity using multiple views has been studied. It has been shown that the problem is NP-hard in general. It has been shown in [140] that a polynomial time algorithm is possible if functional dependencies exist between the different views.

An interesting analysis of the safety of k-anonymization methods has been discussed in [73]. It tries to model the effectiveness of a k-anonymous representation, given that the attacker has some prior knowledge about the data such as a sample of the original data. Clearly, the more similar the sample data is to the true data, the greater the risk. The technique in [73] uses this fact to construct a model in which it calculates the expected number of items identified. This kind of technique can be useful in situations where it is desirable to determine whether or not anonymization should be used as the technique of choice for a particular situation.

2.3.2 Personalized Privacy-Preservation

Not all individuals or entities are equally concerned about their privacy. For example, a corporation may have very different constraints on the privacy of its records as compared to an individual. This leads to the natural problem that we may wish to treat the records in a given data set very differently for anonymization purposes. From a technical point of view, this means that the value of k for anonymization is not fixed but may vary with the record. A condensation-based approach [9] has been proposed for privacy-preserving data mining in the presence of variable constraints on the privacy of the data records. This technique constructs groups of non-homogeneous size from the data, such that it is guaranteed that each record lies in a group whose size is at least equal to its anonymity level. Subsequently, pseudo-data is generated from each group so as to create a synthetic data set with the same aggregate distribution as the original data.

Another interesting model of personalized anonymity is discussed in [132] in which a person can specify the level of privacy for his or her *sensitive values*. This technique assumes that an individual can specify a node of the domain generalization hierarchy in order to decide the level of anonymity that he can work with. This approach has the advantage that it allows for direct protection of the sensitive values of individuals than a vanilla k-anonymity method which is susceptible to different kinds of attacks.

2.3.3 Utility Based Privacy Preservation

The process of privacy-preservation leads to loss of information for data mining purposes. This loss of information can also be considered a loss of *utility* for data mining purposes. Since some negative results [7] on the curse of dimensionality suggest that a lot of attributes may need to be suppressed in order to preserve anonymity, it is extremely important to do this carefully in order to preserve utility. We note that many anonymization methods [18, 50, 83, 126] use cost measures in order to measure the information loss from the anonymization process. examples of such utility measures include generalization height [18], size of anonymized group [83], discernability measures of attribute values [18], and privacy information loss ratio[126]. In addition, a number of metrics such as the classification metric [59] explicitly try to perform the privacy-preservation in such a way so as to tailor the results with use for specific applications such as classification.

The problem of utility-based privacy-preserving data mining was first studied formally in [69]. The broad idea in [69] is to ameliorate the curse of dimensionality by separately publishing marginal tables containing attributes which have utility, but are also problematic for privacy-preservation purposes. The generalizations performed on the marginal tables and the original tables in fact do not need to be the same. It has been shown that this broad approach can preserve considerable utility of the data set without violating privacy.

A method for utility-based data mining using local recoding was proposed in [135]. The approach is based on the fact that different attributes have different utility from an application point of view. Most anonymization methods are *global*, in which a particular tuple value is mapped to the same generalized value globally. In local recoding, the data space is partitioned into a number of regions, and the mapping of the tuple to the generalizes value is local to that region. Clearly, this kind of approach has greater flexibility, since it can tailor the generalization process to a particular region of the data set. In [135], it has been shown that this method can perform quite effectively because of its local recoding strategy.

Another indirect approach to utility based anonymization is to make the privacy-preservation algorithms more aware of the workload [77]. Typically, data recipients may request only a subset of the data in many cases, and the union of these different requested parts of the data set is referred to as the workload. Clearly, a workload in which some records are used more frequently than others tends to suggest a different anonymization than one which is based on the entire data set. In [77], an effective and efficient algorithm has been proposed for workload aware anonymization.

Another direction for utility based privacy-preserving data mining is to anonymize the data in such a way that it remains useful for particular kinds of data mining or database applications. In such cases, the utility measure is often affected by the underlying application at hand. For example, in [50], a method has been proposed for k-anonymization using an information-loss metric as the utility measure. Such an approach is useful for the problem of classification. In [72], a method has been proposed for anonymization, so that the accuracy of the underlying queries is preserved.

2.3.4 Sequential Releases

Privacy-preserving data mining poses unique problems for dynamic applications such as data streams because in such cases, the data is released sequentially. In other cases, different views of the table may be released sequentially. Once a data block is released, it is no longer possible to go back and increase the level of generalization. On the other hand, new releases may sharpen an attacker's view of the data and may make the overall data set more susceptible to attack. For example, when different views of the data are released sequentially, then one may use a join on the two releases [127] in order to sharpen the ability to distinguish particular records in the data. A technique discussed in [127] relies on lossy joins in order to cripple an attack based on global quasi-identifiers. The intuition behind this approach is that if the join is lossy enough, it will reduce the confidence of the attacker in relating the release from previous views to the current release. Thus, the inability to link successive releases is key in preventing further discovery of the identity of records.

While the work in [127] explores the issue of sequential releases from the point of view of adding additional attributes, the work in [134] discusses the same issue when records are added to or deleted from the original data. A new generalization principle called m-invariance is proposed, which effectively limits the risk of privacy-disclosure in re-publication. Another method for handling sequential updates to the data set is discussed in [101]. The broad idea in this approach is to progressively and consistently increase the generalization granularity, so that the released data satisfies the k-anonymity requirement both with respect to the current table, as well as with respect to the previous releases.

2.3.5 The l-diversity Method

The k-anonymity is an attractive technique because of the simplicity of the definition and the numerous algorithms available to perform the anonymization. Nevertheless the technique is susceptible to many kinds of attacks especially when background knowledge is available to the attacker. Some kinds of such attacks are as follows:

- **Homogeneity Attack:** In this attack, all the values for a sensitive attribute within a group of k records are the same. Therefore, even though the data is k-anonymized, the value of the sensitive attribute for that group of k records can be predicted exactly.

- **Background Knowledge Attack:** In this attack, the adversary can use an association between one or more quasi-identifier attributes with the sensitive attribute in order to narrow down possible values of the sensitive field further. An example given in [83] is one in which background knowledge of low incidence of heart attacks among Japanese could be used to narrow down information for the sensitive field of what disease a patient might have. A detailed discussion of the effects of background knowledge on privacy may be found in [88].

Clearly, while k-anonymity is effective in preventing *identification* of a record, it may not always be effective in preventing inference of the sensitive values of the attributes of that record. Therefore, the technique of l-diversity was proposed which not only maintains the minimum group size of k, but also focusses on maintaining the diversity of the sensitive attributes. Therefore, the l-diversity model [83] for privacy is defined as follows:

DEFINITION 2.3 *Let a q^*-block be a set of tuples such that its non-sensitive values generalize to q^*. A q^*-block is l-diverse if it contains l "well represented" values for the sensitive attribute S. A table is l-diverse, if every q^*-block in it is l-diverse.*

A number of different instantiations for the l-diversity definition are discussed in [83]. We note that when there are multiple sensitive attributes, then the l-diversity problem becomes especially challenging because of the curse of dimensionality. Methods have been proposed in [83] for constructing l-diverse tables from the data set, though the technique remains susceptible to the curse of dimensionality [7]. Other methods for creating l-diverse tables are discussed in [133], in which a simple and efficient method for constructing the l-diverse representation is proposed.

2.3.6 The t-closeness Model

The t-closeness model is a further enhancement on the concept of l-diversity. One characteristic of the l-diversity model is that it treats all values of a given attribute in a similar way irrespective of its distribution in the data. This is rarely the case for real data sets, since the attribute values may be very skewed. This may make it more difficult to create feasible l-diverse representations. Often, an adversary may use background knowledge of the global distribution in order to make inferences about sensitive values in the data. Furthermore, not all values of an attribute are equally sensitive. For example, an attribute corresponding to a disease may be more sensitive when the value is positive, rather than when it is negative. In [79], a t-closeness model was proposed which uses the property that the distance between the distribution of the sensitive attribute within an anonymized group should not be different from the global distribution by more than a threshold t. The Earth Mover distance metric is used in order to quantify the distance between the two distributions. Furthermore, the t-closeness approach tends to be more effective than many other privacy-preserving data mining methods for the case of numeric attributes.

2.3.7 Models for Text, Binary and String Data

Most of the work on privacy-preserving data mining is focussed on numerical or categorical data. However, specific data domains such as strings, text, or market basket data may share specific properties with some of these general data domains, but may be different enough to require their own set of techniques for privacy-preservation. Some examples are as follows:

- **Text and Market Basket Data:** While these can be considered a case of text and market basket data, they are typically too high dimensional to work effectively with standard k-anonymization techniques. However, these kinds of data sets have the special property that they are extremely *sparse*. The sparsity property implies that only a few of the attributes are non-zero, and most of the attributes take on zero values. In [11], techniques have been proposed to construct anonymization methods which take advantage of this sparsity. In particular sketch based methods have been used to construct anonymized representations of the data. Variations are proposed to construct anonymizations which may be used at data collection time.

- **String Data:** String Data is considered challenging because of the variations in the lengths of strings across different records. Typically methods for k-anonymity are attribute specific, and therefore constructions of anonymizations for variable length records are quite difficult. In [12], a condensation based method has been proposed for anonymization of string data. This technique creates clusters from the different strings, and then generates synthetic data which has the same aggregate properties as the individual clusters. Since each cluster contains at least k-records, the anonymized data is guaranteed to at least satisfy the definitions of k-anonymity.

2.4 Distributed Privacy-Preserving Data Mining

The key goal in most distributed methods for privacy-preserving data mining is to allow computation of useful aggregate statistics over the entire data set without compromising the privacy of the individual data sets within the different participants. Thus, the participants may wish to collaborate in obtaining aggregate results, but may not fully trust each other in terms of the distribution of their own data sets. For this purpose, the data sets may either be *horizontally partitioned* or be *vertically partitioned*. In horizontally partitioned data sets, the individual records are spread out across multiple entities, each of which have the same set of attributes. In vertical partitioning, the individual entities may have different attributes (or views) of the same set of records. Both kinds of partitioning pose different challenges to the problem of distributed privacy-preserving data mining.

The problem of distributed privacy-preserving data mining overlaps closely with a field in cryptography for determining secure multi-party computations. A broad overview of the intersection between the fields of cryptography and privacy-preserving data mining may be found in [102]. The broad approach to cryptographic methods tends to compute functions over inputs provided by multiple recipients without actually sharing the inputs with one another. For example, in a 2-party setting, Alice and Bob may have two inputs x and y respectively, and may wish to both compute the function $f(x, y)$ without revealing x or y to each other. This problem can also be generalized across k parties by designing the k argument function $h(x_1 \ldots x_k)$. Many data mining algorithms may be viewed in the context of repetitive computations of many such primitive functions such as the scalar dot product, secure sum etc. In order to compute the function $f(x, y)$ or $h(x_1 \ldots, x_k)$, a *protocol* will have to designed for exchanging information in such a way that the function is computed without compromising privacy. We note that the robustness of the protocol depends upon the level of trust one is willing to place on the two participants Alice and Bob. This is because the protocol may be subjected to various kinds of adversarial behavior:

- **Semi-honest Adversaries:** In this case, the participants Alice and Bob are curious and attempt to learn from the information received by them during the protocol, but do not deviate from the protocol themselves. In many situations, this may be considered a realistic model of adversarial behavior.

- **Malicious Adversaries:** In this case, Alice and Bob may vary from the protocol, and may send sophisticated inputs to one another to learn from the information received from each other.

A key building-block for many kinds of secure function evaluations is the 1 out of 2 oblivious-transfer protocol. This protocol was proposed in [45, 105] and involves two parties: a *sender*, and a *receiver*. The sender's input is a pair (x_0, x_1), and the receiver's input is a bit value $\sigma \in \{0, 1\}$. At the end of the process, the receiver learns x_σ only, and the sender learns nothing. A number of simple solutions can be designed for this task. In one solution [45, 53], the receiver generates two random public keys, K_0 and K_1, but the receiver knows only the decryption key for K_σ. The receiver sends these keys to the sender, who encrypts x_0 with K_0, x_1 with K_1, and sends the encrypted data back to the receiver. At this point, the receiver can only decrypt x_σ, since this is the only input for which they have the decryption key. We note that this is a semi-honest solution, since the intermediate steps require an assumption of trust. For example, it is assumed that when the receiver sends two keys to the sender, they indeed know the decryption key to only one of them. In order to deal with the case of malicious adversaries, one must ensure that the sender chooses the public keys according to the protocol. An efficient method for doing so is described in [94]. In [94], generalizations of the 1 out of 2 oblivious transfer protocol to the 1 out N case and k out of N case are described.

Since the oblivious transfer protocol is used as a building block for secure multi-party computation, it may be repeated many times over a given function evaluation. Therefore, the computational effectiveness of the approach is important. Efficient methods for both semi-honest and malicious adversaries are discussed in [94]. More complex problems in this domain include the computation of probabilistic functions over a number of multi-party inputs [137]. Such powerful techniques can be used in order to abstract out the primitives from a number of computationally intensive data mining problems. Many of the above techniques have been described for the 2-party case, though generic solutions also exist for the multiparty case. Some important solutions for the multiparty case may be found in [25].

The oblivious transfer protocol can be used in order to compute several data mining primitives related to vector distances in multi-dimensional space. A classic problem which is often used as a primitive for many other problems is that of computing the scalar dot-product in a distributed environment [58]. A fairly general set of methods in this direction are described in [39]. Many of these techniques work by sending changed or encrypted versions of the inputs to one another in order to compute the function with the different alternative versions followed by an oblivious transfer protocol to retrieve the correct value of the final output. A systematic framework is described in [39] to transform normal data mining problems to secure multi-party computation problems. The problems discussed in [39] include those of clustering, classification, association rule mining, data summarization, and generalization. A second set of methods for distributed privacy-preserving data mining is discussed in [32] in which the secure multi-party computation of a number of important data mining primitives is discussed. These methods include the secure sum, the secure set union, the secure size of set intersection and the scalar product. These techniques can be used as data mining primitives for secure multi-party computation over a variety of horizontally and vertically partitioned data sets. Next, we will discuss algorithms for secure multi-party computation over horizontally partitioned data sets.

2.4.1 Distributed Algorithms over Horizontally Partitioned Data Sets

In horizontally partitioned data sets, different sites contain different sets of records with the same (or highly overlapping) set of attributes which are used for mining purposes. Many of these techniques use specialized versions of the general methods discussed in [32, 39] for various problems. The work in [80] discusses the construction of a popular decision tree induction method called ID3 with the use of approximations of the best splitting attributes. Subsequently, a variety of classifiers have been generalized to the problem of horizontally-partitioned privacy preserving mining including the Naive Bayes Classifier [65], and the SVM Classifier with nonlinear kernels [141].

An extreme solution for the horizontally partitioned case is discussed in [139], in which privacy-preserving classification is performed in a *fully* distributed setting, where each customer has private access to only their own record. A host of other data mining applications have been generalized to the problem of horizontally partitioned data sets. These include the applications of association rule mining [64], clustering [57, 62, 63] and collaborative filtering [104]. Methods for cooperative statistical analysis using secure multi-party computation methods are discussed in [40, 41].

A related problem is that of information retrieval and document indexing in a network of content providers. This problem arises in the context of multiple providers which may need to cooperate with one another in sharing their content, but may essentially be business competitors. In [17], it has been discussed how an adversary may use the output of search engines and content providers in order to reconstruct the documents. Therefore, the level of trust required grows with the number of content providers. A solution to this problem [17] constructs a centralized privacy-preserving index in conjunction with a distributed access control mechanism. The privacy-preserving index maintains strong privacy guarantees even in the face of colluding adversaries, and even if the entire index is made public.

2.4.2 Distributed Algorithms over Vertically Partitioned Data

For the vertically partitioned case, many primitive operations such as computing the scalar product or the secure set size intersection can be useful in computing the results of data mining algorithms. For example, the methods in [58] discuss how to use to scalar dot product computation for frequent itemset counting. The process of counting can also be achieved by using the secure size of set intersection as described in [32]. Another method for association rule mining discussed in [119] uses the secure scalar product over the vertical bit representation of itemset inclusion in transactions, in order to compute the frequency of the corresponding itemsets. This key step is applied repeatedly within the framework of a roll up procedure of itemset counting. It has been shown in [119] that this approach is quite effective in practice.

The approach of vertically partitioned mining has been extended to a variety of data mining applications such as decision trees [122], SVM Classification [142], Naive Bayes Classifier [121], and k-means clustering [120]. A number of theoretical results on the ability to learn different kinds of functions in vertically partitioned databases with the use of cryptographic approaches are discussed in [42].

2.4.3 Distributed Algorithms for k-Anonymity

In many cases, it is important to maintain k-anonymity across different distributed parties. In [60], a k-anonymous protocol for data which is vertically partitioned across two parties is described. The broad idea is for the two parties to agree on the quasi-identifier to generalize to the same value before release. A similar approach is discussed in [128], in which the two parties agree on how the generalization is to be performed before release.

In [144], an approach has been discussed for the case of horizontally partitioned data. The work in [144] discusses an extreme case in which each site is a customer which owns exactly one tuple from the data. It is assumed that the data record has both sensitive attributes and quasi-identifier attributes. The solution uses encryption on the sensitive attributes. The sensitive values can be decrypted only if therefore are at least k records with the same values on the quasi-identifiers. Thus, k-anonymity is maintained.

The issue of k-anonymity is also important in the context of hiding identification in the context of distributed location based services [20, 52]. In this case, k-anonymity of the user-identity is maintained even when the location information is released. Such location information is often released when a user may send a message at any point from a given location.

A similar issue arises in the context of communication protocols in which the anonymity of senders (or receivers) may need to be protected. A message is said to be *sender k-anonymous*, if it is guaranteed that an attacker can at most narrow down the identity of the sender to k individuals. Similarly, a message is said to be *receiver k-anonymous*, if it is guaranteed that an attacker can at most narrow down the identity of the receiver to k individuals. A number of such techniques have been discussed in [56, 135, 138].

2.5 Privacy-Preservation of Application Results

In many cases, the output of applications can be used by an adversary in order to make significant inferences about the behavior of the underlying data. In this section, we will discuss a number of miscellaneous methods for privacy-preserving data mining which tend to preserve the privacy of the end results of applications such as association rule mining and query processing. This problem is related to that of disclosure control [1] in statistical databases, though advances in data mining methods provide increasingly sophisticated methods for adversaries to make inferences about the behavior of the underlying data. In cases, where the commercial data needs to be shared, the association rules may represent sensitive information for target-marketing purposes, which needs to be protected from inference.

In this section, we will discuss the issue of disclosure control for a number of applications such as association rule mining, classification, and query processing. The key goal here is to prevent adversaries from making inferences from the end results of data mining and management applications. A broad discussion of the security and privacy implications of data mining are presented in [33]. We will discuss each of the applications below:

2.5.1 Association Rule Hiding

Recent years have seen tremendous advances in the ability to perform association rule mining effectively. Such rules often encode important target marketing information about a business. Some of the earliest work on the challenges of association rule mining for database security may be found in [16]. Two broad approaches are used for association rule hiding:

- **Distortion:** In distortion [99], the entry for a given transaction is modified to a different value. Since, we are typically dealing with binary transactional data sets, the entry value is flipped.

- **Blocking:** In blocking [108], the entry is not modified, but is left incomplete. Thus, unknown entry values are used to prevent discovery of association rules.

We note that both the distortion and blocking processes have a number of side effects on the non-sensitive rules in the data. Some of the non-sensitive rules may be lost along with sensitive rules, and new *ghost rules* may be created because of the distortion or blocking process. Such side effects are undesirable since they reduce the utility of the data for mining purposes.

A formal proof of the NP-hardness of the distortion method for hiding association rule mining may be found in [16]. In [16], techniques are proposed for changing some of the 1-values to 0-values so that the support of the corresponding sensitive rules is appropriately lowered. The utility of the approach was defined by the number of non-sensitive rules whose support was also lowered by using such an approach. This approach was extended in [34] in which both support and confidence of the appropriate rules could be lowered. In this case, 0-values in the transactional database could also change to 1-values. In many cases, this resulted in spurious association rules (or ghost rules) which was an undesirable side effect of the process. A complete description of the various methods for data distortion for association rule hiding may be found in [124]. Another interesting piece of work which balances privacy and disclosure concerns of sanitized rules may be found in [99].

The broad idea of blocking was proposed in [23]. The attractiveness of the blocking approach is that it maintains the truthfulness of the underlying data, since it replaces a value with an unknown (often represented by '?') rather than a false value. Some interesting algorithms for using blocking for association rule hiding are presented in [109]. The work has been further extended in

[108] with a discussion of the effectiveness of reconstructing the hidden rules. Another interesting set of techniques for association rule hiding with limited side effects is discussed in [131]. The objective of this method is to reduce the loss of non-sensitive rules, or the creation of ghost rules during the rule hiding process.

In [6], it has been discussed how blocking techniques for hiding association rules can be used to prevent discovery of sensitive entries in the data set by an adversary. In this case, certain entries in the data are classified as sensitive, and only rules which disclose such entries are hidden. An efficient depth-first association mining algorithm is proposed for this task [6]. It has been shown that the methods can effectively reduce the disclosure of sensitive entries with the use of such a hiding process.

2.5.2 Downgrading Classifier Effectiveness

An important privacy-sensitive application is that of classification, in which the results of a classification application may be sensitive information for the owner of a data set. Therefore the issue is to modify the data in such a way that the accuracy of the classification process is reduced, while retaining the utility of the data for other kinds of applications. A number of techniques have been discussed in [24, 92] in reducing the classifier effectiveness in context of classification rule and decision tree applications. The notion of *parsimonious downgrading* is proposed [24] in the context of blocking out inference channels for classification purposes while mining the effect to the overall utility. A system called Rational Downgrader [92] was designed with the use of these principles.

The methods for association rule hiding can also be generalized to rule based classifiers. This is because rule based classifiers often use association rule mining methods as subroutines, so that the rules with the class labels in their consequent are used for classification purposes. For a classifier downgrading approach, such rules are sensitive rules, whereas all other rules (with non-class attributes in the consequent) are non-sensitive rules. An example of a method for rule based classifier downgradation is discussed in [95] in which it has been shown how to effectively hide classification rules for a data set.

2.5.3 Query Auditing and Inference Control

Many sensitive databases are not available for public access, but may have a public interface through which *aggregate querying* is allowed. This leads to the natural danger that a smart adversary may pose a sequence of queries through which he or she may infer sensitive facts about the data. The nature of this inference may correspond to *full disclosure*, in which an adversary may determine the exact values of the data attributes. A second notion is that of *partial disclosure* in which the adversary may be able to narrow down the values to a range, but may not be able to guess the exact value. Most work on query auditing generally concentrates on the full disclosure setting.

Two broad approaches are designed in order to reduce the likelihood of sensitive data discovery:

- **Query Auditing:** In query auditing, we deny one or more queries from a sequence of queries. The queries to be denied are chosen such that the sensitivity of the underlying data is preserved. Some examples of query auditing methods include [37, 68, 93, 106].

- **Query Inference Control:** In this case, we perturb the underlying data or the query result itself. The perturbation is engineered in such a way, so as to preserve the privacy of the underlying data. Examples of methods which use perturbation of the underlying data include [3, 26, 90]. Examples of methods which perturb the query result include [22, 36, 42–44].

An overview of classical methods for query auding may be found in [1]. The query auditing problem has an *online* version, in which we do not know the sequence of queries in advance, and an *offline* version, in which we do know this sequence in advance. Clearly, the offline version is open to better optimization from an auditing point of view.

The problem of query auditing was first studied in [37, 106]. This approach works for the online version of the query auditing problem. In these works, the sum query is studied, and privacy is protected by using restrictions on sizes and pairwise overlaps of the allowable queries. Let us assume that the query size is restricted to be at most k, and the number of common elements in pairwise query sets is at most m. Then, if q be the number of elements that the attacker already knows from background knowledge, it was shown that [37, 106] that the maximum number of queries allowed is $(2 \cdot k - (q + 1))/m$. We note that if N be the total number of data elements, the above expression is always bounded above by $2 \cdot N$. If for some constant c, we choose $k = N/c$ and $m = 1$, the approach can only support a constant number of queries, after which all queries would have to be denied by the auditor. Clearly, this is undesirable from an application point of view. Therefore, a considerable amount of research has been devoted to increasing the number of queries which can be answered by the auditor without compromising privacy.

In [67], the problem of sum auditing on sub-cubes of the data cube are studied, where a query expression is constructed using a string of 0, 1, and *. The elements to be summed up are determined by using matches to the query string pattern. In [71], the problem of auditing a database of boolean values is studied for the case of sum and max queries. In [21], and approach for query auditing is discussed which is actually a combination of the approach of denying some queries and modifying queries in order to achieve privacy.

In [68], the authors show that denials to queries depending upon the answer to the current query can leak information. The authors introduce the notion of simulatable auditing for auditing sum and max queries. In [93], the authors devise methods for auditing max queries and bags of max and min queries under the partial and full disclosure settings. The authors also examine the notion of *utility* in the context of auditing, and obtain results for sum queries in the full disclosure setting.

A number of techniques have also been proposed for the offline version of the auditing problem. In [29], a number of variations of the offline auditing problem have been studied. In the offline auditing problem, we are given a sequence of queries which have been truthfully answered, and we need to determine if privacy has been breached. In [29], effective algorithms were proposed for the sum, max, and max and min versions of the problems. On the other hand, the sum and max version of the problem was shown to be NP-hard. In [4], an offline auditing framework was proposed for determining whether a database adheres to its disclosure properties. The key idea is to create an audit expression which specifies sensitive table entries.

A number of techniques have also been proposed for sanitizing or randomizing the data for query auditing purposes. These are fairly general models of privacy, since they preserve the privacy of the data even when the entire database is available. The standard methods for perturbation [2, 5] or k-anonymity [110] can always be used, and it is always guaranteed that an adversary may not derive anything more from the queries than they can from the base data. Thus, since a k-anonymity model guarantees a certain level of privacy even when the entire database is made available, it will continue to do so under any sequence of queries. In [26], a number of interesting methods are discussed for measuring the effectiveness of sanitization schemes in terms of balancing privacy and utility.

Instead of sanitizing the base data, it is possible to use summary constructs on the data, and respond to queries using only the information encoded in the summary constructs. Such an approach preserves privacy, as long as the summary constructs do not reveal sensitive information about the underlying records. A histogram based approach to data sanitization has been discussed in [26, 27]. In this technique the data is recursively partitioned into multi-dimensional cells. The final output is the exact description of the cuts along with the population of each cell. Clearly, this kind of description can be used for approximate query answering with the use of standard histogram query processing methods. In [55], a method has been proposed for privacy-preserving indexing of multi-dimensional data by using bucketizing of the underlying attribute values in conjunction with encryption of identification keys.

We note that a choice of larger bucket sizes provides greater privacy but less accuracy. Similarly, optimizing the bucket sizes for accuracy can lead to reductions in privacy. This tradeoff has been studied in [55], and it has been shown that reasonable query precision can be maintained at the expense of partial disclosure.

In the class of methods which use summarization structures for inference control, an interesting method was proposed by Mishra and Sandler in [90], which uses pseudo-random sketches for privacy-preservation. In this technique sketches are constructed from the data, and the sketch representations are used to respond to user queries. In [90], it has been shown that the scheme preserves privacy effectively, while continuing to be useful from a utility point of view.

Finally, an important class of query inference control methods changes the results of queries in order to preserve privacy. A classical method for aggregate queries such as the sum or relative frequency is that of random sampling [35]. In this technique, a random sample of the data is used to compute such aggregate functions. The random sampling approach makes it impossible for the questioner to precisely control the formation of query sets. The advantage of using a random sample is that the results of large queries are quite robust (in terms of *relative error*), but the privacy of individual records are preserved because of high *absolute error*.

Another method for query inference control is by adding noise to the results of queries. Clearly, the noise should be sufficient that an adversary cannot use small changes in the query arguments in order to infer facts about the base data. In [44], an interesting technique has been presented in which the result of a query is perturbed by an amount which depends upon the underlying sensitivity of the query function. This sensitivity of the query function is defined approximately by the change in the response to the query by changing one argument to the function. An important theoretical result [22, 36, 42, 43] shows that a surprisingly small amount of noise needs to be added to the result of a query, provided that the number of queries is sublinear in the number of database rows. With increasing sizes of databases today, this result provides fairly strong guarantees on privacy. Such queries together with their slightly noisy responses are referred to as the SuLQ primitive.

2.6 Limitations of Privacy: The Curse of Dimensionality

Many privacy-preserving data-mining methods are inherently limited by the curse of dimensionality in the presence of public information. For example, the technique in [7] analyzes the k-anonymity method in the presence of increasing dimensionality. The curse of dimensionality becomes especially important when adversaries may have considerable background information, as a result of which the boundary between pseudo-identifiers and sensitive attributes may become blurred. This is generally true, since adversaries may be familiar with the subject of interest and may have greater information about them than what is publicly available. This is also the motivation for techniques such as l-diversity [83] in which background knowledge can be used to make further privacy attacks. The work in [7] concludes that in order to maintain privacy, a large number of the attributes may need to be suppressed. Thus, the data loses its utility for the purpose of data mining algorithms. The broad intuition behind the result in [7] is that when attributes are generalized into wide ranges, the combination of a large number of generalized attributes is so sparsely populated, that even two anonymity becomes increasingly unlikely. While the method of l-diversity has not been formally analyzed, some observations made in [83] seem to suggest that the method becomes increasingly infeasible to implement effectively with increasing dimensionality.

The method of randomization has also been analyzed in [10]. This paper makes a first analysis of the ability to re-identify data records with the use of maximum likelihood estimates. Consider a d-dimensional record $X = (x_1 \ldots x_d)$, which is perturbed to $Z = (z_1 \ldots z_d)$. For a given public record $W = (w_1 \ldots w_d)$, we would like to find the probability that it could have been perturbed to Z using the perturbing distribution $f_Y(y)$. If this were true, then the set of values given by $(Z - W) = (z_1 - w_1 \ldots z_d - w_d)$ should be all drawn from the distribution $f_Y(y)$. The corresponding log-likelihood fit is given by $-\sum_{i=1}^{d} \log(f_y(z_i - w_i))$. The higher the log-likelihood fit, the greater the probability that the record W corresponds to X. In order to achieve greater anonymity, we would like the perturbations to be large enough, so that some of the spurious records in the data have greater log-likelihood fit to Z than the true record X. It has been shown in [10], that this probability reduces rapidly with increasing dimensionality for different kinds of perturbing distributions. Thus, the randomization technique also seems to be susceptible to the curse of high dimensionality.

We note that the problem of high dimensionality seems to be a fundamental one for privacy preservation, and it is unlikely that more effective methods can be found in order to preserve privacy when background information about a large number of features is available to even a subset of selected individuals. Indirect examples of such violations occur with the use of trail identifications [84, 85], where information from multiple sources can be compiled to create a high dimensional feature representation which violates privacy.

2.7 Applications of Privacy-Preserving Data Mining

The problem of privacy-preserving data mining has numerous applications in homeland security, medical database mining, and customer transaction analysis. Some of these applications such as those involving bio-terrorism and medical database mining may intersect in scope. In this section, we will discuss a number of different applications of privacy-preserving data mining methods.

2.7.1 Medical Databases: The Scrub and Datafly Systems

The scrub system [118] was designed for de-identification of clinical notes and letters which typically occurs in the form of textual data. Clinical notes and letters are typically in the form of text which contain references to patients, family members, addresses, phone numbers or providers. Traditional techniques simply use a global search and replace procedure in order to provide privacy. However clinical notes often contain cryptic references in the form of abbreviations which may only be understood either by other providers or members of the same institution. Therefore traditional methods can identify no more than 30-60% of the identifying information in the data [118]. The Scrub system uses numerous detection algorithms which compete in parallel to determine when a block of text corresponds to a name, address or a phone number. The Scrub System uses local knowledge sources which compete with one another based on the certainty of their findings. It has been shown in [118] that such a system is able to remove more than 99% of the identifying information from the data.

The Datafly System [117] was one of the earliest practical applications of privacy-preserving transformations. This system was designed to prevent identification of the subjects of medical records which may be stored in multi-dimensional format. The multi-dimensional information may include directly identifying information such as the social security number, or indirectly identifying information such as age, sex or zip-code. The system was designed in response to the concern that the process of removing only directly identifying attributes such as social security numbers was not sufficient to guarantee privacy. While the work has a similar motive as the k-anonymity approach of preventing record identification, it does not formally use a k-anonymity model in order to prevent identification through linkage attacks. The approach works by setting a minimum bin size for each field. The anonymity level is defined in Datafly with respect to this bin size. The values in the records are thus generalized to the ambiguity level of a bin size as opposed to exact values. Directly, identifying attributes such as the social-security-number, name, or zip-code are removed from the data. Furthermore, outlier values are suppressed from the data in order to prevent identification. Typically, the user of Datafly will set the anonymity level depending upon the profile of the data recipient in question. The overall anonymity level is defined between 0 and 1, which defines the minimum bin size for each field. An anonymity level of 0 results in Datafly providing the original data, whereas an anonymity level of 1 results in the maximum level of generalization of the underlying data. Thus, these two values provide two extreme values of trust and distrust. We note that these values are set depending upon the recipient of the data. When the records are released to the public, it is desirable to set of higher level of anonymity in order to ensure the maximum amount of protection. The generalizations in the datafly system are typically done independently at the individual attribute level, since the bins are defined independently for different attributes. The Datafly system is one of the earliest systems for anonymization, and is quite simple in its approach to anonymization. A lot of work in the anonymity field has been done since the creation of the Datafly system, and there is considerable scope for enhancement of the Datafly system with the use of these models.

2.7.2 Bioterrorism Applications

In typical bioterrorism applications, we would like to analyze medical data for privacy-preserving data mining purposes. Often a biological agent such as anthrax produces symptoms which are similar to other common respiratory diseases such as the cough, cold and the flu. In the absence of prior knowledge of such an attack, health care providers may diagnose a patient affected by an anthrax attack of have symptoms from one of the more common respiratory diseases. The key is to quickly identify a true anthrax attack from a normal outbreak of a common respiratory disease, In many cases, an unusual number of such cases in a given locality may indicate a bio-terrorism attack. Therefore, in order to identify such attacks it is necessary to track incidences of these common diseases as well. Therefore, the corresponding data would need to be reported to public health agencies. However, the common respiratory diseases are not reportable diseases by law. The solution proposed in [114] is that of "selective revelation" which initially allows only limited access to the data. However, in the event of suspicious activity, it allows a "drill-down" into the underlying data. This provides more identifiable information in accordance with public health law.

2.7.3 Homeland Security Applications

A number of applications for homeland security are inherently intrusive because of the very nature of surveillance. In [113], a broad overview is provided on how privacy-preserving techniques may be used in order to deploy these applications effectively without violating user privacy. Some examples of such applications are as follows:

- **Credential Validation Problem:** In this problem, we are trying to match the subject of the credential to the person presenting the credential. For example, the theft of social security numbers presents a serious threat to homeland security. In the credential validation approach [113], an attempt is made to exploit the semantics associated with the social security number to determine whether the person presenting the SSN credential truly owns it.

- **Identity Theft:** A related technology [115] is to use a more *active* approach to avoid identity theft. The *identity angel* system [115], crawls through cyberspace, and determines people who are at risk from identity theft. This information can be used to notify appropriate parties. We note that both the above approaches to prevention of identity theft are relatively non-invasive and therefore do not violate privacy.

- **Web Camera Surveillance:** One possible method for surveillance is with the use of publicly available webcams [113, 116], which can be used to detect unusual activity. We note that this is a much more invasive approach than the previously discussed techniques because of person-specific information being captured in the webcams. The approach can be made more privacy-sensitive by extracting only *facial count* information from the images and using these in order to detect unusual activity. It has been hypothesized in [116] that unusual activity can be detected only in terms of facial count rather than using more specific information about particular individuals. In effect, this kind of approach uses a domain-specific downgrading of the information available in the webcams in order to make the approach privacy-sensitive.

- **Video-Surveillance:** In the context of sharing video-surveillance data, a major threat is the use of facial recognition software, which can match the facial images in videos to the facial images in a driver license database. While a straightforward solution is to completely black out each face, the result is of limited new, since all facial information has been wiped out. A more balanced approach [96] is to use selective downgrading of the facial information, so that it scientifically limits the ability of facial recognition software to reliably identify faces, while maintaining facial details in images. The algorithm is referred to as k-Same, and the key is to identify faces which are somewhat similar, and then construct new faces which construct combinations of features from these similar faces. Thus, the identity of the underlying individual is anonymized to a certain extent, but the video continues to remain useful. Thus, this approach has the flavor of a k-anonymity approach, except that it creates new synthesized data for the application at hand.

- **The Watch List Problem:** The motivation behind this problem [113] is that the government typically has a list of known terrorists or suspected entities which it wishes to track from the population. The aim is to view transactional data such as store purchases, hospital admissions, airplane manifests, hotel registrations or school attendance records in order to identify or track these entities. This is a difficult problem because the transactional data is private, and the privacy of subjects who do not appear in the watch list need to be protected. Therefore, the transactional behavior of non-suspicious subjects may not be identified or revealed. Furthermore, the problem is even more difficult if we assume that the watch list cannot be revealed to the data holders. The second assumption is a result of the fact that members on the watch list may only be suspected entities and should have some level of protection from identification as suspected terrorists to the general public. The watch list problem is currently an open problem [113].

2.7.4 Genomic Privacy

Recent years have seen tremendous advances in the science of DNA sequencing and forensic analysis with the use of DNA. As result, the databases of collected DNA are growing very fast in the both the medical and law enforcement communities. DNA data is considered extremely sensitive, since it contains almost uniquely identifying information about an individual.

As in the case of multi-dimensional data, simple removal of directly identifying data such as social security number is not sufficient to prevent re-identification. In [86], it has been shown that a software called *CleanGene* can determine the identifiability of DNA entries independent of any other demographic or other identifiable information. The software relies on publicly available medical data and knowledge of particular diseases in order to assign identifications to DNA entries. It was shown in [86] that 98-100% of the individuals are identifiable using this approach. The identification is done by taking the DNA sequence of an individual and then constructing a genetic profile corresponding to the sex, genetic diseases, the location where the DNA was collected etc. This genetic profile has been shown in [86] to be quite effective in identifying the individual to a much smaller group. One way to protect the anonymity of such sequences is with the use of *generalization lattices* [87] which are constructed in such a way that an entry in the modified database cannot be distinguished from at least $(k-1)$ other entities. Another approach discussed in [11] constructs synthetic data which preserves the aggregate characteristics of the original data, but preserves the privacy of the original records. Another method for compromising the privacy of genomic data is that of *trail re-identification*, in which the uniqueness of patient visit patterns [84, 85] is exploited in order to make identifications. The premise of this work is that patients often visit and leave behind genomic data at various distributed locations and hospitals. The hospitals usually separate out the clinical data from the genomic data and make the genomic data available for research purposes. While the data is seemingly anonymous, the visit location pattern of the patients is encoded in the site from which the data is released. It has been shown in [84, 85] that this information may be combined with publicly available data in order to perform unique re-identifications. Some broad ideas for protecting the privacy in such scenarios are discussed in [85].

2.8 Summary

In this paper, we presented a survey of the broad areas of privacy-preserving data mining and the underlying algorithms. We discussed a variety of data modification techniques such as randomization and k-anonymity based techniques. We discussed methods for distributed privacy-preserving mining, and the methods for handling horizontally and vertically partitioned data. We discussed the issue of downgrading the effectiveness of data mining and data management applications such as association rule mining, classification, and query processing. We discussed some fundamental limitations of the problem of privacy-preservation in the presence of increased amounts of public information and background knowledge. Finally, we discussed a number of diverse application domains for which privacy-preserving data mining methods are useful.

References

[1] Adam N., Wortmann J. C.: Security-Control Methods for Statistical Databases: A Comparison Study. *ACM Computing Surveys*, 21(4), 1989.

[2] Agrawal R., Srikant R. Privacy-Preserving Data Mining. *Proceedings of the ACM SIGMOD Conference*, 2000.

[3] Agrawal R., Srikant R., Thomas D. Privacy-Preserving OLAP. *Proceedings of the ACM SIGMOD Conference*, 2005.

[4] Agrawal R., Bayardo R., Faloutsos C., Kiernan J., Rantzau R., Srikant R.: Auditing Compliance via a hippocratic database. *VLDB Conference*, 2004.

[5] Agrawal D. Aggarwal C. C. On the Design and Quantification of Privacy-Preserving Data Mining Algorithms. *ACM PODS Conference*, 2002.

[6] Aggarwal C., Pei J., Zhang B. A Framework for Privacy Preservation against Adversarial Data Mining. *ACM KDD Conference*, 2006.

[7] Aggarwal C. C. On k-anonymity and the curse of dimensionality. *VLDB Conference*, 2005.

[8] Aggarwal C. C., Yu P. S.: A Condensation approach to privacy preserving data mining. *EDBT Conference*, 2004.

[9] Aggarwal C. C., Yu P. S.: On Variable Constraints in Privacy-Preserving Data Mining. *SIAM Conference*, 2005.

[10] Aggarwal C. C.: On Randomization, Public Information and the Curse of Dimensionality. *ICDE Conference*, 2007.

[11] Aggarwal C. C., Yu P. S.: On Privacy-Preservation of Text and Sparse Binary Data with Sketches. *SIAM Conference on Data Mining*, 2007.

[12] Aggarwal C. C., Yu P. S. On Anonymization of String Data. *SIAM Conference on Data Mining*, 2007.

[13] Aggarwal G., Feder T., Kenthapadi K., Motwani R., Panigrahy R., Thomas D., Zhu A.: Anonymizing Tables. *ICDT Conference*, 2005.

[14] Aggarwal G., Feder T., Kenthapadi K., Motwani R., Panigrahy R., Thomas D., Zhu A.: Approximation Algorithms for k-anonymity. *Journal of Privacy Technology*, paper 20051120001, 2005.

[15] Aggarwal G., Feder T., Kenthapadi K., Khuller S., Motwani R., Panigrahy R., Thomas D., Zhu A.: Achieving Anonymity via Clustering. *ACM PODS Conference*, 2006.

[16] Atallah, M., Elmagarmid, A., Ibrahim, M., Bertino, E., Verykios, V.: Disclosure limitation of sensitive rules, *Workshop on Knowledge and Data Engineering Exchange*, 1999.

[17] Bawa M., Bayardo R. J., Agrawal R.: Privacy-Preserving Indexing of Documents on the Network. *VLDB Conference*, 2003.

[18] Bayardo R. J., Agrawal R.: Data Privacy through Optimal k-Anonymization. *Proceedings of the ICDE Conference*, pp. 217–228, 2005.

[19] Bertino E., Fovino I., Provenza L.: A Framework for Evaluating Privacy-Preserving Data Mining Algorithms. *Data Mining and Knowledge Discovery Journal*, 11(2), 2005.

[20] Bettini C., Wang X. S., Jajodia S.: Protecting Privacy against Location Based Personal Identification. *Proc. of Secure Data Management Workshop*, Trondheim, Norway, 2005.

[21] Biskup J., Bonatti P.: Controlled Query Evaluation for Known Policies by Combining Lying and Refusal. *Annals of Mathematics and Artificial Intelligence*, 40(1-2), 2004.

[22] Blum A., Dwork C., McSherry F., Nissim K.: Practical Privacy: The SuLQ Framework. *ACM PODS Conference*, 2005.

[23] Chang L., Moskowitz I.: An integrated framwork for database inference and privacy protection. *Data and Applications Security*. Kluwer, 2000.

[24] Chang L., Moskowitz I.: Parsimonious downgrading and decision trees applied to the inference problem. *New Security Paradigms Workshop*, 1998.

[25] Chaum D., Crepeau C., Damgard I.: Multiparty unconditionally secure protocols. *ACM STOC Conference*, 1988.

[26] Chawla S., Dwork C., McSherry F., Smith A., Wee H.: Towards Privacy in Public Databases, *TCC*, 2005.

[27] Chawla S., Dwork C., McSherry F., Talwar K.: On the Utility of Privacy-Preserving Histograms, *UAI*, 2005.

[28] Chen K., Liu L.: Privacy-preserving data classification with rotation perturbation. *ICDM Conference*, 2005.

[29] Chin F.: Security Problems on Inference Control for SUM, MAX, and MIN Queries. *J. of the ACM*, 33(3), 1986.

[30] Chin F., Ozsoyoglu G.: Auditing for Secure Statistical Databases. *Proceedings of the ACM'81 Conference*, 1981.

[31] Ciriani V., De Capitiani di Vimercati S., Foresti S., Samarati P.: k-Anonymity. *Security in Decentralized Data Management*, ed. Jajodia S., Yu T., Springer, 2006.

[32] Clifton C., Kantarcioglou M., Lin X., Zhu M.: Tools for privacy-preserving distributed data mining. *ACM SIGKDD Explorations*, 4(2), 2002.

[33] Clifton C., Marks D.: Security and Privacy Implications of Data Mining., *Workshop on Data Mining and Knowledge Discovery*, 1996.

[34] Dasseni E., Verykios V., Elmagarmid A., Bertino E.: Hiding Association Rules using Confidence and Support, *4th Information Hiding Workshop*, 2001.

[35] Denning D.: Secure Statistical Databases with Random Sample Queries. *ACM TODS Journal*, 5(3), 1980.

[36] Dinur I., Nissim K.: Revealing Information while preserving privacy. *ACM PODS Conference*, 2003.

[37] Dobkin D., Jones A., Lipton R.: Secure Databases: Protection against User Influence. *ACM Transactions on Databases Systems*, 4(1), 1979.

[38] Domingo-Ferrer J,, Mateo-Sanz J.: Practical data-oriented microaggregation for statistical disclosure control. *IEEE TKDE*, 14(1), 2002.

[39] Du W., Atallah M.: Secure Multi-party Computation: A Review and Open Problems.*CERIAS Tech. Report* 2001-51, Purdue University, 2001.

[40] Du W., Han Y. S., Chen S.: Privacy-Preserving Multivariate Statistical Analysis: Linear Regression and Classification, Proc. SIAM Conf. Data Mining, 2004.

[41] Du W., Atallah M.: Privacy-Preserving Cooperative Statistical Analysis, 17th Annual Computer Security Applications Conference, 2001.

[42] Dwork C., Nissim K.: Privacy-Preserving Data Mining on Vertically Partitioned Databases, *CRYPTO*, 2004.

[43] Dwork C., Kenthapadi K., McSherry F., Mironov I., Naor M.: Our Data, Ourselves: Privacy via Distributed Noise Generation. *EUROCRYPT*, 2006.

[44] Dwork C., McSherry F., Nissim K., Smith A.: Calibrating Noise to Sensitivity in Private Data Analysis, *TCC*, 2006.

[45] Even S., Goldreich O., Lempel A.: A Randomized Protocol for Signing Contracts. *Communications of the ACM*, vol 28, 1985.

[46] Evfimievski A., Gehrke J., Srikant R. Limiting Privacy Breaches in Privacy Preserving Data Mining. *ACM PODS Conference*, 2003.

[47] Evfimievski A., Srikant R., Agrawal R., Gehrke J.: Privacy-Preserving Mining of Association Rules. *ACM KDD Conference*, 2002.

[48] Evfimievski A.: Randomization in Privacy-Preserving Data Mining. *ACM SIGKDD Explorations*, 4, 2003.

[49] Fienberg S., McIntyre J.: Data Swapping: Variations on a Theme by Dalenius and Reiss. *Technical Report, National Institute of Statistical Sciences*, 2003.

[50] Fung B., Wang K., Yu P.: Top-Down Specialization for Information and Privacy Preservation. *ICDE Conference*, 2005.

[51] Gambs S., Kegl B., Aimeur E.: Privacy-Preserving Boosting. *Knowledge Discovery and Data Mining Journal*, to appear.

[52] Gedik B., Liu L.: A customizable k-anonymity model for protecting location privacy, *ICDCS Conference*, 2005.

[53] Goldreich O.: Secure Multi-Party Computation, Unpublished Manuscript, 2002.

[54] Huang Z., Du W., Chen B.: Deriving Private Information from Randomized Data. pp. 37–48, *ACM SIGMOD Conference*, 2005.

[55] Hore B., Mehrotra S., Tsudik B.: A Privacy-Preserving Index for Range Queries. *VLDB Conference*, 2004.

[56] Hughes D, Shmatikov V.: Information Hiding, Anonymity, and Privacy: A modular Approach. *Journal of Computer Security*, 12(1), 3–36, 2004.

[57] Inan A., Saygin Y., Savas E., Hintoglu A., Levi A.: Privacy-Preserving Clustering on Horizontally Partitioned Data. *Data Engineering Workshops*, 2006.

[58] Ioannidis I., Grama A., Atallah M.: A secure protocol for computing dot products in clustered and distributed environments, *International Conference on Parallel Processing*, 2002.

[59] Iyengar V. S.: Transforming Data to Satisfy Privacy Constraints. *KDD Conference*, 2002.

[60] Jiang W., Clifton C.: Privacy-preserving distributed k-Anonymity. *Proceedings of the IFIP 11.3 Working Conference on Data and Applications Security*, 2005.

[61] Johnson W., Lindenstrauss J.: Extensions of Lipshitz Mapping into Hilbert Space, *Contemporary Math.* vol. 26, pp. 189-206, 1984.

[62] Jagannathan G., Wright R.: Privacy-Preserving Distributed k-means clustering over arbitrarily partitioned data. *ACM KDD Conference*, 2005.

[63] Jagannathan G., Pillaipakkamnatt K., Wright R.: A New Privacy-Preserving Distributed k-Clustering Algorithm. *SIAM Conference on Data Mining*, 2006.

[64] Kantarcioglu M., Clifton C.: Privacy-Preserving Distributed Mining of Association Rules on Horizontally Partitioned Data. *IEEE TKDE Journal*, 16(9), 2004.

[65] Kantarcioglu M., Vaidya J.: Privacy-Preserving Naive Bayes Classifier for Horizontally Partitioned Data. *IEEE Workshop on Privacy-Preserving Data Mining*, 2003.

[66] Kargupta H., Datta S., Wang Q., Sivakumar K.: On the Privacy Preserving Properties of Random Data Perturbation Techniques. *ICDM Conference*, pp. 99-106, 2003.

[67] Karn J., Ullman J.: A model of statistical databases and their security. *ACM Transactions on Database Systems*, 2(1):1–10, 1977.

[68] Kenthapadi K.,Mishra N., Nissim K.: Simulatable Auditing, *ACM PODS Conference*, 2005.

[69] Kifer D., Gehrke J.: Injecting utility into anonymized datasets. *SIGMOD Conference*, pp. 217-228, 2006.

[70] Kim J., Winkler W.: Multiplicative Noise for Masking Continuous Data, *Technical Report Statistics 2003-01, Statistical Research Division, US Bureau of the Census*, Washington D.C., Apr. 2003.

[71] Kleinberg J., Papadimitriou C., Raghavan P.: Auditing Boolean Attributes. *Journal of Computer and System Sciences*, 6, 2003.

[72] Koudas N., Srivastava D., Yu T., Zhang Q.: Aggregate Query Answering on Anonymized Tables. *ICDE Conference*, 2007.

[73] Lakshmanan L., Ng R., Ramesh G. To Do or Not To Do: The Dilemma of Disclosing Anonymized Data. *ACM SIGMOD Conference*, 2005.

[74] Liew C. K., Choi U. J., Liew C. J. A data distortion by probability distribution. *ACM TODS*, 10(3):395-411, 1985.

[75] LeFevre K., DeWitt D., Ramakrishnan R.: Incognito: Full Domain K-Anonymity. *ACM SIGMOD Conference*, 2005.

[76] LeFevre K., DeWitt D., Ramakrishnan R.: Mondrian Multidimensional K-Anonymity. *ICDE Conference*, 25, 2006.

[77] LeFevre K., DeWitt D., Ramakrishnan R.: Workload Aware Anonymization. *KDD Conference*, 2006.

[78] Li F., Sun J., Papadimitriou S. Mihaila G., Stanoi I.: Hiding in the Crowd: Privacy Preservation on Evolving Streams through Correlation Tracking. *ICDE Conference*, 2007.

[79] Li N., Li T., Venkatasubramanian S: t-Closeness: Orivacy beyond k-anonymity and l-diversity. *ICDE Conference*, 2007.

[80] Lindell Y., Pinkas B.: Privacy-Preserving Data Mining. *CRYPTO*, 2000.

[81] Liu K., Kargupta H., Ryan J.: Random Projection Based Multiplicative Data Perturbation for Privacy Preserving Distributed Data Mining. *IEEE Transactions on Knowledge and Data Engineering*, 18(1), 2006.

[82] Liu K., Giannella C. Kargupta H.: An Attacker's View of Distance Preserving Maps for Privacy-Preserving Data Mining. *PKDD Conference*, 2006.

[83] Machanavajjhala A., Gehrke J., Kifer D., and Venkitasubramaniam M.: l-Diversity: Privacy Beyond k-Anonymity. *ICDE*, 2006.

[84] Malin B, Sweeney L. Re-identification of DNA through an automated linkage process. *Journal of the American Medical Informatics Association*, pp. 423–427, 2001.

[85] Malin B. Why methods for genomic data privacy fail and what we can do to fix it, *AAAS Annual Meeting*, Seattle, WA, 2004.

[86] Malin B., Sweeney L.: Determining the identifiability of DNA database entries. *Journal of the American Medical Informatics Association*, pp. 537–541, November 2000.

[87] Malin, B. Protecting DNA Sequence Anonymity with Generalization Lattices. *Methods of Information in Medicine*, 44(5): 687-692, 2005.

[88] Martin D., Kifer D., Machanavajjhala A., Gehrke J., Halpern J.: Worst-Case Background Knowledge. *ICDE Conference*, 2007.

[89] Meyerson A., Williams R. On the complexity of optimal k-anonymity. *ACM PODS Conference*, 2004.

[90] Mishra N., Sandler M.: Privacy vis Pseudorandom Sketches. *ACM PODS Conference*, 2006.

[91] Mukherjee S., Chen Z., Gangopadhyay S.: A privacy-preserving technique for Euclidean distance-based mining algorithms using Fourier based transforms, *VLDB Journal*, 2006.

[92] Moskowitz I., Chang L.: A decision theoretic system for information downgrading. *Joint Conference on Information Sciences*, 2000.

[93] Nabar S., Marthi B., Kenthapadi K., Mishra N., Motwani R.: Towards Robustness in Query Auditing. *VLDB Conference*, 2006.

[94] Naor M., Pinkas B.: Efficient Oblivious Transfer Protocols, *SODA Conference*, 2001.

[95] Natwichai J., Li X., Orlowska M.: A Reconstruction-based Algorithm for Classification Rules Hiding. *Australasian Database Conference*, 2006.

[96] Newton E., Sweeney L., Malin B.: Preserving Privacy by De-identifying Facial Images. *IEEE Transactions on Knowledge and Data Engineering, IEEE TKDE*, February 2005.

[97] Oliveira S. R. M., Zaane O.: Privacy Preserving Clustering by Data Transformation, *Proc. 18th Brazilian Symp. Databases*, pp. 304-318, Oct. 2003.

[98] Oliveira S. R. M., Zaiane O.: Data Perturbation by Rotation for Privacy-Preserving Clustering, *Technical Report TR04-17*, Department of Computing Science, University of Alberta, Edmonton, AB, Canada, August 2004.

[99] Oliveira S. R. M., Zaiane O., Saygin Y.: Secure Association-Rule Sharing. *PAKDD Conference*, 2004.

[100] Park H., Shim K. Approximate Algorithms for K-anonymity. *ACM SIGMOD Conference*, 2007.

[101] Pei J., Xu J., Wang Z., Wang W., Wang K.: Maintaining k-Anonymity against Incremental Updates. *Symposium on Scientific and Statistical Database Management*, 2007.

[102] Pinkas B.: Cryptographic Techniques for Privacy-Preserving Data Mining. *ACM SIGKDD Explorations*, 4(2), 2002.

[103] Polat H., Du W.: SVD-based collaborative filtering with privacy. *ACM SAC Symposium*, 2005.

[104] Polat H., Du W.: Privacy-Preserving Top-N Recommendations on Horizontally Partitioned Data. *Web Intelligence*, 2005.

[105] Rabin M. O.: How to exchange secrets by oblivious transfer, *Technical Report* TR-81, Aiken Corporation Laboratory, 1981.

[106] Reiss S.: Security in Databases: A combinatorial Study, *Journal of ACM*, 26(1), 1979.

[107] Rizvi S., Haritsa J.: Maintaining Data Privacy in Association Rule Mining. *VLDB Conference*, 2002.

[108] Saygin Y., Verykios V., Clifton C.: Using Unknowns to prevent discovery of Association Rules, *ACM SIGMOD Record*, 30(4), 2001.

[109] Saygin Y., Verykios V., Elmagarmid A.: Privacy-Preserving Association Rule Mining, *12th International Workshop on Research Issues in Data Engineering*, 2002.

[110] Samarati P.: Protecting Respondents' Identities in Microdata Release. IEEE Trans. Knowl. Data Eng. 13(6): 1010-1027 (2001).

[111] Shannon C. E.: The Mathematical Theory of Communication, University of Illinois Press, 1949.

[112] Silverman B. W.: Density Estimation for Statistics and Data Analysis. *Chapman and Hall*, 1986.

[113] Sweeney L.: Privacy Technologies for Homeland Security. *Testimony before the Privacy and Integrity Advisory Committee of the Deprtment of Homeland Scurity*, Boston, MA, June 15, 2005.

[114] Sweeney L.: Privacy-Preserving Bio-terrorism Surveillance. *AAAI Spring Symposium, AI Technologies for Homeland Security*, 2005.

[115] Sweeney L.: AI Technologies to Defeat Identity Theft Vulnerabilities. *AAAI Spring Symposium, AI Technologies for Homeland Security*, 2005.

[116] Sweeney L., Gross R.: Mining Images in Publicly-Available Cameras for Homeland Security. *AAAI Spring Symposium, AI Technologies for Homeland Security*, 2005.

[117] Sweeney L.: Guaranteeing Anonymity while Sharing Data, the Datafly System. *Journal of the American Medical Informatics Association*, 1997.

[118] Sweeney L.: Replacing Personally Identifiable Information in Medical Records, the Scrub System. *Journal of the American Medical Informatics Association*, 1996.

[119] Vaidya J., Clifton C.: Privacy-Preserving Association Rule Mining in Vertically Partitioned Databases. *ACM KDD Conference*, 2002.

[120] Vaidya J., Clifton C.: Privacy-Preserving k-means clustering over vertically partitioned Data. *ACM KDD Conference*, 2003.

[121] Vaidya J., Clifton C.: Privacy-Preserving Naive Bayes Classifier over vertically partitioned data. *SIAM Conference*, 2004.

[122] Vaidya J., Clifton C.: Privacy-Preserving Decision Trees over vertically partitioned data. *Lecture Notes in Computer Science*, Vol 3654, 2005.

[123] Verykios V. S., Bertino E., Fovino I. N., Provenza L. P., Saygin Y., Theodoridis Y.: State-of-the-art in privacy preserving data mining. *ACM SIGMOD Record*, v.33 n.1, 2004.

[124] Verykios V. S., Elmagarmid A., Bertino E., Saygin Y.,, Dasseni E.: Association Rule Hiding. *IEEE Transactions on Knowledge and Data Engineering*, 16(4), 2004.

[125] Wang K., Yu P., Chakraborty S.: Bottom-Up Generalization: A Data Mining Solution to Privacy Protection. *ICDM Conference*, 2004.

[126] Wang K., Fung B. C. M., Yu P. Template based Privacy -Preservation in classification problems. *ICDM Conference*, 2005.

[127] Wang K., Fung B. C. M.: Anonymization for Sequential Releases. *ACM KDD Conference*, 2006.

[128] Wang K., Fung B. C. M., Dong G.: Integarting Private Databases for Data Analysis. *Lecture Notes in Computer Science*, 3495, 2005.

[129] Warner S. L. Randomized Response: A survey technique for eliminating evasive answer bias. *Journal of American Statistical Association*, 60(309):63–69, March 1965.

[130] Winkler W.: Using simulated annealing for k-anonymity. *Technical Report 7, US Census Bureau.*

[131] Wu Y.-H., Chiang C.-M., Chen A. L. P.: Hiding Sensitive Association Rules with Limited Side Effects. *IEEE Transactions on Knowledge and Data Engineering*, 19(1), 2007.

[132] Xiao X., Tao Y.. Personalized Privacy Preservation. *ACM SIGMOD Conference*, 2006.

[133] Xiao X., Tao Y. Anatomy: Simple and Effective Privacy Preservation. *VLDB Conference*, pp. 139-150, 2006.

[134] Xiao X., Tao Y.: m-Invariance: Towards Privacy-preserving Republication of Dynamic Data Sets. *SIGMOD Conference*, 2007.

[135] Xu J., Wang W., Pei J., Wang X., Shi B., Fu A. W. C.: Utility Based Anonymization using Local Recoding. *ACM KDD Conference*, 2006.

[136] Xu S., Yung M.: k-anonymous secret handshakes with reusable credentials. *ACM Conference on Computer and Communications Security*, 2004.

[137] Yao A. C.: How to Generate and Exchange Secrets. *FOCS Conferemce*, 1986.

[138] Yao G., Feng D.: A new k-anonymous message transmission protocol. *International Workshop on Information Security Applications*, 2004.

[139] Yang Z., Zhong S., Wright R.: Privacy-Preserving Classification of Customer Data without Loss of Accuracy. *SDM Conference*, 2006.

[140] Yao C., Wang S., Jajodia S.: Checking for k-Anonymity Violation by views. *ACM Conference on Computer and Communication Security*, 2004.

[141] Yu H., Jiang X., Vaidya J.: Privacy-Preserving SVM using nonlinear Kernels on Horizontally Partitioned Data. *SAC Conference*, 2006.

[142] Yu H., Vaidya J., Jiang X.: Privacy-Preserving SVM Classification on Vertically Partitioned Data. *PAKDD Conference*, 2006.

[143] Zhang P., Tong Y., Tang S., Yang D.: Privacy-Preserving Naive Bayes Classifier. *Lecture Notes in Computer Science*, Vol 3584, 2005.

[144] Zhong S., Yang Z., Wright R.: Privacy-enhancing k-anonymization of customer data, In Proceedings of the ACM SIGMOD-SIGACT-SIGART Principles of Database Systems, Baltimore, MD. 2005.

[145] Zhu Y., Liu L. Optimal Randomization for Privacy- Preserving Data Mining. *ACM KDD Conference*, 2004.

Chapter 3

A Survey of Inference Control Methods for Privacy-Preserving Data Mining

Josep Domingo-Ferrer*

Rovira i Virgili University of Tarragona †
UNESCO Chair in Data Privacy
Dept. of Computer Engineering and Mathematics
Av. Països Catalans 26, E-43007 Tarragona, Catalonia
josep.domingo@urv.cat

Abstract Inference control in databases, also known as Statistical Disclosure Control (SDC), is about protecting data so they can be published without revealing confidential information that can be linked to specific individuals among those to which the data correspond. This is an important application in several areas, such as official statistics, health statistics, e-commerce (sharing of consumer data), etc. Since data protection ultimately means data modification, the challenge for SDC is to achieve protection with minimum loss of the accuracy sought by database users. In this chapter, we survey the current state of the art in SDC methods for protecting individual data (microdata). We discuss several information loss and disclosure risk measures and analyze several ways of combining them to assess the performance of the various methods. Last but not least, topics which need more research in the area are identified and possible directions hinted.

Keywords: Privacy, inference control, statistical disclosure control, statistical disclosure limitation, statistical databases, microdata.

*This work received partial support from the Spanish Ministry of Science and Education through project SEG2004-04352-C04-01 "PROPRIETAS", the Government of Catalonia under grant 2005 SGR 00446 and Eurostat through the CENEX SDC project. The author is solely responsible for the views expressed in this chapter, which do not necessarily reflect the position of UNESCO nor commit that organization.

† Part of this chapter was written while the author was a Visiting Fellow at Princeton University.

3.1 Introduction

Inference control in statistical databases, also known as Statistical Disclosure Control (SDC) or Statistical Disclosure Limitation (SDL), seeks to protect statistical data in such a way that they can be publicly released and mined without giving away private information that can be linked to specific individuals or entities. There are several areas of application of SDC techniques, which include but are not limited to the following:

- *Official statistics*. Most countries have legislation which compels national statistical agencies to guarantee statistical confidentiality when they release data collected from citizens or companies. This justifies the research on SDC undertaken by several countries, among them the European Union (*e.g.* the CASC project[8]) and the United States.

- *Health information*. This is one of the most sensitive areas regarding privacy. For example, in the U. S., the Privacy Rule of the Health Insurance Portability and Accountability Act (HIPAA,[43]) requires the strict regulation of protected health information for use in medical research. In most western countries, the situation is similar.

- *E-commerce*. Electronic commerce results in the automated collection of large amounts of consumer data. This wealth of information is very useful to companies, which are often interested in sharing it with their subsidiaries or partners. Such consumer information transfer should not result in public profiling of individuals and is subject to strict regulation; see [28] for regulations in the European Union and [77] for regulations in the U.S.

The protection provided by SDC techniques normally entails some degree of data modification, which is an intermediate option between no modification (maximum utility, but no disclosure protection) and data encryption (maximum protection but no utility for the user without clearance).

The challenge for SDC is to modify data in such a way that sufficient protection is provided while keeping at a minimum the information loss, *i.e.* the loss of the accuracy sought by database users. In the years that have elapsed since the excellent survey by [3], the state of the art in SDC has evolved so that now at least three subdisciplines are clearly differentiated:

Tabular data protection This is the oldest and best established part of SDC, because tabular data have been the traditional output of national statistical offices. The goal here is to publish *static* aggregate information, *i.e.* tables, in such a way that no confidential information on specific individuals among those to which the table refers can be inferred. See [79] for a conceptual survey and [36] for a software survey.

Dynamic databases The scenario here is a database to which the user can submit statistical queries (sums, averages, etc.). The aggregate information obtained by a user as a result of successive queries should not allow him to infer information on specific individuals. Since the 80s, this has been known to be a difficult problem, subject to the tracker attack [69]. One possible strategy is to perturb the answers to queries; solutions based on perturbation can be found in [26], [54] and [76]. If perturbation is not acceptable and exact answers are needed, it may become necessary to refuse answers to certain queries; solutions based on query restriction can be found in [9] and [38]. Finally, a third strategy is to provide correct (unperturbed) interval answers, as done in [37] and [35].

Microdata protection This subdiscipline is about protecting static individual data, also called microdata. It is only recently that data collectors (statistical agencies and the like) have been persuaded to publish microdata. Therefore, microdata protection is the youngest subdiscipline and is experiencing continuous evolution in the last years.

Good general works on SDC are [79, 45]. This survey will cover the current state of the art in SDC methods for microdata, the most common data used for data mining. First, the main existing methods will be described. Then, we will discuss several information loss and disclosure risk measures and will analyze several approaches to combining them when assessing the performance of the various methods. The comparison metrics being presented should be used as a benchmark for future developments in this area. Open research issues and directions will be suggested at the end of this chapter.

Plan of This Chapter

Section 3.2 introduces a classification of microdata protection methods. Section 3.3 reviews perturbative masking methods. Section 3.4 reviews non-perturbative masking methods. Section 3.5 reviews methods for synthetic microdata generation. Section 3.6 discusses approaches to trade off information loss for disclosure risk and analyzes their strengths and limitations. Conclusions and directions for future research are summarized in Section 3.7.

3.2 A classification of Microdata Protection Methods

A microdata set **V** can be viewed as a file with n records, where each record contains m attributes on an individual respondent. The attributes can be classified in four categories which are not necessarily disjoint:

- *Identifiers.* These are attributes that *unambiguously* identify the respondent. Examples are the passport number, social security number, name-surname, etc.

- *Quasi-identifiers or key attributes.* These are attributes which identify the respondent with some degree of ambiguity. (Nonetheless, a combination of quasi-identifiers may provide unambiguous identification.) Examples are address, gender, age, telephone number, etc.
- *Confidential outcome attributes.* These are attributes which contain sensitive information on the respondent. Examples are salary, religion, political affiliation, health condition, etc.
- *Non-confidential outcome attributes.* Those attributes which do not fall in any of the categories above.

Since the purpose of SDC is to prevent confidential information from being linked to specific respondents, we will assume in what follows that original microdata sets to be protected have been pre-processed to remove from them all identifiers.

The purpose of microdata SDC mentioned in the previous section can be stated more formally by saying that, given an original microdata set $\mathbf{V}$, the goal is to release a protected microdata set $\mathbf{V}'$ in such a way that:

1 Disclosure risk (*i.e.* the risk that a user or an intruder can use $\mathbf{V}'$ to determine confidential attributes on a specific individual among those in $\mathbf{V}$) is low.

2 User analyses (regressions, means, etc.) on $\mathbf{V}'$ and on $\mathbf{V}$ yield the same or at least similar results.

Microdata protection methods can generate the protected microdata set $\mathbf{V}'$

- either by *masking original data, i.e.* generating $\mathbf{V}'$ a modified version of the original microdata set $\mathbf{V}$;
- or by *generating synthetic data* $\mathbf{V}'$ that preserve some statistical properties of the original data $\mathbf{V}$.

Masking methods can in turn be divided in two categories depending on their effect on the original data [79]:

- *Perturbative.* The microdata set is distorted before publication. In this way, unique combinations of scores in the original dataset may disappear and new unique combinations may appear in the perturbed dataset; such confusion is beneficial for preserving statistical confidentiality. The perturbation method used should be such that statistics computed on the perturbed dataset do not differ significantly from the statistics that would be obtained on the original dataset.
- *Non-perturbative.* Non-perturbative methods do not alter data; rather, they produce partial suppressions or reductions of detail in the original

dataset. Global recoding, local suppression and sampling are examples of non-perturbative masking.

At a first glance, synthetic data seem to have the philosophical advantage of circumventing the re-identification problem: since published records are invented and do not derive from any original record, some authors claim that no individual having supplied original data can complain from having been re-identified. At a closer look, some authors (*e.g.*, [80] and [63]) claim that even synthetic data might contain some records that allow for re-identification of confidential information. In short, synthetic data overfitted to original data might lead to disclosure just as original data would. On the other hand, a clear problem of synthetic data is data utility: only the statistical properties explicitly selected by the data protector are preserved, which leads to the question whether the data protector should not directly publish the statistics he wants preserved rather than a synthetic microdata set. We will return to these issues in Section 3.5.

So far in this section, we have classified microdata protection methods by their operating principle. If we consider the type of data on which they can be used, a different dichotomic classification applies:

- *Continuous.* An attribute is considered continuous if it is numerical and arithmetic operations can be performed with it. Examples are income and age. Note that a numerical attribute does not necessarily have an infinite range, as is the case for age. When designing methods to protect continuous data, one has the advantage that arithmetic operations are possible, and the drawback that every combination of numerical values in the original dataset is likely to be unique, which leads to disclosure if no action is taken.

- *Categorical.* An attribute is considered categorical when it takes values over a finite set and standard arithmetic operations do not make sense. Ordinal and nominal scales can be distinguished among categorical attributes. In ordinal scales the order between values is relevant, whereas in nominal scales it is not. In the former case, $\max$ and $\min$ operations are meaningful while in the latter case only pairwise comparison is possible. The instruction level is an example of ordinal attribute, whereas eye color is an example of nominal attribute. In fact, all quasi-identifiers in a microdata set are normally categorical nominal. When designing methods to protect categorical data, the inability to perform arithmetic operations is certainly inconvenient, but the finiteness of the value range is one property that can be successfully exploited.

3.3 Perturbative Masking Methods

Perturbative methods allow for the release of the entire microdata set, although perturbed values rather than exact values are released. Not all perturbative methods are designed for continuous data; this distinction is addressed further below for each method.

Most perturbative methods reviewed below (including additive noise, rank swapping, microaggregation and post-randomization) are special cases of matrix masking. If the original microdata set is **X**, then the masked microdata set **Z** is computed as

$$\mathbf{Z} = \mathbf{A}\mathbf{X}\mathbf{B} + \mathbf{C}$$

where **A** is a record-transforming mask, **B** is an attribute-transforming mask and **C** is a displacing mask (noise)[27].

Table 3.1 lists the perturbative methods described below. For each method, the table indicates whether it is suitable for continuous and/or categorical data.

3.3.1 Additive Noise

The noise additions algorithms in the literature are:

- *Masking by uncorrelated noise addition.* The vector of observations x_j for the j-th attribute of the original dataset X_j is replaced by a vector

$$z_j = x_j + \epsilon_j$$

  where ϵ_j is a vector of normally distributed errors drawn from a random variable $\varepsilon_j \sim N(0, \sigma^2_{\varepsilon_j})$, such that $Cov(\varepsilon_t, \varepsilon_l) = 0$ for all $t \neq l$. This does not preserve variances nor correlations.

- *Masking by correlated noise addition.* Correlated noise addition also preserves means and additionally allows preservation of correlation coefficients. The difference with the previous method is that the covariance

Table 3.1. Perturbative methods vs data types. "X" denotes applicable and "(X)" denotes applicable with some adaptation

Method	*Continuous data*	*Categorical data*
Additive noise	X	
Microaggregation	X	(X)
Rank swapping	X	X
Rounding	X	
Resampling	X	
PRAM		X
MASSC		X

matrix of the errors is now proportional to the covariance matrix of the original data, *i.e.* $\varepsilon \sim N(0, \Sigma_\varepsilon)$, where $\Sigma_\varepsilon = \alpha\Sigma$.

- *Masking by noise addition and linear transformation.* In [49], a method is proposed that ensures by additional transformations that the sample covariance matrix of the masked attributes is an unbiased estimator for the covariance matrix of the original attributes.

- *Masking by noise addition and nonlinear transformation.* An algorithm combining simple additive noise and nonlinear transformation is proposed in [72]. The advantages of this proposal are that it can be applied to discrete attributes and that univariate distributions are preserved. Unfortunately, as justified in [6], the application of this method is very time-consuming and requires expert knowledge on the data set and the algorithm.

For more details on specific algorithms, the reader can check [5]. In practice, only simple noise addition (two first variants) or noise addition with linear transformation are used. When using linear transformations, a decision has to be made whether to reveal them to the data user to allow for bias adjustment in the case of subpopulations.

With the exception of the not very practical method of [72], additive noise is not suitable to protect categorical data. On the other hand, it is well suited for continuous data for the following reasons:

- It makes no assumptions on the range of possible values for V_i (which may be infinite).

- The noise being added is typically continuous and with mean zero, which suits well continuous original data.

- No exact matching is possible with external files. Depending on the amount of noise added, approximate (interval) matching might be possible.

3.3.2 Microaggregation

Microaggregation is a family of SDC techniques for continous microdata. The rationale behind microaggregation is that confidentiality rules in use allow publication of microdata sets if records correspond to groups of k or more individuals, where no individual dominates (*i.e.* contributes too much to) the group and k is a threshold value. Strict application of such confidentiality rules leads to replacing individual values with values computed on small aggregates (microaggregates) prior to publication. This is the basic principle of microaggregation.

To obtain microaggregates in a microdata set with n records, these are combined to form g groups of size at least k. For each attribute, the average value over each group is computed and is used to replace each of the original averaged values. Groups are formed using a criterion of maximal similarity. Once the procedure has been completed, the resulting (modified) records can be published.

The optimal k-partition (from the information loss point of view) is defined to be the one that maximizes within-group homogeneity; the higher the within-group homogeneity, the lower the information loss, since microaggregation replaces values in a group by the group centroid. The sum of squares criterion is common to measure homogeneity in clustering. The within-groups sum of squares SSE is defined as

$$SSE = \sum_{i=1}^{g} \sum_{j=1}^{n_i} (x_{ij} - \bar{x}_i)'(x_{ij} - \bar{x}_i)$$

The lower SSE, the higher the within group homogeneity. Thus, in terms of sums of squares, the optimal k-partition is the one that minimizes SSE.

For a microdata set consisting of p attributes, these can be microaggregated together or partitioned into several groups of attributes. Also the way to form groups may vary. Several taxonomies are possible to classify the microaggregation algorithms in the literature: i) fixed group size [15, 44, 23] vs variable group size [15, 51, 18, 68, 50, 20]; ii) exact optimal (only for the univariate case, [41, 55]) vs heuristic microaggregation; iii) continuous vs categorical microaggregation [75].

To illustrate, we next give a heuristic algorithm called MDAV (Maximum Distance to Average Vector,[23]) for multivariate fixed group size microaggregation on unprojected continuous data. We designed and implemented MDAV for the μ-Argus package [44].

ALGORITHM 3.1 (MDAV)

1 *Compute the average record $\bar{x}$ of all records in the dataset. Consider the most distant record x_r to the average record $\bar{x}$ (using the squared Euclidean distance).*

2 *Find the most distant record x_s from the record x_r considered in the previous step.*

3 *Form two groups around x_r and x_s, respectively. One group contains x_r and the $k-1$ records closest to x_r. The other group contains x_s and the $k-1$ records closest to x_s.*

4 *If there are at least* $3k$ *records which do not belong to any of the two groups formed in Step 3, go to Step 1 taking as new dataset the previous dataset minus the groups formed in the last instance of Step 3.*

5 *If there are between* $3k-1$ *and* $2k$ *records which do not belong to any of the two groups formed in Step 3: a) compute the average record* $\bar{x}$ *of the remaining records; b) find the most distant record* x_r *from* $\bar{x}$*; c) form a group containing* x_r *and the* $k-1$ *records closest to* x_r*; d) form another group containing the rest of records. Exit the Algorithm.*

6 *If there are less than* $2k$ *records which do not belong to the groups formed in Step 3, form a new group with those records and exit the Algorithm.*

The above algorithm can be applied independently to each group of attributes resulting from partitioning the set of attributes in the dataset.

3.3.3 Data Wapping and Rank Swapping

Data swapping was originally presented as an SDC method for databases containing only categorical attributes [11]. The basic idea behind the method is to transform a database by exchanging values of confidential attributes among individual records. Records are exchanged in such a way that low-order frequency counts or marginals are maintained.

Even though the original procedure was not very used in practice (see [32]), its basic idea had a clear influence in subsequent methods. In [59] and [58] data swapping was introduced to protect continuous and categorical microdata, respectively. Another variant of data swapping for microdata is *rank swapping*, which will be described next in some detail.

Although originally described only for ordinal attributes [40], rank swapping can also be used for any numerical attribute [53]. First, values of an attribute X_i are ranked in ascending order, then each ranked value of X_i is swapped with another ranked value randomly chosen within a restricted range (*e.g.* the rank of two swapped values cannot differ by more than $p\%$ of the total number of records, where p is an input parameter). This algorithm is independently used on each original attribute in the original data set.

It is reasonable to expect that multivariate statistics computed from data swapped with this algorithm will be less distorted than those computed after an unconstrained swap. In earlier empirical work by these authors on continuous microdata protection [21], rank swapping has been identified as a particularly well-performing method in terms of the tradeoff between disclosure risk and information loss (see Example 3.4 below). Consequently, it is one of the techniques that have been implemented in the $\mu - Argus$ package [44].

Table 3.2. Example of rank swapping. Left, original file; right, rankswapped file

1	K	3.7	4.4	1	H	3.0	4.8
2	L	3.8	3.4	2	L	4.5	3.2
3	N	3.0	4.8	3	M	3.7	4.4
4	M	4.5	5.0	4	N	5.0	6.0
5	L	5.0	6.0	5	L	4.5	5.0
6	H	6.0	7.5	6	F	6.7	9.5
7	H	4.5	10.0	7	K	3.8	11.0
8	F	6.7	11.0	8	H	6.0	10.0
9	D	8.0	9.5	9	C	10.0	7.5
10	C	10.0	3.2	10	D	8.0	3.4

EXAMPLE 3.2 *In Table 3.2, we can see an original microdata set on the left and its rankswapped version on the right. There are four attributes and ten records in the original dataset; the second attribute is alphanumeric, and the standard alphabetic order has been used to rank it. A value of* $p = 10\%$ *has been used for all attributes.* □

3.3.4 Rounding

Rounding methods replace original values of attributes with rounded values. For a given attribute X_i, rounded values are chosen among a set of rounding points defining a *rounding set* (often the multiples of a given base value). In a multivariate original dataset, rounding is usually performed one attribute at a time (*univariate* rounding); however, multivariate rounding is also possible [79, 10]. The operating principle of rounding makes it suitable for continuous data.

3.3.5 Resampling

Originally proposed for protecting tabular data [42, 17], resampling can also be used for microdata. Take t independent samples $S_1, \cdots, S_t$ of the values of an original attribute X_i. Sort all samples using the same ranking criterion. Build the masked attribute Z_i as $\bar{x}_1, \cdots, \bar{x}_n$, where n is the number of records and $\bar{x}_j$ is the average of the j-th ranked values in $S_1, \cdots, S_t$.

3.3.6 PRAM

The Post-RAndomization Method (PRAM, [39]) is a probabilistic, perturbative method for disclosure protection of categorical attributes in microdata files. In the masked file, the scores on some categorical attributes for certain records in the original file are changed to a different score according to a prescribed probability mechanism, namely a Markov matrix. The Markov approach makes PRAM very general, because it encompasses noise addition, data suppression and data recoding.

PRAM information loss and disclosure risk largely depend on the choice of the Markov matrix and are still (open) research topics [14].

The PRAM matrix contains a row for each possible value of each attribute to be protected. This rules out using the method for continuous data.

3.3.7 MASSC

MASSC [71] is a masking method whose acronym summarizes its four steps: Micro Agglomeration, Substitution, Subsampling and Calibration. We briefly recall the purpose of those four steps:

1 Micro agglomeration is applied to partition the original dataset into risk strata (groups of records which are at a similar risk of disclosure). These strata are formed using the key attributes, *i.e.* the quasi-identifiers in the records. The idea is that those records with rarer combinations of key attributes are at a higher risk.

2 Optimal probabilistic substitution is then used to perturb the original data.

3 Optimal probabilistic subsampling is used to suppress some attributes or even entire records.

4 Optimal sampling weight calibration is used to preserve estimates for outcome attributes in the treated database whose accuracy is critical for the intended data use.

MASSC in interesting in that, to the best of our knowledge, it is the first attempt at designing a perturbative masking method in such a way that disclosure risk can be analytically quantified. Its main shortcoming is that its disclosure model simplifies reality by considering only disclosure resulting from linkage of key attributes with external sources. Since key attributes are typically categorical, the risk of disclosure can be analyzed by looking at the probability that a sample unique is a population unique; however, doing so ignores the fact that continuous outcome attributes can also be used for respondent re-identification via record linkage. As an example, if respondents are companies and turnover is one outcome attribute, everyone in a certain industrial sector knows which is the company with largest turnover. Thus, in practice, MASSC is a method only suited when continuous attributes are not present.

3.4 Non-perturbative Masking Methods

Non-perturbative methods do not rely on distortion of the original data but on partial suppressions or reductions of detail. Some of the methods are usable

Table 3.3. Non-perturbative methods vs data types

Method	*Continuous data*	*Categorical data*
Sampling		X
Global recoding	X	X
Top and bottom coding	X	X
Local suppression		X

on both categorical and continuous data, but others are not suitable for continuous data. Table 3.3 lists the non-perturbative methods described below. For each method, the table indicates whether it is suitable for continuous and/or categorical data.

3.4.1 Sampling

Instead of publishing the original microdata file, what is published is a sample S of the original set of records [79].

Sampling methods are suitable for categorical microdata, but for continuous microdata they should probably be combined with other masking methods. The reason is that sampling alone leaves a continuous attribute V_i unperturbed for all records in S. Thus, if attribute V_i is present in an external administrative public file, unique matches with the published sample are very likely: indeed, given a continuous attribute V_i and two respondents o_1 and o_2, it is highly unlikely that V_i will take the same value for both o_1 and o_2 unless $o_1 = o_2$ (this is true even if V_i has been truncated to represent it digitally).

If, for a continuous identifying attribute, the score of a respondent is only approximately known by an attacker (as assumed in [78]), it might still make sense to use sampling methods to protect that attribute. However, assumptions on restricted attacker resources are perilous and may prove definitely too optimistic if good quality external administrative files are at hand.

3.4.2 Global Recoding

This method is also sometimes known as generalization [67, 66]. For a categorical attribute V_i, several categories are combined to form new (less specific) categories, thus resulting in a new V_i' with $|D(V_i')| < |D(V_i)|$ where $|\cdot|$ is the cardinality operator. For a continuous attribute, global recoding means replacing V_i by another attribute V_i' which is a discretized version of V_i. In other words, a potentially infinite range $D(V_i)$ is mapped onto a finite range $D(V_i')$. This is the technique used in the μ-Argus SDC package [44].

This technique is more appropriate for categorical microdata, where it helps disguise records with strange combinations of categorical attributes. Global recoding is used heavily by statistical offices.

EXAMPLE 3.3 *If there is a record with "Marital status = Widow/er" and "Age = 17", global recoding could be applied to "Marital status" to create a broader category "Widow/er or divorced", so that the probability of the above record being unique would diminish. Global recoding can also be used on a continuous attribute, but the inherent discretization leads very often to an unaffordable loss of information. Also, arithmetical operations that were straightforward on the original* V_i *are no longer easy or intuitive on the discretized* V_i'. □

3.4.3 Top and Bottom Coding

Top and bottom coding is a special case of global recoding which can be used on attributes that can be ranked, that is, continuous or categorical ordinal. The idea is that top values (those above a certain threshold) are lumped together to form a new category. The same is done for bottom values (those below a certain threshold). See [44].

3.4.4 Local Suppression

Certain values of individual attributes are suppressed with the aim of increasing the set of records agreeing on a combination of key values. Ways to combine local suppression and global recoding are discussed in [16] and implemented in the μ-Argus SDC package [44].

If a continuous attribute V_i is part of a set of key attributes, then each combination of key values is probably unique. Since it does not make sense to systematically suppress the values of V_i, we conclude that local suppression is rather oriented to categorical attributes.

3.5 Synthetic Microdata Generation

Publication of synthetic — *i.e.* simulated — data was proposed long ago as a way to guard against statistical disclosure. The idea is to randomly generate data with the constraint that certain statistics or internal relationships of the original dataset should be preserved.

We next review some approaches in the literature to synthetic data generation and then proceed to discuss the global pros and cons of using synthetic data.

3.5.1 Synthetic Data by Multiple Imputation

More than twenty years ago, it was suggested in [65] to create an entirely synthetic dataset based on the original survey data and multiple imputation. Rubin's proposal was more completely developed in [57]. A simulation study of it was given in [60]. In [64] inference on synthetic data is discussed and in [63] an application is given.

We next sketch the operation of the original proposal by Rubin. Consider an original microdata set X of size n records drawn from a much larger population of N individuals, where there are background attributes A, non-confidential attributes B and confidential attributes C. Background attributes are observed and available for all N individuals in the population, whereas B and C are only available for the n records in the sample X. The first step is to construct from X a multiply-imputed population of N individuals. This population consists of the n records in X and M (the number of multiple imputations, typically between 3 and 10) matrices of (B, C) data for the $N - n$ non-sampled individuals. The variability in the imputed values ensures, theoretically, that valid inferences can be obtained on the multiply-imputed population. A model for predicting (B, C) from A is used to multiply-impute (B, C) in the population. The choice of the model is a nontrivial matter. Once the multiply-imputed population is available, a sample Z of n' records can be drawn from it whose structure looks like the one a sample of n' records drawn from the original population. This can be done M times to create M replicates of (B, C) values. The result are M multiply-imputed synthetic datasets. To make sure no original data are in the synthetic datasets, it is wise to draw the samples from the multiply-imputed population excluding the n original records from it.

3.5.2 Synthetic Data by Bootstrap

Long ago, [30] proposed generating synthetic microdata by using bootstrap methods. Later, in [31] this approach was used for categorical data.

The bootstrap approach bears some similarity to the data distortion by probability distribution and the multiple-imputation methods described above. Given an original microdata set X with p attributes, the data protector computes its empirical p-variate cumulative distribution function (c.d.f.) F. Now, rather than distorting the original data to obtain masked data (as done by the masking methods in Sections 3.3 and 3.4), the data protector alters (or "smoothes") the c.d.f. F to derive a similar c.d.f. F'. Finally, F' is sampled to obtain a synthetic microdata set Z.

3.5.3 Synthetic Data by Latin Hypercube Sampling

Latin Hypercube Sampling (LHS) appears in the literature as another method for generating multivariate synthetic datasets. In [46], the LHS updated technique of [33] was improved, but the proposed scheme is still time-intensive even for a moderate number of records. In [12], LHS is used along with a rank correlation refinement to reproduce both the univariate (*i.e.* mean and covariance) and multivariate structure (in the sense of rank correlation) of the original dataset. In a nutshell, LHS-based methods rely on iterative refinement, are time-intensive and their running time does not only depend on the number of values to be reproduced, but on the starting values as well.

3.5.4 Partially Synthetic Data by Cholesky Decomposition

Generating plausible synthetic values for all attributes in a database may be difficult in practice. Thus, several authors have considered mixing actual and synthetic data.

In [7], a non-iterative method for generating continuous synthetic microdata is proposed. It consists of three methods sketched next. Informally, suppose two sets of attributes X and Y, where the former are the confidential outcome attributes and the latter are quasi-identifier attributes. Then X are taken as independent and Y as dependent attributes. Conditional on the specific confidential attributes x_i, the quasi-identifier attributes Y_i are assumed to follow a multivariate normal distribution with covariance matrix $\Sigma = \{\sigma_{jk}\}$ and a mean vector $x_i B$, where B is a matrix of regression coefficients.

Method A computes a multiple regression of Y on X and fitted Y'_A attributes. Finally, attributes X and Y'_A are released in place of X and Y.

If a user fits a multiple regression model to (y'_A, x), she will get estimates $\hat{B}_A$ and $\hat{\Sigma}_A$ which, in general, are different from the estimates $\hat{B}$ and $\hat{\Sigma}$ obtained when fitting the model to the original data (y, x). IPSO Method B modifies y'_A into y'_B in such a way that the estimate $\hat{B}_B$ obtained by multiple linear regression from (y'_B, x) satisfies $\hat{B}_B = \hat{B}$.

A more ambitious goal is to come up with a data matrix y'_C such that, when a multivariate multiple regression model is fitted to (y'_C, x), *both* sufficient statistics $\hat{B}$ and $\hat{\Sigma}$ obtained on the original data (y, x) are preserved. This is achieved by IPSO Method C.

3.5.5 Other Partially Synthetic and Hybrid Microdata Approaches

The multiple imputation approach described in [65] for creating entirely synthetic microdata can be extended for partially synthetic microdata. As a result multiply-imputed, partially synthetic datasets are obtained that contain a mix of actual and imputed (synthetic) values. The idea is to multiply-impute confidential values and release non-confidential values without perturbation. This approach was first applied to protect the Survey of Consumer Finances [47, 48]. In Abowd and Woodcock [1, 2], this technique was adopted to protect longitudinal linked data, that is, microdata that contain observations from two or more related time periods (successive years, etc.). Methods for valid inference on this kind of partial synthetic data were developed in [61] and a non-parametric method was presented in [62] to generate multiply-imputed, partially synthetic data.

Closely related to multiply imputed, partially synthetic microdata is model-based disclosure protection [34, 56]. In this approach, a set of confidential continuous outcome attributes is regressed on a disjoint set of non-confidential attributes; then the fitted values are released for the confidential attributes instead of the original values.

A different approach called hybrid masking was proposed in [13]. The idea is to compute masked data as a combination of original and synthetic data. Such a combination allows better control than purely synthetic data over the individual characteristics of masked records. For hybrid masking to be feasible, a rule must be used to pair one original data record with one synthetic data record. An option suggested in [13] is to go through all original data records and pair each original record with the nearest synthetic record according to some distance. Once records have been paired, [13] suggest two possible ways for combining one original record X with one synthetic record X_s: additive combination and multiplicative combination. Additive combination yields

$$Z = \alpha X + (1 - \alpha)X_s$$

and multiplicative combination yields

$$Z = X^{\alpha} \cdot X_s^{(1-\alpha)}$$

where α is an input parameter in $[0, 1]$ and Z is the hybrid record. [13] present empirical results comparing the hybrid approach with rank swapping and microaggregation masking (the synthetic component of hybrid data is generated using Latin Hypercube Sampling [12]).

Another approach to combining original and synthetic microdata is proposed in [70]. The idea here is to first mask an original dataset using a masking method (see Sections 3.3 and 3.4 above). Then a hill-climbing optimization heuristic is run which seeks to modify the masked data to preserve the first and second-order moments of the original dataset as much as possible without increasing the disclosure risk with respect to the initial masked data. The optimization heuristic can be modified to preserve higher-order moments, but this significantly increases computation. Also, the optimization heuristic can take as initial dataset a random dataset instead of a masked dataset; in this case, the output dataset is purely synthetic.

3.5.6 Pros and Cons of Synthetic Microdata

As pointed out in Section 3.2, synthetic data are appealing in that, at a first glance, they seem to circumvent the re-identification problem: since published records are invented and do not derive from any original record, it might be concluded that no individual can complain from having been re-identified. At a closer look this advantage is less clear. If, by chance, a published synthetic record matches a particular citizen's non-confidential attributes (age, marital status, place of residence, etc.) and confidential attributes (salary, mortgage, etc.), re-identification using the non-confidential attributes is easy and that citizen may feel that his confidential attributes have been unduly revealed. In that case, the citizen is unlikely to be happy with or even understand the explanation that the record was synthetically generated.

On the other hand, limited data utility is another problem of synthetic data. Only the statistical properties explicitly captured by the model used by the data protector are preserved. A logical question at this point is why not directly publish the statistics one wants to preserve rather than release a synthetic microdata set.

One possible justification for synthetic microdata would be if valid analyses could be obtained on a number of subdomains, *i.e.* similar results were obtained in a number of subsets of the original dataset and the corresponding subsets of the synthetic dataset. Partially synthetic or hybrid microdata are more likely to succeed in staying useful for subdomain analysis. However, when using partially synthetic or hybrid microdata, we lose the attractive feature of purely synthetic data that the number of records in the protected (synthetic) dataset is independent from the number of records in the original dataset.

3.6 Trading off Information Loss and Disclosure Risk

Sections 3.2 through 3.5 have presented a plethora of methods to protect microdata. To complicate things further, most of such methods are parametric (*e.g.*, in microaggregation, one parameter is the minimum number of records in a cluster), so the user must go through two choices rather than one: a primary choice to select a method and a secondary choice to select parameters for the method to be used. To help reducing the *embarras du choix*, some guidelines are needed.

3.6.1 Score Construction

The mission of SDC to modify data in such a way that sufficient protection is provided at minimum information loss suggests that a good SDC method is one achieving a good tradeoff between disclosure risk and information loss.

Following this idea, [21] proposed a score for method performance rating based on the average of information loss and disclosure risk measures. For each method M and parameterization P, the following score is computed:

$$Score(\mathbf{V}, \mathbf{V}') = \frac{IL(\mathbf{V}, \mathbf{V}') + DR(\mathbf{V}, \mathbf{V}')}{2}$$

where IL is an information loss measure, DR is a disclosure risk measure and $\mathbf{V}'$ is the protected dataset obtained after applying method M with parameterization P to an original dataset $\mathbf{V}$.

In [21] and [19] IL and DR were computed using a weighted combination of several information loss and disclosure risk measures. With the resulting score, a ranking of masking methods (and their parameterizations) was obtained. In [81] the line of the above two papers was followed to rank a different set of methods using a slightly different score.

To illustrate how a score can be constructed, we next describe the particular score used in [21].

EXAMPLE 3.4 *Let X and X' be matrices representing original and protected datasets, respectively, where all attributes are numerical. Let V and R be the covariance matrix and the correlation matrix of X, respectively; let $\bar{X}$ be the vector of attribute averages for X and let S be the diagonal of V. Define V', R', $\bar{X}'$, and S' analogously from X'. The Information Loss (IL) is computed by averaging the mean variations of $X - X'$, $\bar{X} - \bar{X}'$, $V - V'$, $S - S'$, and the mean absolute error of $R - R'$ and multiplying the resulting average by 100. Thus, we obtain the following expression for information loss:*

$$IL = \frac{100}{5}\Bigg(\frac{\sum_{j=1}^{p}\sum_{i=1}^{n}\frac{|x_{ij}-x'_{ij}|}{|x_{ij}|}}{np} + \frac{\sum_{j=1}^{p}\frac{|\bar{x}_j-\bar{x}'_j|}{|\bar{x}_j|}}{p} +$$
$$\frac{\sum_{j=1}^{p}\sum_{1\le i\le j}\frac{|v_{ij}-v'_{ij}|}{|v_{ij}|}}{\frac{p(p+1)}{2}} + \frac{\sum_{j=1}^{p}\frac{|v_{jj}-v'_{jj}|}{|v_{jj}|}}{p} + \frac{\sum_{j=1}^{p}\sum_{1\le i\le j}|r_{ij}-r'_{ij}|}{\frac{p(p-1)}{2}}\Bigg)$$

The expression of the overall score is obtained by combining information loss and information risk as follows:

$$Score = \frac{IL + \frac{(0.5DLD+0.5PLD)+ID}{2}}{2}$$

Here, DLD (Distance Linkage Disclosure risk) is the percentage of correctly linked records using distance-based record linkage [19], PLD (Probabilistic Linkage Record Disclosure risk) is the percentage of correctly linked records using probabilistic linkage [29], ID (Interval Disclosure) is the percentage of original records falling in the intervals around their corresponding masked values and IL is the information loss measure defined above.

Based on the above score, [21] found that, for the benchmark datasets and the intruder's external information they used, two good performers among the set of methods and parameterizations they tried were: i) rankswapping with parameter p around 15 (see description above); ii) multivariate microaggregation on unprojected data taking groups of three attributes at a time (Algorithm 3.1 with partitioning of the set of attributes). □

Using a score permits to regard the selection of a masking method and its parameters as an optimization problem. This idea was first used in the above-mentioned contribution [70]. In that paper, a masking method was applied to the original data file and then a post-masking optimization procedure was applied to decrease the score obtained.

On the negative side, no specific score weighting can do justice to all methods. Thus, when ranking methods, the values of all measures of information loss and disclosure risk should be supplied along with the overall score.

3.6.2 R-U Maps

A tool which may be enlightening when trying to construct a score or, more generally, optimize the tradeoff between information loss and disclosure risk is a graphical representation of pairs of measures (disclosure risk, information loss) or their equivalents (disclosure risk, data utility). Such maps are called R-U confidentiality maps [24, 25]. Here, R stands for disclosure risk and U for data utility. According to [25], "in its most basic form, an R-U confidentiality map is the set of paired values (R, U), of disclosure risk and data utility that correspond to various strategies for data release" (*e.g.*, variations on a parameter). Such (R, U) pairs are typically plotted in a two-dimensional graph, so that the user can easily grasp the influence of a particular method and/or parameter choice.

3.6.3 k-anonymity

A different approach to facing the conflict between information loss and disclosure risk is suggested by Samarati and Sweeney [67, 66, 73, 74]. A protected dataset is said to satisfy k-anonymity for $k > 1$ if, for each combination of quasi-identifier values (*e.g.* address, age, gender, etc.), at least k records exist in the dataset sharing that combination. Now if, for a given k, k-anonymity is assumed to be enough protection, one can concentrate on minimizing information loss with the only constraint that k-anonymity should be satisfied. This is a clean way of solving the tension between data protection and data utility. Since k-anonymity is usually achieved via generalization (equivalent to global recoding, as said above) and local suppression, minimizing information loss usually translates to reducing the number and/or the magnitude of suppressions.

k-anonymity bears some resemblance to the underlying principle of microaggregation and is a useful concept because quasi-identifiers are usually categorical or can be categorized, *i.e.* they take values in a finite (and ideally reduced) range. However, re-identification is not necessarily based on categorical quasi-identifiers: sometimes, numerical outcome attributes —which are continuous and often cannot be categorized— give enough clues for re-identification (see discussion on the MASSC method above). Microaggregation was suggested in [23] as a possible way to achieve k-anonymity for numerical, ordinal and nominal attributes. A similar idea called data condensation had also been independently proposed by [4] to achieve k-anonymity for the specific case of numerical attributes.

Another connection between k-anonymity and microaggregation is the NP-hardness of solving them optimally. Satisfying k-anonymity with minimal data modification has been shown to be NP-hard in [52], which is parallel to the NP-hardness of optimal multivariate microaggregation proven in [55].

3.7 Conclusions and Research Directions

Inference control methods for privacy-preserving data mining are a hot research topic progressing very fast. There are still many open issues, some of which can be hopefully solved with further research and some which are likely to stay open due to the inherent nature of SDC.

We first list some of the issues that we feel can be and should be settled in the near future:

- Identifying a comprehensive listing of data uses (*e.g.* regression models, association rules, etc.) that would allow the definition of data use-specific information loss measures broadly accepted by the community; those new measures could complement and/or replace the generic measures currently used. Work in this line has been started in Europe in 2006 under the CENEX SDC project sponsored by Eurostat.

- Devising disclosure risk assessment procedures which are as universally applicable as record linkage while being less greedy in computational terms.

- Identifying the external data sources that intruders can typically access in order to attempt re-identification for each domain of application. This would help data protectors figuring out in more realistic terms which are the disclosure scenarios they should protect data against.

- Creating one or several benchmarks to assess the performance of SDC methods. Benchmark creation is currently hampered by the confidentiality of the original datasets to be protected. Data protectors should agree on a collection of non-confidential original-looking data sets (financial datasets, population datasets, etc.) which can be used by anybody to compare the performance of SDC methods. The benchmark should also incorporate state-of-the-art disclosure risk assessment methods, which requires continuous update and maintenance.

There are other issues which, in our view, are less likely to be resolved in the near future, due to the very nature of SDC methods. As pointed out in [22], if an intruder knows the SDC algorithm used to create a protected data set, he can mount algorithm-specific re-identification attacks which can disclose more confidential information than conventional data mining attacks. Keeping secret the SDC algorithm used would seem a solution, but in many cases the protected dataset itself gives some clues on the SDC algorithm used to produce it. Such is the case for a rounded, microaggregated or partially suppressed microdata set. Thus, it is unclear to what extent the SDC algorithm used can be kept secret.

Other data security areas where slightly distorted data are sent to a recipient who is legitimate but untrusted also share the same concerns about the secrecy of protection algorithms in use. This is the case of watermarking. Teaming up with those areas sharing similar problems is probably one clever line of action for SDC.

References

[1] J. M. Abowd and S. D. Woodcock. Disclosure limitation in longitudinal linked tables. In P. Doyle, J. I. Lane, J. J. Theeuwes, and L. V. Zayatz, editors, *Confidentiality, Disclosure and Data Access: Theory and Practical Applications for Statistical Agencies*, pages 215–278, Amsterdam, 2001. North-Holland.

[2] J. M. Abowd and S. D. Woodcock. Multiply-imputing confidential characteristics and file links in longitudinal linked data. In J. Domingo-Ferrer and V. Torra, editors, *Privacy in Statistical Databases*, volume 3050 of *Lecture Notes in Computer Science*, pages 290–297, Berlin Heidelberg, 2004. Springer.

[3] N. R. Adam and J. C. Wortmann. Security-control for statistical databases: a comparative study. *ACM Computing Surveys*, 21(4):515–556, 1989.

[4] C. C. Aggarwal and P. S. Yu. A condensation approach to privacy preserving data mining. In E. Bertino, S. Christodoulakis, D. Plexousakis, V. Christophides, M. Koubarakis, K. Böhm, E. Ferrari, editors, *Advances in Database Technology - EDBT 2004*, vol. 2992 of *Lecture Notes in Computer Science*, pages 183-199, Berlin Heidelberg, 2004. Springer.

[5] R. Brand. Microdata protection through noise addition. In J. Domingo-Ferrer, editor, *Inference Control in Statistical Databases*, volume 2316 of *Lecture Notes in Computer Science*, pages 97–116, Berlin Heidelberg, 2002. Springer.

[6] R. Brand. Tests of the applicability of sullivan's algorithm to synthetic data and real business data in official statistics, 2002. European Project IST-2000-25069 CASC, Deliverable 1.1-D1, http://neon.vb.cbs.nl/casc.

[7] J. Burridge. Information preserving statistical obfuscation. *Statistics and Computing*, 13:321–327, 2003.

[8] CASC. Computational aspects of statistical confidentiality, 2004. European project IST-2000-25069 CASC, 5th FP, 2001-2004, http://neon.vb.cbs.nl/casc.

[9] F. Y. Chin and G. Ozsoyoglu. Auditing and inference control in statistical databases. *IEEE Transactions on Software Engineering*, SE-8:574–582, 1982.

[10] L. H. Cox and J. J. Kim. Effects of rounding on the quality and confidentiality of statistical data. In J. Domingo-Ferrer and L. Franconi, editors, *Privacy in Statistical Databases-PSD 2006*, volume 4302 of *Lecture Notes in Computer Science*, pages 48–56, Berlin Heidelberg, 2006.

[11] T. Dalenius and S. P. Reiss. Data-swapping: a technique for disclosure control (extended abstract). In *Proc. of the ASA Section on Survey Research Methods*, pages 191–194, Washington DC, 1978. American Statistical Association.

[12] R. Dandekar, M. Cohen, and N. Kirkendall. Sensitive micro data protection using latin hypercube sampling technique. In J. Domingo-Ferrer, editor, *Inference Control in Statistical Databases*, volume 2316 of *Lecture Notes in Computer Science*, pages 245–253, Berlin Heidelberg, 2002. Springer.

[13] R. Dandekar, J. Domingo-Ferrer, and F. Sebé. Lhs-based hybrid microdata vs rank swapping and microaggregation for numeric microdata protection. In J. Domingo-Ferrer, editor, *Inference Control in Statistical Databases*, volume 2316 of *Lecture Notes in Computer Science*, pages 153–162, Berlin Heidelberg, 2002. Springer.

[14] P.-P. de Wolf. Risk, utility and pram. In J. Domingo-Ferrer and L. Franconi, editors, *Privacy in Statistical Databases-PSD 2006*, volume 4302 of *Lecture Notes in Computer Science*, pages 189–204, Berlin Heidelberg, 2006.

[15] D. Defays and P. Nanopoulos. Panels of enterprises and confidentiality: the small aggregates method. In *Proc. of 92 Symposium on Design and Analysis of Longitudinal Surveys*, pages 195–204, Ottawa, 1993. Statistics Canada.

[16] A. G. DeWaal and L. C. R. J. Willenborg. Global recodings and local suppressions in microdata sets. In *Proceedings of Statistics Canada Symposium'95*, pages 121–132, Ottawa, 1995. Statistics Canada.

[17] J. Domingo-Ferrer and J. M. Mateo-Sanz. On resampling for statistical confidentiality in contingency tables. *Computers & Mathematics with Applications*, 38:13–32, 1999.

[18] J. Domingo-Ferrer and J. M. Mateo-Sanz. Practical data-oriented microaggregation for statistical disclosure control. *IEEE Transactions on Knowledge and Data Engineering*, 14(1):189–201, 2002.

[19] J. Domingo-Ferrer, J. M. Mateo-Sanz, and V. Torra. Comparing sdc methods for microdata on the basis of information loss and disclosure risk. In *Pre-proceedings of ETK-NTTS'2001 (vol. 2)*, pages 807–826, Luxemburg, 2001. Eurostat.

[20] J. Domingo-Ferrer, F. Sebé, and A. Solanas. A polynomial-time approximation to optimal multivariate microaggregation. *Computers & Mathematics with Applications*, 2007. (To appear).

[21] J. Domingo-Ferrer and V. Torra. A quantitative comparison of disclosure control methods for microdata. In P. Doyle, J. I. Lane, J. J. M. Theeuwes, and L. Zayatz, editors, *Confidentiality, Disclosure and Data Access: Theory and Practical Applications for Statistical Agencies*, pages 111–134, Amsterdam, 2001. North-Holland. http://vneumann.etse.urv.es/publications/bcpi.

[22] J. Domingo-Ferrer and V. Torra. Algorithmic data mining against privacy protection methods for statistical databases. *manuscript*, 2004.

[23] J. Domingo-Ferrer and V. Torra. Ordinal, continuous and heterogenerous k-anonymity through microaggregation. *Data Mining and Knowledge Discovery*, 11(2):195–212, 2005.

[24] G. T. Duncan, S. E. Fienberg, R. Krishnan, R. Padman, and S. F. Roehrig. Disclosure limitation methods and information loss for tabular data. In P. Doyle, J. I. Lane, J. J. Theeuwes, and L. V. Zayatz, editors, *Confidentiality, Disclosure and Data Access: Theory and Practical Applications for Statistical Agencies*, pages 135–166, Amsterdam, 2001. North-Holland.

[25] G. T. Duncan, S. A. Keller-McNulty, and S. L Stokes. Disclosure risk vs. data utility: The r-u confidentiality map, 2001.

[26] G. T. Duncan and S. Mukherjee. Optimal disclosure limitation strategy in statistical databases: deterring tracker attacks through additive noise. *Journal of the American Statistical Association*, 95:720–729, 2000.

[27] G. T. Duncan and R. W. Pearson. Enhancing access to microdata while protecting confidentiality: prospects for the future. *Statistical Science*, 6:219–239, 1991.

[28] E.U.Privacy. European privacy regulations, 2004. http://europa.eu.int/comm/internal_market/privacy/law_en.htm.

[29] I. P. Fellegi and A. B. Sunter. A theory for record linkage. *Journal of the American Statistical Association*, 64(328):1183–1210, 1969.

[30] S. E. Fienberg. A radical proposal for the provision of micro-data samples and the preservation of confidentiality. Technical Report 611, Carnegie Mellon University Department of Statistics, 1994.

[31] S. E. Fienberg, U. E. Makov, and R. J. Steele. Disclosure limitation using perturbation and related methods for categorical data. *Journal of Official Statistics*, 14(4):485–502, 1998.

[32] S. E. Fienberg and J. McIntyre. Data swapping: variations on a theme by dalenius and reiss. In J. Domingo-Ferrer and V. Torra, editors, *Privacy in Statistical Databases*, volume 3050 of *Lecture Notes in Computer Science*, pages 14–29, Berlin Heidelberg, 2004. Springer.

[33] A. Florian. An efficient sampling scheme: updated latin hypercube sampling. *Probabilistic Engineering Mechanics*, 7(2):123–130, 1992.

[34] L. Franconi and J. Stander. A model based method for disclosure limitation of business microdata. *Journal of the Royal Statistical Society D - Statistician*, 51:1–11, 2002.

[35] R. Garfinkel, R. Gopal, and D. Rice. New approaches to disclosure limitation while answering queries to a database: protecting numerical confidential data against insider threat based on data and algorithms, 2004. Manuscript. Available at http://www-eio.upc.es/seminar/04/garfinkel.pdf.

[36] S. Giessing. Survey on methods for tabular data protection in argus. In J. Domingo-Ferrer and V. Torra, editors, *Privacy in Statistical Databases*, volume 3050 of *Lecture Notes in Computer Science*, pages 1–13, Berlin Heidelberg, 2004. Springer.

[37] R. Gopal, R. Garfinkel, and P. Goes. Confidentiality via camouflage: the cvc approach to disclosure limitation when answering queries to databases. *Operations Research*, 50:501–516, 2002.

[38] R. Gopal, P. Goes, and R. Garfinkel. Interval protection of confidential information in a database. *INFORMS Journal on Computing*, 10:309–322, 1998.

[39] J. M. Gouweleeuw, P. Kooiman, L. C. R. J. Willenborg, and P.-P. DeWolf. Post randomisation for statistical disclosure control: Theory and implementation, 1997. Research paper no. 9731 (Voorburg: Statistics Netherlands).

[40] B. Greenberg. Rank swapping for ordinal data, 1987. Washington, DC: U. S. Bureau of the Census (unpublished manuscript).

[41] S. L. Hansen and S. Mukherjee. A polynomial algorithm for optimal univariate microaggregation. *IEEE Transactions on Knowledge and Data Engineering*, 15(4):1043–1044, 2003.

[42] G. R. Heer. A bootstrap procedure to preserve statistical confidentiality in contingency tables. In D. Lievesley, editor, *Proc. of the International Seminar on Statistical Confidentiality*, pages 261–271, Luxemburg, 1993. Office for Official Publications of the European Communities.

[43] HIPAA. Health insurance portability and accountability act, 2004. http://www.hhs.gov/ocr/hipaa/.

[44] A. Hundepool, A. Van de Wetering, R. Ramaswamy, L. Franconi, A. Capobianchi, P.-P. DeWolf, J. Domingo-Ferrer, V. Torra, R. Brand, and S. Giessing. *μ-ARGUS version 4.0 Software and User's Manual.* Statistics Netherlands, Voorburg NL, may 2005. http://neon.vb.cbs.nl/casc.

[45] A. Hundepool, J. Domingo-Ferrer, L. Franconi, S. Giessing, R. Lenz, J. Longhurst, E. Schulte-Nordholt, G. Seri, and P.-P. DeWolf. *Handbook on Statistical Disclosure Control (version 1.0).* Eurostat (CENEX SDC Project Deliverable), 2006.

[46] D. E. Huntington and C. S. Lyrintzis. Improvements to and limitations of latin hypercube sampling. *Probabilistic Engineering Mechanics*, 13(4):245–253, 1998.

[47] A. B. Kennickell. Multiple imputation and disclosure control: the case of the 1995 survey of consumer finances. In *Record Linkage Techniques*, pages 248–267, Washington DC, 1999. National Academy Press.

[48] A. B. Kennickell. Multiple imputation and disclosure protection: the case of the 1995 survey of consumer finances. In J. Domingo-Ferrer, editor, *Statistical Data Protection*, pages 248–267, Luxemburg, 1999. Office for Official Publications of the European Communities.

[49] J. J. Kim. A method for limiting disclosure in microdata based on random noise and transformation. In *Proceedings of the Section on Survey Research Methods*, pages 303–308, Alexandria VA, 1986. American Statistical Association.

[50] M. Laszlo and S. Mukherjee. Minimum spanning tree partitioning algorithm for microaggregation. *IEEE Transactions on Knowledge and Data Engineering*, 17(7):902–911, 2005.

[51] J. M. Mateo-Sanz and J. Domingo-Ferrer. A method for data-oriented multivariate microaggregation. In J. Domingo-Ferrer, editor, *Statistical Data Protection*, pages 89–99, Luxemburg, 1999. Office for Official Publications of the European Communities.

[52] A. Meyerson and R. Williams. General k-anonymization is hard. Technical Report 03-113, Carnegie Mellon School of Computer Science (USA), 2003.

[53] R. Moore. Controlled data swapping techniques for masking public use microdata sets, 1996. U. S. Bureau of the Census, Washington, DC, (unpublished manuscript).

[54] K. Muralidhar, D. Batra, and P. J. Kirs. Accessibility, security and accuracy in statistical databases: the case for the multiplicative fixed data perturbation approach. *Management Science*, 41:1549–1564, 1995.

[55] A. Oganian and J. Domingo-Ferrer. On the complexity of optimal microaggregation for statistical disclosure control. *Statistical Journal of the United Nations Economic Comission for Europe*, 18(4):345–354, 2001.

[56] S. Polettini, L. Franconi, and J. Stander. Model based disclosure protection. In J. Domingo-Ferrer, editor, *Inference Control in Statistical Databases*, volume 2316 of *Lecture Notes in Computer Science*, pages 83–96, Berlin Heidelberg, 2002. Springer.

[57] T. J. Raghunathan, J. P. Reiter, and D. Rubin. Multiple imputation for statistical disclosure limitation. *Journal of Official Statistics*, 19(1):1–16, 2003.

[58] S. P. Reiss. Practical data-swapping: the first steps. *ACM Transactions on Database Systems*, 9:20–37, 1984.

[59] S. P. Reiss, M. J. Post, and T. Dalenius. Non-reversible privacy transformations. In *Proceedings of the ACM Symposium on Principles of Database Systems*, pages 139–146, Los Angeles, CA, 1982. ACM.

[60] J. P. Reiter. Satisfying disclosure restrictions with synthetic data sets. *Journal of Official Statistics*, 18(4):531–544, 2002.

[61] J. P. Reiter. Inference for partially synthetic, public use microdata sets. *Survey Methodology*, 29:181–188, 2003.

[62] J. P. Reiter. Using cart to generate partially synthetic public use microdata, 2003. Duke University working paper.

[63] J. P. Reiter. Releasing multiply-imputed, synthetic public use microdata: An illustration and empirical study. *Journal of the Royal Statistical Society, Series A*, 168:185–205, 2005.

[64] J. P. Reiter. Significance tests for multi-component estimands from multiply-imputed, synthetic microdata. *Journal of Statistical Planning and Inference*, 131(2):365–377, 2005.

[65] D. B. Rubin. Discussion of statistical disclosure limitation. *Journal of Official Statistics*, 9(2):461–468, 1993.

[66] P. Samarati. Protecting respondents' identities in microdata release. *IEEE Transactions on Knowledge and Data Engineering*, 13(6):1010–1027, 2001.

[67] P. Samarati and L. Sweeney. Protecting privacy when disclosing information: k-anonymity and its enforcement through generalization and suppression. Technical report, SRI International, 1998.

[68] G. Sande. Exact and approximate methods for data directed microaggregation in one or more dimensions. *International Journal of Uncertainty, Fuzziness and Knowledge-Based Systems*, 10(5):459–476, 2002.

[69] J. Schlörer. Disclosure from statistical databases: quantitative aspects of trackers. *ACM Transactions on Database Systems*, 5:467–492, 1980.

[70] F. Sebé, J. Domingo-Ferrer, J. M. Mateo-Sanz, and V. Torra. Post-masking optimization of the tradeoff between information loss and disclosure risk in masked microdata sets. In J. Domingo-Ferrer, editor, *Inference Control in Statistical Databases*, volume 2316 of *Lecture Notes in Computer Science*, pages 163–171, Berlin Heidelberg, 2002. Springer.

[71] A. C. Singh, F. Yu, and G. H. Dunteman. Massc: A new data mask for limiting statistical information loss and disclosure. In H. Linden, J. Riecan, and L. Belsby, editors, *Work Session on Statistical Data Confidentiality 2003*, Monographs in Official Statistics, pages 373–394, Luxemburg, 2004. Eurostat.

[72] G. R. Sullivan. *The Use of Added Error to Avoid Disclosure in Microdata Releases*. PhD thesis, Iowa State University, 1989.

[73] L. Sweeney. Achieving k-anonymity privacy protection using generalization and suppression. *International Journal of Uncertainty, Fuzziness and Knowledge Based Systems*, 10(5):571–588, 2002.

[74] L. Sweeney. k-anonimity: A model for protecting privacy. *International Journal of Uncertainty, Fuzziness and Knowledge Based Systems*, 10(5):557–570, 2002.

[75] V. Torra. Microaggregation for categorical variables: a median based approach. In J. Domingo-Ferrer and V. Torra, editors, *Privacy in Statistical Databases*, volume 3050 of *Lecture Notes in Computer Science*, pages 162–174, Berlin Heidelberg, 2004. Springer.

[76] J. F. Traub, Y. Yemini, and H. Wozniakowski. The statistical security of a statistical database. *ACM Transactions on Database Systems*, 9:672–679, 1984.

[77] U.S.Privacy. U. s. privacy regulations, 2004. http://www.media-awareness.ca/english/issues/privacy/us_legislation_privacy.cfm.

[78] L. Willenborg and T. DeWaal. *Statistical Disclosure Control in Practice*. Springer-Verlag, New York, 1996.

[79] L. Willenborg and T. DeWaal. *Elements of Statistical Disclosure Control*. Springer-Verlag, New York, 2001.

[80] W. E. Winkler. Re-identification methods for masked microdata. In J. Domingo-Ferrer and V. Torra, editors, *Privacy in Statistical Databases*, volume 3050 of *Lecture Notes in Computer Science*, pages 216–230, Berlin Heidelberg, 2004. Springer.

[81] W. E. Yancey, W. E. Winkler, and R. H. Creecy. Disclosure risk assessment in perturbative microdata protection. In J. Domingo-Ferrer, editor, *Inference Control in Statistical Databases*, volume 2316 of *Lecture Notes in Computer Science*, pages 135–152, Berlin Heidelberg, 2002. Springer.

Chapter 4

Measures of Anonymity

Suresh Venkatasubramanian
School of Computing, University of Utah
suresh@cs.utah.edu

Abstract To design a privacy-preserving data publishing system, we must first quantify the very notion of privacy, or information loss. In the past few years, there has been a proliferation of measures of privacy, some based on statistical considerations, others based on Bayesian or information-theoretic notions of information, and even others designed around the limitations of bounded adversaries. In this chapter, we review the various approaches to capturing privacy. We will find that although one can define privacy from different standpoints, there are many structural similarities in the way different approaches have evolved. It will also become clear that the notions of privacy and utility (the useful information one can extract from published data) are intertwined in ways that are yet to be fully resolved.

Keywords: Measures of privacy, statistics, Bayes inference, information theory, cryptography.

4.1 Introduction

In this chapter, we survey the various approaches that have been proposed to measure privacy (and the loss of privacy). Since most privacy concerns (especially those related to health-care information [44]) are raised in the context of legal concerns, it is instructive to view privacy from a legal perspective, rather than from purely technical considerations.

It is beyond the scope of this survey[1] to review the legal interpretations of privacy [11]. However, one essay on privacy that appears directly relevant (and has inspired at least one paper surveyed here) is the view of privacy in terms of access that others have to us and our information, presented by Ruth Gavison [23]. In her view, a general definition of privacy must be one that is measurable, of value, and actionable. The first property needs no explanation; the second means that the entity being considered private must be valuable, and the third property argues that from a legal perspective, only those losses of privacy are interesting that can be prosecuted.

This survey, and much of the research on privacy, concerns itself with the measuring of privacy. The second property is implicit in most discussion of measures of privacy: authors propose basic data items that are valuable and must be protected (fields in a record, background knowledge about a distribution, and so on). The third aspect of privacy is of a legal nature and is not directly relevant to our discussion here.

4.1.1 What is Privacy?

To measure privacy, we must define it. This, in essence, is the hardest part of the problem of measuring privacy, and is the reason for the plethora of proposed measures. Once again, we turn to Gavison for some insight. In her paper, she argues that there are three inter-related kinds of privacy: secrecy, anonymity, and solitude. Secrecy concerns information that others may gather about us. Anonymity addresses how much "in the public gaze" we are, and solitude measures the degree to which others have physical access to us. From the perspective of protecting information, solitude relates to the physical protection of data, and is again beyond the purview of this article. Secrecy and anonymity are useful ways of thinking about privacy, and we will see that measures of privacy preservation can be viewed as falling mostly into one of these two categories.

If we think of privacy as secrecy (of our information), then a loss of privacy is leakage of that information. This can measured through various means: the probability of a data item being accessed, the change in knowledge of an adversary upon seeing the data, and so on. If we think in terms of anonymity, then privacy leakage is measured in terms of the size of the blurring accompanying the release of data: the more the blurring, the more anonymous the data.

Privacy versus Utility. It would seem that the most effective way to preserve privacy of information would be to encrypt it. Users wishing to access the data could be given keys, and this would summarily solve all privacy issues. Unfortunately, this approach does not work in a *data publishing* scenario, which is the primary setting for much work on privacy preservation.

The key notion here is one of *utility*: the goal of privacy preservation measures is to secure access to confidential information *while at the same time releasing aggregate information to the public.* One common example used is that of the U.S. Census. The U.S Census wishes to publish survey data from the census so that demographers and other public policy experts can analyze trends in the general population. On the other hand, they wish to avoid releasing information that could be used to infer facts about specific individuals; the case of the AOL search query release [34] indicates the dangers of releasing data without adequately anonymizing it.

It is this idea of utility that makes cryptographic approaches to privacy preservation problematic. As Dwork points out in her overview of differential privacy [16], a typical cryptographic scenario involves two communicating parties and an adversary attempting to eavesdrop. In the scenarios we consider, the adversary is the same as the recipient of the message, making security guarantees much harder to prove.

Privacy and utility are fundamentally in tension with each other. We can achieve perfect privacy by not releasing any data, but this solution has no utility. Thus, any discussion of privacy measures is incomplete without a corresponding discussion of utility measures. Traditionally, the two concepts have been measured using different yardsticks, and we are now beginning to see attempts to unify the two notions along a common axis of measurement.

A Note on Terminology. Various terms have been used in the literate to describe privacy and privacy loss. *Anonymization* is a popular term, often used to describe methods like k-anonymity and its successors. *Information loss* is used by some of the information-theoretic methods, and *privacy leakage* is another common expression describing the loss of privacy. We will use these terms interchangeably.

4.1.2 Data Anonymization Methods

The measures of anonymity we discuss here are usually defined with respect to a particular data anonymization method. There are three primary methods in use today, *random perturbation*, *generalization* and *suppression*. In what follows, we discuss these methods.

Perhaps the most natural way of anonymizing numerical data is to perturb it. Rather than reporting a value x for an attribute, we report the value $\tilde{x} = x + r$, where r is a random value drawn from an appropriate (usually bias-free) distribution. One must be careful with this approach however; if the value r is chosen independently each time x is queried, then simple averaging will eliminate its effect. Since introducing bias would affect any statistical analysis one might wish to perform on the data, a preferred method is to fix the perturbations in advance.

If the attribute x has a domain other than $\mathbb{R}$, then perturbation is more complex. As long as the data lies in a continuous metric space (like $\mathbb{R}^d$ for instance), then a perturbation is well defined. If the data is categorical however, other methods, such as deleting items and inserting other, randomly chosen items, must be employed. We will see more of such methods below.

It is often useful to distinguish between two kinds of perturbation. *Input perturbation* is the process of perturbing the source data itself, and returning correct answers to queries on this perturbed data. *Output perturbation* on the other hand perturbs the answers sent to a query, rather than modifying the input itself.

The other method for anonymizing data is generalization, which is often used in conjunction with suppression. Suppose the data domain possesses a natural hierarchical structure. For example, ZIP codes can be thought of as the leaves of a hierarchy, where $8411*$ is the parent of 84117, and $84*$ is an ancestor of $8411*$, and so on. In the presence of such a hierarchy, attributes can be generalized by replacing their values with that of their (common) parent. Again returning to the ZIP code example, ZIP codes of the form $84117, 84118, 84120$ might all be replaced by the generic ZIP $841*$. The degree of perturbation can then be measured in terms of the height of the resulting generalization above the leaf values.

Data suppression, very simply, is the omission of data. For example, a set of database tuples might all have ZIP code fields of the form 84117 or 84118, with the exception of a few tuples that have a ZIP code field value of 90210. In this case, the outlier tuples can be *suppressed* in order to construct valid and compact generalization. Another way of performing data suppression is to replace a field with a generic identifier for that field. In the above example, the ZIP code field value of 90210 might be replaced by a null value $\perp_{\text{ZIP}}$.

Another method of data anonymization that was proposed by Zhang *et al.* [50] is to *permute* the data. Given a table consisting of sensitive and identifying attributes, their approach is to permute the projection of the table consisting of the sensitive attributes; the purpose of doing this is to retain the aggregate properties of the table, while destroying the link between identifying and sensitive attributes that could lead to a privacy leakage.

4.1.3 A Classification of Methods

Broadly speaking, methods for measuring privacy can be divided into three distinct categories. Early work on statistical databases measured privacy in terms of the variance of key perturbed variables: the larger the variance, the better the privacy of the perturbed data. We refer to these approaches as *statistical methods*.

Much of the more recent work on privacy measures starts with the observation that statistical methods are unable to quantify the idea of *background information* that an adversary may possess. As a consequence, researchers have employed tools from information theory and Bayesian analysis to quantify more precisely notions of information transfer and loss. We will describe these methods under the general heading of *probabilistic methods*.

Almost in parallel with the development of probabilistic methods, some researchers have attacked the problem of privacy from a computational angle.

In short, rather than relying on statistical or probabilistic estimates for the amount of information leaked, these measures start from the idea of a resource-bounded adversary, and measure privacy in terms of the amount of information accessible by such an adversary. This approach is reminiscent of cryptographic approaches, but for the reasons outlined above is substantially more difficult.

An Important Omission: Secure Multiparty Computation. One important technique for preserving data privacy is the approach from cryptography called *secure multi-party computation* (SMC). The simplest version of this framework is the so-called 'Millionaires Problem' [49]:

> Two millionaires wish to know who is richer; however, they do not want to find out inadvertently any additional information about each others wealth. How can they carry out such a conversation?

In general, an SMC scenario is described by N clients, each of whom owns some private data, and a public function $f(x_1, \ldots x_N)$ that needs to be computed from the shared data without any of the clients revealing their private information.

Notice that in an SMC setting, the clients are trusted, and do not trust the central server to preserve their information (otherwise they could merely transmit the required data to the server). In all the privacy-preservation settings we will consider in this article, it is the server that is trusted, and queries to the server emanate from untrusted clients. We will not address SMC-based privacy methods further.

4.2 Statistical Measures of Anonymity

4.2.1 Query Restriction

Query restriction was one of the first methods for preserving anonymity in data [22, 25, 21, 40]. For a database of size N, and a fixed parameter k, all queries that returned either fewer than k or more than $N - k$ records were rejected. Query restriction anticipates k-anonymity, in that the method for preserving anonymity is by returning a large set of records for any query. Contrast this with data suppression; rather than deleting records, the procedure deletes queries.

It was pointed out later [13, 12, 41, 10, 41] that query restriction could be subverted by requesting a specific sequence of queries, and then combining them using simple Boolean operators, in a construction referred to as a *tracker*. Thus, this mechanism is not very effective.

4.2.2 Anonymity via Variance

Here, we start with randomly perturbed data $\tilde{x} = x + r$, as described in Section 4.1.2. Intuitively, the larger the perturbation, the more blurred, and thus more protected the value is. Thus, we can measure anonymity by measuring the variance of the perturbed data. The larger the variance, the better the guarantee of anonymity, and thus one proposal by Duncan *et al.* [15] is to lower bound the variance for estimators of sensitive attributes. An alternative approach, used by Agrawal and Srikant [3], is to fix a confidence level and measure the length of the interval of values of the estimator that yields this confidence bound; the longer the interval, the more successful the anonymization.

Under this model, utility can be measured in a variety of ways. The Duncan *et al.* paper measures utility by combining the perturbation scheme with a query restriction method, and measuring the fraction of queries that are permitted after perturbation. Obviously, the larger the perturbation (measured by the variance σ^2), the larger the fraction of queries that return sets of high cardinality. This presents a natural tradeoff between privacy (increased by increasing σ^2) and utility (increased by increasing the fraction of permitted queries).

The paper by Agrawal and Srikant implicitly measures utility in terms of how hard it is to reconstruct the original data distribution. They use many iterations of a Bayesian update procedure to perform this reconstruction; however the reconstruction itself provides no guarantees (in terms of distance to the true data distribution).

4.2.3 Anonymity via Multiplicity

Perturbation-based privacy works by changing the values of data items. In generalization-based privacy, the idea is to "blur" the data via generalization. The hope here is that the blurred data set will continue to provide the statistical utility that the original data provided, while preventing access to individual tuples.

The measure of privacy here is a combinatorial variant of the length-of-interval measure used in [3]. A database is said to be k-anonymous [42] if there is no query that can extract fewer than k records from it. This is achieved by aggregating tuples along a generalization hierarchy: for example, by aggregating zip codes upto to the first three digits, and so on. k-anonymity was first defined in the context of *record linkage*: can tuples from multiple databases be joined together to infer private information inaccessible from the individual sources?

The k-anonymity requirement means such access cannot happen, since no query returns fewer than k records, and so cannot be used to isolate a single tuple containing the private information. As a method for blocking record linkage, k-anonymity is effective, and much research has gone into optimizing the computations, investigating the intrinsic hardness of computing it, and generalizing it to multiple dimensions.

4.3 Probabilistic Measures of Anonymity

Upto this point, an information leak has been defined as the revealing of specific data in a tuple. Often though, information can be leaked even if the adversary does not gain access to a specific data item. Such attacks usually rely on knowing aggregate information about the (perturbed) source database, as well as the method of perturbation used when modifying the data.

Suppose we attempt to anonymize an attribute X by perturbing it with a random value chosen uniformly from the interval $[-1, 1]^2$. Fixing a confidence level of 100%, and using the measure of privacy from [3], we infer that the privacy achieved by this perturbation is 2 (the length of the interval $[-1, 1]$). Suppose however that a distribution on the values of X is revealed: namely, X takes a value in the range $[0, 1]$ with probability $1/2$, and a value in the range $[4, 5]$ with probability $1/2$. In this case, no matter what the actual value of X is, an adversary can infer from the perturbed value $\tilde{X}$ which of the two intervals of length 1 the true value of X really lies in, reducing the effective privacy to at most 1.

Incorporating background information changes the focus of anonymity measurements. Rather than measuring the likelihood of some data being released, we now have to measure a far more nebulous quantity: the "amount of new information learned by an adversary" relative to the background. In order to do this, we need more precise notions of information leakage than the variance of a perturbed value.

This analysis applies irrespective of whether we do anonymization based on random perturbation or generalization. We first consider measures of anonymization that are based on perturbation schemes, following this with an examination of measures based on generalization. In both settings, the measures are probabilistic: they compute functions of distributions defined on the data.

4.3.1 Measures Based on Random Perturbation

Using Mutual Information The paper by Agrawal and Aggarwal [2] proposes the use of mutual information to measure leaked information. We can use the entropy $H(A)$ to encode the amount of uncertainty (and therefore the degree of privacy) in a random variable A. $H(A|B)$, the *conditional entropy of A given B*, can be interpreted as the amount of privacy "left" in A after B is revealed. Since entropy is usually expressed in terms of *bits* of information, we will use the expression $2^{H(A)}$ to represent the measure of privacy in A. Using this measure, the fraction of privacy leaked by an adversary who knows B can be written as

$$\mathcal{P}(A|B) = 1 - 2^{H(A|B)}/2^{H(A)} = 1 - 2^{-I(A;B)}$$

where $I(A; B) = H(A) - H(A|B)$ is the mutual information between the random variables A and B.

They also develop a notion of utility measured by the statistical distance between the source distribution of data and the perturbed distribution. They also demonstrate an EM-based method for reconstructing the maximum likelihood estimate of the source distribution, and show that it converges to the correct answer (they do not address the issue of rate of convergence).

Handling Categorical Values The above schemes rely on the source data being numerical. For data mining applications, the relevant source data is usually categorical, consisting of collections of transactions, each transaction defined as a set of items. For example, in the typical market-basket setting, a transaction consists of a set of items purchased by a customer.

Such sets are typically represented by binary characteristic vectors. The elementary datum that requires anonymity is membership: does item i belong to transaction t? The questions requiring utility, on the other hand, are of the form, "which patterns have reasonable support and confidence"? In such a setting, the only possible perturbation is to flip an item's membership in a transaction, but not so often as to change the answers to questions about patterns in any significant way.

There are two ways of measuring privacy in this setting. The approach taken by Evfimievski *et al.* [20] is to evaluate whether an anonymization scheme leaves clues for an adversary with high probability. Specifically, the define a privacy breach one in which the probability of some property of the input data is high, conditioned on the output perturbed data having certain properties.

DEFINITION 4.3.1 *An itemset* A causes a privacy breach of level ρ *if for some item* $a \in A$ *and some* $i \in 1 \ldots N$ *we have* $\mathbf{P}[a \in t_i | A \subseteq t'_i] \geq \rho$.

Here, the event "$A \subseteq t_i$" is leaking information about the event "$a \in t_i$". Note that this measure is absolute, regardless of what the prior probability of $a \in t_i$ might have been. The perturbation method is based on randomly sampling some items of the transaction t_i to keep, and buffering with elements $a \notin t_i$ at random.

The second approach, taken by Rizvi and Haritsa [38], is to measure privacy in terms of the probability of correctly reconstructing the original bit, given a perturbed bit. This can be calculated using Bayes' Theorem, and is parametrized by the probability of flipping a bit (which they set to a constant p). Privacy is then achieved by setting p to a value that minimizes the reconstruction probability; the authors show that a wide range of values for p yields acceptable privacy thresholds.

Both papers then frame utility as the problem of reconstructing itemset frequencies accurately. [20] establishes a tradeoff between utility more precisely, in terms of the probabilities $p[l \to l'] = \mathbf{P}[\#(t' \cap A) = l' | \#(t \cap A) = l]$.

For privacy, we have to ensure that (for example) if we fix an element $a \in t$, then the set of tuples t that do contain a are not overly represented in the modified itemset. Specifically, in terms of an average over the size of tuple sets returned, we obtain a condition on the $p[l \to l']$. In essence, the probabilities $p[l \to l']$ encode the tradeoff between utility (or ease of reconstruction) and privacy.

Measuring Transfer of Information Both the above papers have the same weakness that plagued the original statistics-based anonymization works: they ignore the problem of the background knowledge attack. A related, and yet subtlely different problem is that ignoring the source data distribution may yield meaningless results. For example, suppose the probability of an item occurring any particular transaction is very high. Then the probability of reconstructing its value correctly is also high, but this would not ordinarily be viewed as a leak of information. A more informative approach would be to measure the level of "surprise": namely whether the probability $P[a \in t_i]$ increases (or decreases) dramatically, conditioned on seeing the event $A \subseteq t'_i$.

Notice that this idea is the motivation for [2]; in their paper, the mutual information $I(A; B)$ measures the transfer of information between the source and anonymized data. Evfimievski *et al.* [19], in a followup to [20], develop a slightly different notion of information transfer, motivated by the idea that mutual information is an "averaged" measure and that for privacy preservation, worst-case bounds are more relevant.

Formally, information leakage is measured by estimating the change in probability of a property from source to distorted data. For example, given a property $Q(X)$ of the data, they say that there is a privacy breach after perturbing the data by function $R(X)$ if for some y,

$$\mathbf{P}[Q(X)] \leq \rho_1, \mathbf{P}[Q(X)|R(X) = y] \geq \rho_2$$

where $\rho_1 \ll \rho_2$.

However, ensuring that this property holds is computationally intensive. The authors show that a sufficient condition for guaranteeing no (ρ_1, ρ_2) privacy breach is to bound the difference in probability between two different x_i being mapped to a particular y. Formally, they propose perturbation schemes such that

$$\frac{p[x_1 \to y]}{p[x_2 \to y]} \leq \gamma$$

Intuitively, this means that if we look back from y, there is no easy way of telling whether the source was x_1 or x_2. The formal relation to (ρ_1, ρ_2)-privacy is established via this intuition.

Formally, we can rewrite

$$I(X;Y) = \sum_y p(y)\mathrm{KL}(p(X|Y=y)|p(X))$$

The function $\mathrm{KL}(p(X|Y=y)|p(X))$ measures the transfer distance; it asks how different the induced distribution $p(X|Y=y)$ is from the source distribution $p(X)$. The more the difference is, the less the privacy breach is. The authors propose replacing the averaging in the above expression by a max, yielding a modified notion

$$I_w(X;Y) = \max_y p(y)\mathrm{KL}(p(X|Y=y)|p(X))$$

They then show that a (ρ_1, ρ_2)-privacy breach yields a lower bound on the worst-case mutual information $I_w(X;Y)$, which is what we would expect.

More general perturbation schemes All of the above described perturbation schemes are *local*: perturbations are applied independently to data items. Kargupta *et al.* [27] showed that the lack of correlation between perturbations can be used to attack such a privacy-preserving mechanism. Their key idea is a spectral filtering method based on computing principal components of the data transformation matrix.

Their results suggest that for more effective privacy preservation, one should consider more general perturbation schemes. It is not hard to see that a natural generalization of these perturbation schemes is a Markov-chain based approach, where an item x is perturbed to item y based on a transition probability $p(y|x)$. FRAPP [4] is one such scheme based on this idea. The authors show that they can express the notion of a (ρ_1, ρ_2)-privacy breach in terms of properties of the Markov transition matrix. Moreover, they can express the utility of this scheme in terms of the condition number of the transition matrix.

4.3.2 Measures Based on Generalization

It is possible to mount a 'background knowledge' attack on k-anonymity. For example, it is possible that all the k records returned from a particular query share the same value of some attribute. Knowing that the desired tuple is one of the k tuples, we have thus extracted a value from this tuple without needing to isolate it.

The first approach to address this problem was the work on ℓ-diversity [32]. Here, the authors start with the now-familiar idea that the privacy measure should capture the *change* in the adversary's world-view upon seeing the data.

However, they execute this idea with an approach that is absolute. They require that the *distribution* of sensitive values in an aggregate have high entropy (at least $\log \ell$). This subsumes k-anonymity, since we can think of the probability of leakage of a single tuple in k-anonymity as $1/k$, and so the "entropy" of the aggregate is $\log k$. Starting with this idea, they introduce variants of ℓ-diversity that are more relaxed about disclosure, or allow one to distinguish between positive and negative disclosure, or even allow for multi-attribute disclosure measurement.

Concurrently published, the work on p-sensitive-k-anonymity [43] attempts to do the same thing, but in a more limited way, by requiring at least p distinct sensitive values in each generalization block, instead of using entropy. A variant of this idea was proposed by Wong *et al.* [47]; in their scheme, termed (α, k)-anonymity, the additional constraint imposed on the generalization is that the fractional frequency of each value in a generalization is no more than α. Note that this approach automatically lower bounds the entropy of the generalization by $\log(1/\alpha)$.

Machanavajjhala *et al.* [32] make the point that it is difficult to model the adversary's background knowledge; they use this argument to justify the ℓ-diversity measure. One way to address this problem is to assume that the adversary has access to global statistics of the sensitive attribute in question. In this case, the goal is to make the sensitive attribute "blend in"; its distribution in the generalization should mimic its distribution in the source data.

This is the approach taken by Li, Li and the author [31]. They define a measure called t-closeness that requires that the "distance" between the distribution of a sensitive attribute in the generalized and original tables is at most t.

A natural distance measure to use would be the KL-distance from the generalized to the source distribution. However, for numerical attributes, the notion of closeness must incorporate the notion of a metric on the attribute. For example, suppose that a salary field in a table is generalized to have three distinct values $(20000, 21000, 22000)$. One might reasonably argue that this generalization leaks more information than a generalization that has the three distinct values $(20000, 50000, 80000)$.

Computing the distance between two distributions where the underlying domains inhabit a metric space can be performed using the metric known as the earth-mover distance [39], or the Monge-Kantorovich transportation distance [24]. Formally, suppose we have two distributions p, q defined over the elements X of a metric space (X, d). Then the earth-mover distance between p and q is

$$d_E(p, q) = \inf_{P[x'|x]} \sum_{x,x'} d(x, x') P[x'|x] p(x)$$

subject to the constraint $\sum_x P[x'|x] p(x) = q(x')$.

Intuitively, this distance is defined as the value that minimizes the transportation cost of transforming one distribution to the other, where transportation cost is measured in terms of the distance in the underlying metric space. Note that since any underlying metric can be used, this approach can be used to integrate numerical and categorical attributes, by imposing any suitable metric (based on domain generalization or other methods) on the categorical attributes.

The idea of extending the notion of diversity to numerical attributes was also considered by Zhang *et al.* [50]. In this paper, the notion of distance for numerical attributes is extended in a different way: the goal for the k-anonymous blocks is that the "diameter" of the range of sensitive attributes is larger than a parameter e. Such a generalization is said to be (k, e)-anonymous. Note that this condition makes utility difficult. If we relate this to the ℓ-diversity condition of having at least ℓ distinct values, this represents a natural generalization of the approach. As stated however, the approach appears to require defining a total order on the domain of the attribute; this would prevent it from being used for higher dimensional attributes sets.

Another interesting feature of the Zhang *et al.* method is that it considers the down-stream problem of answering aggregate queries on an anonymized database, and argues that rather than performing generalization, it might be better to perform a permutation of the data. They show that this permutation-based anonymization can answer aggregate queries more accurately than generalization-based anonymization.

Anonymizing Inferences. In all of the above measures, the data being protected is an attribute of a record, or some distributional characteristic of the data. Another approach to anonymization is to protect the possible inferences that can be made from the data; this is akin to the approach taken by Evfimievski *et al.* [19, 20] for perturbation-based privacy. Wang *et al.* [45] investigate this idea in the context of generalization and suppression. A *privacy template* is an inference on the data, coupled with a confidence bound, and the requirement is that in the anonymized data, this inference not be valid with a confidence larger than the provided bound. In their paper, they present a scheme based on data suppression (equivalent to using a unit height generalization hierarchy) to ensure that a given set of privacy templates can be preserved.

Clustering as k-anonymity. Viewing attributes as elements of metric space and defining privacy accordingly has not been studied extensively. However, from the perspective of generalization, many papers ([7, 30, 35]) have pointed out that generalization along a domain generalization hierarchy is only one way of aggregating data. In fact, if we endow the attribute space with a metric, then the process of generalization can be viewed in general as a clustering problem on this metric space, where the appropriate measure of anonymity is applied to each *cluster*, rather than to each generalized group.

Such an approach has the advantage of placing different kinds of attributes on an equal footing. When anonymizing categorical attributes, generalization proceeds along a generalization hierarchy, which can be interpreted as defining a tree metric. Numerical attributes are generalized along ranges, and t-closeness works with attributes in a general metric space. By lifting all such attributes to a general metric space, generalization can happen in a uniform manner, measured in terms of the diameters of the clusters.

Strictly speaking, these methods do not introduce a new notion of privacy; however, they do extend the applicability of generalization-based privacy measures like k-anonymity and its successors.

Measuring utility in generalization-based anonymity The original k-anonymity work defines the utility of a generalized table as follows. Each cell is the result of generalizing an attribute up a certain number of levels in a generalization hierarchy. In normalized form, the "height" of a generalization ranges from 0 if the original value is used, to 1 if a completely generalized value is used (in the scheme proposed, a value of 1 corresponds to value suppression, since that is the top level of all hierarchies). The *precision* of a generalization scheme is then 1 - the average height of a generalization (measured over all cells). The precision is 1 if there is no generalization and is 0 if all values are generalized.

Bayardo and Agrawal ([5]) define a different utility measure for k-anonymity. In their view, a tuple that inhabits a generalized equivalence class E of size $|E| = j, j > k$ incurs a "cost" of j. A tuple that is suppressed entirely incurs a cost of D, where D is the size of the entire database. Thus, the cost incurred by an anonymization is given by

$$C = \sum_{|E|\geq k} |E|^2 + \sum_{|E|<k} |D||E|$$

This measure is known as the *discernability* metric. One can also compute the average size of a generalized class as a measure of utility [32].

Another cost measure proposed by Iyengar [26] is a *misclassification metric*: As above, consider the equivalence class produced by an anonymization, and charge one unit of cost for each tuple in a *minority class* with respect to the collection of classes. Ignore all suppressed tuples. Again, averaging this over all tuples returns the total penalty.

Once again, when we introduce a metric structure on numeric attributes, utility has to be measured differently. Zhang *et al.* [50] propose measuring utility by ensuring that the range of sensitive values in each group is as small as possible, subject to the privacy constraints.

4.3.3 Utility vs Privacy

The problems of utility and anonymity ask the same kind of question: "how much information does the anonymized data distribution reveal about the source?". For an attribute to be anonymized, we wish this quantity to be small, but for a useful attribute, we want this quantity to be large !

Most of the schemes for ensuring data anonymity focus their effort on defining measures of anonymity, while using *ad hoc* measures of utility. A more balanced treatment of the two notions would use similar measures for utility and anonymity, or quantify the tradeoff that must exist between the two. In the next section, we will see how this can be performed in the context of computational approaches to anonymization.

However, even in the probabilistic context, some principled approaches to utility measurement have been developed. One paper that attempts this in the context of generalization-based anonymization is the work by Kifer and Gehrke [28]. In this paper, after performing a standard anonymization, they publish carefully chosen marginals of the source data. From these marginals, they then construct a consistent maximum entropy distribution, and measure utility as the KL-distance between this distribution and the source. The remainder of the paper is devoted to methods for constructing good marginals, and reconstructing the maximum entropy extension.

Switching to perturbation-based methods, the paper by Rastogi *et al.* [37] provides strong tradeoffs between privacy and utility. In this work, the authors define a measure of utility in terms of the discrepancy between the value of a counting query returned by an estimator, and the true value of the counting query. Privacy is measured using the framework of Evfimievski *et al.* [19], in terms of the conditional probability of a tuple being present in the anonymized data, relative to the prior probability of tuple being present in the source. One of the main results in their paper is an impossibility result limiting the tradeoff between these measures of privacy and utility.

4.4 Computational Measures of Anonymity

We now turn to measures of anonymity that are defined computationally: privacy statements are phrased in terms of the power of an adversary, rather than the amount of background knowledge they possess. Such an approach is attractive for a variety of reasons: measuring privacy in terms of a distance between distributions does not tell us what kinds of attacks a resource-bounded adversary can mount: in this sense, privacy measures that rely on distributional distances are overly conservative. On the other hand, it is difficult to define precisely what kind of background knowledge an adversary has, and in the absence of such information, *any* privacy scheme based on background information attacks is susceptible to information leakage.

The first study of anonymity in the presence of computationally bounded adversaries was carried out by Dinur and Nissim [14]. In their framework, a database consists of a sequence of bits (this is without loss of generality), and a query q consists of a subset of bit positions, with the output a_q being the number of 1s in the subset. Such a query can be thought of as abstracting standard aggregation queries. The anonymization procedure is represented by an algorithm that returns the (possibly modified) answer $\tilde{a}_q$ to query q. The utility of the anonymization is measured by a parameter $\mathcal{E}$: an anonymization is said to be within $\mathcal{E}$ perturbation if $|a_q - \tilde{a}_q| \leq \mathcal{E}$, for all q.

An adversary is a Turing machine that can reconstruct a constant fraction of the bits in the database with high probability, using only invocations of the query algorithm. This reconstruction can be measured by the Hamming distance between the reconstructed database and the original database; the adversary succeeds if this distance is at most ϵn.

Rather than define privacy, the authors define "non-privacy": they say a database is $t(n)$-non-private if for all $\epsilon > 0$, there is some adversary running in time $t(n)$ that can succeed with high probability.

In this model, adversaries are surprisingly strong. The authors show that even with almost-linear perturbation, an adversary permitted to run in exponential time can break privacy. Restricting the adversary to run in polynomial time helps, but only slightly; any perturbation $\mathcal{E} = o\sqrt{n}$ is not enough to preserve privacy, and this is tight.

Feasibility results are hard to prove in this model: as the authors point out, an adversary, with one query, can distinguish between the databases 1^n and 0^n if it has background knowledge that these are the only two choices. A perturbation of $n/2$ would be needed to hide the database contents. One way of circumventing this is to assume that the database itself is generated from some distribution, and that the adversary is required to reveal the value of a specific bit (say, the ith bit) after making an arbitrary number of queries, and after being given *all bits of the database* except the ith bit.

In this setting, privacy is defined as the condition that the adversary's reconstruction probability is at most $1/2 + \delta$. In this setting, they show that a $\sqrt{T(n)}$-perturbed database is private against all adversaries that run in time $T(n)$.

Measuring Anonymity Via Information Transfer As before, in the case of probabilistic methods, we can reformulate the anonymity question in terms of information transfer; how much does the probability of a bit being 1 (or 0) change upon anonymization ?

Dwork and Nissim [18] explore this idea in the context of computationally bounded adversaries. Starting with a database d represented as a Boolean *matrix* and drawn from a distribution $\mathcal{D}$, we can define the prior probability $p_0^{ij} = P[d_{ij} = 1]$. Once an adversary asks T queries to the anonymized database as above, and all other values of the database are provided, we can now define the posterior probability p_T^{ij} of d_{ij} taking the value 1. The change in belief can be quantified by the expression $\Delta = |c(p_T^{ij}) - c(p_0^{ij})|$, where $c(x) = \log(x/(1-x))$ is a monotonically increasing function of x.

This is the simplified version of their formulation. In general, we can replace the event $d_{ij} = 1$ by the more general $f(d_{i1}, d_{i2}, \ldots d_{ik}) = 1$, where f is some k-ary Boolean function. All the above definitions translate to this more general setting. We can now define $(\delta, T(n))$-privacy as the condition that for all distributions over databases, all functions f, and all adversaries making T queries, the probability that the maximum change of belief is more than δ is negligibly small.

As with [14], the authors show a natural tradeoff between the degree of perturbation needed, and the level of privacy achieved. Specifically, the authors show that a previously proposed algorithm SuLQ [6] achieves $(\delta, T(n))$ privacy with a perturbation $\mathcal{E} = O(\sqrt{T(n)}/\delta)$. They then go on to show that under such conditions, it is possible to perform efficient and accurate data mining on the anonymized database to estimate probabilities of the form $P[\beta|\alpha]$, where α, β are two attributes.

Indistinguishability Although the above measures of privacy develop precise notions of information transfer with respect to a bounded adversary, they still require some notion of a distribution on the input databases, as well as a specific protocol followed by an adversary. To abstract the ideas underlying privacy further, Dwork *et al.* [17] formulate a definition of privacy inspired by Dalenius [16]: *A database is private if anything learnable from it can be learned in the absence of the database.*

In order to do this, they distinguish between non-interactive privacy mechanisms, where the data publisher anonymizes the data and publishes it (*input perturbation*), and interactive mechanisms, in which the output to queries are perturbed (*output perturbation*). Dwork [16] shows that in a non-interactive setting, it is impossible to achieve privacy under this definition; in other words, it is always possible to design an adversary and an auxiliary information generator such that the adversary, combining the anonymized data and the auxiliary information, can effect a privacy breach far more often than an adversary lacking access to the database can.

In the interactive setting, we can think of the interaction between the database and the adversary as a *transcript*. The idea of indistinguishability is that if two databases are very similar, then their transcripts with respect to an adversary should also be similar. Intuitively, this means that if an individual adds their data to a database (causing a small change), the nominal loss in privacy is very small.

The main consequence of this formulation is that it is possible to design perturbation schemes that depend only on the query functions and the error terms, and are *independent of the database*. Informally, the amount of perturbation required depends on the sensitivity of the query functions: the more the function can change when one input is perturbed slightly, the more perturbation the database must incur. The details of these procedures are quite technical: the reader is referred to [16, 17] for more details.

4.4.1 Anonymity via Isolation

Another approach to anonymization is taken by [8, 9]. The underlying principle here is *isolation*: a record is private if it cannot be singled out from its neighbors. Formally, they define an adversary as an algorithm that takes an anonymized database and some auxiliary information, and outputs a single point q. The adversary succeeds if a small ball around q does not contain too many points of the database. In this sense, the adversary has *isolated* some points of the database[3].

Under this definition of a privacy breach, they then develop methods for anonymizing a database. Like the papers above, they use a differential model of privacy: an anonymization is successful if the adversary, combining the anonymization with auxiliary information, can do no better at isolation than a weaker adversary with no access to the anonymized data.

One technical problem with the idea of isolation, which the authors acknowledge, is that it can be attacked in the same way that methods like k-anonymity are attacked. If the anonymization causes many points with similar characteristics to cluster together, then even though the adversary cannot isolate a single point, it can determine some special characteristics of the data from the clustering that might not have otherwise been inferred.

4.5 Conclusions and New Directions

The evolution of measures of privacy, irrespective of the specific method of perturbation or class of measure, has proceeded along a standard path. The earliest measures are absolute in nature, defining an intuitive notion of privacy in terms of a measure of obfuscation. Further development occurs when the notion of background information is brought in, and this culminates in the idea of a change in adversarial information before and after the anonymized data is presented.

From the perspective of theoretical rigor, computational approaches to privacy are the most attractive. They rely on few to no modelling assumptions about adversaries, and their cryptographic flavor reinforces our belief in their overall reliability as measures of privacy. Although the actual privacy preservation methods proposed in this space are fairly simple, they do work from very simple models of the underlying database, and one question that so far remains unanswered is the degree to which these methods can be made practically effective when dealing with the intricacies of actual databases.

The most extensive attention has been paid to the probabilistic approaches to privacy measurements. k-anonymity and its successors have inspired numerous works that study not only variants of the basic measures, but systems for managing privacy, extensions to higher dimensional spaces, as well as better methods for publishing data tables. The challenge in dealing with methods deriving from k-anonymity is the veritable alphabet soup of approaches that have been proposed, all varying subtlety in the nature of the assumptions used. The work by Wong *et al.* [46] illustrates the subtleties of modelling background information; their m-confidentiality measure attempts to model adversaries who exploit the desire of k-anonymizing schemes to generate a minimal anonymization. This kind of background information is very hard to formalize and argue rigorously about, even when we consider the general framework for analyzing background information proposed by Martin *et al.* [33].

4.5.1 New Directions

There are two recent directions in the area of privacy preservation measures that are quite interesting and merit further study. The first addresses the problem noted earlier: the imbalance in the study of utility versus privacy. The computational approaches to privacy preservation, starting with the work of Dinur and Nissim [14], provide formal tradeoffs between utility and privacy, for bounded adversaries. The work of Kifer *et al.* [28] on injecting utility into privacy-preservation allows for a more general measure of utility as a distance between distributions, and Rastogi *et al.* [37] examine the tradeoff between privacy and utility rigorously in the perturbation framework.

With a few exceptions, all of the above measures of privacy are *global*: they assume a worst-case (or average-case) measure of privacy over the entire input, or prove privacy guarantees that are independent of the specific instance of a database being anonymized. It is therefore natural to consider *personalized* privacy, where the privacy guarantee need only be accurate with respect to the specific instance being considered, or can be tuned depending on auxiliary inputs.

The technique for anonymizing inferences developed in [45] can be viewed as such a scheme: the set of inferences needing protection are supplied as part of the input, and other inferences need not be protected. In the context of k-anonymity, Xiao and Tao [48] propose a technique that takes as input user preferences about the level of generalization they desire for their sensitive attributes, and adapts the k-anonymity method to satisfy these preferences. The work on worst-case background information modelling by Martin *et al.* [33] assumes that the specific background knowledge possessed by an adversary is an input to the privacy-preservation algorithm. Recent work by Nissim *et al.* [36] revisits the indistinguishability measure [17] (which is oblivious of the specific database instance) by designing an *instance-based* property of the query function that they use to anonymize a given database.

Notes

1. ...and the expertise of the author!
2. This example is taken from [2].
3. This bears a strong resemblance to k-anonymity, but is more general.

References

[1] *Proceedings of the 23rd International Conference on Data Engineering, ICDE 2007, April 15-20, 2007, The Marmara Hotel, Istanbul, Turkey* (2007), IEEE.

[2] AGRAWAL, D., AND AGGARWAL, C. C. On the design and quantification of privacy preserving data mining algorithms. In *Proceedings of the twentieth ACM SIGMOD-SIGACT-SIGART symposium on Principles of Database Systems* (Santa Barbara, CA, 2001), pp. 247–255.

[3] AGRAWAL, R., AND SRIKANT, R. Privacy preserving data mining. In *Proceedings of the ACM SIGMOD Conference on Management of Data* (Dallas, TX, May 2000), pp. 439–450.

[4] AGRAWAL, S., AND HARITSA, J. R. FRAPP: A framework for high-accuracy privacy-preserving mining. In *ICDE '05: Proceedings of the 21st International Conference on Data Engineering (ICDE'05)* (Washington, DC, USA, 2005), IEEE Computer Society, pp. 193–204.

[5] BAYARDO, JR., R. J., AND AGRAWAL, R. Data privacy through optimal k-anonymization. In *ICDE* (2005), IEEE Computer Society, pp. 217–228.

[6] BLUM, A., DWORK, C., MCSHERRY, F., AND NISSIM, K. Practical privacy: the sulq framework. In *PODS '05: Proceedings of the twenty-fourth ACM SIGMOD-SIGACT-SIGART symposium on Principles of database systems* (New York, NY, USA, 2005), ACM Press, pp. 128–138.

[7] BYUN, J.-W., KAMRA, A., BERTINO, E., AND LI, N. Efficient -anonymization using clustering techniques. In *DASFAA* (2007), K. Ramamohanarao, P. R. Krishna, M. K. Mohania, and E. Nantajeewarawat, Eds., vol. 4443 of *Lecture Notes in Computer Science*, Springer, pp. 188–200.

[8] CHAWLA, S., DWORK, C., MCSHERRY, F., SMITH, A., AND WEE, H. Toward privacy in public databases. In *TCC* (2005), J. Kilian, Ed., vol. 3378 of *Lecture Notes in Computer Science*, Springer, pp. 363–385.

[9] CHAWLA, S., DWORK, C., MCSHERRY, F., AND TALWAR, K. On privacy-preserving histograms. In *UAI* (2005), AUAI Press.

[10] DE JONGE, W. Compromising statistical databases responding to queries about means. *ACM Trans. Database Syst. 8*, 1 (1983), 60–80.

[11] DECEW, J. Privacy. In *The Stanford Encyclopedia of Philosophy*, E. N. Zalta, Ed. Fall 2006.

[12] DENNING, D. E., DENNING, P. J., AND SCHWARTZ, M. D. The tracker: A threat to statistical database security. *ACM Trans. Database Syst. 4*, 1 (1979), 76–96.

[13] DENNING, D. E., AND SCHLÖRER, J. A fast procedure for finding a tracker in a statistical database. *ACM Trans. Database Syst. 5*, 1 (1980), 88–102.

[14] DINUR, I., AND NISSIM, K. Revealing information while preserving privacy. In *PODS '03: Proceedings of the twenty-second ACM SIGMOD-SIGACT-SIGART symposium on Principles of database systems* (New York, NY, USA, 2003), ACM Press, pp. 202–210.

[15] DUNCAN, G. T., AND MUKHERJEE, S. Optimal disclosure limitation strategy in statistical databases: Deterring tracker attacks through additive noise. *Journal of the American Statistical Association 95*, 451 (2000), 720.

[16] DWORK, C. Differential privacy. In *Proc. 33rd Intnl. Conf. Automata, Languages and Programming (ICALP)* (2006), pp. 1–12. Invited paper.

[17] DWORK, C., MCSHERRY, F., NISSIM, K., AND SMITH, A. Calibrating noise to sensitivity in private data analysis. In *TCC* (2006), S. Halevi and T. Rabin, Eds., vol. 3876 of *Lecture Notes in Computer Science*, Springer, pp. 265–284.

[18] DWORK, C., AND NISSIM, K. Privacy-preserving datamining on vertically partitioned databases. In *CRYPTO* (2004), M. K. Franklin, Ed., vol. 3152 of *Lecture Notes in Computer Science*, Springer, pp. 528–544.

[19] EVFIMEVSKI, A., GEHRKE, J., AND SRIKANT, R. Limiting privacy breaches in privacy preserving data mining. In *Proceedings of the ACM SIGMOD/PODS Conference* (San Diego, CA, June 2003), pp. 211–222.

[20] EVFIMIEVSKI, A., SRIKANT, R., AGRAWAL, R., AND GEHRKE, J. Privacy preserving mining of association rules. In *KDD '02: Proceedings of the eighth ACM SIGKDD international conference on Knowledge discovery and data mining* (New York, NY, USA, 2002), ACM Press, pp. 217–228.

[21] FELLEGI, I. P. On the question of statistical confidentiality. *J. Am. Stat. Assoc 67*, 337 (1972), 7–18.

[22] FRIEDMAN, A. D., AND HOFFMAN, L. J. Towards a fail-safe approach to secure databases. In *Proc. IEEE Symp. Security and Privacy* (1980).

[23] GAVISON, R. Privacy and the limits of the law. *The Yale Law Journal 89*, 3 (January 1980), 421–471.

[24] GIVENS, C. R., AND SHORTT, R. M. A class of Wasserstein metrics for probability distributions. *Michigan Math J. 31* (1984), 231–240.

[25] HOFFMAN, L. J., AND MILLER, W. F. Getting a personal dossier from a statistical data bank. *Datamation 16*, 5 (1970), 74–75.

[26] IYENGAR, V. S. Transforming data to satisfy privacy constraints. In *KDD '02: Proceedings of the eighth ACM SIGKDD international conference on Knowledge discovery and data mining* (New York, NY, USA, 2002), ACM Press, pp. 279–288.

[27] KARGUPTA, H., DATTA, S., WANG, Q., AND SIVAKUMAR, K. On the privacy preserving properties of random data perturbation techniques. In *Proceedings of the IEEE International Conference on Data Mining* (Melbourne, FL, November 2003), p. 99.

[28] KIFER, D., AND GEHRKE, J. Injecting utility into anonymized datasets. In *SIGMOD '06: Proceedings of the 2006 ACM SIGMOD international conference on Management of data* (New York, NY, USA, 2006), ACM Press, pp. 217–228.

[29] KOCH, C., GEHRKE, J., GAROFALAKIS, M. N., SRIVASTAVA, D., ABERER, K., DESHPANDE, A., FLORESCU, D., CHAN, C. Y., GANTI, V., KANNE, C.-C., KLAS, W., AND NEUHOLD, E. J., Eds. *Proceedings of the 33rd International Conference on Very Large Data Bases, University of Vienna, Austria, September 23-27, 2007* (2007), ACM.

[30] LEFEVRE, K., DEWITT, D. J., AND RAMAKRISHNAN, R. Mondrian multidimensional k-anonymity. In *ICDE '06: Proceedings of the 22nd International Conference on Data Engineering (ICDE'06)* (Washington, DC, USA, 2006), IEEE Computer Society, p. 25.

[31] LI, N., LI, T., AND VENKATASUBRAMANIAN, S. t-closeness: Privacy beyond k-anonymity and ℓ-diversity. In *IEEE International Conference on Data Engineering (this proceedings)* (2007).

[32] MACHANAVAJJHALA, A., GEHRKE, J., KIFER, D., AND VENKITASUBRAMANIAM, M. l-diversity: Privacy beyond k-anonymity. In *Proceedings of the 22nd International Conference on Data Engineering (ICDE'06)* (2006), p. 24.

[33] MARTIN, D. J., KIFER, D., MACHANAVAJJHALA, A., GEHRKE, J., AND HALPERN, J. Y. Worst-case background knowledge for privacy-preserving data publishing. In *ICDE* [1], pp. 126–135.

[34] NAKASHIMA, E. AOL Search Queries Open Window Onto Users' Worlds. *The Washington Post* (August 17 2006).

[35] NERGIZ, M. E., AND CLIFTON, C. Thoughts on k-anonymization. In *ICDE Workshops* (2006), R. S. Barga and X. Zhou, Eds., IEEE Computer Society, p. 96.

[36] NISSIM, K., RASKHODNIKOVA, S., AND SMITH, A. Smooth sensitivity and sampling in private data analysis. In *STOC '07: Proceedings of the thirty-ninth annual ACM symposium on Theory of computing* (New York, NY, USA, 2007), ACM Press, pp. 75–84.

[37] RASTOGI, V., HONG, S., AND SUCIU, D. The boundary between privacy and utility in data publishing. In Koch et al. [29], pp. 531–542.

[38] RIZVI, S. J., AND HARITSA, J. R. Maintaining data privacy in association rule mining. In *VLDB '2002: Proceedings of the 28th international conference on Very Large Data Bases* (2002), VLDB Endowment, pp. 682–693.

[39] RUBNER, Y., TOMASI, C., AND GUIBAS, L. J. The earth mover's distance as a metric for image retrieval. *Int. J. Comput. Vision 40*, 2 (2000), 99–121.

[40] SCHLÖRER, J. Identification and retrieval of personal records from a statistical data bank. *Methods Info. Med. 14*, 1 (1975), 7–13.

[41] SCHWARTZ, M. D., DENNING, D. E., AND DENNING, P. J. Linear queries in statistical databases. *ACM Trans. Database Syst. 4*, 2 (1979), 156–167.

[42] SWEENEY, L. Achieving k-anonymity privacy protection using generalization and suppression. *Int. J. Uncertain. Fuzziness Knowl.-Based Syst. 10*, 5 (2002), 571–588.

[43] TRUTA, T. M., AND VINAY, B. Privacy protection: p-sensitive k-anonymity property. In *ICDEW '06: Proceedings of the 22nd International Conference on Data Engineering Workshops (ICDEW'06)* (Washington, DC, USA, 2006), IEEE Computer Society, p. 94.

[44] U. S. DEPARTMENT OF HEALTH AND HUMAN SERVICES. Office for Civil Rights - HIPAA. `http://www.hhs.gov/ocr/hipaa/`.

[45] WANG, K., FUNG, B. C. M., AND YU, P. S. Handicapping attacker's confidence: an alternative to k-anonymization. *Knowl. Inf. Syst. 11*, 3 (2007), 345–368.

[46] WONG, R. C.-W., FU, A. W.-C., WANG, K., AND PEI, J. Minimality attack in privacy preserving data publishing. In Koch et al. [29], pp. 543–554.

[47] WONG, R. C.-W., LI, J., FU, A. W.-C., AND WANG, K. (α, k)-anonymity: an enhanced k-anonymity model for privacy preserving data publishing. In *KDD '06: Proceedings of the 12th ACM SIGKDD international conference on Knowledge discovery and data mining* (New York, NY, USA, 2006), ACM Press, pp. 754–759.

[48] XIAO, X., AND TAO, Y. Personalized privacy preservation. In *SIGMOD '06: Proceedings of the 2006 ACM SIGMOD international conference on Management of data* (New York, NY, USA, 2006), ACM Press, pp. 229–240.

[49] YAO, A. C. Protocols for secure computations. In *Proc. IEEE Foundations of Computer Science* (1982), pp. 160–164.

[50] ZHANG, Q., KOUDAS, N., SRIVASTAVA, D., AND YU, T. Aggregate query answering on anonymized tables. In *ICDE* [1], pp. 116–125.

Chapter 5

k-Anonymous Data Mining: A Survey

V. Ciriani, S. De Capitani di Vimercati, S. Foresti, and P. Samarati
DTI - Università degli Studi di Milano
26013 Crema - Italy
{ciriani, decapita, foresti, samarati} @dti.unimi.it

Abstract Data mining technology has attracted significant interest as a means of identifying patterns and trends from large collections of data. It is however evident that the collection and analysis of data that include personal information may violate the privacy of the individuals to whom information refers. Privacy protection in data mining is then becoming a crucial issue that has captured the attention of many researchers.

In this chapter, we first describe the concept of k-anonymity and illustrate different approaches for its enforcement. We then discuss how the privacy requirements characterized by k-anonymity can be violated in data mining and introduce possible approaches to ensure the satisfaction of k-anonymity in data mining.

Keywords: k-anonymity, data mining, privacy.

5.1 Introduction

The amount of data being collected every day by private and public organizations is quickly increasing. In such a scenario, *data mining techniques* are becoming more and more important for assisting decision making processes and, more generally, to extract hidden knowledge from massive data collections in the form of patterns, models, and trends that hold in the data collections. While not explicitly containing the original actual data, data mining results could potentially be exploited to infer information - contained in the original data - and not intended for release, then potentially breaching the privacy of the parties to whom the data refer. Effective application of data mining can take place only if proper guarantees are given that the privacy of the underlying data is not compromised. The concept of privacy preserving data mining has been proposed in response to these privacy concerns [6]. Privacy preserving data mining aims at providing a trade-off between sharing information for data mining analysis, on the one side, and protecting information to preserve the privacy of the involved parties on the other side. Several privacy preserving data mining approaches have been proposed, which usually protect data by modifying them to mask or erase the original sensitive data that should not be revealed [4, 6, 13]. These approaches typically are based on the concepts of: *loss of privacy*, measuring the capacity of estimating the original data from the modified data, and *loss of information*, measuring the loss of accuracy in the data. In general, the more the privacy of the respondents to which the data refer, the less accurate the result obtained by the miner and vice versa. The main goal of these approaches is therefore to provide a trade-off between privacy and accuracy. Other approaches to privacy preserving data mining exploit cryptographic techniques for preventing information leakage [20, 30]. The main problem of cryptography-based techniques is, however, that they are usually computationally expensive.

Privacy preserving data mining techniques clearly depend on the definition of privacy, which captures what information is sensitive in the original data and should therefore be protected from either direct or indirect (via inference) disclosure. In this chapter, we consider a specific aspect of privacy that has been receiving considerable attention recently, and that is captured by the notion of *k-anonymity* [11, 26, 27]. k-anonymity is a property that models the protection of released data against possible re-identification of the respondents to which the data refer. Intuitively, k-anonymity states that each release of data must be such that every combination of values of released attributes that are also externally available and therefore exploitable for linking can be indistinctly matched to at least k respondents. k-anonymous data mining has been recently introduced as an approach to ensuring privacy-preservation when releasing data mining results. Very few, preliminary, attempts have been presented looking at different aspects in guaranteeing k-anonymity in data mining. We discuss possible threats to k-anonymity posed by data mining and sketch possible approaches to their counteracting, also briefly illustrating some preliminary results existing in the current literature. After recalling the concept of k-anonymity (Section 5.2) and some proposals for its enforcement (Section 5.3), we discuss possible threats to k-anonymity to which data mining results are exposed (Section 5.4). We then illustrate (Section 5.5) possible approaches combining k-anonymity and data mining, distinguishing them depending on whether k-anonymity is enforced directly on the private data (before mining) or on the mined data themselves (either as a post-mining sanitization process or by the mining process itself). For each of the two approaches (Section 5.6 and 5.7, respectively) we discuss possible ways to capture k-anonymity violations to the aim, on the one side, of defining when mined results respect k-anonymity of the original data and, on the other side, of identifying possible protection techniques for enforcing such a definition of privacy.

5.2 k-Anonymity

k-anonymity [11, 26, 27] is a property that captures the protection of released data against possible re-identification of the respondents to whom the released data refer. Consider a private table PT, where data have been de-identified by removing explicit identifiers (e.g., `SSN` and `Name`). However, values of other released attributes, such as `ZIP`, `Date_of_birth`, `Marital_status`, and `Sex` can also appear in some external tables jointly with the individual respondents' identities. If some combinations of values for these attributes are such that their occurrence is unique or rare, then parties observing the data can determine the identity of the respondent to which the data refer or reduce the uncertainty over a limited set of respondents. *k-anonymity demands that every tuple in the private table being released be indistinguishably related to no fewer than k respondents.* Since it seems impossible, or highly impractical and limiting, to make assumptions on which data are known to a potential attacker and can be used to (re-)identify respondents, k-anonymity takes a safe approach requiring that, in the released table itself, the respondents be indistinguishable (within a given set of individuals) with respect to the set of attributes, called *quasi-identifier*, that can be exploited for linking. In other words, k-anonymity requires that if a combination of values of quasi-identifying attributes appears in the table, then it appears with at least k occurrences.

To illustrate, consider a private table reporting, among other attributes, the marital status, the sex, the working hours of individuals, and whether they suffer from hypertension. Assume attributes `Marital_status`, `Sex`, and `Hours` are the attributes jointly constituting the quasi-identifier. Figure 5.1 is a simplified representation of the projection of the private table over the quasi-identifier. The representation has been simplified by collapsing tuples with the same quasi-identifying values into a single tuple. The numbers at the right hand side of the table report, for each tuple, the number of actual occurrences, also specifying how many of these occurrences have values `Y` and `N`, respectively, for attribute `Hypertension`. For simplicity, in the following we use such a simplified table as our table PT.

The private table PT in Figure 5.1 guarantees k-anonymity only for $k \leq 2$. In fact, the table has only two occurrences of divorced (fe)males working 35 hours. If such a situation is satisfied in a particular correlated external table as well, the uncertainty of the identity of such respondents can be reduced to two specific individuals. In other words, a data recipient can infer that any

Marital_status	*Sex*	*Hours*	*#tuples (Hyp. values)*
divorced	M	35	2 (0Y, 2N)
divorced	M	40	17 (16Y, 1N)
divorced	F	35	2 (0Y, 2N)
married	M	35	10 (8Y, 2N)
married	F	50	9 (2Y, 7N)
single	M	40	26 (6Y, 20N)

Figure 5.1. Simplified representation of a private table

information appearing in the table for such divorced (fe)males working 35 hours, actually pertains to one of two specific individuals.

It is worth pointing out a simple but important observation (to which we will come back later in the chapter): if a tuple has k occurrences, then any of its sub-tuples must have at least k-occurrences. In other words, the existence of k occurrences of any sub-tuple is a necessary (not sufficient) condition for having k occurrences of a super-tuple. For instance, with reference to our example, k-anonymity over quasi-identifier {`Marital_status`, `Sex`, `Hours`} requires that each value of the individual attributes, as well as of any sub-tuple corresponding to a combination of them, appears with at least k occurrences. This observation will be exploited later in the chapter to assess the non satisfaction of a k-anonymity constraint for a table based on the fact that a sub-tuple of the quasi-identifier appears with less than k occurrences. Again with reference to our example, the observation that there are only two tuples referring to divorced females allows us to assert that the table will certainly not satisfy k-anonymity for $k > 2$ (since the two occurrences will remain at most two when adding attribute `Hours`).

Two main techniques have been proposed for enforcing k-anonymity on a private table: *generalization* and *suppression*, both enjoying the property of preserving the truthfulness of the data.

Generalization consists in replacing attribute values with a generalized version of them. Generalization is based on a domain generalization hierarchy and a corresponding value generalization hierarchy on the values in the domains. Typically, the domain generalization hierarchy is a total order and the corresponding value generalization hierarchy a tree, where the parent/child relationship represents the direct generalization/specialization relationship. Figure 5.2 illustrates an example of possible domain and value generalization hierarchies for the quasi-identifying attributes of our example.

Generalization can be applied at the level of single cell (substituting the cell value with a generalized version of it) or at the level of attribute (generalizing all the cells in the corresponding column). It is easy to see how generalization can enforce k-anonymity: values that were different in the private table can be generalized to a same value, whose number of occurrences would be

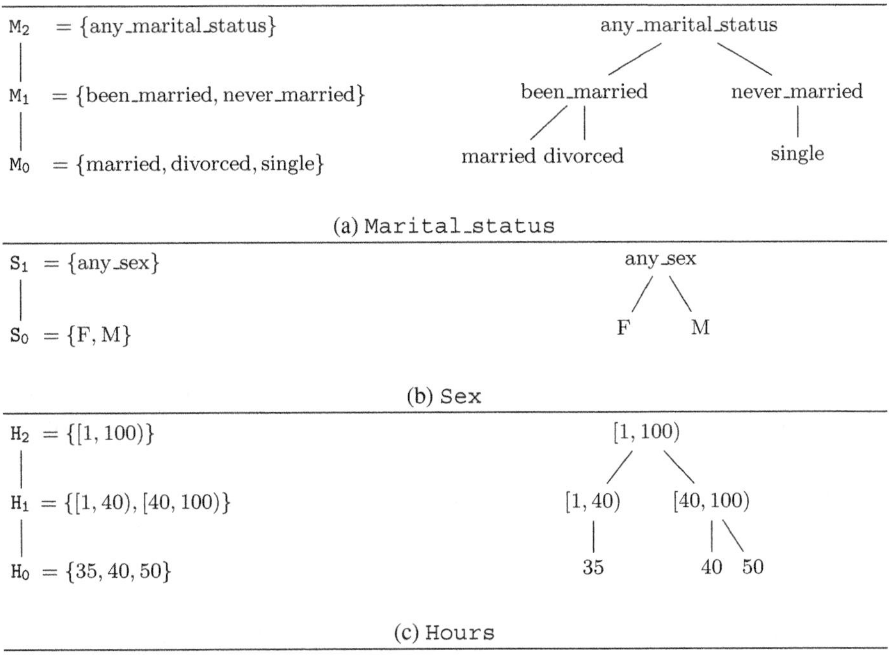

Figure 5.2. An example of domain and value generalization hierarchies

the sum of the number of occurrences of the values that have been generalized to it. The same reasoning extends to tuples. Figure 5.11(d) reports the result of a generalization over attribute `Sex` on the table in Figure 5.1, which resulted, in particular, in divorced people working 35 hours to be collapsed to the same tuple {`divorced`, `any_sex`, `35`}, with 4 occurrences. The table in Figure 5.11(d) satisfies k-anonymity for any $k \leq 4$ (since there are no less than 4 respondents for each combination of values of quasi-identifying attributes). Note that 4-anonymity could be guaranteed also by only generalizing (to `any_sex`) the sex value of divorced people (males and females) working 35 hours while leaving the other tuples unaltered, since for all the other tuples not satisfying this condition there are already at least 4 occurrences in the private table. This cell generalization approach has the advantage of avoiding generalizing all values in a column when generalizing only a subset of them suffices to guarantee k-anonymity. It has, however, the disadvantage of not preserving the homogeneity of the values appearing in the same column.

Suppression consists in protecting sensitive information by removing it. Suppression, which can be applied at the level of single cell, entire tuple, or entire column, allows reducing the amount of generalization to be enforced to achieve k-anonymity. Intuitively, if a limited number of outliers would force a large amount of generalization to satisfy a k-anonymity constraint, then

	Suppression			
Generalization	*Tuple*	*Attribute*	*Cell*	*None*
Attribute	AG_TS	AG_AS ≡ AG_	AG_CS	AG_ ≡ AG_AS
Cell	CG_TS not applicable	CG_AS not applicable	CG_CS ≡ CG_	CG_ ≡ CG_CS
None	_TS	_AS	_CS	_ not interesting

Figure 5.3. Classification of k-anonymity techniques [11]

such outliers can be removed from the table thus allowing satisfaction of k-anonymity with less generalization (and therefore reducing the loss of information).

Figure 5.3 summarizes the different combinations of generalization and suppression at different granularity levels (including combinations where one of the two techniques is not adopted), which correspond to different approaches and solutions to the k-anonymity problem [11]. It is interesting to note that the application of generalization and suppression at the same granularity level is equivalent to the application of generalization only (AG_≡AG_AS and CG_≡CG_CS), since suppression can be modeled as a generalization to the top element in the value generalization hierarchy. Combinations CG_TS (cell generalization, tuple suppression) and CG_AS (cell generalization, attribute suppression) are not applicable since the application of generalization at the cell level implies the application of suppression at that level too.

5.3 Algorithms for Enforcing k-Anonymity

The application of generalization and suppression to a private table PT produces less precise (more general) and less complete (some values are suppressed) tables that provide protection of the respondents' identities. It is important to maintain under control, and minimize, the information loss (in terms of loss of precision and completeness) caused by generalization and suppression. Different definitions of minimality have been proposed in the literature and the problem of finding minimal k-anonymous tables, with attribute generalization and tuple suppression, has been proved to be computationally hard [2, 3, 22].

Within a given definition of minimality, more generalized tables, all ensuring minimal information loss, may exist. While existing approaches typically aim at returning any of such solutions, different criteria could be devised according to which a solution should be preferred over the others. This aspect is particularly important in data mining, where there is the need to maximize the usefulness of the data with respect to the goal of the data mining process (see Section 5.6). We now describe some algorithms proposed in literature for producing k-anonymous tables.

Samarati's Algorithms. The first algorithm for AG_TS (i.e., generalization over quasi-identifier attributes and tuple suppression) was proposed in conjunction with the definition of k-anonymity [26]. Since the algorithm operates on a set of attributes, the definition of domain generalization hierarchy is extended to refer to tuples of domains. The domain generalization hierarchy of a domain tuple is a lattice, where each vertex represents a generalized table that is obtained by generalizing the involved attributes according to the corresponding domain tuple and by suppressing a certain number of tuples to fulfill the k-anonymity constraint. Figure 5.4 illustrates an example of domain generalization hierarchy obtained by considering `Marital_status` and `Sex` as quasi-identifying attributes, that is, by considering the domain tuple $\langle M_0, S_0 \rangle$. Each path in the hierarchy corresponds to a generalization strategy according to which the original private table PT can be generalized. The main goal of the algorithm is to find a k-minimal generalization that suppresses less tuples. Therefore, given a threshold MaxSup specifying the maximum number of tuples that can be suppressed, the algorithm has to compute a generalization that satisfies k-anonymity within the MaxSup constraint. Since going up in the hierarchy the number of tuples that must be removed to guarantee k-anonymity decreases, the algorithm performs a binary search on the hierarchy. Let h be the height of the hierarchy. The algorithm first evaluates all the solutions at height $\lfloor h/2 \rfloor$. If there is at least a k-anonymous table that satisfies the MaxSup threshold, the algorithm checks solutions at height $\lfloor h/4 \rfloor$; otherwise it evaluates solutions at height $\lfloor 3h/4 \rfloor$, and so on, until it finds the lowest height where there is a solution that satisfies the k-anonymity constraint. As an example, consider the private table in Figure 5.1 with QI=$\{$`Marital_status`, `Sex`$\}$, the domain and value generalization hierarchies in Figure 5.2, and the generalization hierarchy in Figure 5.4. Suppose also that $k = 4$ and MaxSup$= 1$. The algorithm first evaluates solutions at height $\lfloor 3/2 \rfloor$, that is, $\langle M_0, S_1 \rangle$ and

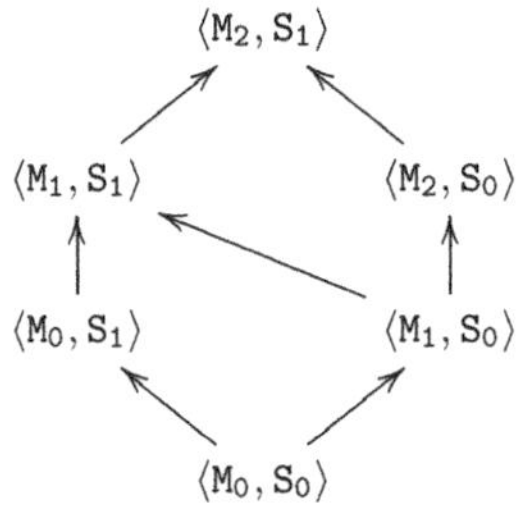

Figure 5.4. Generalization hierarchy for QI=$\{$`Marital_status`, `Sex`$\}$

$\langle \mathtt{M}_1, \mathtt{S}_0 \rangle$. Since both the solutions are 4-anonymous within the MaxSup constraint, the algorithm evaluates solutions at height $\lfloor 3/4 \rfloor$, that is, $\langle \mathtt{M}_0, \mathtt{S}_0 \rangle$. Solution $\langle \mathtt{M}_0, \mathtt{S}_0 \rangle$ corresponds to the original table that is not 4-anonymous and violates the MaxSup constraint since 4-anonymity requires to suppress the two tuples $\langle \mathtt{divorced}, \mathtt{F} \rangle$. Consequently, the two solutions $\langle \mathtt{M}_0, \mathtt{S}_1 \rangle$ and $\langle \mathtt{M}_1, \mathtt{S}_0 \rangle$ are considered as minimal.

Bayardo-Agrawal's Algorithm. Bayardo and Agrawal [10] proposed another algorithm for AG_TS, called *k-Optimize*. Given a private table PT, and an ordered set QI=$\{A_1, \ldots, A_n\}$ of quasi-identifying attributes, *k-Optimize* assumes that each attribute $A_i \in$ QI is defined over a totally ordered domain D_i. An attribute generalization of A on D consists in partitioning D into a set of ordered intervals $\{I_1, \ldots, I_m\}$ such that $\bigcup_{i=1}^{m} I_i = D$ and $\forall v_i \in I_i, \forall v_j \in I_j$ if $i < j$, then $v_i < v_j$. The approach associates an integer, called *index*, with each interval in any domain of the quasi-identifying attributes. The index assignment reflects the total order relationship over intervals in the domains and among quasi-identifier attributes. As an example, consider the private table in Figure 5.1 where the quasi-identifying attributes are Marital_status and Sex. Suppose that the order between the quasi-identifying attributes is Marital_status followed by Sex, and the order among values inside each attribute domain is married, divorced, single for Marital_status, and F, M for Sex. Figure 5.5 represents the index assignment obtained when no generalization is applied, that is, when each attribute value represents an interval.

A generalization is represented through the union of generalized sets for each attribute domain. Since the least value from each attribute domain must appear in any valid generalization for the attribute domain, it can be omitted. With respect to our example in Figure 5.5, the least values are 1 (Marital_status=married) and 4 (Sex=F). As an example, consider now the index list $\{3, 5\}$. After adding the least values, we obtain the generalizer sets $\{1,3\}$ for attribute Marital_status and $\{4, 5\}$ for attribute Sex, which in turn correspond to the following intervals of domain values: $\langle$[married or divorced], [single]$\rangle$ and $\langle$[F], [M]$\rangle$. The empty set $\{\ \}$ represents the generalization where, for each domain, all values in the domain are generalized to the most general value. In our example, $\{\ \}$ corresponds to index values $\{1\}$ for

	Marital_status		Sex	
⟨[married]	[divorced]	[single]⟩	⟨[F]	[M]⟩
1	2	3	4	5

Figure 5.5. Index assignment to attributes Marital_status and Sex

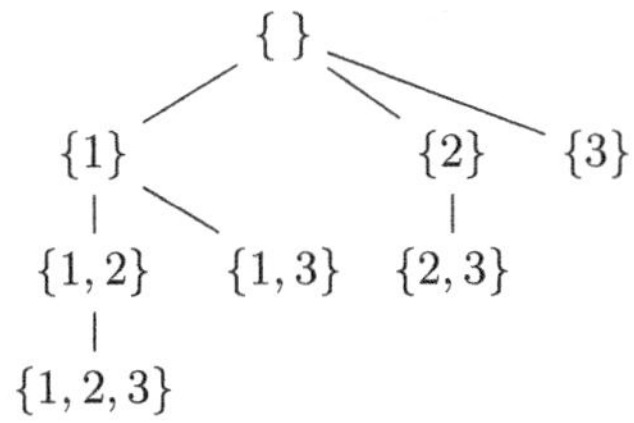

Figure 5.6. An example of set enumeration tree over set $\mathcal{I} = \{1, 2, 3\}$ of indexes

`Marital_status` and $\{4\}$ for `Sex`, which in turn correspond to $\langle$[married or divorced or single]$\rangle$ and $\langle$[F or M]$\rangle$ generalized domain values.

The k-Optimize algorithm builds a *set enumeration tree* over the set $\mathcal{I}$ of index values, which is a tree representing all possible subsets of $\mathcal{I}$, without duplications. The children of a node n correspond to the sets that can be formed by appending a single element of $\mathcal{I}$ to n, with the restriction that this single element must follow every element already in n according to the total order previously defined. Figure 5.6 illustrates an example of set enumeration tree over $\mathcal{I} = \{1, 2, 3\}$. Since each node in the tree represents how to generalize the original table PT, the visit of the set enumeration tree is equivalent to the evaluation of each possible solution to the k-anonymity problem. At each node n in the tree, the algorithm computes the cost (as determined by some cost metric) associated with the table that can be obtained by applying the generalization represented by n. This cost is then compared against the best cost found until that point. If the cost is lower than the best cost found until that point, it becomes the new best cost and node n is retained. Since a complete visit of the tree may however be impractical (the tree contains $2^{|\mathcal{I}|}$ nodes), k-Optimize proposes an heuristic pruning strategy. Intuitively, a node n can be pruned when the cost associated with its descendants cannot be optimal. To this purpose, the algorithm computes a lower bound on the cost that can be obtained by any node in the subtree rooted at n. If this lower bound is greater than the current best cost, node n is pruned. Note that *k*-Optimize can also be exploited as an heuristic algorithm, by stopping in advance the visit of the tree.

Incognito. Incognito, proposed by LeFevre, DeWitt and Ramakrishnan [18], is an algorithm for AG_TS based on the observation that k-anonymity with respect to any subset of QI is a necessary (not sufficient) condition for k-anonymity with respect to QI. Consequently, given a generalization hierarchy over QI, the generalizations that are not k-anonymous with respect to a subset QI$'$ of QI can be discarded along with all their descendants in the hierarchy.

Exploiting this observation, at each iteration i, for $i = 1, \ldots, |\,\text{QI}\,|$, Incognito builds the generalization hierarchies for all subsets of the quasi-identifying

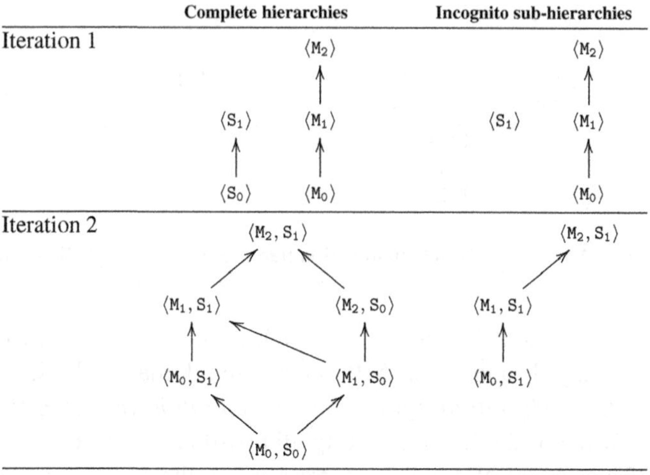

Figure 5.7. Sub-hierarchies computed by Incognito for the table in Figure 5.1

attributes of size i. It then visits each node of the hierarchies discarding the generalizations that do not satisfy k-anonymity with respect to the considered set of attributes. Note that if a node of a generalization hierarchy satisfies k-anonymity, also all its generalizations satisfy k-anonymity and therefore they are not checked in the subsequent visits of the hierarchy. The algorithm then constructs the generalization hierarchies for all subsets of the quasi-identifying attributes of size $i + 1$ by considering only the generalizations of size i that satisfy the k-anonymity constraint at iteration i. Incognito terminates when the whole set of attributes in QI has been considered.

As an example, consider the table PT in Figure 5.1 and suppose that QI = {`Marital_status`, `Sex`} and $k = 12$. The first iteration of Incognito finds that the original table is 12-anonymous with respect to $\mathtt{M}_0$, and $\mathtt{S}_1$. Note that since PT is 12-anonymous with respect to $\mathtt{M}_0$, the table is 12-anonymous also with respect to $\mathtt{M}_1$ and $\mathtt{M}_2$ and therefore they are not checked. The algorithm then builds the generalization hierarchy on the ⟨`Marital_status`, `Sex`⟩ pair by considering only the generalizations $\mathtt{M}_0$, $\mathtt{M}_1$, $\mathtt{M}_2$ and $\mathtt{S}_1$ that are 12-anonymous. The algorithm finds that the table is 12-anonymous with respect to $\langle \mathtt{M}_0, \mathtt{S}_1 \rangle$. Consequently, all generalizations of $\langle \mathtt{M}_0, \mathtt{S}_1 \rangle$ (i.e., $\langle \mathtt{M}_1, \mathtt{S}_1 \rangle$ and $\langle \mathtt{M}_2, \mathtt{S}_1 \rangle$) are 12-anonymous and the search terminates. Figure 5.7 illustrates on the left-hand side the complete domain generalization hierarchies and on the right-hand side the sub-hierarchies computed by Incognito at each iteration.

Mondrian. The Mondrian algorithm, proposed by LeFevre, DeWitt and Ramakrishnan [19], is based on the *multidimensional global recoding* technique.

A private table PT is represented as a set of points in a multidimensional space, where each attribute represents one dimension. To the aim of computing a k-anonymous table, the multidimensional space is partitioned in regions that have to contain at least k points. All points in a given region are then generalized to the same value for QI. Note that tuples in different regions can be generalized in different ways. It is proved that any multidimensional space partition contains at most $2d(k-1)+m$ points, where $d = |\mathsf{QI}|$ and m is the maximum number of tuples with the same quasi-identifier value in PT.

Since the computation of a multidimensional partitioning that minimizes information loss is a NP-hard problem, the authors propose a greedy algorithm that works as follows. Given a space region r, at each iteration the algorithm chooses a dimension d (if such a dimension exists) and splits the region at the median value x of d: all points such that $d > x$ will belong to one of the resulting regions, while all points with $d \leq x$ will belong to the other region. Note that this splitting operation is allowed only if there are more than k points within any region. The algorithm terminates when there are no more splitting operations allowed. The tuples within a given region are then generalized to a unique tuple of summary statistics for the considered region. For each quasi-identifying attribute, a summary statistic may simply be a static value (e.g., the average value) or the pair of maximum and minimum values for the attribute in the region. As an example, consider the private table PT in Figure 5.1 and suppose that QI = {`Marital_status`, `Sex`} and $k = 10$. Figure 5.8(a) illustrates the two dimensional representation of the table for the `Marital_status` and `Sex` quasi-identifying attributes, where the number associated with each point corresponds to the occurrences of the quasi-identifier value in PT. Suppose to perform a split operation on the `Marital_status` dimension. The resulting two regions illustrated in Figure 5.8(b) are 10-anonymous. The bottom region can be further partitioned along the `Sex` dimension, as represented in Figure 5.8(c). Another splitting operation along the `Marital_status` dimension can be performed on the region containing the points that correspond to the quasi-identifying values ⟨`married`, `M`⟩ and ⟨`divorced`, `M`⟩. Figure 5.8(d) illustrates the final solution.

The experimental results [19] show that the Mondrian multidimensional method obtains good solutions for the k-anonymity problem, also compared with k-Optimize and Incognito.

Approximation Algorithms. Since the majority of the exact algorithms proposed in literature have computational time exponential in the number of the attributes composing the quasi-identifier, approximation algorithms have been also proposed. Approximation algorithms for _CS and CG_ have been

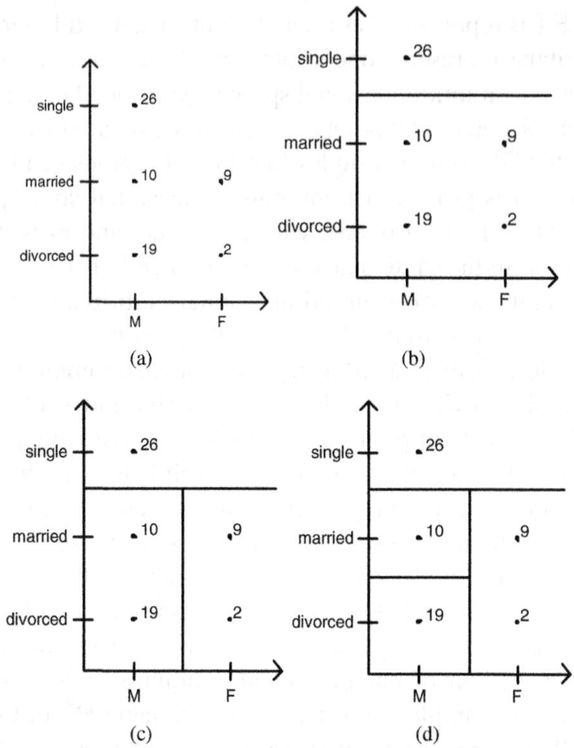

Figure 5.8. Spatial representation (a) and possible partitioning (b)-(d) of the table in Figure 5.1

presented, both for general and specific values of k (e.g., 1.5-approximation[1] for 2-anonymity, and 2-approximation for 3-anonymity [3]).

The first approximation algorithm for _CS was proposed by Meyerson and Williams [22] and guarantees a $O(k \log(k))$-approximation. The best-known approximation algorithm for _CS is described in [2] and guarantees a $O(k)$-approximate solution. The algorithm constructs a complete weighted graph from the original private table PT. Each vertex in the graph corresponds to a tuple in PT, and the edges are weighted with the number of different attribute values between the two tuples represented by extreme vertices. The algorithm then constructs, starting from the graph, a forest composed of trees containing at least k vertices, which represents the clustering for k-anonymization. Some cells in the vertices are suppressed to obtain that all the tuples in the same tree have the same quasi-identifier value. The cost of a vertex is evaluated as the number of cells suppressed, and the cost of a tree is the sum of the weights of

[1]In a minimization framework, a p-approximation algorithm guarantees that the cost C of its solution is such that $C/C^* \leq p$, where C^* is the cost of an optimal solution [17].

its vertices. The cost of the final solution is equal to the sum of the costs of its trees. In constructing the forest, the algorithm limits the maximum number of vertices in a tree to be $3k-3$. Partitions with more than $3k-3$ elements are decomposed, without increasing the total solution cost. The construction of trees with no more than $3k-3$ vertices guarantees a $O(k)$-approximate solution.

An approximation algorithm for CG_ is described in [3] as a direct extension of the approximation algorithm for _CS presented in [2]. For taking into account the generalization hierarchies, each edge has a weight that is computed as follows. Given two tuples i and j and an attribute a, the generalization cost $h_{i,j}(a)$ associated with a is the lowest level of the value generalization hierarchy of a such that tuples i and j have the same generalized value for a. The weight $w(e)$ of the edge $e = (i, j)$ is therefore $w(e) = \Sigma_a h_{i,j}(a)/l_a$, where l_a is the number of levels in the value generalization hierarchy of a. The solution of this algorithm is guaranteed to be a $O(k)$-approximation.

Besides algorithms that compute k-anonymized tables for any value of k, ad-hoc algorithms for specific values of k have also been proposed. For instance, to find better results for Boolean attributes, in the case where $k = 2$ or $k = 3$, an ad-hoc approach has been provided in [3]. The algorithm for $k = 2$ exploits the minimum-weight $[1, 2]$-factor built on the graph constructed for the 2-anonymity. The $[1, 2]$-factor for graph G is a spanning subgraph of G built using only vertices with no more than 2 outgoing edges. Such a subgraph is a vertex-disjoint collection of edges and pairs of adjacent vertices and can be computed in polynomial time. Each component in the subgraph is treated as a cluster, and a 2-anonymized table is obtained by suppressing each cell, for which the vectors in the cluster differ in value. This procedure is a 1.5-approximation algorithm. The approximation algorithm for $k = 3$ is similar and guarantees a 2-approximation solution.

5.4 k-Anonymity Threats from Data Mining

Data mining techniques allow the extraction of information from large collections of data. Data mined information, even if not explicitly including the original data, is built on them and can therefore allow inferences on original data to be withdrawn, possibly putting privacy constraints imposed on the original data at risk. This observation holds also for k-anonymity. The desire to ensure k-anonymity of the data in the collection may therefore require to impose restrictions on the possible output of the data mining process. In this section, we discuss possible threats to k-anonymity that can arise from performing mining on a collection of data maintained in a private table PT subject to k-anonymity constraints.

We discuss the problems for the two main classes of data mining techniques, namely *association rule* mining and *classification* mining.

5.4.1 Association Rules

The classical association rule mining operates on a set of transactions, each composed of a set of items, and produce association rules of the form $X \rightarrow Y$, where X and Y are sets of items. Intuitively, rule $X \rightarrow Y$ expresses the fact that transactions that contain items X tend to also contain items Y. Each rule has a *support* and a *confidence*, in the form of percentage. The support expresses the percentage of transactions that contain both X and Y, while the confidence expresses the percentage of transactions, among those containing X, that also contain Y. Since the goal is to find common patterns, typically only those rules that have support and confidence greater than some predefined thresholds are considered of interest [5, 28, 31].

Translating association rule mining over a private table PT on which k-anonymity should be enforced, we consider the values appearing in the table as items, and the tuples reporting respondents' information as transactions. For simplicity, we assume here that the domains of the attributes are disjoint. Also, we assume support and confidence to be expressed in absolute values (in contrast to percentage). The reason for this assumption, which is consistent with the approaches in the literature, is that k-anonymity itself is expressed in terms of absolute numbers. Note, however, that this does not imply that the release itself will be made in terms of absolute values.

Association rule mining over a private table PT allows then the extraction of rules expressing combination of values common to different respondents. For instance, with reference to the private table in Figure 5.1, rule {divorced} → {M} with support 19, and confidence $\frac{19}{21}$ states that 19 tuples in the table refer to divorced males, and among the 21 tuples referring to divorced people 19 of them are male. If the quasi-identifier of table PT contains both attributes Marital_status and Sex, it is easy to see that such a rule violates any k-anonymity for $k > 19$, since it reflects the existence of 19 respondents who are divorced male (being Marital_status and Sex included in the quasi-identifier, this implies that no more than 19 indistinguishable tuples can exist for divorced male respondents). Less trivially, the rule above violates also k-anonymity for any $k > 2$, since it reflects the existence of 2 respondents who are divorced and not male; again, being Marital_status and Sex included in the quasi-identifier, this implies that no more than 2 indistinguishable tuples can exist for non male divorced respondents.

5.4.2 Classification Mining

In classification mining, a set of database tuples, acting as a training sample, are analyzed to produce a model of the data that can be used as a predictive classification method for classifying new data into classes. Goal of the classification process is to build a model that can be used to further classify tuples being inserted and that represents a descriptive understanding of the table content [25].

One of the most popular classification mining techniques is represented by *decision trees*, defined as follows. Each internal node of a decision tree is associated with an attribute on which the classification is defined (excluding the classifying attributes, which in our example is `Hypertension`). Each outgoing edge is associated with a split condition representing how the data in the training sample are partitioned at that tree node. The form of a split condition depends on the type of the attribute. For instance, for a numerical attribute A, the split condition may be of the form $A \leq v$, where v is a possible value for A. Each node contains information about the number of samples at that node and how they are distributed among the different class values.

As an example, the private table PT in Figure 5.1 can be used as a learning set to build a decision tree for predicting if people are likely to suffer from hypertension problems, based on their marital status, if they are male, and on their working hours, if they are female. A possible decision tree for such a case performing the classification based on some values appearing in quasi-identifier attributes is illustrates in Figure 5.9. The quasi-identifier attributes correspond to internal (splitting) nodes in the tree, edges are labeled with (a subset of) attribute values instead of reporting the complete split condition, and nodes simply contain the number of respondents classified by the node values, distinguishing between people suffering (`Y`) and not suffering (`N`) of hypertension.

While the decision tree does not directly release the data of the private table, it indeed allows inferences on them. For instance, Figure 5.9 reports the existence of 2 females working 35 hours (node reachable from path $\langle$`F,35`$\rangle$). Again, since `Sex` and `Hours` belong to the quasi-identifier, this information

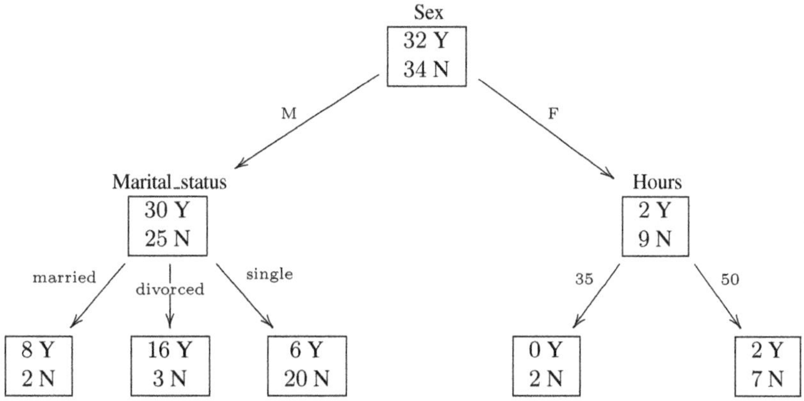

Figure 5.9. An example of decision tree

reflects the existence of no more than two respondents for such occurrences of values, thus violating k-anonymity for any $k > 2$. Like for association rules, threats can also be possible by combining classifications given by different nodes along the same path. For instance, considering the decision tree in Figure 5.9, the combined release of the nodes reachable from paths $\langle \texttt{F} \rangle$ (with 11 occurrences) and $\langle \texttt{F}, 50 \rangle$ (with 9 occurrences) allows to infer that there are 2 female respondents in PT who do not work 50 hours per week.

5.5 k-Anonymity in Data Mining

Section 5.4 has illustrated how data mining results can compromise the k-anonymity of a private table, even if the table itself is not released. Since proper privacy guarantees are a must for enabling information sharing, it is then important to devise solutions ensuring that data mining does not open the door to possible privacy violations. With particular reference to k-anonymity, we must ensure that k-anonymity for the original table PT be not violated.

There are two possible approaches to guarantee k-anonymity in data mining.

- *Anonymize-and-Mine*: anonymize the private table PT and perform mining on its k-anonymous version.
- *Mine-and-Anonymize*: perform mining on the private table PT and anonymize the result. This approach can be performed by executing the two steps independently or in combination.

Figure 5.10 provides a graphical illustration of these approaches, reporting, for the Mine-and-Anonymize approach, the two different cases: one step or two

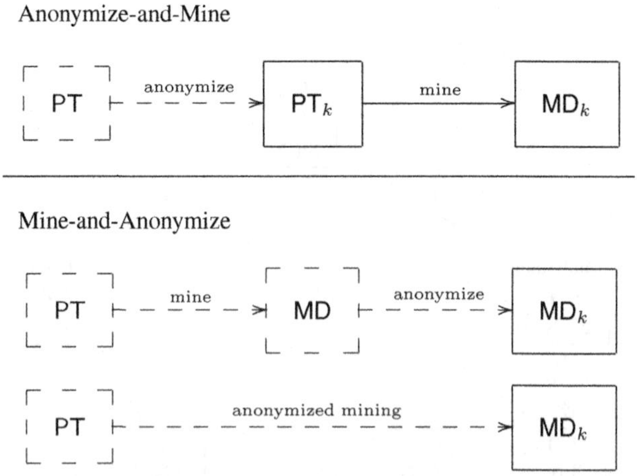

Figure 5.10. Different approaches for combining k-anonymity and data mining

steps. In the figure, boxes represent data, while arcs represent processes producing data from data. The different data boxes are: PT, the private table; PT_k, an anonymized version of PT; MD, a result of a data mining process (without any consideration of k-anonymity constraints); and MD_k, a result of a data mining process that respects the k-anonymity constraint for the private table PT. Dashed lines for boxes and arcs denote data and processes, respectively, reserved to the data holder, while continuous lines denote data and processes that can be viewed and executed by other parties (as their visibility and execution does not violate the k-anonymity for PT).

Let us then discuss the two approaches more in details and their trade-offs between applicability and efficiency of the process on the one side, and utility of data on the other side.

Anonymize-and-Mine (AM) This approach consists in applying a k-anonymity algorithm on the original private table PT and releasing then a table PT_k that is a k-anonymized version of PT. Data mining is performed, by the data holder or even external parties, on PT_k. The advantage of such an approach is that it allows the decoupling of data protection from mining, giving a double benefit. First, it guarantees that data mining is safe: since data mining is executed on PT_k (and not on PT), by definition the data mining results cannot violate k-anonymity for PT. Second, it allows data mining to be executed by others than the data holder, enabling different data mining processes and different uses of the data. This is convenient, for example, when the data holder may not know a priori how the recipient may analyze and classify the data. Moreover, the recipient may have application-specific data mining algorithms and she may want to directly define parameters (e.g., accuracy and interpretability) and decide the mining method only after examining the data. On the other hand, the possible disadvantages of performing mining on anonymized data is that mining operates on less specialized and complete data, therefore usefulness and significance of the mining results can be compromised. Since classical k-anonymity approaches aim at satisfying k-anonymity minimizing information loss (i.e., minimizing the amount of generalization and suppression adopted), a k-anonymity algorithm may produce a result that is not suited for mining purposes. As a result, classical k-anonymity algorithms may hide information that is highly useful for data mining purposes. Particular care must then be taken in the k-anonymization process to ensure maximal utility of the k-anonymous table PT_k with respect to the goals of the data mining process that has to be executed. In particular, the aim of k-anonymity algorithms operating on data intended for data mining should not be the mere minimization of information loss, but the optimization of a measure suitable for data mining purposes. A further limitation of the Anonymize-and-Mine approach is that it is not applicable when the input data can be accessed only once (e.g., when the data source is a stream). Also, it may be overall less efficient, since the anonymization process may be quite expensive with respect to the mining one, especially in case of sparse and large databases [1]. Therefore, performing k-anonymity before data mining is likely to be more expensive than doing the contrary.

Mine-and-Anonymize (MA) This approach consists in mining original non-k-anonymous data, performing data mining on the original table PT, and then applying an anonymization process on the data mining result. Data mining can then be performed by the data holder only, and only the sanitized data mining results (MD_k) are released to other parties. The definition of k-anonymity must then be adapted to the output of the data mining phase. Intuitively, no inference should be possible on the mined data allowing violating k-anonymity for the original table PT. This does not mean that the table PT must be k-anonymous, but that if it was not, it should not be known and the effect of its non being k-anonymous be not visible in the mined results. In the Mine-and-Anonymize approach, k-anonymity constraints can be taken into consideration after data mining is complete (*two-step* Mine-and-Anonymize) or within the mining process itself (*one-step* Mine-and-Anonymize). In *two-step* Mine-and-Anonymize the result needs to be sanitized removing from MD all data that would compromise k-anonymity for PT. In *one-step* Mine-and-Anonymize the data mining algorithm needs to be modified so to ensure that only results that would not compromise k-anonymity for PT are computed (MD_k). The two possible implementations (one step vs two steps) provide different trade-offs between applicability and efficiency: *two-step* Mine-and-Anonymize does not require any modification to the mining process and therefore can use any data mining tool available (provided that results are then anonymized); *one-step* Mine-and-Anonymize requires instead to redesign data mining algorithms and tools to directly enforce k-anonymity, combining the two steps can however result in a more efficient process giving then performance advantages. Summarizing, the main drawback of Mine-and-Anonymize is that it requires mining to be executed only by the data holder (or parties authorized to access the private table PT). This may therefore impact applicability. The main advantages are efficiency of the mining process and quality of the results: performing mining before, or together with, anonymization can in fact result more efficient and allow to keep data distortion under control to the goal of maximizing the usefulness of the data.

5.6 Anonymize-and-Mine

The main objective of classical k-anonymity techniques is the minimization of information loss. Since a private table may have more than one minimal k-anonymous generalization, different preference criteria can be applied in choosing a minimal generalization, such as minimum absolute distance, minimum relative distance, maximum distribution, or minimum suppression [26]. In fact, the strategies behind heuristics for k-anonymization can be typically based on preference criteria or even user policies (e.g., the discourage of the generalization of some given attributes).

In the context of data mining, the main goal is retaining useful information for data mining, while determining a k-anonymization that protects the respondents against linking attacks. However, it is necessary to define k-anonymity algorithms that guarantee data usefulness for subsequent mining operations. A possible solution to this problem is the use of existing k-anonymizing algorithms, choosing the maximization of the usefulness of the data for classification as a preference criteria.

Recently, two approaches that anonymize data before mining have been presented for classification (e.g., decision trees): a top-down [16] and a bottom-up [29] technique. These two techniques aim at releasing a k-anonymous table $T(A_1, \ldots, A_m, class)$ for modeling classification of attribute *class* considering the quasi-identifier QI = $\{A_1, \ldots, A_m\}$. k-anonymity is achieved with cell generalization and cell suppression (CG_), that is, different cells of the same attribute may have values belonging to different generalized domains. The aim of *preserving anonymity for classification* is then to satisfy the k-anonymity constraint while preserving the classification structure in the data.

The top-down approach starts from a table containing the most general values for all attributes and tries to *refine* (i.e., specialize) some values. For instance, the table in Figure 5.11(a) represents a completely generalized table for the table in Figure 5.1. The bottom-up approach starts from a private table and tries to generalize the attributes until the k-anonymity constraint is satisfied.

In the top-down technique a refinement is performed only if it has some suitable properties for guaranteeing both anonymity and good classification. For this purpose, a selection criterion is described for guiding the top-down refinement process to heuristically maximize the classification goal. The refinement has two opposite effects: it increases the information of the table for classification and it decreases its anonymity. The algorithm is guided by the functions *InfoGain(v)* and *AnonyLoss(v)* measuring the information gain and the anonymity loss, respectively, where v is the attribute value (cell) candidate for refinement. A good candidate v is such that *InfoGain(v)* is large, and *AnonyLoss(v)* is small. Thus, the selection criterion for choosing the candidate v to be refined maximizes function $Score(v) = \frac{InfoGain(v)}{AnonyLoss(v)+1}$. Function *Score(v)* is computed for each value v of the attributes in the table. The value with the highest score is then specialized to its children in the value generalization hierarchy.

An attribute value v, candidate for specialization, is considered useful to obtain a good classification if the frequencies of the class values are not uniformly distributed for the specialized values of v. The entropy of a value in a table measures the dominance of the majority: the more dominating the majority value in the class is, the smaller the entropy is. *InfoGain(v)* then measures the reduction of entropy after refining v (for a formal definition of *InfoGain(v)* see [16]). A good candidate is a value v that reduces the entropy of the table. For instance, with reference to the private table in Figure 5.1 and its generalized version in Figure 5.11(a), *InfoGain*(`any_marital_status`) is high since for `been_married` we have 14 N and 26 Y, with a difference of 12, and for `never_married` we have 20 N and 6 Y, with a difference of 14 (see

Marital_status	*Sex*	*Hours*	*#tuples (Hyp. values)*
any_marital_status	any_sex	[1,100)	66 (32Y, 34N)

(a) Step 1: the most general table

Marital_status	*Sex*	*Hours*	*#tuples (Hyp. values)*
been_married	any_sex	[1,100)	40 (26Y, 14N)
never_married	any_sex	[1,100)	26 (6Y, 20N)

(b) Step 2

Marital_status	*Sex*	*Hours*	*#tuples (Hyp. values)*
divorced	any_sex	[1,100)	21 (16Y, 5N)
married	any_sex	[1,100)	19 (10Y, 9N)
never_married	any_sex	[1,100)	26 (6Y, 20N)

(c) Step 3

Marital_status	*Sex*	*Hours*	*#tuples (Hyp. values)*
divorced	any_sex	35	4 (0Y, 4N)
divorced	any_sex	40	17 (16Y, 1N)
married	any_sex	35	10 (8Y, 2N)
married	any_sex	50	9 (2Y, 7N)
single	any_sex	40	26 (6Y, 20N)

(d) Final table (after 7 steps)

Figure 5.11. An example of top-down anonymization for the private table in Figure 5.1

Figure 5.11(b)). On the contrary, *InfoGain*($[1, 100)$) is low since for $[0, 40)$ we have 8 Y and 6 N, with a difference of 2, and for $[40, 100)$ we have 24 Y and 28 N, with a difference of 2. Thus `Marital_status` is more useful for classification than `Hours`.

Let us define the anonymity degree of a table as the maximum k for which the table is k-anonymous. The loss of anonymity, defined as *AnonyLoss(v)*, is the difference between the degrees of anonymity of the table before and after refining v. For instance, the degrees of the tables in Figures 5.11(b) and 5.11(c) are 26 (tuples containing: `never_married, any_sex, [1,100)`) and 19 (tuples containing: `married, any_sex, [1,100)`), respectively. Since the table in Figure 5.11(c) is obtained by refining the value `been_married` of the table in Figure 5.11(b), *AnonyLoss*(`been_married`) is 7.

The algorithm terminates when any further refinement would violate the k-anonymity constraint.

EXAMPLE 5.1 *Consider the private table in Figure 5.1, and the value generalization hierarchies in Figure 5.2. Let us suppose* QI = {`Marital_status`, `Sex`, `Hours`} *and* $k = 4$. *The algorithm starts from the most generalized table in Figure 5.11(a), and computes the scores: Score*(`any_marital_status`), *Score*(`any_sex`), *and Score*($[1, 100)$).

Since the maximum score corresponds to value `any_marital_status`, *this value is refined, producing the table in Figure 5.11(b). The remaining tables computed by the algorithm are shown in Figures 5.11(c), and 5.11(d). Figure 5.11(d) illustrates the final table since the only possible refinement (*`any_sex` *to* `M` *and* `F`*) violates 4-anonymity. Note that the final table is 4-anonymous with respect to* QI = {`Marital_status`, `Sex`, `Hours`}.

The bottom-up approach is the dual of the top-down approach. Starting from the private table, the objective of the bottom-up approach is to generalize the values in the table to determine a k-anonymous table preserving good qualities for classification and minimizing information loss. The effect of generalization is thus measured by a function involving anonymity gain (instead of anonymity loss) and information loss.

Note that, since these methods compute a minimal k-anonymous table suitable for classification with respect to *class* and QI, the computed table PT_k is optimized only if classification is performed using the entire set QI. Otherwise, the obtained table PT_k could be too general. For instance, consider the table in Figure 5.1, the table in Figure 5.11(d) is a 4-anonymization for it considering QI = {`Marital_status, Sex, Hours`}. If classification is to be done with respect to a subset QI′ = {`Marital_status, Sex`} of QI, such a table would be too general. As a matter of fact, a 4-anonymization for PT with respect to QI′ can be obtained from PT by simply generalizing `divorced` and `married` to `been_married`. This latter generalization would generalize only 40 cells, instead of the 66 cells (M and F to any_sex) generalized in the table in Figure 5.11(d).

5.7 Mine-and-Anonymize

The Mine-and-Anonymize approach performs mining on the original table PT. Anonymity constraints must therefore be enforced with respect to the mined results to be returned. Regardless of whether the approach is executed in one or two steps (see Section 5.5), the problem to be solved is to translate k-anonymity constraints for PT over the mined results. Intuitively, the mined results should not allow anybody to infer the existence of sets of quasi-identifier values that have less than k occurrences in the private table PT. Let us then discuss what this implies for association rules and for decision trees.

5.7.1 Enforcing k-Anonymity on Association Rules

To discuss k-anonymity for association rules it is useful to distinguish the two different phases of association rule mining:

1 find all combinations of items whose support (i.e., the number of joint occurrences in the records) is greater than a minimum threshold σ (frequent itemsets mining);

2 use the frequent itemsets to generate the desired rules.

The consideration of these two phases conveniently allows expressing k-anonymity constraints with respect to observable itemsets instead of association rules. Intuitively, k-anonymity for PT is satisfied if the observable itemsets do not allow inferring (the existence of) sets of quasi-identifier values that have less than k occurrences in the private table. It is trivial to see that any itemset X that includes only values on quasi-identifier attributes and with a support lower than k is clearly *unsafe*. In fact, the information given by the itemset corresponds to stating that there are less than k respondents with occurrences of values as in X, thus violating k-anonymity. Besides trivial itemsets such as this, also the combination of itemsets with support greater than or equal to k can breach k-anonymity.

As an example, consider the private table in Figure 5.1, where the quasi-identifier is {Marital_status, Sex, Hours} and suppose 3-anonymity must be guaranteed. All itemsets with support lower than 3 clearly violate the constraint. For instance, itemset {divorced, F} with support 2, which holds in the table, cannot be released. Figure 5.12 illustrates some examples of itemsets with support greater than or equal to 19 (assuming lower supports are not of interest). While one may think that releasing these itemsets guarantees any k-anonymity for $k \leq 19$, it is not so. Indeed, the combination of the two itemsets {divorced, M}, with support 19, and {divorced}, with support 21,

Itemset	*Support*
{∅}	66
{M}	55
{M, 40}	43
{single, M, 40}	26
{divorced}	21
{divorced, M}	19
{married}	19

Figure 5.12. Frequent itemsets extracted from the table in Figure 5.1

clearly violates it. In fact, from their combination we can infer the existence of two tuples in the private table for which the condition '`Marital_status` = `divorced` ∧ ¬(`Sex` = `M`)' is satisfied. Being `Marital_status` and `Sex` included in the quasi-identifier, this implies that no more than 2 indistinguishable tuples can exist for divorced non male respondents, thus violating k-anonymity for $k > 2$. In particular, since `Sex` can assume only two values, the two itemsets above imply the existence of (not released) itemset {`divorced, F`} with support 2. Note that, although both itemsets ({`divorced`}, 21) and ({`divorced, M`}, 19) cannot be released, there is no reason to suppress both, since each of them individually taken is safe.

The consideration of inferences such as those, and of possible solutions for suppressing itemsets to block the inferences while maximizing the utility of the released information, bring some resembling with the primary and secondary suppression operations in statistical data release [12]. It is also important to note that suppression is not the only option that can be applied to sanitize a set of itemsets so that no unsafe inferences violating k-anonymity are possible. Alternative approaches can be investigated, including adapting classical statistical protection strategies [12, 14]. For instance, itemsets can be combined, essentially providing a result that is equivalent to operating on *generalized* (in contrast to specific) data. Another possible approach consists in introducing *noise* in the result, for example, *modifying the support* of itemsets in such a way that their combination never allows inferring itemsets (or patterns of them) with support lower than the specified k.

A first investigation of translating the k-anonymity property of a private table on itemsets has been carried out in [7–9] with reference to private tables where all attributes are defined on binary domains. The identification of unsafe itemsets bases on the concept of *pattern*, which is a boolean formula of items, and on the following observation. Let X and $X \cup \{A_i\}$ be two itemsets. The support of pattern $X \wedge \neg A_i$ can be obtained by subtracting the support of itemset $X \cup \{A_i\}$ from the support of X. By generalizing this observation, we can conclude that given two itemsets $X = \{A_{x_1} \ldots A_{x_n}\}$

and $Y = \{A_{x_1} \ldots A_{x_n}, A_{y_1} \ldots A_{y_m}\}$, with $X \subset Y$, the support of pattern $A_{x_1} \wedge \ldots \wedge A_{x_n} \wedge \neg A_{y_1} \wedge \ldots \wedge \neg A_{y_m}$ (i.e., the number of tuples in the table containing X but not $Y - X$) can be inferred from the support of X, Y, and all itemsets Z such that $X \subset Z \subset Y$. This observation allows stating that a set of itemsets satisfies k-anonymity only if all itemsets, as well as the patterns derivable from them, have support greater than or equal to k.

As an example, consider the private table PT in Figure 5.13(a), where all attributes can assume two distinct values. This table can be transformed into the binary table T in Figure 5.13(b), where A corresponds to 'Marital_status = been_married', B corresponds to 'Sex = M', and C corresponds to 'Hours = [40,100)'. Figure 5.14 reports the lattice of all itemsets derivable from T together with their support. Assume that all itemsets with support greater than or equal to the threshold $\sigma = 40$, represented in Figure 5.15(a), are of interest, and that $k = 10$. The itemsets in Figure 5.15(a) present two inference channels. The first inference is obtained through itemsets $X_1 = \{C\}$ with support 52, and $Y_1 = \{BC\}$ with support 43. According to

Marital_status	*Sex*	*Hours*	*#tuples*
been_married	M	[1-40)	12
been_married	M	[40-100)	17
been_married	F	[1-40)	2
been_married	F	[40-100)	9
never_married	M	[40-100)	26

(a) PT

A	*B*	*C*	*#tuples*
1	1	0	12
1	1	1	17
1	0	0	2
1	0	1	9
0	1	1	26

(b) T

Figure 5.13. An example of binary table

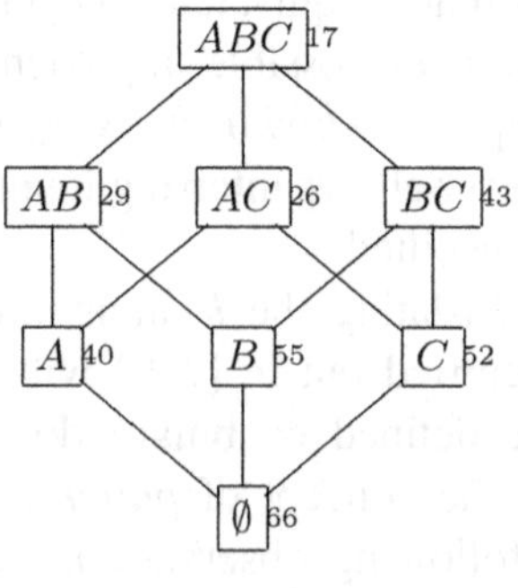

Figure 5.14. Itemsets extracted from the table in Figure 5.13(b)

Figure 5.15. Itemsets with support at least equal to 40 (a) and corresponding anonymized itemsets (b)

the observation previously mentioned, since $X_1 \subset Y_1$, we can infer that pattern $C \wedge \neg B$ has support $52 - 43 = 9$. The second inference channel is obtained through itemsets $X_2 =\{\emptyset\}$ with support 66, $Y_2 = \{BC\}$ with support 43, and all itemsets Z such that $X_2 \subset Z \subset Y_2$, that is, itemsets $\{B\}$ with support 55, and $\{C\}$ with support 52. The support of pattern $\neg B \wedge \neg C$ can then be obtained by applying again the observation previously mentioned. Indeed, from $\{BC\}$ and $\{B\}$ we infer pattern $B \wedge \neg C$ with support $55 - 43 = 12$, and from $\{BC\}$ and $\{C\}$ we infer pattern $\neg B \wedge C$ with support $52 - 43 = 9$. Since the support of itemset $\{\emptyset\}$ corresponds to the total number of tuples in the binary table, the support of $\neg B \wedge \neg C$ is computed by subtracting the support of $B \wedge \neg C$ (12), $\neg B \wedge C$ (9), and $B \wedge C$ (43) from the support of $\{\emptyset\}$, that is, $66 - 12 - 9 - 43 = 2$. The result is that release of the itemsets in Figure 5.15(a) would not satisfy k-anonymity for any $k > 2$.

In [9] the authors present an algorithm for detecting inference channels that is based on a classical data mining solution for concisely representing all frequent itemsets (closed itemsets [24]) and on the definition of maximal inference channels. In the same work, the authors propose to block possible inference channels violating k-anonymity by modifying the support of involved itemsets. In particular, an inference channel due to a pair of itemsets $X = \{A_{x_1} \dots A_{x_n}\}$ and $Y = \{A_{x_1} \dots A_{x_n}, A_{y_1} \dots A_{y_m}\}$ is blocked by increasing the support of X by k. In addition, to avoid contradictions among the released itemsets, also the support of all subsets of X is increased by k. For instance, with respect to the previous two inference channels, since k is equal to 10, the support of itemset $\{C\}$ is increased by 10 and the support of $\{\emptyset\}$ is increased by 20, because $\{\emptyset\}$ is involved in the two channels. Figure 5.15(b) illustrates the resulting anonymized itemsets. Another possible strategy for blocking channels consists in decreasing the support of the involved itemsets to zero. Note that this corresponds basically to removing some tuples in the original table.

5.7.2 Enforcing k-Anonymity on Decision Trees

Like for association rules, a decision tree satisfies k-anonymity for the private table PT from which the tree has been built if no information in the tree allows inferring quasi-identifier values that have less than k occurrences in the private table PT. Again, like for association rules, k-anonymity breaches can be caused by individual pieces of information or by combination of apparently anonymous information. In the following, we briefly discuss the problem distinguishing two cases depending on whether the decision tree reports frequencies information for the internal nodes also or for the leaves only.

Let us first consider the case where the tree reports frequencies information for all the nodes in the tree. An example of such a tree is reported in Figure 5.9. With a reasoning similar to that followed for itemsets, given a k, all nodes with a number of occurrences lower than k are *unsafe* as they breach k-anonymity. For instance, the fourth leaf (reachable through path $\langle$F,35$\rangle$) is unsafe for any k-anonymity higher than 2. Again, with a reasoning similar to that followed for itemsets, also combinations of nodes that allow inferring patterns of tuples containing quasi-identifying attributes with a number of occurrences lower than k breach k-anonymity for the given k. For instance, nodes corresponding to paths $\langle$F$\rangle$ and to $\langle$F,50$\rangle$, which taken individually would appear to satisfy any k-anonymity constraint for $k \leq 9$, considered in combination would violate any k-anonymity for $k > 2$ since their combination allows inferring that there are no more than two tuples in the table referring to females working a number of hours different from 50. It is interesting to draw a relationship between decision trees and itemsets. In particular, any node in the tree corresponds to an itemset dictated by the path to reach the node. For instance, with reference to the tree in Figure 5.9, the nodes correspond to itemsets: {}, {M}, {M,married}, {M,divorced}, {M,single}, {F}, {F,35}, {F,40}, {F,50}, where the support of each itemset is the sum of the Ys and Ns in the corresponding node. This observation can be exploited for translating approaches for sanitizing itemsets for the sanitization of decision trees (or viceversa). With respect to blocking inference channels, different approaches can be used to anonymize decision trees, including suppression of unsafe nodes as well as other nodes as needed to block combinations breaching anonymity (secondary suppression). To illustrate, suppose that 3-anonymity is to be guaranteed. Figure 5.16 reports a 3-anonymized version of the tree in Figure 5.9. Here, besides suppressing node $\langle$F,35$\rangle$, its sibling $\langle$F,50$\rangle$ has been suppressed to block the inference channel described above.

Let us now consider the case where the tree reports frequencies information only for the leaf nodes. Again, there is an analogy with the itemset problem with the additional consideration that, in this case, itemsets are such that none

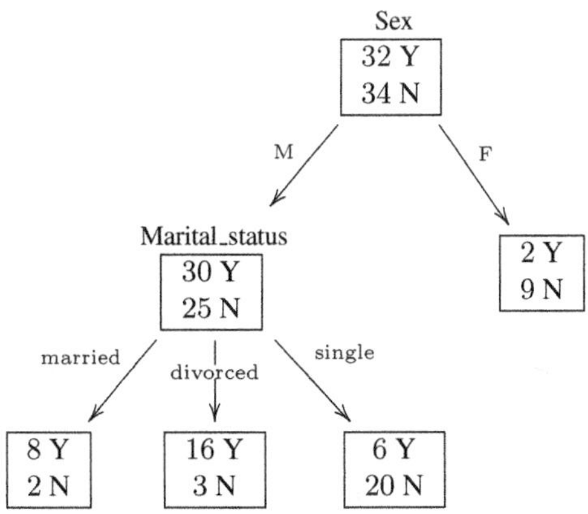

Figure 5.16. 3-anonymous version of the tree of Figure 5.9

of them is a subset of another one. It is therefore quite interesting to note that the set of patterns of tuples identified by the tree nodes directly corresponds to a generalized version of the private table PT, where some values are suppressed (CG_). This property derives from the fact that, in this case, every tuple in PT satisfies exactly one pattern (path to a leaf). To illustrate, consider the decision tree in Figure 5.17, obtained from the tree in Figure 5.9 by suppressing occurrences in non-leaf nodes. Each leaf in the tree corresponds to a generalized tuple reporting the value given by the path (for attributes appearing in the path). The number of occurrences of such a generalized tuple is reported in the leaf. If a quasi-identifier attribute does not appear along the path, then its value is set to $*$. As a particular case, if every path in the tree contains all the quasi-identifier attributes and puts conditions on specific values, the generalization coincides with the private table PT. For instance, Figure 5.18 reports the table containing tuple patterns that can be derived from the tree in Figure 5.17, and which corresponds to a generalization of the original private table PT in Figure 5.1. The relationship between trees and generalized tables is very important as it allows us to express the protection enjoyed of a decision tree in terms of the generalized table corresponding to it, with the advantage of possibly exploiting classical k-anonymization approaches referred to the private table. In particular, this observation allows us to identify as unsafe *all and only* those nodes corresponding to tuples whose number of occurrences is lower than k. In other words, in this case (unlike for the case where frequencies of internal nodes values are reported) there is no risk that combination of nodes, each with occurrences higher than or equal to k, can breach k-anonymity.

Again, different strategies can be applied to protect decision trees in this case, including exploiting the correspondence just withdrawn, translating on

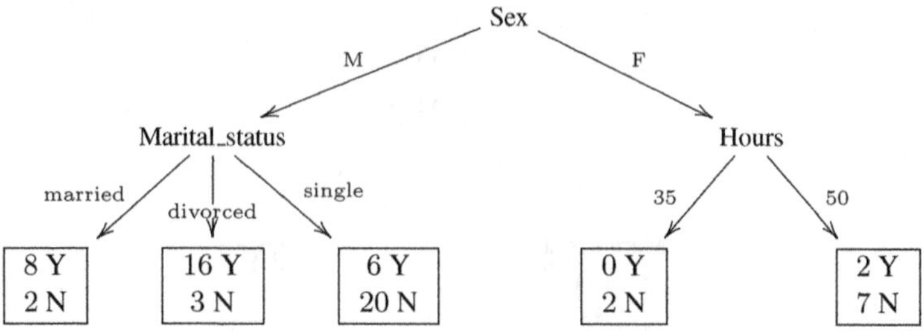

Figure 5.17. Suppression of occurrences in non-leaf nodes in the tree in Figure 5.9

Marital_status	*Sex*	*Hours*	*#tuples (Hyp. values)*
divorced	M	*	19 (16Y, 3N)
*	F	35	2 (0Y, 2N)
married	M	*	10 (8Y, 2N)
*	F	50	9 (2Y, 7N)
single	M	*	26 (6Y, 20N)

Figure 5.18. Table inferred from the decision tree in Figure 5.17

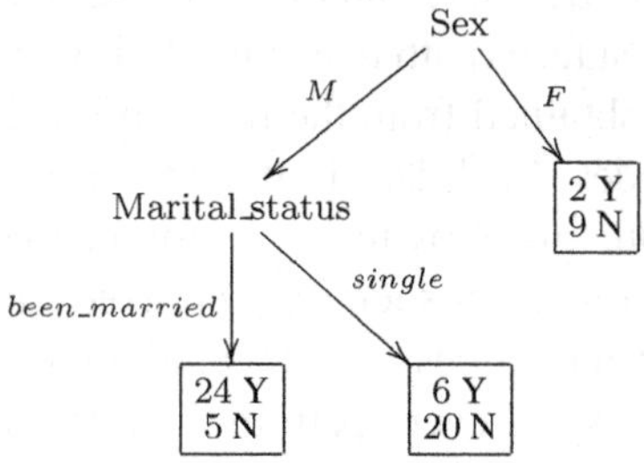

Figure 5.19. 11-anonymous version of the tree in Figure 5.17

the tree the generalization and suppression operations that could be executed on the private table. To illustrate, consider the tree in Figure 5.17, the corresponding generalized table is in Figure 5.18, which clearly violates any k-anonymity for $k > 2$. Figure 5.19 illustrates a sanitized version of the tree for guaranteeing 11-anonymity obtained by suppressing the splitting node `Hours` and combining nodes ⟨`M,married`⟩ and ⟨`M,divorced`⟩ into a single node. Note how the two operations have a correspondence with reference to the starting table in Figure 5.18 with an attribute generalization over `Hours` and a cell generalization over `Marital_status`, respectively. Figure 5.20 illustrates the table corresponding to the tree in Figure 5.19.

The problem of sanitizing decision trees has been studied in the literature by Friedman et al. [15, 16], who proposed a method for directly building a

Marital_status	*Sex*	*Hours*	*#tuples (Hyp. values)*
been_married	M	*	29 (24Y, 5N)
*	F	*	11 (2Y, 9N)
single	M	*	26 (6Y, 20N)

Figure 5.20. Table inferred from the decision tree in Figure 5.19

k-anonymous decision tree from a private table PT. The proposed algorithm is basically an improvement of the classical decision tree building algorithm, combining mining and anonymization in a single process. At initialization time, the decision tree is composed of a unique root node, representing all the tuples in PT. At each step, the algorithm inserts a new splitting node in the tree, by choosing the attribute in the quasi-identifier that is more useful for classification purposes, and updates the tree accordingly. If the tree obtained is non-k-anonymous, then the node insertion is rolled back. The algorithm stops when no node can be inserted without violating k-anonymity, or when the classification obtained is considered satisfactory.

5.8 Conclusions

A main challenge in data mining is to enable the legitimate usage and sharing of mined information while at the same time guaranteeing proper protection of the original sensitive data. In this chapter, we have discussed how k-anonymity can be combined with data mining for protecting the identity of the respondents to whom the data being mined refer. We have described the possible threats to k-anonymity that can arise from performing mining on a collection of data and characterized two main approaches to combine k-anonymity in data mining. We have also discussed different methods that can be used for detecting k-anonymity violations and consequently eliminate them in association rule mining and classification mining.

k-anonymous data mining is however a recent research area and many issues are still to be investigated such as: the combination of k-anonymity with other possible data mining techniques; the investigation of new approaches for detecting and blocking k-anonymity violations; and the extension of current approaches to protect the released data mining results against attribute, in contrast to identity, disclosure [21].

Acknowledgements

This work was supported in part by the European Union under contract IST-2002-507591, by the Italian Ministry of Research Fund for Basic Research (FIRB) under project "RBNE05FKZ2", and by the Italian MIUR under project 2006099978.

References

[1] Charu C. Aggarwal. On k-anonymity and the curse of dimensionality. In *Proc. of the 31th VLDB Conference*, Trondheim, Norway, September 2005.

[2] Gagan Aggarwal, Tomas Feder, Krishnaram Kenthapadi, Rajeev Motwani, Rina Panigrahy, Dilys Thomas, and An Zhu. Anonymizing tables. In *Proc. of the 10th International Conference on Database Theory (ICDT'05)*, Edinburgh, Scotland, January 2005.

[3] Gagan Aggarwal, Tomas Feder, Krishnaram Kenthapadi, Rajeev Motwani, Rina Panigrahy, Dilys Thomas, and An Zhu. Approximation algorithms for k-anonymity. *Journal of Privacy Technology*, November 2005.

[4] Dakshi Agrawal and Charu C. Aggarwal. On the design and quantification of privacy preserving data mining algorithms. In *Proc. of the 20th ACM SIGMOD-SIGACT-SIGART symposium on Principles of database systems*, Santa Barbara, California, June 2001.

[5] Rakesh Agrawal and Ramakrishnan Srikant. Fast algorithms for mining association rules. In *Proc. of the 20th VLDB Conference*, Santiago, Chile, September 1994.

[6] Rakesh Agrawal and Ramakrishnan Srikant. Privacy-preserving data mining. In *Proc. of the ACM SIGMOD Conference on Management of Data*, Dallas, Texas, May 2000.

[7] Maurizio Atzori, Francesco Bonchi, Fosca Giannotti, and Dino Pedreschi. Blocking anonymity threats raised by frequent itemset mining. In *Proc. of the 5th IEEE International Conference on Data Mining (ICDM 2005)*, Houston, Texas, November 2005.

[8] Maurizio Atzori, Francesco Bonchi, Fosca Giannotti, and Dino Pedreschi. k-anonymous patterns. In *Proc. of the 9th European Conference on Principles and Practice of Knowledge Discovery in Databases (PKDD)*, Porto, Portugal, October 2005.

[9] Maurizio Atzori, Francesco Bonchi, Fosca Giannotti, and Dino Pedreschi. Anonymity preserving pattern discovery. *VLDB Journal*, November 2006.

[10] Roberto J. Bayardo and Rakesh Agrawal. Data privacy through optimal k-anonymization. In *Proc. of the International Conference on Data Engineering (ICDE'05)*, Tokyo, Japan, April 2005.

[11] Valentina Ciriani, Sabrina De Capitani di Vimercati, Sara Foresti, and Pierangela Samarati. k-anonymity. In T. Yu and S. Jajodia, editors, *Security in Decentralized Data Management*. Springer, Berlin Heidelberg, 2007.

[12] Valentina Ciriani, Sabrina De Capitani di Vimercati, Sara Foresti, and Pierangela Samarati. Microdata protection. In T. Yu and S. Jajodia, editors, *Security in Decentralized Data Management.* Springer, Berlin Heidelberg, 2007.

[13] Alexandre Evfimievski, Ramakrishnan Srikant, Rakesh Agrawal, and Johannes Gehrke. Privacy preserving mining of association rules. In *Proc. of the 8th ACM SIGKDD International Conference on Knowledge Discovery and Data Mining*, Edmonton, Alberta, Canada, July 2002.

[14] Federal Committee on Statistical Methodology. Statistical policy working paper 22, May 1994. Report on Statistical Disclosure Limitation Methodology.

[15] Arik Friedman, Assaf Schuster, and Ran Wolff. Providing k-anonymity in data mining. *VLDB Journal.* Forthcoming.

[16] Benjamin C.M. Fung, Ke Wang, and Philip S. Yu. Anonymizing classification data for privacy preservation. *IEEE Transactions on Knowledge and Data Engineering*, 19(5):711–725, May 2007.

[17] Michael R. Garey and David S. Johnson *Computers and Intractability.* W. H. Freeman & Co., New York, NY, USA, 1979.

[18] Kristen LeFevre, David J. DeWitt, and Raghu Ramakrishnan. Incognito: efficient full-domain k-anonymity. In *Proc. of the ACM SIGMOD Conference on Management of Data*, Baltimore, Maryland, June 2005.

[19] Kristen LeFevre, David J. DeWitt, and Raghu Ramakrishnan. Mondrian multidimensional k-anonymity. In *Proc. of the International Conference on Data Engineering (ICDE'06)*, Atlanta, Georgia, April 2006.

[20] Yehuda Lindell and Benny Pinkas. Privacy preserving data mining. *Journal of Cryptology*, 15(3):177–206, June 2002.

[21] Ashwin Machanavajjhala, Johannes Gehrke, and Daniel Kifer. ℓ-density: Privacy beyond k-anonymity. In *Proc. of the International Conference on Data Engineering (ICDE'06)*, Atlanta, Georgia, April 2006.

[22] Adam Meyerson and Ryan Williams On the complexity of optimal k-anonymity. In *Proc. of the 23rd ACM SIGMOD-SIGACT-SIGART Symposium on Principles of Database Systems*, Paris, France, June 2004.

[23] Hyoungmin Park and Kyuseok Shim. Approximate algorithms for k-anonymity. In *Proc. of the ACM SIGMOD Conference on Management of Data*, Beijing, China, June 2007.

[24] Nicolas Pasquier, Yves Bastide, Rafik Taouil, and Lotfi Lakhal. Discovering frequent closed itemsets for association rules. In *Proc. of the 7th International Conference on Database Theory (ICDT '99)*, Jerusalem, Israel, January 1999.

[25] Rajeev Rastogi and Kyuseok Shim. PUBLIC: A decision tree classifier that integrates building and pruning. In *Proc. of the 24th VLDB Conference*, New York, September 1998.

[26] Pierangela Samarati. Protecting respondents' identities in microdata release. *IEEE Transactions on Knowledge and Data Engineering*, 13(6):1010–1027, November 2001.

[27] Pierangela Samarati and Latanya Sweeney. Generalizing data to provide anonymity when disclosing information (abstract). In *Proc. of the 17th ACM-SIGMOD-SIGACT-SIGART Symposium on the Principles of Database Systems*, page 188, Seattle, WA, 1998.

[28] Ramakrishnan Srikant and Rakesh Agrawal. Mining generalized association rules. In *Proc. of the 21th VLDB Conference*, Zurich, Switzerland, September 1995.

[29] Ke Wang, Philip S. Yu, and Sourav Chakraborty. Bottom-up generalization: A data mining solution to privacy protection. In *Proc. of the 4th IEEE International Conference on Data Mining (ICDM 2004)*, Brighton, UK, November 2004.

[30] Zhiqiang Yang, Sheng Zhong, and Rebecca N. Wright. Privacy-preserving classification of customer data without loss of accuracy. In *Proc. of the 5th SIAM International Conference on Data Mining*, Newport Beach, California, April 2005.

[31] Mohammed J. Zaki and Ching-Jui Hsiao. Charm: An efficient algorithm for closed itemset mining. In *Proc. of the 2nd SIAM International Conference on Data Mining*, Arlington, Virginia, April 2002.

Chapter 6

A Survey of Randomization Methods for Privacy-Preserving Data Mining

Charu C. Aggarwal
IBM T. J. Watson Research Center
Hawthorne, NY 10532
charu@us.ibm.com

Philip S. Yu
University of Illinois at Chicago
Chicago, IL 60607
psyu@us.ibm.com

Abstract A well known method for privacy-preserving data mining is that of randomization. In randomization, we add noise to the data so that the behavior of the individual records is masked. However, the aggregate behavior of the data distribution can be reconstructed by subtracting out the noise from the data. The reconstructed distribution is often sufficient for a variety of data mining tasks such as classification. In this chapter, we will provide a survey of the randomization method for privacy-preserving data mining.

Keywords: Randomization, privacy quantification, perturbation.

6.1 Introduction

In the randomization method, we add noise to the data in order to mask the values of the records. The noise added is sufficiently large so that the individual values of the records can no longer be recovered. However, the probability distribution of the aggregate data can be recovered and subsequently used for privacy-preservation purposes. The earliest work on randomization may be found in [16, 12], in which it has been used in order to eliminate evasive answer bias. In [3] it has been shown how the reconstructed distributions may be used for data mining. The specific problem which has been discussed in [3] is that of classification, though the approach can be easily extended to a variety of other problems such as association rule mining [8, 24].

The method of randomization can be described as follows. Consider a set of data records denoted by $X = \{x_1 \ldots x_N\}$. For record $x_i \in X$, we add a noise component which is drawn from the probability distribution $f_Y(y)$. These noise components are drawn independently, and are denoted $y_1 \ldots y_N$. Thus, the new set of distorted records are denoted by $x_1 + y_1 \ldots x_N + y_N$. We denote this new set of records by $z_1 \ldots z_N$. In general, it is assumed that the variance of the added noise is large enough, so that the original record values cannot be easily guessed from the distorted data. Thus, the original records cannot be recovered, but the distribution of the original records can be recovered. We note that the addition of X and Y creates a new distribution Z. We know N instantiations of this new distribution, and can therefore estimate it approximately. Furthermore, since the distribution of Y is publicly known, we can estimate the distribution obtained by subtracting Y from Z. In a later section, we will discuss more accurate strategies for distribution estimation. Furthermore, the above-mentioned technique is an *additive strategy* for randomization. In the *multiplicative strategy*, it is possible to multiply the records with random vectors in order yo provide the final representation of the data. Thus, this approach uses a random projection kind of approach in order to perform the privacy-preserving transformation. The resulting data can be be re-constructed within a certain variance depending upon the number of components of the multiplicative perturbation.

We note that methods such as randomization add or multiply the noise to the records in a *data-independent* way. In other methods such as k-anonymity [25], the overall behavior of the records is leveraged in the anonymization process. This is very useful from a practical point of view, since it means that the randomization can be performed at *data-collection* time. Thus, a trusted server is not required (as in k-anonymization) in order to perform the transformations on the records. This is a key advantage of randomization methods, though it comes at the expense that there are no guarantees against re-identification of the data in the presence of public information. Another key property of the randomization method is that the original records are not used after the transformation. Rather, the data mining algorithms use aggregate distributions of the data in order to perform the mining process.

This paper is organized as follows. In the next section, we will discuss a number of reconstruction methods for randomization. We will also discuss the issue of optimaility and utility of randomization methods. In section 3, we will discuss a number of applications of randomization. We will show how the approach can be used for a number of applications such as classification and association rule mining. In section 4, we will discuss issues surrounding the quantification of privacy-preserving data mining algorithms. In section 5, we will discuss a number of adversarial attacks on the randomization method. In section 6, we discuss applications of the randomization method to the case of time series data. In section 7, we discuss the method of multiplicative perturbations and its applications to a variety of data mining algorithms. The conclusions and summary are presented in section 8.

6.2 Reconstruction Methods for Randomization

In this section, we will discuss reconstruction algorithms for the randomization method. We note that the perturbed data distribution Z can be obtained by adding the distributions of the original data X and that of the perturbation Y. Therefore, we have:

$$Z = X + Y$$
$$X = Z - Y$$

We note that only the distribution of Y is known explicit. The distribution of X is unknown, and N instantiations of the probability distribution Z are known. These N instantiations can be used to construct an estimate of the probability distribution Z. When the value of N is large, this estimate can be quite accurate. Once Z is known, we can subtract Y from it in order to obtain the probability distribution of X. For modest values of N, the errors in the estimation of Z can be quite large, and these errors may get magnified on subtraction of Y. Therefore, a more indirect method is desirable in order to estimate the probability distribution of X.

A pair of closely related iterative methods have been discussed in [3, 5] for approximation of the corresponding probability distributions. The method in [3] uses the Bayes rule for distribution approximation, whereas that in [5] uses the EM method for distribution approximation. In this section, we will describe both methods. First, we will discuss the method in [3] for distribution reconstruction.

6.2.1 The Bayes Reconstruction Method

Let f' and F' be the estimated density functions and cumulative density functions with the use of the reconstructed distributions. The, we can use the bayes formula in order to derive an estimate for f', using the first observed value z_1:

$$F'(a) = \int_{-\infty}^{a} f_{X_1}(w|X_1 + Y_1 = z_1)dw \qquad (6.1)$$

We can expand the above expression using the Bayes rule (in conjunction with the independence of the random variables Y and X) in order to construct the following expression for $F'(a)$.

$$F'(a) = \frac{\int_{-\infty}^{a} f_X(z_1 - w) \cdot f_X(w)dw}{\int_{-\infty}^{\infty} f_X(z_1 - w) \cdot f_X(w)dw} \tag{6.2}$$

We note that the above expression for $F'(a)$ was derived using a single observation z_1. In practice, the average distribution of multiple observations $z_1 \dots z_N$ can be used in order to construct the estimated cumulative distribution $F'(a)$. Thus, we can construct the estimated distribution as follows:

$$F'(a) = (1/N) \cdot \sum_{i=1}^{N} \frac{\int_{-\infty}^{a} f_X(z_i - w) \cdot f_X(w)dw}{\int_{-\infty}^{\infty} f_X(z_i - w) \cdot f_X(w)dw} \tag{6.3}$$

The corresponding density distribution can be obtained by differentiating $F'(a)$. This differentiation results in the removal of the integral sign from the numerator, and the corresponding instantiation of w to a. Therefore, we have:

$$f'(a) = (1/N) \cdot \sum_{i=1}^{N} \frac{f_X(z_i - a) \cdot f_X(a)}{\int_{-\infty}^{\infty} f_X(z_i - w) \cdot f_X(w)dw} \tag{6.4}$$

We note that it is tricky to compute $f(\cdot)$ using the above equation, since we do not know the distribution for f on the right hand side. This suggests an iterative method for computing the distribution f. We start of by setting f as the uniform distribution, and iteratively update it using the equation above. The algorithm for computing $f(a)$ for a particular value of a is described as follows:

Set f to be the uniform distribution;
repeat
Update $f(a) = (1/N) \cdot \sum_{i=1}^{N} \frac{f_X(z_i - a) \cdot f_X(a)}{\int_{-\infty}^{\infty} f_X(z_i - w) \cdot f_X(w)dw}$
until convergence

We note that we cannot compute the value of $f(a)$ over all possible (infinite number of) values of a in a continuous domain. Therefore, we partition the domain of X into a number of intervals $[l_1, u_1] \dots [l_n, u_n]$, and assume that the function is uniform over each interval. For each interval $[l_i, u_i]$, the value of a in the above equation is picked to be $(l_i + u_i)/2$. Thus, in each iteration, we use n different values of a corresponding to each of the intervals. We note that the density functions on the right hand sides can be computed using the mean values over the corresponding intervals.

We note that the algorithm is terminated when the distribution does not change significantly over successive steps of the algorithm. A χ^2 test was used to compare the two distributions. The implementation in [3] terminated the algorithm when the difference between successive estimates was given by 1% of the threshold of the χ^2 test. While this algorithm is known to perform effectively in practice, the work in [3] does not prove this algorithm to be a provably convergent solution. In [5], an Expectation Maximization (EM) algorithm has been proposed which converges to a provably optimal solution. It is also shown in [5] that the Bayes algorithm of [3] is actually an approximation of the Expectation Maximization algorithm proposed in [5]. This is one of the reasons why the Bayes method proposed in [3] is so robust in practice.

6.2.2 The EM Reconstruction Method

In this subsection, we will discuss the EM algorithm for distribution reconstruction. Since the function $f_X(x)$ is defined over a continuous domain, we need to parameterize and discretize it for the purpose of any numerical estimation method. We assume that the data domain Ω_X can be discretized into K intervals $\Omega_1 \ldots \Omega_K$, where $\cup_{i=1}^{k} \Omega_i = \Omega_X$. Let $m_i = m(\Omega_i)$ be the length of the interval Ω_i. We assume that $f_X(x)$ is constant over Ω_i and the corresponding density function value is equal to θ_i. Thus, such a form will restrict $f_X(x)$ to a class parameterized by the finite set of parameters $\Theta = \{\theta_1, \theta_2, \ldots, \theta_K\}$. In order to explicitly denote the parametric dependence of the density function on Θ we will use the notation $f_{X;\Theta}(x)$ for the density function of X. Therefore, we have $f_{X;\boldsymbol{\Theta}}(x) = \sum_{i=1}^{K} \theta_i I_{\Omega_i}(x)$. Here $I_{\Omega_i}(x) = 1$ if $x \in \Omega_i$ and 0 otherwise. Since $f_{X;\boldsymbol{\Theta}}(x)$ is a density, it follows that $\sum_{i=1}^{K} \theta_i m(\Omega_i) = 1$. By choosing K large enough, density functions of the form discussed above can approximate any density function with arbitrary precision.

After this parameterization, the algorithm will proceed to estimate Θ, and thereby determine $\hat{f}_{X;\boldsymbol{\Theta}}(x)$. Let $\widehat{\Theta} = \{\hat{\theta}_1, \hat{\theta}_2, \ldots, \hat{\theta}_K\}$ be the estimate of these parameters produced by the reconstruction algorithm.

Given a set of observations $\mathbf{Z} = \mathbf{z}$, we would ideally like to find the *maximum-likelihood* (ML) estimate $\widehat{\boldsymbol{\Theta}}_{ML} = \text{argmax}_{\boldsymbol{\Theta}} \ln f_{\mathbf{Z};\boldsymbol{\Theta}}(\mathbf{z})$. The ML estimate has many attractive properties such as consistency, asymptotic unbiasedness, and asymptotic minimum variance among unbiased estimates. However, it is not always be possible to find $\widehat{\boldsymbol{\Theta}}_{ML}$ directly, and this turns out to be the case with the $f_{\mathbf{Z};\boldsymbol{\Theta}}(\mathbf{z})$ given above.

In order to achieve this goal, we will derive a reconstruction algorithm which fits into the broad framework of Expectation Maximization (EM) algorithms. The algorithm proceeds as if a more comprehensive set of data, say $\mathbf{D} = \mathbf{d}$ is observable and maximizes $\ln f_{\mathbf{D};\boldsymbol{\Theta}}(\mathbf{d})$ over all values of $\boldsymbol{\Theta}$ (M-step). Since $\mathbf{d}$ is in fact unavailable, it replaces $\ln f_{\mathbf{D};\boldsymbol{\Theta}}(\mathbf{d})$ by its conditional expected value given $\mathbf{Z} = \mathbf{z}$ and the current estimate of $\boldsymbol{\Theta}$ (E-Step). The $\mathbf{D}$ is chosen to make E-step and M-step easy to compute.

In this paper, we propose the use of $\mathbf{X} = \mathbf{x}$ as the more comprehensive set of data. As shown in the next section, this choice results in a computationally efficient algorithm. More formally, we define a Q function as follows:

$$Q(\mathbf{\Theta}, \widehat{\mathbf{\Theta}}) = E\Big[\ln f_{\mathbf{X};\mathbf{\Theta}}(\mathbf{X}) \,\Big|\, \mathbf{Z} = \mathbf{z}; \widehat{\mathbf{\Theta}}\Big] \tag{6.5}$$

Thus, $Q(\mathbf{\Theta}, \widehat{\mathbf{\Theta}})$ is the expected value of $\ln f_{\mathbf{X};\mathbf{\Theta}}(\mathbf{X})$ computed with respect $f_{\mathbf{X}|\mathbf{Z}=\mathbf{z};\widehat{\mathbf{\Theta}}}$, the density of $\mathbf{X}$ given $\mathbf{Z} = \mathbf{z}$ and parameter vector $\widehat{\mathbf{\Theta}}$. After the initialization of $\mathbf{\Theta}$ to a nominal value $\mathbf{\Theta}^0$, the EM algorithm will iterate over the following two steps:

1 E-step: Compute $Q(\mathbf{\Theta}, \mathbf{\Theta}^k)$.

2 M-step: Update $\mathbf{\Theta}^{k+1} = \text{argmax}_{\mathbf{\Theta}}\, Q(\mathbf{\Theta}, \mathbf{\Theta}^k)$.

The above discussion provides the general framework of EM algorithms; the actual details of the E-step and M-steps require a derivation which is problem specific. Similarly, the precise convergence properties of an EM algorithm are rather sensitive to the problem and its corresponding derivation. In the next subsection, we will derive the EM algorithm for the reconstruction problem and show that the resulting EM-algorithm has desirable convergence properties. The values of $Q(\mathbf{\Theta}, \widehat{\mathbf{\Theta}})$ during the E-step and the M-step of the reconstruction algorithm are discussed in [5].

THEOREM 6.1 *The value of* $Q(\mathbf{\Theta}, \widehat{\mathbf{\Theta}})$ *during the E-step of the reconstruction algorithm is given by:* $Q(\mathbf{\Theta}, \widehat{\mathbf{\Theta}}) = \sum_{i=1}^{K} \psi_i(\mathbf{z}; \widehat{\mathbf{\Theta}}) \ln \theta_i$, *where* $\psi_i(\mathbf{z}; \widehat{\mathbf{\Theta}}) = \hat{\theta}_i \sum_{j=1}^{N} \frac{Pr(Y \in z_j - \Omega_i)}{f_{Z;\widehat{\mathbf{\Theta}}}(z_j)}$.

In the next proposition, we calculate the value of $\mathbf{\Theta}$ that maximizes $Q(\mathbf{\Theta}, \widehat{\mathbf{\Theta}})$.

THEOREM 6.2 *The value of* $\mathbf{\Theta}$ *which maximizes* $Q(\mathbf{\Theta}, \widehat{\mathbf{\Theta}})$ *during the M-step of the reconstruction algorithm is given by:* $\theta_i = \frac{\psi_i(\mathbf{z}; \widehat{\mathbf{\Theta}})}{m_i N}$, *where* $\psi_i(\mathbf{z}; \widehat{\mathbf{\Theta}}) = \hat{\theta}_i \sum_{j=1}^{N} \frac{Pr(Y \in z_j - \Omega_i)}{f_{Z;\widehat{\mathbf{\Theta}}}(z_j)}$.

Now, we are in a position to describe the EM algorithm for the reconstruction problem.
1. Initialize $\theta_i^0 = \frac{1}{K}$, $i = 1, 2, \ldots, K$; $k = 0$;
2. Update $\mathbf{\Theta}$ as follows: $\theta_i^{(k+1)} = \frac{\psi_i(\mathbf{z}; \mathbf{\Theta}^k)}{m_i N}$;
3. $k = k + 1$;
4. If *not termination-criterion* then return to Step 2.

One key observation is that the EM algorithm is actually a refined version of the Bayes method discussed in [3]. The key difference between the two methods is in how the approximation of the values within an interval is treated.

While the Bayes method uses the crude estimate of the midpoint of the interval, the EM algorithm is more refined about it. While the Bayes method has not been shown to provably converge, it has been known to always empirically converge. On the other hand, our argument below shows that the EM algorithm does converge to a provably optimal solution. The close relationship between the two methods is the reason that the Bayes method is always known to empirically converge to an approximately optimal solution. The termination criterion for this method is based on how much $\boldsymbol{\Theta}^k$ has changed since the last iteration. It has been shown in [5] that the EM algorithm converges to the true distribution of the random variable X. We summarize the result as follows:

THEOREM 6.3 *The EM sequence* $\{\Theta^{(k)}\}$ *for the reconstruction algorithm converges to the unique Maximum Likelihood Estimate* $\widehat{\Theta}_{ML}$.

The above results lead to the following desirable property of the EM Algorithm.

OBSERVATION 6.2.1 *When there is a very large number of data observations, then the EM algorithm provides zero information loss.*

This is because as the number of observations increases, $\widehat{\Theta}_{ML} \Rightarrow \Theta$. Therefore, the original and estimated distribution become the same (subject to the discretization needed for any numerical estimation algorithm), resulting in zero information loss.

6.2.3 Utility and Optimality of Randomization Models

We note that the use of different perturbing distributions results in a different level of effectiveness of the randomization scheme. A key issue is how the randomization may be performed in order to optimize the tradeoff between privacy and accuracy. Clearly, the provision of a higher level of accuracy for the same privacy level is desirable from the point of view of maintaining greater utility of the randomized data. In order to achieve this goal, the work in [30] defines a randomization scheme in which the noise added to a given observation depends upon the value of the underlying data record as well as a user-defined parameter. Thus, in this case, the noise is *conditional* on the value of the record itself. This is a more general and flexible model for the randomization process. We note that this approach still does not depend upon the behavior of the other records, and can therefore be performed at data collection time. Methods are defined in [30] in order to perform reconstruction of the data with the use of this kind of randomization. The reconstruction methods proposed in [30] are designed with the use of kernel estimators or iterative EM methods. In [30] a number of information loss and interval metrics are used to quantify the tradeoff between privacy and optimality. The approach explores the issue of optimizing the information loss within a privacy constraint, or optimizing the privacy within an information loss constraint. A number of simulations have been presented in [30] to illustrate the effectiveness of the approach.

6.3 Applications of Randomization

The randomization method has been extended to a variety of data mining problems. In [3], it was discussed how to use the approach for classification. A number of other techniques [29, 30] have also been proposed which seem to work well over a variety of different classifiers. Techniques have also been proposed for privacy-preserving methods of improving the effectiveness of classifiers. For example, the work in [10] proposes methods for privacy-preserving boosting of classifiers. Methods for privacy-preserving mining of association rules have been proposed in [8, 24]. The problem of association rules is especially challenging because of the discrete nature of the attributes corresponding to presence or absence of items. In order to deal with this issue, the randomization technique needs to be modified slightly. Instead of adding quantitative noise, random items are dropped or included with a certain probability. The perturbed transactions are then used for aggregate association rule mining. This technique has shown to be extremely effective in [8]. The randomization approach has also been extended to other applications such as OLAP [4], and SVD based collaborative filtering [22]. We will discuss details of many of these techniques below.

We note that a variety of other randomization schemes exist for privacy-preserving data mining. The above-mentioned scheme uses a single perturbing distribution in order to perform the randomization over the entire data. The randomization scheme can be tailored much more effectively by using mixture models [30] in order to perform the privacy-preservation. The work in [30] shows that this approach has a number of optimality properties in terms of the quality of the perturbation.

6.3.1 Privacy-Preserving Classification with Randomization

A number of methods have been proposed for privacy-preserving classification with randomization. In [3], a method has been discussed for decision tree classification with the use of the aggregate distributions reconstructed from the randomized distribution. The key idea is to construct the distributions separately for the different classes. Then, the splitting condition for the decision tree uses the relative presence of the different classes which is derived from the aggregate distributions. It has been shown in [3] that such an approach can be used in order to design very effective classifiers.

Since the probabilistic behavior is encoded in aggregate data distributions, it can be used to construct a naive Bayes classifier. In such a classifier [29], the approach of randomized response with partial hiding is used in order to perform the classification. It has been shown in [29] that this approach is effective both empirically and analytically.

6.3.2 Privacy-Preserving OLAP

In [4], a randomization algorithm for distributed privacy-preserving OLAP is discussed. In this approach, each client independently perturbs their data before sending it to a centralized server. The technique uses *local perturbation techniques* in which the perturbation added to an element depends upon its initial value. A variety of reconstruction techniques are discussed in order to respond to different kinds of queries. The key in such queries is to develop effective algorithms for estimating counts of different subcubes in the data. Such queries are typical in most OLAP applications. The approach has been shown in [4] to satisfy a number of privacy-breach guarantees.

The method in [4] uses an interesting technique called *retention replacement perturbation*. In retention replacement perturbation, each element from column j is retained with probability p_j, or replaced with an element from the selected pdf. It has been shown in [4] that approximate probabilistic reconstructability is possible when a least a certain number of rows are present in the data. Methods have also been devised in [4] to express the estimated query results on the perturbed table as a function of the query results on the perturbed table. Methods are devised in [4] to reconstruct the original distributed, single column aggregates, and multiple column aggregates.

Techniques have also been devised on [4] for perturbation of categorical data sets. In this case, the retention-replacement approach needs to be modified appropriately. In this case, the replacement approach is to use a random element to replace an element which is not retained.

6.3.3 Collaborative Filtering

A variety of collaborative filtering techniques have been discussed in [22, 23]. The collaborative filtering problem is used in the context of electronic commerce when users choose to leave quantitative feedback (or ratings) about the products which they may like. In the collaborative filtering problem, we wish to make predictions of ratings of products for a particular user with the use of ratings of users with similar profiles. Such ratings are useful for making recommendations that the user may like. In [23], a correlation based collaborative filtering technique with randomization was proposed. In [22], an SVD based collaborative filtering method was proposed using randomized perturbation techniques. Since the collaborative filtering technique is inherently one in which ratings from multiple users are incorporated, we use a client-server mechanism in order to perform the perturbation. The broad approach of SVD-based collaborative filtering technique is as follows:

- The server decides on the nature (eg. uniform or Gaussian) of the perturbing distribution along with the corresponding parameters. These parameters are transmitted to each user.
- Each user computes the mean and z-number for their ratings. The entries which are not rated are substituted with the mean for the corresponding ratings and a z-number of 0.
- Each user then adds random number to all the ratings, and sends the disguised ratings to the server.
- The server receives the ratings from the different users and uses SVD on the disguised matrix in order to make predictions.

6.4 The Privacy-Information Loss Tradeoff

The quantity used to measure privacy should indicate how closely the original value of an attribute can be estimated. The work in [3] uses a measure that defines privacy as follows: If the original value can be estimated with $c\%$ confidence to lie in the interval $[\alpha_1, \alpha_2]$, then the interval width $(\alpha_2 - \alpha_1)$ defines the amount of privacy at $c\%$ confidence level. For example, if the perturbing additive is uniformly distributed in an interval of width 2α, then α is the amount of privacy at confidence level 50% and 2α is the amount of privacy at confidence level 100%. However, this simple method of determining privacy can be subtly incomplete in some situations. This can be best explained by the following example.

EXAMPLE 6.4 *Consider an attribute X with the density function $f_X(x)$ given by:*

$$f_X(x) = \begin{cases} 0.5 & 0 \leq x \leq 1 \\ 0.5 & 4 \leq x \leq 5 \\ 0 & \text{otherwise} \end{cases}$$

Assume that the perturbing additive Y is distributed uniformly between $[-1, 1]$. Then according to the measure proposed in [3], the amount of privacy is 2 at confidence level 100%.

However, after performing the perturbation and subsequent reconstruction, the density function $f_X(x)$ will be approximately revealed. Let us assume for a moment that a large amount of data is available, so that the distribution function is revealed to a high degree of accuracy. Since the (distribution of the) perturbing additive is publicly known, the two pieces of information can *be combined to determine that if* $Z \in [-1, 2]$, *then* $X \in [0, 1]$; *whereas if* $Z \in [3, 6]$ *then* $X \in [4, 5]$.

Thus, in each case, the value of X *can be localized to an interval of length 1. This means that the actual amount of privacy offered by the perturbing additive* Y *is* at most *1 at confidence level 100%. We use the qualifier 'at most' since* X *can often be localized to an interval of length less than one. For example, if the value of* Z *happens to be* -0.5, *then the value of* X *can be localized to an even smaller interval of* $[0, 0.5]$. ∎

This example illustrates that the method suggested in [3] does not take into account the distribution of original data. In other words, the (aggregate) reconstruction of the attribute value also provides a certain level of knowledge which can be used to guess a data value to a higher level of accuracy. To accurately quantify privacy, we need a method which takes such side-information into account.

A key privacy measure [5] is based on the *differential entropy* of a random variable. The differential entropy $h(A)$ of a random variable A is defined as follows:

$$h(A) = -\int_{\Omega_A} f_A(a) \log_2 f_A(a)\, da \tag{6.6}$$

where Ω_A is the domain of A. It is well-known that $h(A)$ is a measure of uncertainty inherent in the value of A [111]. It can be easily seen that for a random variable U distributed uniformly between 0 and a, $h(U) = \log_2(a)$. For $a = 1$, $h(U) = 0$.

In [5], it was proposed that $2^{h(A)}$ is a measure of privacy inherent in the random variable A. This value is denoted by $\Pi(A)$. Thus, a random variable U distributed uniformly between 0 and a has privacy $\Pi(U) = 2^{log_2(a)} = a$. For a general random variable A, $\Pi(A)$ denote the length of the interval, over which a uniformly distributed random variable has the same uncertainty as A.

Given a random variable B, the *conditional* differential entropy of A is defined as follows:

$$h(A|B) = -\int_{\Omega_{A,B}} f_{A,B}(a, b) \log_2 f_{A|B=b}(a)\, da\, db \tag{6.7}$$

Thus, the average conditional privacy of A given B is $\Pi(A|B) = 2^{h(A|B)}$. This motivates the following metric $\mathcal{P}(A|B)$ for the conditional privacy loss of A, given B:

$$\mathcal{P}(A|B) = 1 - \Pi(A|B)/\Pi(A) = 1 - 2^{h(A|B)}/2^{h(A)} = 1 - 2^{-I(A;B)}.$$

where $I(A; B) = h(A) - h(A|B) = h(B) - h(B|A)$. $I(A; B)$ is also known as the *mutual information* between the random variables A and B. Clearly, $\mathcal{P}(A|B)$ is the fraction of privacy of A which is lost by revealing B.

As an illustration, let us reconsider Example 6.4 given above. In this case, the differential entropy of X is given by:

$$h(X) = -\int_{\Omega_X} f_X(x) \log_2 f_X(x)\, dx =$$

$$= -\int_0^1 0.5 \log_2 0.5\, dx - \int_4^5 0.5 \log_2 0.5\, dx = 1$$

Thus the privacy of X, $\Pi(X) = 2^1 = 2$. In other words, X has as much privacy as a random variable distributed uniformly in an interval of length 2. The density function of the perturbed value Z is given by $f_Z(z) = \int_{-\infty}^{\infty} f_X(\nu) f_Y(z - \nu)\, d\nu$.

Using $f_Z(z)$, we can compute the differential entropy $h(Z)$ of Z. It turns out that $h(Z) = 9/4$. Therefore, we have:

$$I(X; Z) = h(Z) - h(Z|X) = 9/4 - h(Y) = 9/4 - 1 = 5/4$$

Here, the second equality $h(Z|X) = h(Y)$ follows from the fact that X and Y are independent and $Z = X + Y$. Thus, the fraction of privacy loss in this case is $\mathcal{P}(X|Z) = 1 - 2^{-5/4} = 0.5796$. Therefore, after revealing Z, X has privacy $\Pi(X|Z) = \Pi(X) \times (1 - \mathcal{P}(X|Z)) = 2 \times (1.0 - 0.5796) = 0.8408$. This value is less than 1, since X can be localized to an interval of length less than one for many values of Z. Given the perturbed values $z_1, z_2, \ldots, z_N$, it is (in general) not possible to reconstruct the original density function $f_X(x)$ with an arbitrary precision. The greater the variance of the perturbation, the lower the precision in estimating $f_X(x)$. This constitutes the classic tradeoff between privacy and information loss. We refer the lack of precision in estimating $f_X(x)$ as *information loss*. Clearly, the lack of precision is estimating the true distribution will degrade the accuracy of the application that such a distribution is used for. The work in [3] uses an application dependent approach to measure the information loss. For example, for a classification problem, the inaccuracy in distribution reconstruction is measured by examining the effects on the mis-classification rate. The work in [5] uses a more direct approach to measure the information loss.

Let $\hat{f}_X(x)$ denote the density function of X as estimated by a reconstruction algorithm.

We propose the metric $\mathcal{I}(f_X, \hat{f}_X)$ to measure the information loss incurred by a reconstruction algorithm in estimating $f_X(x)$:

$$\mathcal{I}(f_X, \hat{f}_X) = \frac{1}{2} E\left[\int_{\Omega_X} \left| f_X(x) - \hat{f}_X(x) \right| dx\right] \tag{6.8}$$

Thus the proposed metric equals half the expected value of L_1-norm between the original distribution $f_X(x)$ and its estimate $\hat{f}_X(x)$. Note that information

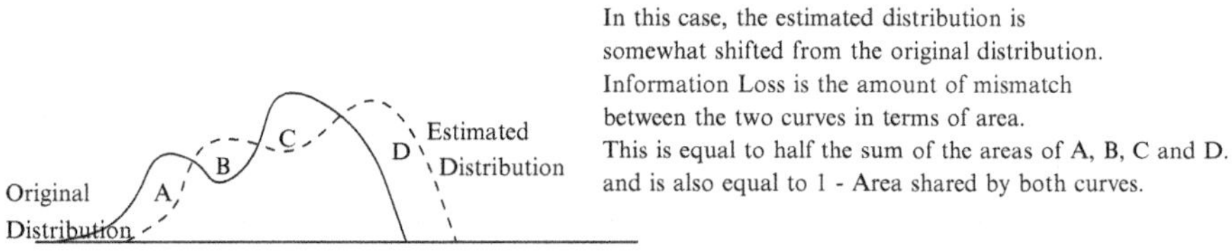
Original
Distribution
A
B
C
D
Estimated
Distribution
In this case, the estimated distribution is
somewhat shifted from the original distribution.
Information Loss is the amount of mismatch
between the two curves in terms of area.
This is equal to half the sum of the areas of A, B, C and D.
and is also equal to 1 - Area shared by both curves.

Figure 6.1. Illustration of the Information Loss Metric

loss $\mathcal{I}(f_X, \hat{f}_X)$ lies between 0 and 1; $\mathcal{I}(f_X, \hat{f}_X) = 1$ implies perfect reconstruction of $f_X(x)$ and $\mathcal{I}(f_X, \hat{f}_X) = 0$ implies that there is no overlap between $f_X(x)$ and its estimate $\hat{f}_X(x)$ (see Figure 6.1). The proposed metric is *universal* in the sense that it can be applied to any reconstruction algorithm since it depends only on the original density $f_X(x)$, and its estimate $\hat{f}_X(x)$. We advocate the use of a universal metric since it is independent of the particular data mining task at hand, and therefore facilitates absolute comparisons between disparate reconstruction algorithms.

6.5 Vulnerabilities of the Randomization Method

In the earlier section on privacy quantification, we illustrated an example in which the reconstructed distribution on the data can be used in order to reduce the privacy of the underlying data record. In general, a systematic approach can be used to do this in multi-dimensional data sets with the use of spectral filtering or PCA based techniques [11, 14]. The broad idea in techniques such as PCA [11] is that the correlation structure in the original data can be estimated fairly accurately (in larger data sets) even after noise addition. This is because the noise is added to each dimension independently, and it does not affect the expected covariance between different pairs of attributes. Only the variance of the attributes is affected, and the change in variance can be estimated accurately from the public information about the perturbing distribution. To understand this point, consider the case when the noise variable Y_1 is added to the first column X_1, and the noise variable Y_2 is added to the second column X_2. Then, we have:

$$covariance((X_1 + Y_1) \cdot (X_2 + Y_2)) = covariance(X_1 \cdot X_2)$$
$$variance((X_1 + Y_1)) = variance(X_1) + variance(Y_1)$$

Both results can be derived by expanding the expressions and using the fact that the covariance between either of $\{X_1, X_2\}$ with either of $\{Y_1, Y_2\}$ is zero and that $covariance(Y_1, Y_2) = 0$. This is because it is assumed that the noise is added independently to each dimension. Therefore, the covariance of Y_1 and Y_2 with each other or the original data columns is zero. Furthermore, the variances of Y_1 and Y_2 are known, since the corresponding distributions are publicly known. This means that the covariance matrix of the perturbed data can be used to derive the covariance matrix of the original data by simply modifying the diagonal entries. Once the covariance matrix of the original data has been estimated, one can then try to remove the noise in the data in such a way that it fits the aggregate correlation structure of the data. For example, the data is expected to be distributed along the eigenvectors of this covariance matrix, so that the variance along these eigenvectors are given by the corresponding eigenvalues. Since real data usually shows considerable skew in the eigenvalue structure, it is often the case that the entire data set of a few hundred dimensions can be captured on a plane containing less than 20 to 30 eigenvectors. In such cases, it is apparent that the data points which deviate significantly from this much lower dimensional plane need to be projected back onto it in order to derive the original data. It has been shown in [11] that such an approach can reconstruct the data quite accurately. Furthermore, we note that the accuracy of this kind of approach increases with the size of the data set, and the relationship of the intrinsic dimensionality to the full dimensionality of the data set. A related method in [14] uses spectral filtering in order to reconstruct the data accurately. It has been shown that such techniques can reduce the privacy of the perturbation process significantly since the noise removal results in values which are fairly close to their original values [11, 14]. The approach is particularly effective in cases where the data is embedded in a much lower intrinsic dimensionality as compared to its true dimensionality. It has been shown in [11] that the addition of noise along the eigenvectors of the data is safer from the point of view of privacy-preservation. This is because the discrepancy between the behavior of individual randomized points with the correlation structure of the data may no longer be used for reconstruction. Some other discussions on limiting breaches of privacy in the randomization method may be found in [7].

A second kind of adversarial attack is with the use of public information [1]. While the PCA-approach is good for *value-reconstruction*, it does not say much about *identification of the subject* of a record. Both value-reconstruction and subject-identification are required in adversarial attacks. For this purpose, it is possible to use public data in order to try to determine the identity of the subject. Consider a record $X = (x_1 \ldots x_d)$, which is perturbed to $Z = (z_1 \ldots z_d)$. Then, since the distribution of the perturbations is known, we can try to use a maximum likelihood fit of the *potential perturbation* of Z to a public record. Consider the publicly public record $W = (w_1 \ldots w_d)$. Then, the *potential perturbation* of Z with respect to W is given by $(Z - W) = (z_1 - w_1 \ldots z_d - w_d)$. Each of these values $(z_i - w_i)$ should fit the distribution $f_Y(y)$. The corresponding log-likelihood fit is given by $-\sum_{i=1}^{d} \log(f_y(z_i - w_i))$. The higher the log-likelihood fit, the greater the probability that the record W corresponds to X. If it is known that the public data set always includes X, then the maximum likelihood fit can provide a high degree of certainty in identifying the correct record, especially in cases where d is large. Another result in [10] suggests that the use of different perturbing distributions can have significant effects on the privacy of the underlying data. For example, the use of uniform perturbations is experimentally shown to be more effective in the low dimensional case. However, for the high dimensional case, gaussian perturbations are more effective. The work in [10] characterizes the amount of perturbation required for a particular dimensionality with each kind of perturbing distribution. For the case of gaussian distributions, the standard deviation of the perturbation needs to increases with the square-root of the implicit dimensionality, and for the case of uniform distributions, the standard deviation of the perturbation increases at least linearly with the implicit dimensionality. In either case, both kinds of perturbations tend to become ineffective with increasing dimensionality.

6.6 Randomization of Time Series Data Streams

The randomization approach is particularly well suited to privacy-preserving data mining of streams, since the noise added to a given record is independent of the rest of the data. However, streams provide a particularly vulnerable target for adversarial attacks with the use of PCA based techniques [11] because of the large volume of the data available for analysis. In addition, there are typically auto-correlations among the different components of a series. Such auto-correlations can also be used for reconstruction purposes. In [28], an interesting technique for randomization has been proposed which uses the correlations and auto-correlations in different time series while deciding the noise to be added to any particular value. The key idea for the case of correlated noise is to use a similar idea as in [11] in order to use principal component analysis to determine the directions in which the second order correlations are zero. These principal components are the eigenvectors of the covariance matrix for the data. Then, the noise is added along these principal components (or eigenvectors) rather than the original space. This ensures that it is extremely difficult to reconstruct the data using correlation analysis. This approach is effective for the case of correlations across multiple streams, but not auto-correlations within a single stream. In the case of dynamic auto-correlations, we are dealing with the case when there are *correlations within a single stream at different local time instants*. Such correlations can also be removed by treating a window of the stream at one time, and performing the principal components analysis on all the components of the window. Thus, we are using essentially the same idea, except that we are using multiple time instants of the sane stream to construct the co-variance matrix. The ideas can in fact be combined when there are both correlations and auto-correlations by using multiple time-instants from all streams, in order to create one covariance matrix. This will also capture correlations between different streams at slightly displaced time instants. Such situations are are referred to as *lag correlations*, and are quite common in data streams when slight changes in one stream precede changes in another because of the same cause.

In many cases, the directions of correlations may change over time. If a static approach is used for randomization, then the changes in the correlation structure will result in a risk of the data becoming exposed over time, when the principal components have changed sufficiently. Therefore, the technique in [28] is designed to dynamically adjust the directions of correlation as more and more points from the data stream are received. It has been shown in [28] that such an approach is more robust since the noise correlates with the stream behavior, and it is more difficult to create effective adversarial attacks with the use of correlation analysis techniques.

6.7 Multiplicative Noise for Randomization

The most common method of randomization is that of additive perturbations. However, multiplicative perturbations can also be used to good effect for privacy-preserving data mining. Many of these techniques derive their roots in the work of [13] which shows how to use multi-dimensional projections in order to reduce the dimensionality of the data. This technique preserves the inter-record distances approximately, and therefore the transformed records can be used in conjunction with a variety of distance-intensive data mining applications. In particular, the approach is discussed in detail in [20, 21], in which it is shown how to use the method for privacy-preserving clustering. The technique can also be applied to the problem of classification as discussed in [28]. We note that both clustering and classification are locality specific problems, and are therefore particularly well suited to the multiplicative perturbation technique. One key difference between the use of additive and multiplicative perturbations is that in the former case, we can reconstruct only aggregate distributions, whereas in the latter case more record-specific information (eg. distances) are preserved. Therefore, the latter technique is often more friendly to different kinds of data mining techniques.

Multiplicative perturbations can also be used for distributed privacy-preserving data mining. Details can be found in [17]. In [17], a number of key assumptions have also been discussed, which ensure that privacy is preserved. These assumptions discuss the level of privacy when the attacker knows partial characateristics about the algorithm used to perform the transformation, or other statistics associated with the transformation. The effects of using special kinds of data (eg. boolean data) are also discussed.

A number of techniques for multiplicative perturbation in the context of masking census data may be found in [15]. A variation on this theme may be implemented with the use of distance preserving fourier transforms, which work effectively for a variety of cases [19].

6.7.1 Vulnerabilities of Multiplicative Randomization

As in the case of additive perturbations, multiplicative perturbations are not entirely safe from adversarial attacks. In general, if the attacker has no prior knowledge of the data, then it is relatively difficult to attack the privacy of the transformation. However, with some prior knowledge, two kinds of attacks are possible [18]:

- **Known Input-Output Attack:** In this case, the attacker knows some linearly independent collection of records, and their corresponding perturbed version. In such cases, linear algebra techniques can be used to reverse-engineer the nature of the privacy preserving transformation. The number of records required depends upon the dimensionality of the data and the available records. The probability of a privacy breach with a given sample size is characterized in [18].

- **Known Sample Attack:** In this case, the attacker has a collection of independent data samples from the same distribution from which the original data was drawn. In such cases, principal component analysis techniques can be used in order to reconstruct the behavior of the original data. Then, one can try to determine how the current random projection of the data relates to this principal component analysis. This can provide an approximate idea of the corresponding geometric transformation.

One observation is that both the above mentioned techniques require much more samples (or background knowledge) to work effectively in the high dimensional case. Thus, random projection techniques should generally be used for the case of high dimensional data, and only a smaller number of projections should be retained in order to preserve privacy. Thus, as with the additive perturbation technique, the multiplicative technique is not completely secure from attacks. A key research direction is to use a combination of additive and multiplicative perturbation techniques in order to construct more robust privacy-preservation techniques.

6.7.2 Sketch Based Randomization

A closely related case to the use of multiplicative perturbations is the use of sketch-based randomization. In sketch based randomization [2], we use sketches in order to construct the randomization from the data set. We note that sketches are a special case of multiplicative perturbation techniques in the sense that the individual components of the multiplicative vector are drawn from $\{-1, +1\}$. Sketches are particularly useful for the case of *sparse* data such as text or binary data in which most components are zero and only a few components are non-zero. Furthermore, sketches are designed in such a way that many aggregate properties such as the dot product can be estimate very accurately from a small number of constant components. Since text and market basket data are both high-dimensional, the use of random projections is particularly effective from the point of view of adversarial attacks. In [11], it as been shown how the method of sketches can be used in order to perform effective privacy-preserving data mining of text and market basket data.

It is possible to use sketches to create a scheme which is similar to randomization in the sense that the transformation of a given record can be performed at data collection time. It is possible to control the anonymization in such a way so that the absolute variance of the randomization scheme is preserved. If desired, it is also possible to use sketches to add noise so that records cannot be distinguished easily from their k-nearest neighbors. This is a similar model to the k-anonymity model, but comes at the expense of using a trusted server for anonymization.

6.8 Conclusions and Summary

In this chapter, we discussed the randomization method for privacy-preserving data mining. We discussed a number of different algorithms for randomization, such as the Bayes method and the EM reconstruction technique. The EM-reconstruction algorithm also exhibits a number of optimality properties with respect to its convergence to the maximum likelihood estimate of the data distribution. We also discussed a number of variants of the perturbation technique such as the method of multiplicative perturbations. A number of applications of the randomization method were discussed over a variety of data mining problems.

References

[1] Aggarwal C. C.: On Randomization, Public Information and the Curse of Dimensionality. *ICDE Conference*, 2007.

[2] Aggarwal C. C., Yu P. S.: On Privacy-Preservation of Text and Sparse Binary Data with Sketches. *SIAM Conference on Data Mining*, 2007.

[3] Agrawal R., Srikant R. Privacy-Preserving Data Mining. *Proceedings of the ACM SIGMOD Conference*, 2000.

[4] Agrawal R., Srikant R., Thomas D. Privacy-Preserving OLAP. *Proceedings of the ACM SIGMOD Conference*, 2005.

[5] Agrawal D. Aggarwal C. C. On the Design and Quantification of Privacy-Preserving Data Mining Algorithms. *ACM PODS Conference*, 2002.

[6] Chen K., Liu L.: Privacy-preserving data classification with rotation perturbation. *ICDM Conference*, 2005.

[7] Evfimievski A., Gehrke J., Srikant R. Limiting Privacy Breaches in Privacy Preserving Data Mining. *ACM PODS Conference*, 2003.

[8] Evfimievski A., Srikant R., Agrawal R., Gehrke J.: Privacy-Preserving Mining of Association Rules. *ACM KDD Conference*, 2002.

[9] Fienberg S., McIntyre J.: Data Swapping: Variations on a Theme by Dalenius and Reiss. *Technical Report, National Institute of Statistical Sciences*, 2003.

[10] Gambs S., Kegl B., Aimeur E.: Privacy-Preserving Boosting. *Knowledge Discovery and Data Mining Journal*, to appear.

[11] Huang Z., Du W., Chen B.: Deriving Private Information from Randomized Data. pp. 37–48, *ACM SIGMOD Conference*, 2005.

[12] Warner S. L. Randomized Response: A survey technique for eliminating evasive answer bias. *Journal of American Statistical Association*, 60(309):63–69, March 1965.

[13] Johnson W., Lindenstrauss J.: Extensions of Lipshitz Mapping into Hilbert Space, *Contemporary Math.* vol. 26, pp. 189–206, 1984.

[14] Kargupta H., Datta S., Wang Q., Sivakumar K.: On the Privacy Preserving Properties of Random Data Perturbation Techniques. *ICDM Conference*, pp. 99–106, 2003.

[15] Kim J., Winkler W.: Multiplicative Noise for Masking Continuous Data, *Technical Report Statistics 2003-01, Statistical Research Division, US Bureau of the Census*, Washington D.C., Apr. 2003.

[16] Liew C. K., Choi U. J., Liew C. J. A data distortion by probability distribution. *ACM TODS*, 10(3):395–411, 1985.

[17] Liu K., Kargupta H., Ryan J.: Random Projection Based Multiplicative Data Perturbation for Privacy Preserving Distributed Data Mining. *IEEE Transactions on Knowledge and Data Engineering*, 18(1), 2006.

[18] Liu K., Giannella C., Kargupta H.: An Attacker's View of Distance Preserving Maps for Privacy-Preserving Data Mining. *PKDD Conference*, 2006.

[19] Mukherjee S., Chen Z., Gangopadhyay S.: A privacy-preserving technique for Euclidean distance-based mining algorithms using Fourier based transforms, *VLDB Journal*, 2006.

[20] Oliveira S. R. M., Zaane O.: Privacy Preserving Clustering by Data Transformation, *Proc. 18th Brazilian Symp. Databases*, pp. 304–318, Oct. 2003.

[21] Oliveira S. R. M., Zaiane O.: Data Perturbation by Rotation for Privacy-Preserving Clustering, *Technical Report TR04-17*, Department of Computing Science, University of Alberta, Edmonton, AB, Canada, August 2004.

[22] Polat H., Du W.: SVD-based collaborative filtering with privacy. *ACM SAC Symposium*, 2005.

[23] Polat H., Du W.: Privacy-preserving collaborative filtering with randomized perturbation techniques. *ICDM Conference*, 2003.

[24] Rizvi S., Haritsa J.: Maintaining Data Privacy in Association Rule Mining. *VLDB Conference*, 2002.

[25] Samarati P.: Protecting Respondents' Identities in Microdata Release. IEEE Trans. Knowl. Data Eng. 13(6): 1010–1027 (2001).

[26] Shannon C. E.: The Mathematical Theory of Communication, University of Illinois Press, 1949.

[27] Silverman B. W.: Density Estimation for Statistics and Data Analysis. *Chapman and Hall*, 1986.

[28] Li F., Sun J., Papadimitriou S., Mihaila G., Stanoi I.: Hiding in the Crowd: Privacy Preservation on Evolving Streams through Correlation Tracking. *ICDE Conference*, 2007.

[29] Zhang P., Tong Y., Tang S., Yang D.: Privacy-Preserving Naive Bayes Classifier. *Lecture Notes in Computer Science*, Vol 3584, 2005.

[30] Zhu Y., Liu L. Optimal Randomization for Privacy- Preserving Data Mining. *ACM KDD Conference*, 2004.

Chapter 7

A Survey of Multiplicative Perturbation for Privacy-Preserving Data Mining

Keke Chen
College of Computing
Georgia Institute of Technology
kekechen@cc.gatech.edu

Ling Liu
College of Computing
Georgia Institute of Technology
lingliu@cc.gatech.edu

Abstract The major challenge of data perturbation is to achieve the desired balance between the level of privacy guarantee and the level of data utility. Data privacy and data utility are commonly considered as a pair of conflicting requirements in privacy-preserving data mining systems and applications. Multiplicative perturbation algorithms aim at improving data privacy while maintaining the desired level of data utility by selectively preserving the mining task and model specific information during the data perturbation process. By preserving the task and model specific information, a set of "transformation-invariant data mining models" can be applied to the perturbed data directly, achieving the required model accuracy. Often a multiplicative perturbation algorithm may find multiple data transformations that preserve the required data utility. Thus the next major challenge is to find a good transformation that provides a satisfactory level of privacy guarantee. In this chapter, we review three representative multiplicative perturbation methods: rotation perturbation, projection perturbation, and geometric perturbation, and discuss the technical issues and research challenges. We first describe the mining task and model specific information for a class of data mining models, and the transformations that can (approximately) preserve the information. Then we discuss the design of appropriate privacy evaluation models for multiplicative perturbations, and give an overview of how we use the privacy evaluation model to measure the level of privacy guarantee in the context of different types of attacks.

Keywords: Multiplicative perturbation, random projection, sketches.

7.1 Introduction

Data perturbation refers to a data transformation process typically performed by the data owners before publishing their data. The goal of performing such data transformation is two-fold. On one hand, the data owners want to change the data in a certain way in order to disguise the sensitive information contained in the published datasets, and on the other hand, the data owners want the transformation to best preserve those domain-specific data properties that are critical for building meaningful data mining models, thus maintaining mining task specific data utility of the published datasets.

Data perturbation techniques are one of the most popular models for privacy preserving data mining. It is especially useful for applications where data owners want to participate in cooperative mining but at the same time want to prevent the leakage of privacy-sensitive information in their published datasets. Typical examples include publishing micro data for research purpose or outsourcing the data to the third party data mining service providers. Several perturbation techniques have been proposed to date [4–1, 8, 3, 13, 14, 26, 35], among which the most popular one is the randomization approach that focuses on single-dimensional perturbation and assumes independency between data columns [4, 13]. Only recently, the data management community has shown some development on multi-dimensional data perturbation techniques, such as the condensation approach using k-nearest neighbor (kNN) method [1], the multi-dimensional K-anonymization using kd-tree [24], and the multiplicative data perturbation techniques [31, 8, 28, 9]. Compared to single-column-based data perturbation techniques that assume data columns to be independent and focus on developing single-dimensional perturbation techniques, multi-dimensional data perturbation aims at perturbing the data while preserving the *multi-dimensional information* with respect to inter-column dependency and distribution.

In this chapter, we will discuss multiplicative data perturbations. This category includes three types of particular perturbation techniques: Rotation Perturbation, Projection Perturbation, and Geometric Perturbation. Comparing to other multi-dimensional data perturbation methods, these perturbations exhibit unique properties for privacy preserving data classification and data clustering. They all preserve (or approximately preserve) distance or inner product, which are important to many classification and clustering models. As a result, the classification and clustering mining models based on the perturbed data through multiplicative data perturbation show similar accuracy to those based on the original data. The main challenge for multiplicative data perturbations thus is how to maximize the desired data privacy. In contrast, many other data perturbation techniques focus on seeking for the better trade-off between the level of data utility and accuracy preserved and the level of data privacy guaranteed.

7.1.1 Data Privacy vs. Data Utility

Perturbation techniques are often evaluated with two basic metrics: level of privacy guarantee and level of model-specific data utility preserved, which is often measured by the loss of accuracy for data classification and data clustering. An ultimate goal for all data perturbation algorithms is to optimize the data transformation process by maximizing both data privacy and data utility achieved. However, the two metrics are typically representing two conflicting goals in many existing perturbation techniques [4, 3, 12–1].

Data privacy is commonly measured by the difficulty level in estimating the original data from the perturbed data. Given a data perturbation technique, the higher level of difficulty in which the original values can be estimated from the perturbed data, the higher level of data privacy this technique supports. In [4], the variance of the added random noise is used as the level of difficulty for estimating the original values as traditionally used in statistical data distortion [23]. However, recent research [12, 3] reveals that variance of the noise is not an effective indicator for random noise addition. In addition, [22] shows that the level of data privacy guaranteed is also bounded to the types of special attacks that can reconstruct the original data from the perturbed data and noise distribution. k-Anonymization is another popular way of measuring the level of privacy, originally proposed for relational databases [34], by enabling the effective estimation of the original data record to a k-record group, assuming that each record in the k-record group is equally protected. However, recent study [29] shows that the privacy evaluation of k-Anonymized records is far more complicated than this simple k-anonymization assumption.

Data utility typically refers to the amount of mining-task/model specific critical information preserved about the dataset after perturbation. Different data mining tasks, such as classification mining task vs. association rule mining, or different models for the same task, such as decision tree model vs. k-Nearest-Neighbor (kNN) classifier for classification, typically utilize different sets of data properties about the dataset. For example, the task of building decision trees primarily concerns the column distribution. Hence, the quality of preserving column distribution should be the key data utility to be maintained in perturbation techniques for decision tree model, as shown in the randomization approach [4]. In comparison, the kNN model relies heavily on the distance relationship, which is quite different from the column distribution. Furthermore, such task/model-specific information is often multidimensional. Many classification models typically concern the multidimensional information rather than single column distribution. Multi-dimensional perturbation techniques with the focus on preserving the model-specific multidimensional information will be more effective for these models.

It is also interesting to note that the data privacy metric and the data utility metric are often contradictory rather than complimentary in many existing data perturbation techniques [4, 3, 12–1]. Typically data perturbation algorithms that aim at maximizing the level of data privacy often have to bear with higher information loss. The intrinsic correlation between the data privacy and the data utility raises a number of important issues regarding how to find a right balance between the two measures.

In summary, we identify three important design principles for multiplicative data perturbations. First, preserving the mining task and model-specific data properties is critical for providing better quality guarantee on both privacy and model accuracy. Second, it is beneficial if data perturbation can effectively preserve the task/model-specific data utility information, and avoid the need for developing special mining algorithms that can use the perturbed data as random noise addition requires. Third and most importantly, if one can develop a data perturbation technique that does not induce any lost of mining-task/model specific data utility, this will enable us to focus on optimizing perturbation algorithms by maximizing the level of data privacy against attacks, which ultimately leads to better overall quality of both data privacy and data utility.

7.1.2 Outline

In the remaining of the chapter we will first give the definition of multiplicative perturbation in Section 7.2. Specifically, we categorize multiplicative perturbations into three categories: rotation perturbation, projection perturbation, and geometric perturbation. Rotation perturbation is often criticized not resilient to attacks, while geometric perturbation is a direct enhancement to rotation perturbation by adding more components, such as translation perturbation and noise addition, to the original rotation perturbation. Both rotation perturbation and geometric Perturbation keep the dimensionality of dataset unchanged, while projection perturbation reduces the dimensionality, and thus incurs more errors in distance or inner product calculation.

One of the unique features that distinguish multiplicative perturbations from other perturbations is that it provides high guarantee on data utility in terms of data classification and clustering. Since many data mining models utilize distance or inner product, as long as such information is preserved, models trained on perturbed data will have similar accuracy to those trained on the original data. In Section 7.3, we define transformation-invariant classifiers and clustering models, the representative models to which multiplicative perturbations are applied.

Evaluation of privacy guarantee for perturbations is an important component in the analysis of multiplicative perturbation. In Section 7.4, we review a set of privacy metrics specifically designed for multiplicative perturbations. We argue that in multidimensional perturbation, the values of multiple columns should be perturbed together and the evaluation metrics should be unified for all columns. We also describe a general framework for privacy evaluation of multiplicative data perturbation by incorporating attack analysis.

We argue that attack analysis is a necessary step in order to accurately evaluate the privacy guarantee of any particular perturbation. In Section 7.5, we review a selection of known attacks to multiplicative perturbations based on different levels of attack's knowledge about the original dataset. By incorporating attack analysis under the general framework of privacy evaluation, a randomized perturbation optimization is developed and described in Section 7.5.5.

7.2 Definition of Multiplicative Perturbation

We will first describe the notations used in this chapter, and then describe three categories of multiplicative perturbations and their basic characteristics.

7.2.1 Notations

In privacy-preserving data mining, either a portion of or the entire data set will be perturbed and then exported. For example, in classification, the training data is exported and the testing data might be exported, too, while in clustering, the entire data for clustering is exported. Suppose that X is the exported dataset consisting of N data rows (records) and d columns (attributes, or dimensions). For presentation convenience, we use $X_{d\times N}$, $X = [\mathbf{x}_1 \ldots \mathbf{x}_N]$, to denote the dataset, where a column $\mathbf{x}_i$ ($1 \leq i \leq N$) is a data tuple, representing a vector in the real space $\mathbb{R}^d$. In classification, each of such data tuples $\mathbf{x}_i$ also belongs to a predefined class, which is indicated by the class label attribute y_i. The class label can be nominal (or continuous for regression), and is public, i.e., privacy-insensitive.

For clear presentation, we can also consider X is a sample dataset from the d-dimension random vector $\mathbf{X} = [\mathbf{X}_1, \mathbf{X}_2, \ldots, \mathbf{X}_d]^T$. As a convention, we use bold lower case to represent vectors, bold upper case to represent random variables, and upper case to represent matrices or datasets.

7.2.2 Rotation Perturbation

This category does not cover traditional "rotations" only, but literally, it includes all orthonormal perturbations. A rotation perturbation is defined as following $G(X)$:

$$G(X) = RX$$

The matrix $R_{d \times d}$ is an orthonormal matrix [32], which has following properties. Let R^T represent the transpose of R, r_{ij} represent the (i, j) element of R, and I be the identity matrix. The rows and columns of R are orthonormal, i.e., for any column j, $\sum_{i=1}^{d} r_{ij}^2 = 1$, and for any two columns j and k, $j \neq k$, $\sum_{i=1}^{d} r_{ij} r_{ik} = 0$. A similar property is held for rows. This definition infers that

$$R^T R = R R^T = I$$

It also implies that by changing the order of the rows or columns of an orthogonal matrix, the resulting matrix is still orthogonal. A random orthonormal matrix can be efficiently generated following the Haar distribution [33].

A key feature of rotation transformation is that it preserve the Euclidean distance of multi-dimensional points during the transformation. Let $\mathbf{x}^T$ represent the transpose of vector $\mathbf{x}$, and $||\mathbf{x}|| = \mathbf{x}^T\mathbf{x}$ represent the length of a vector $\mathbf{x}$. By the definition of rotation matrix, we have

$$||R\mathbf{x}|| = ||\mathbf{x}||$$

Similarly, inner product is also invariant to rotation. Let $\langle \mathbf{x}, \mathbf{y} \rangle = \mathbf{x}^T\mathbf{y}$ represent the inner product of $\mathbf{x}$ and $\mathbf{y}$. We have

$$\langle R\mathbf{x}, R\mathbf{y} \rangle = \mathbf{x}^T R^T R\mathbf{y} = \langle \mathbf{x}, \mathbf{y} \rangle$$

In general, rotation also preserves the geometric shapes such as hyperplane and hyper curved surface in the multidimensional space [7]. We observed that since many classifiers look for geometric decision boundary, such as hyperplane and hyper surface, rotation transformation will preserve the most critical information for many classification models.

There are two ways to apply rotation perturbation. We can either apply it to the whole dataset X [8], or group columns to pairs and apply different rotation perturbations to different pairs of columns [31].

7.2.3 Projection Perturbation

Projection perturbation refers to the technique of projecting a set of data points from a high-dimensional space to a randomly chosen lower-dimensional subspace. Let $P_{k \times d}$ be a projection matrix.

$$G(X) = PX$$

Why can it also be used for perturbation? The rationale is based on the Johnson-Lindenstrauss Lemma [21].

THEOREM 1 *For any* $0 < \epsilon < 1$ *and any integer* n, *let* k *be a positive integer such that* $k \geq \frac{4 \ln n}{\epsilon^2/2 - \epsilon^3/3}$. *Then, for any set* $\mathbf{S}$ *of* n *data points in* d *dimensional space* $\mathbb{R}^d$, *there is a map* $f: \mathbb{R}^d \rightarrow \mathbb{R}^k$ *such that, for all* $\mathbf{x} \in \mathbf{S}$,

$$(1-\epsilon)\|\mathbf{x}-\mathbf{x}\|^2 \leq \|f(\mathbf{x})-f(\mathbf{x})\|^2 \leq (1+\epsilon)\|\mathbf{x}-\mathbf{x}\|^2$$

where $\|\cdot\|$ *denotes the vector 2-norm.*

This lemma shows that any set of n points in d-dimensional Euclidean space could be embedded into a $O(\frac{\log n}{\epsilon^2})$ -dimensional space, such that the pair-wise distance of any two points are maintained with small error. With large n (large dataset) and small ϵ (high accuracy in distance preservation), the ideal dimensionality might be large and may not be practical for the perturbation purpose. Furthermore, although this lemma implies that we can always find one good projection that approximately preserves distances for a particular dataset, the geometric decision boundary might still be distorted and thus the model accuracy is reduced. Due to the different distributions of dataset and particular properties of data mining models, it is challenging to develop an algorithm that can find random projections that preserves model accuracy well for any given dataset.

In paper [28] a method is used to generate random projection matrix. The process can be briefly described as follows. Let P be the projection matrix. Each entry $r_{i,j}$ of P is independent and identically chosen from some distribution with mean zero and variance σ^2. A row-wise projection is defined as

$$G(X) = \frac{1}{\sqrt{k}\sigma} PX$$

Let $\mathbf{x}$ and $\mathbf{y}$ be two points in the original space, and $\mathbf{u}$ and $\mathbf{v}$ be their projections. The statistical properties of inner product under projection perturbation can be shown as follows.

$$E[\mathbf{u}^t\mathbf{v} - \mathbf{x}^t\mathbf{y}] = 0$$

and

$$Var[\mathbf{u}^t\mathbf{v} - \mathbf{x}^t\mathbf{y}] = \frac{1}{k}(\sum_i x_i^2 \sum_i y_i^2 + (\sum_i x_i y_i)^2)$$

Since $\mathbf{x}$ and $\mathbf{y}$ are not normalized by rows, but by columns in practice, with large dimensionality d and relatively small k, the variance is substantial. Similarly, the conclusion can be extended to the distance relationship. Therefore, projection perturbation does not strictly guarantee the preservation of distance/inner product as rotation or geometric perturbation does, which may significantly downgrade the model accuracy.

7.2.4 Sketch-based Approach

Sketch-based approach is primarily proposed to perturb high-dimensional sparse data [2], such as the datasets in text mining and market basket mining. A sketch of the original record $\mathbf{x} = (x_1, \ldots, x_d)$ is defined by a r dimensional vector $\mathbf{s} = (s_1, \ldots, s_r)$, $r \ll d$, where

$$s_j = \sum_{i=1}^{d} x_i r_{ij}$$

The random variable r_{ij} is drawn from {-1,+1} with a mean of 0, and is generated from a pseudo-random number generator [5], which produces 4-wise independent values for the variable r_{ij}.

Note that the sketch based approach defers from projection perturbation with the following two features. First, the number of components for each sketch, i.e., r, can vary across different records, and is carefully controlled so as to provide a uniform measure of privacy guarantee across different records. Second, for each record, r_{ij} is different – there is no fixed projection matrix across records.

The sketch based approach has a few statistical properties that enable approximate calculation of dot product of the original data records with their sketches. Let $\mathbf{s}$ and $\mathbf{t}$ with the same number of components r, be the sketches of the original records $\mathbf{x}$ and $\mathbf{y}$, respectively. The expected dot product $\mathbf{x}$ and $\mathbf{y}$ is given by the following.

$$E[\langle \mathbf{x}, \mathbf{y} \rangle] = \langle \mathbf{s}, \mathbf{t} \rangle / r$$

and the variance of the above estimation is determined by the few non-zeros entries in the sparse original vectors

$$Var(\langle \mathbf{s}, \mathbf{t} \rangle / r) = (\sum_{i=1}^{d} \sum_{l=1}^{d} x_i^2 y_l^2 - (\sum_{i=1}^{d} x_i y_i)^2)/r \tag{7.1}$$

On the other side, the original value x_k in the vector $\mathbf{x}$ can also be estimated by privacy attackers, the precision of which is determined by its variance $(\sum_{i=1}^{d} x_i^2 - x_k^2)/r, k = 1 \ldots d$. The larger the variance is, the better the original value is protected. Therefore, by decreasing r the level of privacy guarantee is possibly increased. However, the precision of dot-product estimation (Eq. 7.1) is decreased. This typical tradeoff has to be carefully controlled in practice [2].

7.2.5 Geometric Perturbation

Geometric perturbation is an enhancement to rotation perturbation by incorporating additional components such as random translation perturbation and noise addition to the basic form of multiplicative perturbation $Y = R \times X$. We show that by adding random translation perturbation and noise addition, Geometric perturbation exhibits more robustness in countering attacks than simple rotation based perturbation [9]. Let $\mathbf{t}_{d\times 1}$ represent a random vector. We define a *translation matrix* as follows.

DEFINITION 1 *Ψ is a translation matrix if $\Psi = [\mathbf{t}, \mathbf{t}, \ldots, \mathbf{t}]_{d\times n}$, i.e., $\Psi_{d\times n} = \mathbf{t}_{d\times 1}\mathbf{1}^T_{N\times 1}$.*

where $\mathbf{1}_{N\times 1}$ is the vector of N '1's. Let $\Delta_{d\times N}$ be a random noise matrix, where each element is Independently and Identically Distributed (iid) variable ε_{ij}, e.g., a Gaussian noise $N(0, \sigma^2)$.

The definition of geometric perturbation is given by a function $G(X)$,

$$G(X) = RX + \Psi + \Delta$$

Clearly, translation perturbation does not change distance, as for any pair of points $\mathbf{x}$ and $\mathbf{y}$, $\|(\mathbf{x}+\mathbf{t}) - (\mathbf{y}+\mathbf{t})\| = \|\mathbf{x}-\mathbf{y}\|$. Comparing with rotation perturbation, it protects the rotation center from attacks and adds additional difficulty to ICA-based attacks. However, translation perturbation does not preserve inner product.

In [9], it shows that by adding an appropriate level of noise Δ, one can effectively prevent knowledgeable attackers from distance-based data reconstruction, since noise addition perturbs distances, which protects perturbation from distance-inference attacks. For example, the experiments in [9] shows that a Gaussian noise $N(0, \sigma^2)$ is effective to counter the distance-inference attacks. Although noise addition prevents from fully preserving distance information, a low intensity noise will not change class boundary or cluster membership much.

In addition, the noise component is optional – if the data owner makes sure that the original data records are secure and no people except the data owner knows any record in the original dataset, the noise component can be removed from geometric perturbation.

7.3 Transformation Invariant Data Mining Models

By using multiplicative perturbation algorithms, we can mine the the perturbed data directly with a set of existing "transformation-invariant data mining models", instead of developing new data mining algorithms to mine the perturbed data [4]. In this section, we will define the concept of transformation-invariant mining models with the example of "transformation-invariant classifiers", and then we extend our discussion to the transformation-invariant models in data classification and data clustering.

7.3.1 Definition of Transformation Invariant Models

Generally speaking, a transformation invariant model, if trained or mined on the transformed data, performs as good as the model based on the original data. We take the classification problem as an example. A classification problem is also a function approximation problem – classifiers are the functions learned from the training data [16]. In the following discussion, we use functions to represent classifiers. Let $\hat{f}_X$ represent a classifier $\hat{f}$ trained with dataset X and $\hat{f}_X(Y)$ be the classification result on the dataset Y. Let $T(X)$ be any transformation function, which transforms the dataset X to another dataset X_T. We use $Err(\hat{f}_X(Y))$ to denote the error rate of classifier $\hat{f}_X$ on testing data Y and let ε be some small real number, $|\varepsilon| < 1$.

DEFINITION 2 *A classifier $\hat{f}$ is invariant to a transformation T if and only if $Err(\hat{f}_X(Y)) = Err(\hat{f}_{T(X)}(T(Y))) + \varepsilon$ for any training dataset X and testing dataset Y.*

With the strict condition $\hat{f}_X(Y) \equiv \hat{f}_{T(X)}(T(Y))$, we get the Proposition 2.

PROPOSITION 2 *In particular, if $\hat{f}_X(Y) \equiv \hat{f}_{T(X)}(T(Y))$ is satisfied for any training dataset X and testing dataset Y, the classifier is invariant to the transformation $T(X)$.*

For instance, if a classifier $\hat{f}$ is invariant to rotation transformation, we call it *rotation-invariant classifier*. Similar definition applies to *translation-invariant classifier*.

In subsequent sections, we will list some examples of transformation invariant models for classification and clustering. Some detailed proofs can be found in [7].

7.3.2 Transformation-Invariant Classification Models

kNN Classifiers and Kernel Methods

A k-Nearest-Neighbor (kNN) classifier determines the class label of a point by looking at the labels of its k nearest neighbors in the training dataset and classifies the point to the class that most of its neighbors belong to. Since the distances between any points are not changed with rotation and translation transformation, the k nearest neighbors are not changed and thus the classification result is not changed either.

Since kNN classifier is a special case of kernel methods, we can also extend our conclusion to kernel methods. Here, we refer kernel methods to the traditional local methods [16]. In general, since the kernels are dependent on the local points, the locality of which is evaluated by distance, transformations that preserve distance will make kernel methods invariant.

Support Vector Machines

Support Vector Machine (SVM) classifier also utilizes kernel functions in training and classification. However, it has an explicit training procedure, which differentiates itself from the traditional kernel methods we just discussed. We can use a two-step procedure to prove that a SVM classifier is invariant to a transformation. 1) Training with the transformed dataset generates the same set of model parameters; 2) the classification function with the model parameters is also invariant to the transformation. The detailed proof will involve the quadratic optimization procedure for SVM. We have demonstrated that SVM classifiers with typical kernels are invariant to rotation transformation [7]. It turns out that if a transformation makes the kernel invariant, then the SVM classifier is also invariant to the transformation.

There are the three popular choices for the kernels discussed in the SVM literature [10, 16].

$$\begin{aligned}
\text{d-th degree polynomial:} \quad & K(\mathbf{x}, \mathbf{x}') = (1 + \langle \mathbf{x}, \mathbf{x}' \rangle)^d, \\
\text{radial basis:} \quad & K(\mathbf{x}, \mathbf{x}') = \exp(-\|\mathbf{x} - \mathbf{x}'\|/c), \\
\text{neural network:} \quad & K(\mathbf{x}, \mathbf{x}') = \tanh(\kappa_1 \langle \mathbf{x}, \mathbf{x}' \rangle + \kappa_2)
\end{aligned}$$

Apparently, all of the three are invariant to rotation transformation. Since translation does not preserve inner product, it is not straightforward to prove that SVMs with polynomial and neural network kernels are invariant to translation perturbation. However, experiments [9] showed that these classifiers are also invariant to translation perturbation.

Linear Classifiers

Linear classification models are popular methods due to their simplicity. In linear classification models, the classification boundary is modeled as a hyperplane, which is clearly a geometric concept. It is easy to understand that distance-preserving transformations, such as rotation and translation, will still make the classes separated if they are originally separated. There is also a detailed proof showing that a typical linear classifier, perceptron, is invariant to rotation transformation [7].

7.3.3 Transformation-Invariant Clustering Models

Most clustering models are based on Euclidean distance such as the popular k-means algorithm [16]. Many are focused on the density property, which is derived from Euclidean distance, such as DBSCAN [11], DENCLUE [17] and OPTICS [6]. All of these clustering models are invariant to Euclidean-distance-preserving transformations, such as rotation and translation.

There are other clustering models, which employ different distance metrics [19], such as linkage based clustering and cosine-distance based clustering. As long as we can find a transformation preserving the particular distance metric, the corresponding clustering model will be invariant to this transformation.

7.4 Privacy Evaluation for Multiplicative Perturbation

The goal of data perturbation is twofold: preserving the accuracy of specific data mining models (data utility), and preserving the privacy of original data (data privacy). The discussion about transformation-invariant data mining models has shown that multiplicative perturbations can theoretically guarantee zero-loss of accuracy for a number of data mining models. The challenge is to find one that maximizes the privacy guarantee in terms of potential attacks.

We dedicate this section to discuss how good a multiplicative perturbation is in terms of preserving privacy under a set of privacy attacks. We first define a multi-column (or multidimensional) privacy measure for evaluating the privacy quality of a multiplicative perturbation over a given dataset. Then, we introduce a framework of privacy evaluation, which can incorporate different attack analysis into the evaluation of privacy guarantee. We show that using this framework, we can employ certain optimization methods (Section 7.5.5) to find a good perturbation among a bunch of randomly generated perturbations, which is locally optimal for the given dataset.

7.4.1 A Conceptual Multidimensional Privacy Evaluation Model

In practice, different columns (or dimensions, or attributes) may have different privacy concern. Therefore, we advocate that the general-purpose privacy metric Φ defined for an entire dataset should be based on **column privacy metric**, rather than point-based privacy metrics, such distance-based metrics. A conceptual privacy model is defined as $\Phi = \Phi(\mathbf{p}, \mathbf{w})$, where $\mathbf{p}$ denotes the column privacy metric vector $\mathbf{p} = [p_1, p_2, \ldots, p_d]$ of a given dataset X, and $\mathbf{w} = (w_1, w_2, \ldots, w_d)$ denote **privacy weights** associated to the d columns respectively. The column privacy p_i itself is defined by a function, which we will discuss later. In summary, the model suggests that the column-wise privacy metric should be calculated first and then use Φ to generate a composite metric. We will first describe some basic designs to the components in function Φ. Then, we dedicate another subsection to the concrete design of the function for generating $\mathbf{p}$.

The first design idea is to take the column importance into unification of different column privacy. Intuitively, the more important the column is, the higher level of privacy guarantee will be required for the perturbed data column. Since $\mathbf{w}$ is used to denote the importance of columns in terms of preserving privacy, we use p_i/w_i to represent the *weighted column privacy* of column i.

The second concept is the *minimum privacy guarantee* and the *average privacy guarantee* among all columns. Normally, when we measure the privacy guarantee of a multidimensional perturbation, we need to pay more attention to the column that has the lowest weighted column privacy, because such a column could become the weakest link of privacy protection. Hence, the first composition function is the minimum privacy guarantee.

$$\Phi_1 = \min_{i=1}^{d}\{p_i/w_i\}$$

Similarly, the *average privacy guarantee* of the multi-column perturbation is defined by $\Phi_2 = \frac{1}{d}\sum_{i=1}^{d} p_i/w_i$, which could be another interesting measure. Note that these two functions assume that p_i should be comparable crossing columns, which is one of the important requirement in the following discussion.

7.4.2 Variance of Difference as Column Privacy Metric

After defining the conceptual privacy model, we move to the design of column-wise privacy metric. Intuitively, for a data perturbation approach, the quality of preserved privacy can be understood as the difficulty level of estimating the original data from the perturbed data. Therefore, how statistically different the *estimated data* is from the original data could be an intuitive measure. We use a variance-of-difference (VoD) based approach, which has a similar form to the naive variance-based evaluation [4], but with very different semantics.

Let the difference between the original column data and the estimated data be a random variable $\mathbf{D}_i$. Without any knowledge about the original data, the mean and variance of the difference present the quality of the estimation. The perfect estimation will have zero mean and variance. Since the mean of difference, i.e., the bias of estimation, can be easily removed if the attacker knows the original distribution of column, we use only the variance of the difference (VoD) as the primary metric to determine the level of difficulty in estimating the original data.

VoD is formally defined as follows. Let $\mathbf{X}_i$ be a random variable representing the column i, $\mathbf{X}_i'$ be the *estimated result*[1] of $\mathbf{X}_i$, and $\mathbf{D}_i$ be $\mathbf{D}_i = \mathbf{X}_i' - \mathbf{X}_i$. Let $E[\mathbf{D}_i]$ and $Var(\mathbf{D}_i)$ denote the mean and the variance of $\mathbf{D}$ respectively. Then VoD for column i is $Var(\mathbf{D}_i)$. Let an estimate of certain value, say x_i, be x_i', $\sigma = \sqrt{Var(\mathbf{D}_i)}$, and c denote confidence parameter depending on both the distribution of $\mathbf{D}_i$ and the corresponding confidence level. The corresponding

[1] It would not be appropriate to use only the perturbed data for privacy estimation, if we consider the potential attacks.

original value x_i in $\mathbf{X}_i$ is located in the range defined below:

$$[x'_i - E[\mathbf{D}_i] - c\sigma, x'_i - E[\mathbf{D}_i] + c\sigma]$$

By removing the effect of $E[\mathbf{D}_i]$, the width of the estimation range, $2c\sigma$, presents the quality of estimating the original value, which proportionally reflects the level of privacy guarantee. The smaller range means better estimation, i.e., a lower level of privacy guarantee. For simplicity, we often use σ to represent the privacy level.

VoD only defines the privacy guarantee for a single column. However, we usually need to evaluate the privacy level of all perturbed columns together if a multiplicative perturbation is applied. The single-column VoD does not work across different columns since different column value ranges may result in very different VoDs. For example, the VoD of age may be much smaller than VoD of salary. Therefore, a same amount of VoD is not equally effective for columns with different value ranges. One straightforward method to unify the different value ranges is via *normalization* over the original dataset and the perturbed dataset. Normalization can be done with various ways, such as max/min normalization or standardized normalization [30]. After normalization, the level of privacy guarantee for each column should be approximately comparable. Note that normalization after VoD calculation, such as relative variance $VoD_i/Var(\mathbf{X}_i)$ is not appropriate, since small $Var(\mathbf{X}_i)$ will inappropriately increase the value.

7.4.3 Incorporating Attack Evaluation

Privacy evaluation has to consider the resilience to attacks as well. The VoD evaluation has a unique advantage in incorporating attack analysis in privacy evaluation. In general, let X be the normalized original dataset, P be the perturbed dataset, and O be the estimated/observed dataset through "attack simulation". We can calculate $VoD(\mathbf{X}_i, \mathbf{O}_i)$ for the column i in terms of different attacks. For example, the attacks to rotation perturbation can be evaluated by following steps. Details will be discussed shortly.

1 Naive Estimation: $O \equiv P$;

2 ICA-based Reconstruction: Independent Component Analysis (ICA) is used to estimate R. Let $\hat{R}$ be the estimate of R, and the estimated data $\hat{R}^{-1}P$ aligned with the known column statistics to get the dataset O;

3 Distance-based Inference: knowing a set of special points in X that can be mapped to certain set of points in P, so that the mapping helps to get the estimated rotation $\hat{R}$, and then $O = \hat{R}^{-1}P$.

7.4.4 Other Metrics

Other metrics include distance-based risk of privacy breach, which was used to evaluate the level of privacy breach when a few pairs of original data points and their maps in perturbed data are known [27]. Assume $\hat{\mathbf{x}}$ is the estimate of an original point $\mathbf{x}$. An ϵ-*privacy breach* occurs if

$$||\hat{\mathbf{x}} - \mathbf{x}|| \le \epsilon ||\mathbf{x}||$$

This roughly represents that, if the estimate is within an arbitrarily small local area around the original point, then the risk of privacy breach is high. However, even though the estimated point is distant from the original point, the estimation can still be effective – large distance may only be determined by the difference between a few columns, while other columns may be very similar. That is the reason why we should consider column-wise privacy metrics.

7.5 Attack Resilient Multiplicative Perturbations

Attack analysis is the essential component in privacy evaluation of multiplicative perturbation. The previous section has set up an evaluation model that can conveniently incorporate attack analysis through "attack simulation". Namely, privacy attacks to multiplicative perturbations are the methods for estimating original points (or values of particular columns) from the perturbed data, with certain level of additional knowledge about the original data. As the perturbed data goes public, the level of effectiveness is solely determined by the additional knowledge the attacker may have. In the following sections, we describe some potential inference attacks to multiplicative perturbations, primarily focused on rotation perturbation.

These attacks are organized according to the different levels of knowledge that an attacker may have. We hope that, from this section the interested readers will have more ideas about the attacks to general multiplicative perturbations and are able to apply appropriate tools to counter attacks. Most content of this section can be found in the paper [9], and we will just present the basic ideas here.

7.5.1 Naive Estimation to Rotation Perturbation

When the attacker knows no additional information, we call attacks under such circumstance as naive estimation, which simply estimates the original data from perturbed data. In this case, an appropriate rotation perturbation is enough to achieve high level of privacy guarantee. With the VoD metric over the normalized data, we can formally analyze the privacy guarantee provided by the rotation perturbed data. Let X be the normalized dataset, X' be the rotation of X, and I_d be the d-dimensional identity matrix. VoD of column i

can be evaluated by

$$\begin{aligned} Cov(\mathbf{X}' - \mathbf{X})_{(i,i)} &= Cov(R\mathbf{X} - \mathbf{X})_{(i,i)} \\ &= ((R - I_d)Cov(\mathbf{X})(R - I_d)^T)_{(i,i)} \end{aligned} \tag{7.2}$$

Let r_{ij} represent the element (i, j) in the matrix R, and c_{ij} be the element (i, j) in the covariance matrix of $\mathbf{X}$. The VoD for ith column is computed as follows.

$$Cov(\mathbf{X}' - \mathbf{X})_{(i,i)} = \sum_{j=1}^{d}\sum_{k=1}^{d} r_{ij}r_{ik}c_{kj} - 2\sum_{j=1}^{d} r_{ij}c_{ij} + c_{ii} \tag{7.3}$$

When the random rotation matrix is generated following the Haar distribution, a considerable number of matrix entries are approximately independent normal distribution $N(0, 1/d)$ [20]. For simplicity and easy understanding, we assume that all entries in random rotation matrix approximately follow independent normal distribution $N(0, 1/d)$. Therefore, random rotations will make VoD_i changing around the mean value c_{ii} as shown in the following equation.

$$E[VoD_i] \sim \sum_{j=1}^{d}\sum_{k=1}^{d} E[r_{ij}]E[r_{ik}]c_{kj} - 2\sum_{j=1}^{d} E[r_{ij}]c_{ij} + c_{ii} = c_{ii}$$

It means that the original column variance could substantially influence the result of random rotation. However, the expectation of VoDs is not the only factor determining the final privacy guarantee. We should also look at the variance of VoDs. If the variance of $VoDs$ is considerably large, we still get great chance to find a rotation with high VoDs in a set of sample random rotations, and the larger the $Var(VoD_i)$ is, the more likely the randomly generated rotation matrices can provide a high privacy level. With the approximately independency assumption, we have

$$\begin{aligned} Var(VoD_i) &\sim \sum_{i=1}^{d}\sum_{j=1}^{d} Var(r_{ij})Var(r_{ik})c_{ij}^2 \\ &\quad +4\sum_{j=1}^{d} Var(r_{ij})c_{ij}^2 \\ &\sim O(1/d^2\sum_{i=1}^{d}\sum_{j=1}^{d} c_{ij}^2 + 4/d\sum_{j=1}^{d} c_{ij}^2). \end{aligned}$$

The above result shows that $Var(VoD_i)$ seems approximately related to the average of the squared covariance entries, with more influence from the row

i of covariance matrix. Therefore, by looking at the covariance matrix of the original dataset and estimate the $Var(VoD_i)$, we can estimate the chance of finding a random rotation that can give high privacy guarantee.

Rotation Center. The basic rotation perturbation uses the origin as the rotation center. Therefore, the points around the origin will be still close to the origin after the perturbation, which leads to weaker privacy protection over these points. The attack to rotation center can be regarded as another kind of naive estimation. This problem is addressed by random translation perturbation, which hides the rotation center. More sophisticated attacks to the combination of rotation and translation would have to utilize the ICA technique with sufficient additional knowledge, which will be described shortly.

7.5.2 ICA-Based Attacks

In this section, we introduce a high-level attack based on data reconstruction. The basic method for reconstructing X from the perturbed data RX would be Independent Component Analysis (ICA) technique, derived from the research of signal processing [18].

The ICA technique can be applied to estimate the independent components (the row vectors in our definition) of the original dataset X from the perturbed data, if the following conditions are satisfied:

1 The source row vectors are independent;

2 All source row vectors should be non-Gaussian with possible exception of one row;

3 The number of observed row vectors must be at least as large as the independent source row vectors.

4 The transformation matrix R must be of full column rank.

For rotation matrices, the 3rd and 4th conditions are always satisfied. However, the first two conditions although practical for signal processing, are often not satisfied in data classification or clustering. Furthermore, there are a few more difficulties in applying direct ICA-based attack. First of all, even ICA can be done successfully, the order of the original independent components cannot be preserved or determined through only ICA [18]. Formally, any permutation matrix P and its inverse P^{-1} can be substituted in the model to give $X' = RP^{-1}PX$. ICA could possibly give the estimate for some permutated source PX. Thus, we cannot identify the particular column without more knowledge about the original data. Second, even if the ordering of columns can be identified, ICA reconstruction does not guarantee to preserve the variance of the original signal – the estimated signal is often scaled up but we do not know how much the scaling is unless we know the original value range of the column. Therefore, without knowing the basic statistics of original columns, ICA-attack is not effective.

However, such basic column statistics are not impossible to get in some cases. Now, we assume that attackers know the basic statistics, including the column max/min values and the probability density function (PDF), or empirical PDF of each column. An enhanced ICA-based attack can be described as follows.

1 Run ICA algorithm to get a reconstructed dataset;

2 For each pair of $(\mathbf{O}_i, \mathbf{X}_j)$, where $\mathbf{O}_i$ is a reconstructed column and $\mathbf{X}_i$ is an original column, scale $\mathbf{O}_i$ with the max/min values of $\mathbf{X}_j$;

3 Compare the PDFs of the scaled $\mathbf{O}_i$ and $\mathbf{X}_j$ to find the closest match among all possible combinations.

Note the the PDFs should be aligned before comparison. [9] gives one method to align it.

The above procedure describes how to use ICA and additional knowledge about the original dataset to precisely reconstruct the original dataset. Note if the four conditions for effective ICA are exactly satisfied and the basic statistics and PDFs are all known distinct from each other, the basic rotation perturbation will be totally broken by the enhanced ICA-based attack. In practice, we can test if the first two conditions for effective ICA are satisfied to decide whether we can safely use rotation perturbation, when the column distributional information is released. If ICA-based attacks can be effectively done, it is also trivial to reveal an additional translation perturbation, which is used to protect the rotation center.

If the first and second conditions are not satisfied, as for most datasets in data classification and clustering, precise ICA reconstruction cannot be achieved. Under this circumstance, different rotation perturbations may result in different levels of privacy guarantee and the goal is to find one perturbation that is resilient to the enhanced ICA-based attacks.

For projection perturbation [28], the third condition of effective ICA is not satisfied either. Although overcomplete ICA is available for this particular case [25], it is generally ineffective to break projection perturbation with ICA-based attacks. The major concern of projection perturbation is to find one that preserves the utility of perturbed data.

7.5.3 Distance-Inference Attacks

In the previous sections, we have discussed naive estimation and ICA-based attacks. In the following discussion, we assume that, besides the information necessary to perform the discussed attacks, the attacker manages to get more knowledge about the original dataset. We assume two scenarios:

1) s/he also knows at least $d+1$ linearly independent original data records, $X = \{\mathbf{x}_1, \mathbf{x}_2, \ldots, \mathbf{x}_{d+1}\}$; or 2) s/he can only get less then d linearly independent points. S/he then tries to find the mapping between these points and their images in the perturbed dataset, denoted by $O = \{\mathbf{o}_1, \mathbf{o}_2, \ldots, \mathbf{o}_{d+1}\}$, to break rotation perturbation and possible also translation perturbation.

For both scenarios, it is possible to find the images of the known points in the perturbed data. Particularly, if a few original points are highly distinguishable, such as "outliers", their images in the perturbed data can be correctly identified with high probability for low-dimensional small datasets (< 4 dimensions). With considerable cost, it is not impossible for higher dimensional and larger datasets by simple exhaustive search, although the probability to get the exact images is relatively low. For scenario 1), with the known mapping, the rotation R and translation $\mathbf{t}$ can be precisely calculated if the incomplete geometric perturbation $G(X) = RX + \Psi$ is applied. Therefore, the threat will be substantial to any other data point in the original dataset.

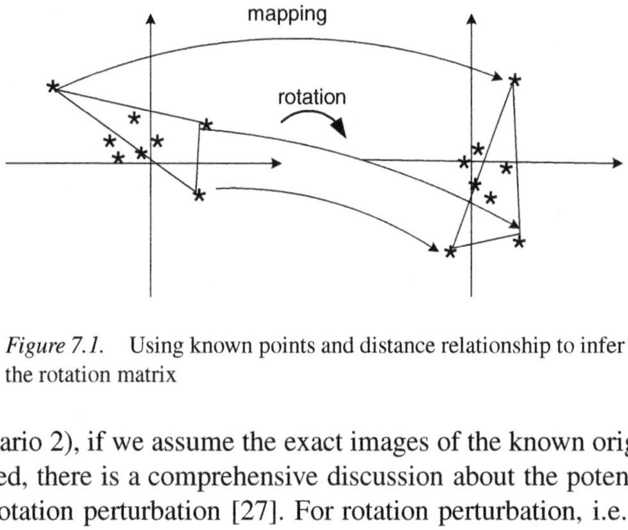

Figure 7.1. Using known points and distance relationship to infer the rotation matrix

For scenario 2), if we assume the exact images of the known original points are identified, there is a comprehensive discussion about the potential privacy breach to rotation perturbation [27]. For rotation perturbation, i.e., $O = RX$ between the known points X and their images O, if X consists of less than d points, there are numerous estimates of R, denoted by $\hat{R}$, satisfying the relationship between X and O. The weakest points, except the known points X, are those around X. Paper [27] gives some estimation to the risk of privacy breach for certain point $\mathbf{x}$ if a set of points X and their image O are known. The definition is based on ϵ-privacy breach (Section 7.4.1). The probability of ϵ -privacy breach, $\rho(\mathbf{x}, \epsilon)$, for any $\mathbf{x}$ in the original dataset can be estimated as follows. Let $d(\mathbf{x}, X)$ be the distance between $\mathbf{x}$ and X.

$$\rho(\mathbf{x}, \epsilon) = \frac{2}{\pi} \arcsin(\frac{\epsilon\|\mathbf{x}\|}{2d(\mathbf{x}, X)}), \text{if} \quad \epsilon\|\mathbf{x}\| < 2d(\mathbf{x}, X); \quad 1 \quad \text{otherwise}.$$

Note that ϵ-privacy breach is not sufficient to column-wise privacy evaluation. Thus, the above definition may not be sufficient as well.

In order to protect from distance-inference attack for both scenarios, an additional noise component Δ is introduced to form the complete version of geometric perturbation $G(X) = RX + \Psi + \Delta$, where $\Delta = [\delta_1, \delta_2, \ldots, \delta_N]$, and δ_i is a d-dimensional Gaussian random vector. The Δ component reduces the probability of getting exact images and the precision of estimation to R and Ψ, which significantly increases the resilience to distance-inference attacks.

Assume the attacker still knows enough pairs of independent (point, image). Now, with the additional noise component, the most effective way to estimate the rotation/translation component is linear regression. The steps include 1) filtering out the translation component first; 2) applying linear regression to estimate R; 3) plugging the estimate $\hat{R}$ back to estimate the translation component; 4) estimating the original data with $\hat{R}$ and $\hat{\Psi}$. There is a detailed procedure in [9]. We can simulate the procedure to estimate the resilience of a perturbation.

Note that the additional noise component also implies that we have to sacrifice some model accuracy for gaining the stronger privacy protection. An empirical study has been performed on a bunch of datasets to evaluate the relationship between noise intensity, resilience to attacks and model accuracy [9]. In general, a low-intense noise component will be enough to reduce the risk of being attacked, while still preserving model accuracy. However, the noise component is required only when the data owner is sure that a small part of the original data is released.

7.5.4 Attacks with More Prior Knowledge

There are also extreme cases that may not happen in practice, which assume the attacker knows a considerable amount of original data points and these points form a sample set that the higher-order statistical properties of the original dataset, like the covariance matrix, are approximately estimated from the sample set. By using the sample statistics and the sample points, the attacker can have more effective attacks.

Note that, in general, if the attacker has known so much information about the original data, its privacy may already be breached. It should not be advised to publish more original data. Further discussion about perturbations will make less sense. However, the techniques developed in these attacks, such as PCA-based attack [27] and AK-ICA attack [15] might be eventually utilized in other aspects to enhance multiplicative perturbations in the future. We will not give detailed description about these attacks due to the space limitation. Instead, they will be covered by another dedicated chapter.

7.5.5 Finding Attack-Resilient Perturbations

We have discussed the unified privacy metric for evaluating the quality of a random geometric perturbation. Some known inference attacks have been analyzed under the framework of multi-column privacy evaluation, which allows us to design an algorithm to choose a good geometric perturbation in terms of these attacks – if the attacker knows considerable amount of original data, it is advised not to release the perturbed dataset, however. A deterministic algorithm in optimizing the perturbation may also provide extra clue to privacy attackers. Therefore, it is also expected to have certain level of randomization in the perturbation optimization.

A randomized perturbation-optimization algorithm for geometric perturbation was proposed in [9]. We briefly describe it as follows. Algorithm 1 is a hill-climbing method, which runs in a given number of iterations to find a geometric perturbation that maximizes the minimum privacy guarantee as possible. Initially, a random translation is selected, which needs not optimization at all. In each iteration, the algorithm randomly generates a rotation matrix. Local maximization of VoD [9] is applied to find a better rotation matrix in terms of naive estimation, which is then tested by the ICA reconstruction with the algorithm described in section 7.5.2. The rotation matrix is accepted as the currently best perturbation if it provides higher minimum privacy guarantee than the previous perturbations. After the iterations, if necessary, a noise component is appended to the perturbation, so that the distance-inference attack cannot reduce the privacy guarantee to a safety level ϕ, e.g., $\phi = 0.2$. Algorithm 1 outputs the rotation matrix R_t, the random translation matrix Ψ, the noise level σ^2, and the corresponding privacy guarantee (we use minimum privacy guarantee in the following algorithm) in terms of the known attacks. If the final privacy guarantee is lower than the expected threshold, the data owner can select not to release the data. This algorithm provides a framework, in which any discovered attacks can be simulated and evaluated.

7.6 Conclusion

We have reviewed the multiplicative perturbation method as an alternative method to privacy preserving data mining. The design of this category of perturbation algorithms is based on an important principle: by developing perturbation algorithms that can always preserve the mining task and model specific data utility, one can focus on finding a perturbation that can provide higher level of privacy guarantee. We described three representative multiplicative perturbation methods – rotation perturbation, projection perturbation, and geometric perturbation. All aim at preserving the distance relationship in the original data, thus achieving good data utility for a set of classification and clustering models. Another important advantage of using these multiplicative

Algorithm 1 Finding_a_resilient_perturbation ($X_{d\times N}$, **w**, m)

Input: $X_{d\times N}$:the original dataset, **w**: weights for attributes in privacy evaluation, m: the number of iterations.
Output: R_t: the selected rotation matrix, Ψ: the random translation, σ^2: the noise level, p: privacy quality
calculate the covariance matrix C of X;
$p = 0$, and randomly generate the translation Ψ;
for Each iteration **do**
  randomly generate a rotation matrix R;
  swapping the rows of R to get R', which maximizes $min_{1\le i\le d}\{\frac{1}{w_i}(Cov(R'X - X)_{(i,i)}\}$;
  p_0 = the privacy guarantee of R', $p_1 = 0$;
  if $p_0 > p$ **then**
    generate $\hat{X}$ with ICA;
    $\{(1),(2),\ldots,(d)\} = argmin_{\{(1),(2),\ldots,(d)\}} \sum_{i=1}^{d} \Delta PDF(X_i, O_{(i)})$
    $p_1 = min_{1\le k\le d}\frac{1}{w_k} VoD(X_k, O_{(k)})$
  end if
  if $p < \min(p_0, p_1)$ **then**
    $p = \min(p_0, p_1)$, $R_t = R'$;
  end if
end for
p_2 = the privacy guarantee to the distance-inference attack with the perturbation $G(X) = R_tX + \Psi + \Delta$.
Tune the noise level σ^2, so that $p_2 \ge p$ if $p < \phi$ or $p_2 > \phi$ if $p < \phi$.

perturbation methods is the fact that we are not required to re-design the existing data mining algorithms in order to perform data mining over the perturbed data.

Privacy evaluation and attack analysis are the major challenging issues for multiplicative perturbations. We reviewed the multi-column variance of difference (VoD) based evaluation method and the distance-based method. Since column distribution information has high probability to be released publicly, in principle it is necessary to evaluate privacy guarantee based on columns. Although this chapter does not intend to enumerate all possible attacks, as we know, attack analysis to multiplicative perturbation is still a very active area, we describe several types of attacks and organize the discussion according to the level of knowledge that the attacker may have about the original data. We also outlined some techniques developed to date for addressing these attacks. Based on attack analysis and the VoD-based evaluation method, we show how to find the perturbations that locally optimize the level of privacy guarantee in terms of various attacks.

Acknowledgment

This work is partially supported by grants from NSF CISE CyberTrust program, IBM faculty award 2006, and an AFOSR grant.

References

[1] AGGARWAL, C. C., AND YU, P. S. A condensation approach to privacy preserving data mining. *Proc. of Intl. Conf. on Extending Database Technology (EDBT) 2992* (2004), 183–199.

[2] AGGARWAL, C. C., AND YU, P. S. On privacy-preservation of text and sparse binary data with sketches. *SIAM Data Mining Conference* (2007).

[3] AGRAWAL, D., AND AGGARWAL, C. C. On the design and quantification of privacy preserving data mining algorithms. *Proc. of ACM PODS Conference* (2002).

[4] AGRAWAL, R., AND SRIKANT, R. Privacy-preserving data mining. *Proc. of ACM SIGMOD Conference* (2000).

[5] ALON, N., MATIAS, Y., AND SZEGEDY, M. The space complexity of approximating the frequency moments. *Proc. of ACM PODS Conference* (1996).

[6] ANKERST, M., BREUNIG, M. M., KRIEGEL, H.-P., AND SANDER, J. OPTICS: Ordering points to identify the clustering structure. *Proc. of ACM SIGMOD Conference* (1999), 49–60.

[7] CHEN, K., AND LIU, L. A random geometric perturbation approach to privacy-preserving data classification. *Technical Report, College of Computing, Georgia Tech* (2005).

[8] CHEN, K., AND LIU, L. A random rotation perturbation approach to privacy preserving data classification. *Proc. of Intl. Conf. on Data Mining (ICDM)* (2005).

[9] CHEN, K., AND LIU, L. Towards attack-resilient geometric data perturbation. *SIAM Data Mining Conference* (2007).

[10] CRISTIANINI, N., AND SHAWE-TAYLOR, J. *An Introduction to Support Vector Machines and Other Kernel-based Learning Methods.* Cambridge University Press, 2000.

[11] ESTER, M., KRIEGEL, H.-P., SANDER, J., AND XU, X. A density-based algorithm for discovering clusters in large spatial databases with noise. *Second International Conference on Knowledge Discovery and Data Mining* (1996), 226–231.

[12] EVFIMIEVSKI, A., GEHRKE, J., AND SRIKANT, R. Limiting privacy breaches in privacy preserving data mining. *Proc. of ACM PODS Conference* (2003).

[13] EVFIMIEVSKI, A., SRIKANT, R., AGRAWAL, R., AND GEHRKE, J. Privacy preserving mining of association rules. *Proc. of ACM SIGKDD Conference* (2002).

[14] FEIGENBAUM, J., ISHAI, Y., MALKIN, T., NISSIM, K., STRAUSS, M., AND WRIGHT, R. N. Secure multiparty computation of approximations. In *ICALP '01: Proceedings of the 28th International Colloquium on Automata, Languages and Programming,* (2001), Springer-Verlag, pp. 927–938.

[15] GUO, S., AND WU, X. Deriving private information from arbitrarily projected data. In *Proceedings of the 11th European Conference on Principles and Practice of Knowledge Discovery in Databases (PKDD07)* (Warsaw, Poland, Sept 2007).

[16] HASTIE, T., TIBSHIRANI, R., AND FRIEDMANN, J. *The Elements of Statistical Learning.* Springer-Verlag, 2001.

[17] HINNEBURG, A., AND KEIM, D. A. An efficient approach to clustering in large multimedia databases with noise. *Proc. of ACM SIGKDD Conference* (1998), 58–65.

[18] HYVARINEN, A., KARHUNEN, J., AND OJA, E. *Independent Component Analysis.* Wiley-Interscience, 2001.

[19] JAIN, A. K., AND DUBES, R. C. Data clustering: A review. *ACM Computing Surveys 31* (1999), 264–323.

[20] JIANG, T. How many entries in a typical orthogonal matrix can be approximated by independent normals. *To appear in The Annals of Probability* (2005).

[21] JOHNSON, W. B., AND LINDENSTRAUSS, J. Extensions of lipshitz mapping into hilbert space. *Contemporary Mathematics 26* (1984).

[22] KARGUPTA, H., DATTA, S., WANG, Q., AND SIVAKUMAR, K. On the privacy preserving properties of random data perturbation techniques. *Proc. of Intl. Conf. on Data Mining (ICDM)* (2003).

[23] KIM, J. J., AND WINKLER, W. E. Multiplicative noise for masking continuous data. Tech. Rep. Statistics #2003-01, Statistical Research Division, U.S. Bureau of the Census, Washington D.C., April 2003.

[24] LEFEVRE, K., DEWITT, D. J., AND RAMAKRISHNAN, R. Mondrain multidimensional k-anonymity. *Proc. of IEEE Intl. Conf. on Data Eng. (ICDE)* (2006).

[25] LEWICKI, M. S., AND SEJNOWSKI, T. J. Learning overcomplet representations. *Neural Computation 12*, 2 (2000).

[26] LINDELL, Y., AND PINKAS, B. Privacy preserving data mining. *Journal of Cryptology 15*, 3 (2000), 177–206.

[27] LIU, K., GIANNELLA, C., AND KARGUPTA, H. An attacker's view of distance preserving maps for privacy preserving data mining. In *Proceedings of the 10th European Conference on Principles and Practice of*

Knowledge Discovery in Databases (PKDD'06) (Berlin, Germany, September 2006).

[28] LIU, K., KARGUPTA, H., AND RYAN, J. Random projection-based multiplicative data perturbation for privacy preserving distributed data mining. *IEEE Transactions on Knowledge and Data Engineering (TKDE) 18*, 1 (January 2006), 92–106.

[29] MACHANAVAJJHALA, A., GEHRKE, J., KIFER, D., AND VENKITASUBRAMANIAM, M. l-diversity: Privacy beyond k-anonymity. *Proc. of IEEE Intl. Conf. on Data Eng. (ICDE)* (2006).

[30] NETER, J., KUTNER, M. H., NACHTSHEIM, C. J., AND WASSERMAN, W. *Applied Linear Statistical Methods.* WCB/McGraw-Hill, 1996.

[31] OLIVEIRA, S. R. M., AND ZAÏANE, O. R. Privacy preservation when sharing data for clustering. In *Proceedings of the International Workshop on Secure Data Management in a Connected World* (Toronto, Canada, August 2004), pp. 67–82.

[32] SADUN, L. *Applied Linear Algebra: the Decoupling Principle.* Prentice Hall, 2001.

[33] STEWART, G. The efficient generation of random orthogonal matrices with an application to condition estimation. *SIAM Journal on Numerical Analysis 17* (1980).

[34] SWEENEY, L. k-anonymity: a model for protecting privacy. *International Journal on Uncertainty, Fuzziness and Knowledge-based Systems 10*, 5 (2002).

[35] VAIDYA, J., AND CLIFTON, C. Privacy preserving k-means clustering over vertically partitioned data. *Proc. of ACM SIGKDD Conference* (2003).

Chapter 8

A Survey of Quantification of Privacy Preserving Data Mining Algorithms

Elisa Bertino
Department of Computer Science
Purdue University
bertino@cs.purdue.edu

Dan Lin
Department of Computer Science
Purdue University
lindan@cs.purdue.edu

Wei Jiang
Department of Computer Science
Purdue University
wjiang@cs.purdue.edu

Abstract The aim of privacy preserving data mining (PPDM) algorithms is to extract relevant knowledge from large amounts of data while protecting at the same time sensitive information. An important aspect in the design of such algorithms is the identification of suitable evaluation criteria and the development of related benchmarks. Recent research in the area has devoted much effort to determine a trade-off between the right to privacy and the need of knowledge discovery. It is often the case that no privacy preserving algorithm exists that outperforms all the others on all possible criteria. Therefore, it is crucial to provide a comprehensive view on a set of metrics related to existing privacy preserving algorithms so that we can gain insights on how to design more effective measurement and PPDM algorithms. In this chapter, we review and summarize existing criteria and metrics in evaluating privacy preserving techniques.

Keywords: Privacy metric.

8.1 Introduction

Privacy is one of the most important properties that an information system must satisfy. For this reason, several efforts have been devoted to incorporating privacy preserving techniques with data mining algorithms in order to prevent the disclosure of sensitive information during the knowledge discovery. The existing privacy preserving data mining techniques can be classified according to the following five different dimensions [32]: (i) data distribution (centralized or distributed); (ii) the modification applied to the data (encryption, perturbation, generalization, and so on) in order to sanitize them; (iii) the data mining algorithm which the privacy preservation technique is designed for; (iv) the data type (single data items or complex data correlations) that needs to be protected from disclosure; (v) the approach adopted for preserving privacy (heuristic or cryptography-based approaches). While heuristic-based techniques are mainly conceived for centralized datasets, cryptography-based algorithms are designed for protecting privacy in a distributed scenario by using encryption techniques. Heuristic-based algorithms recently proposed aim at hiding sensitive raw data by applying perturbation techniques based on probability distributions. Moreover, several heuristic-based approaches for hiding both raw and aggregated data through a hiding technique (k-anonymization, adding noises, data swapping, generalization and sampling) have been developed, first, in the context of association rule mining and classification and, more recently, for clustering techniques.

Given the number of different privacy preserving data mining (PPDM) techniques that have been developed in these years, there is an emerging need of moving toward standardization in this new research area, as discussed by Oliveira and Zaiane [23]. One step toward this essential process is to provide a quantification approach for PPDM algorithms to make it possible to evaluate and compare such algorithms. However, due to the variety of characteristics of PPDM algorithms, it is often the case that no privacy preserving algorithm exists that outperforms all the others on all possible criteria. Rather, an algorithm may perform better than another one on specific criteria like privacy level, data quality. Therefore, it is important to provide users with a comprehensive set of privacy preserving related metrics which will enable them to select the most appropriate privacy preserving technique for the data at hand, with respect to some specific parameters they are interested in optimizing [6].

For a better understanding of PPDM related metrics, we next identify a proper set of criteria and the related benchmarks for evaluating PPDM algorithms. We then adopt these criteria to categorize the metrics. First, we need to be clear with respect to the concept of "privacy" and the general goals of a PPDM algorithm. In our society the *privacy* term is overloaded, and can, in general, assume a wide range of different meanings. For example, in the context of the HIPAA[1] Privacy Rule, *privacy* means the individual's ability to control who has the access to personal health care information. From the organizations point of view, *privacy* involves the definition of policies stating which information is collected, how it is used, and how customers are informed and involved in this process. Moreover, there are many other definitions of privacy that are generally related with the particular environment in which the privacy has to be guaranteed. What we need is a more generic definition, that can be instantiated to different environments and situations. From a philosophical point of view, Schoeman [26] and Walters [33] identify three possible definitions of privacy:

- Privacy as the right of a person to determine which personal information about himself/herself may be communicated to others.
- Privacy as the control over access to information about oneself.
- Privacy as limited access to a person and to all the features related to the person.

In three definitions, what is interesting from our point of view is the concept of "Controlled Information Release". From this idea, we argue that a definition of privacy that is more related with our target could be the following: "*The right of an individual to be secure from unauthorized disclosure of information about oneself that is contained in an electronic repository*". Performing a final tuning of the definition, we consider privacy as "*The right of an entity to be secure from unauthorized disclosure of sensible information that are contained in an electronic repository or that can be derived as aggregate and complex information from data stored in an electronic repository*". The last generalization is due to the fact that the concept of individual privacy does not even exist. As in [23] we consider two main scenarios.

The first is the case of a Medical Database where there is the need to provide information about diseases while preserving the patient identity. Another scenario is the classical "Market Basket" database, where the transactions related to different client purchases are stored and from which it is possible to extract some information in form of association rules like "If a client buys a product X, he/she will purchase also Z with y% probability". The first is an example where individual privacy has to be ensured by protecting from unauthorized disclosure sensitive information in form of specific data items related to specific individuals. The second one, instead, emphasizes how not only the raw data contained into a database must be protected, but also, in some cases, the high level information that can be derived from non sensible raw data need

[1] Health Insurance Portability and Accountability Act

to protected. Such a scenario justifies the final generalization of our privacy definition. In the light of these considerations, it is, now, easy to define which are the main goals a PPDM algorithm should enforce:

1 A PPDM algorithm should have to prevent the discovery of sensible information.

2 It should be resistant to the various data mining techniques.

3 It should not compromise the access and the use of non sensitive data.

4 It should not have an exponential computational complexity.

Correspondingly, we identify the following set of criteria based on which a PPDM algorithm can be evaluated.

- *Privacy level* offered by a privacy preserving technique, which indicates how closely the sensitive information, that has been hidden, can still be estimated.

- *Hiding failure*, that is, the portion of sensitive information that is not hidden by the application of a privacy preservation technique;

- *Data quality* after the application of a privacy preserving technique, considered both as the quality of data themselves and the quality of the data mining results after the hiding strategy is applied;

- *Complexity*, that is, the ability of a privacy preserving algorithm to execute with good performance in terms of all the resources implied by the algorithm.

For the rest of the chapter, we first present details of each criteria through analyzing existing PPDM techniques. Then we discuss how to select proper metric under a specified condition. Finally, we summarize this chapter and outline future research directions.

8.2 Metrics for Quantifying Privacy Level

Before presenting different metrics related to privacy level, we need to take into account two aspects: (i) sensitive or private information can be contained in the original dataset; and (ii) private information that can be discovered from data mining results. We refer to the first one as data privacy and the latter as result privacy.

8.2.1 Data Privacy

In general, the quantification used to measure data privacy is the degree of uncertainty, according to which original private data can be inferred. The higher the degree of uncertainty achieved by a PPDM algorithm, the better the data privacy is protected by this PPDM algorithm. For various types of PPDM algorithms, the degree of uncertainty is estimated in different ways. According to the adopted techniques, PPDM algorithms can be classified into two main categories: heuristic-based approaches and cryptography-based approaches. Heuristic-based approaches mainly include four sub-categories: additive noise, multiplicative noise, k-anonymization, and statistical disclosure control based approaches. In what follows, we survey representative works of each category of PPDM algorithms and review the metrics used by them.

Additive-Noise-based Perturbation Techniques. The basic idea of the additive-noise-based perturbation technique is to add random noise to the actual data. In [2], Agrawal and Srikant uses an additive-noise-based technique to perturb data. They then estimate the probability distribution of original numeric data values in order to build a decision tree classifier from perturbed training data. They introduce a quantitative measure to evaluate the amount of privacy offered by a method and evaluate the proposed method against this measure. The privacy is measured by evaluating how closely the original values of a modified attribute can be determined. In particular, if the perturbed value of an attribute can be estimated, with a confidence c, to belong to an interval $[a, b]$, then the privacy is estimated by $(b-a)$ with confidence c. However, this metric does not work well because it does not take into account the distribution of the original data along with the perturbed data. Therefore, a metric that considers all the informative content of data available to the user is needed. Agrawal and Aggarwal [1] address this problem by introducing a new privacy metric based on the concept of information entropy. More specifically, they propose an Expectation Maximization (EM) based algorithm for distribution reconstruction, which converges to the maximum likelihood estimate of the original distribution on the perturbed data. The measurement of privacy given by them considers the fact that both the perturbed individual record and the reconstructed distribution are available to the user as well as the perturbing distribution, as it is specified in [10]. This metric defines the average conditional privacy of an attribute A given other information, modeled with a random variable B, as $2^{h(A|B)}$, where $h(A|B)$ is the conditional differential entropy of A given B representing a measure of uncertainty inherent in the value of A, given the value of B.

Another additive-noise-based perturbation technique is by Rivzi and Haritsa [24]. They propose a distortion method to pre-process the data before executing the mining process. Their privacy measure deals with the probability with which the user's distorted entries can be reconstructed. Their goal is to ensure privacy at the level of individual entries in each customer tuple. In other words, the authors estimate the probability that a given 1 or 0 in the true matrix representing the transactional database can be reconstructed, even if for many applications the 1's and 0's values do not need the same level of privacy.

Evfimievski et al. [11] propose a framework for mining association rules from transactions consisting of categorical items, where the data has been randomized to preserve privacy of individual transactions, while ensuring at the same time that only true associations are mined. They also provide a formal definition of privacy breaches and a class of randomization operators that are much more effective in limiting breaches than uniform randomization. According to Definition 4 from [11], an itemset A results in a privacy breach of level ρ if the probability that an item in A belongs to a non randomized transaction, given that A is included in a randomized transaction, is greater than or equal to ρ. In some scenarios, being confident that an item not present in the original transaction may also be considered a privacy breach. In order to evaluate the privacy breaches, the approach taken by Evfimievski et al. is to count the occurrences of an itemset in a randomized transaction and in its sub-items in the corresponding non randomized transaction. Out of all sub-items of an itemset, the item causing the worst privacy breach is chosen. Then, for each combination of transaction size and itemset size, the worst and the average value of this breach level are computed over all frequent itemsets. The itemset size giving the worst value for each of these two values is selected.

Finally, we introduce a universal measure of data privacy level, proposed by Bertino et al. in [6]. The measure is developed based on [1]. The basic concept used by this measure is information entropy, which is defined by Shannon [27]: let X be a random variable which takes on a finite set of values according to a probability distribution $p(x)$. Then, the entropy of this probability distribution is defined as follows:

$$h(X) = -\sum p(x) \log_2(p(x)) \tag{8.1}$$

or, in the continuous case:

$$h(X) = -\int f(x) \log_2(f(x))dx \tag{8.2}$$

where $f(x)$ denotes the density function of the continuous random variable x. Information entropy is a measure of how much "choice" is involved in the selection of an event or how uncertain we are of its outcome. It can be used for quantifying the amount of information associated with a set of data. The concept of "information associated with data" can be useful in the evaluation of the privacy achieved by a PPDM algorithm. Because the entropy represents the information content of a datum, the entropy after data sanitization should be higher than the entropy before the sanitization. Moreover the entropy can be assumed as the evaluation of the uncertain forecast level of an event which in our context is evaluation of the right value of a datum. Consequently, the level of privacy inherent in an attribute X, given some information modeled by Y, is defined as follows:

$$\Pi(X|Y) = 2^{-\int f_{X,Y}(x,y) \log_2 f_{X|Y=y}(x))dxdy} \tag{8.3}$$

The privacy level defined in equation 8.3 is very general. In order to use it in the different PPDM contexts, it needs to be refined in relation with some characteristics like the type of transactions, the type of aggregation and PPDM methods. In [6], an example of instantiating the entropy concept to evaluate the privacy level in the context of "association rules" is presented.

However, it is worth noting that the value of the privacy level depends not only on the PPDM algorithm used, but also on the knowledge that an attacker has about the data before the use of data mining techniques and the relevance of this knowledge in the data reconstruction operation. This problem is underlined, for example, in [29, 30]. In [6], this aspect is not considered, but it is possible to introduce assumptions on attacker knowledge by properly modeling Y.

Multiplicative-Noise-based Perturbation Techniques. According to [16], additive random noise can be filtered out using certain signal processing techniques with very high accuracy. To avoid this problem, random projection-based multiplicative perturbation techniques has been proposed in [19]. Instead of adding some random values to the actual data, random matrices are used to project the set of original data points to a randomly chosen lower-dimensional space. However, the transformed data still preserves much statistical aggregates regarding the original dataset so that certain data mining tasks (e.g., computing inner product matrix, linear classification, K-means clustering and computing Euclidean distance) can be performed on the transformed data in a distributed environment (data are either vertically partitioned or horizontally partitioned) with small errors.

In addition, this approach provides a high degree of privacy regarding the original data. As analyzed in the paper, even if the random matrix (i.e., the multiplicative noise) is disclosed, it is impossible to find the exact values of the original dataset, but finding approximation of the original data is possible. The variance of the approximated data is used as privacy measure.

Oliveira and Zaiane [22] also adopt a multiplicative-noise-based perturbation technique to perform a clustering analysis while ensuring at the same time privacy preservation. They have introduced a family of geometric data transformation methods where they apply a noise vector to distort confidential numerical attributes. The privacy ensured by such techniques is measured as the variance difference between the actual and the perturbed values. This measure is given by $Var(X - Y)$, where X represents a single original attribute and Y

the distorted attribute. This measure can be made scale invariant with respect to the variance of X by expressing security as $Sec = Var(X - Y)/Var(X)$.

k-Anonymization Techniques. The concept of k-anonymization is introduced by Samarati and Sweeney in [25, 28]. A database is k-anonymous with respect to quasi-identifier attributes (a set of attributes that can be used with certain external information to identify a specific individual) if there exist at least k transactions in the database having the same values according to the quasi-identifier attributes. In practice, in order to protect sensitive dataset T, before releasing T to the public, T is converted into a new dataset T^* that guarantees the k-anonymity property for a sensible attribute by performing some value generalizations on quasi-identifier attributes. Therefore, the degree of uncertainty of the sensitive attribute is at least $1/k$.

Statistical-Disclosure-Control-based Techniques. In the context of statistical disclosure control, a large number of methods have been developed to preserve individual privacy when releasing aggregated statistics on data. To anonymize the released statistics from those data items such as person, household and business, which can be used to identify an individual, not only features described by the statistics but also related information publicly available need to be considered [35]. In [7] a description of the most relevant perturbation methods proposed so far is presented. Among these methods specifically designed for continuous data, the following masking techniques are described: additive noise, data distortion by probability distribution, resampling, microaggregation, rank swapping, etc. For categorical data both perturbative and non-perturbative methods are presented. The top-coding and bottom-coding techniques are both applied to ordinal categorical variables; they recode, respectively, the first/last p values of a variable into a new category. The global-recoding technique, instead, recodes the p lowest frequency categories into a single one.

The privacy level of such method is assessed by using the *disclosure risk*, that is, the risk that a piece of information be linked to a specific individual. There are several approaches to measure the disclosure risk. One approach is based on the computation of the distance-based record linkage. An intruder is assumed to try to link the masked dataset with the external dataset using the key variables. The distance between records in the original and the masked datasets is computed. A record in the masked dataset is labelled as "linked" or "linked to 2nd nearest" if the nearest or 2nd nearest record in the original dataset turns out to be the corresponding original record. Then the disclosure risk is computed as the percentage of "linked" and "linked to 2nd nearest". The second approach is based on the computation of the probabilistic record linkage. The linear sum assignment model is used to 'pair' records in the original file and the masked file. The percentage of correctly paired records is a measure of disclosure risk. Another approach computes rank intervals for the records in the masked dataset. The proportion of original values that fall into the interval centered around their corresponding masked value is a measure of disclosure risk.

Cryptography-based Techniques. The cryptography-based technique usually guarantees very high level of data privacy. In [14], Kantarcioglu and Clifton address the problem of secure mining of association rules over horizontally partitioned data, using cryptographic techniques to minimize the information shared. Their solution is based on the assumption that each party first encrypts its own itemsets using commutative encryption, then the already encrypted itemsets of every other party. Later on, an initiating party transmits its frequency count, plus a random value, to its neighbor, which adds its frequency count and passes it on to other parties. Finally, a secure comparison takes place between the final and initiating parties to determine if the final result is greater than the threshold plus the random value.

Another cryptography-based approach is described in [31]. Such approach addresses the problem of association rule mining in vertically partitioned data. In other words, its aim is to determine the item frequency when transactions are split across different sites, without revealing the contents of individual transactions. The security of the protocol for computing the scalar product is analyzed.

Though cryptography-based techniques can well protect data privacy, they may not be considered good with respect to other metrics like efficiency that will be discussed in later sections.

8.2.2 Result Privacy

So far, we have seen privacy metrics related to the data mining process. Many data mining tasks produce aggregate results, such as Bayesian classifiers. Although it is possible to protect sensitive data when a classifier is constructed, can this classifier be used to infer sensitive data values? In other words, do data mining results violate privacy? This issue has been analyzed and a framework is proposed in [15] to test if a classifier C creates an inference channel that could be adopted to infer sensitive data values.

The framework considers three types of data: public data (P), accessible to every one including the adversary; private/sensitive data (S), must be protected and unknown to the adversary; unknown data (U), not known to the adversary, but the release of this data might cause privacy violation. The framework assumes that S depends only on P and U, and the adversary has at most t data samples of the form (p_i, s_i). The approach to determine whether an inference channel exists is comprised of two steps. First, a classifier C_1 is built on the t data samples. To evaluate the impact of C, another classifier C_2 is built based on the same t data samples plus the classifier C. If the accuracy of C_2 is significantly better than C_1, we can say that C provides an inference channel for S.

Classifier accuracy is measured based on Bayesian classification error. Suppose we have a dataset $\{x_1, \ldots, x_n\}$, and we want to classify x_i into m classes labelled as $\{1, \ldots, m\}$. Given a classifier C:

$$C : x_i \rightarrow C(x_i) \in \{1, \ldots, m\}, \quad i = 1, \ldots, n$$

The classifier accuracy for C is defined as:

$$\sum_{j=1}^{m} Pr(C(x_i) \neq j | z = j) Pr(z = j)$$

where z is the actual class label of x_i. Since cryptography-based PPDM techniques usually produce the same results as those mined from the original dataset, analyzing privacy implications from the mining results is particular important to this class of techniques.

8.3 Metrics for Quantifying Hiding Failure

The percentage of sensitive information that is still discovered, after the data has been sanitized, gives an estimate of the *hiding failure* parameter. Most of the developed privacy preserving algorithms are designed with the goal of obtaining zero hiding failure. Thus, they hide all the patterns considered sensitive. However, it is well known that the more sensitive information we hide, the more non-sensitive information we miss. Thus, some PPDM algorithms have been recently developed which allow one to choose the amount of sensitive data that should be hidden in order to find a balance between privacy and knowledge discovery. For example, in [21], Oliveira and Zaiane define the *hiding failure* (HF) as the percentage of restrictive patterns that are discovered from the sanitized database. It is measured as follows:

$$HF = \frac{\#R_P(D')}{\#R_P(D)} \tag{8.4}$$

where $\#R_P(D)$ and $\#R_P(D')$ denote the number of restrictive patterns discovered from the original data base D and the sanitized database D' respectively. Ideally, HF should be 0. In their framework, they give a specification of a *disclosure threshold* ϕ, representing the percentage of sensitive transactions that are not sanitized, which allows one to find a balance between the hiding failure and the number of misses. Note that ϕ does not control the *hiding failure* directly, but indirectly by controlling the proportion of sensitive transactions to be sanitized for each restrictive pattern.

Moreover, as pointed out in [32], it is important not to forget that intruders and data terrorists will try to compromise information by using various data mining algorithms. Therefore, a PPDM algorithm developed against a particular data mining techniques that assures privacy of information, may not attain similar protection against all possible data mining algorithms. In order to provide for a complete evaluation of a PPDM algorithm, we need to measure its hiding failure against data mining techniques which are different from the technique that the PPDM algorithm has been designed for. The evaluation needs the consideration of a class of data mining algorithms which are significant for our test. Alternatively, a formal framework can be developed that upon testing of a PPDM algorithm against pre-selected data sets, we can transitively prove privacy assurance for the whole class of PPDM algorithms.

8.4 Metrics for Quantifying Data Quality

The main feature of the most PPDM algorithms is that they usually modify the database through insertion of false information or through the blocking of data values in order to hide sensitive information. Such perturbation techniques cause the decrease of the data quality. It is obvious that the more the changes are made to the database, the less the database reflects the domain of interest. Therefore, data quality metrics are very important in the evaluation of PPDM techniques. Since the data is often sold for making profit, or shared with others in the hope of leading to innovation, data quality should have an acceptable level according also to the intended data usage. If data quality is too degraded, the released database is useless for the purpose of knowledge extraction.

In existing works, several data quality metrics have been proposed that are either generic or data-use-specific. However, currently, there is no metric that is widely accepted by the research community. Here we try to identify a set of possible measures that can be used to evaluate different aspects of data quality. In evaluating the data quality after the privacy preserving process, it can be useful to assess both the *quality of the data* resulting from the PPDM process and the *quality of the data mining results*. The quality of the data themselves can be considered as a general measure evaluating the state of the individual items contained in the database after the enforcement of a privacy preserving technique. The quality of the data mining results evaluates the alteration in the information that is extracted from the database after the privacy preservation process, on the basis of the intended data use.

8.4.1 Quality of the Data Resulting from the PPDM Process

The main problem with data quality is that its evaluation is relative [18], in that it usually depends on the context in which data are used. In particular, there are some aspects related to data quality evaluation that are heavily related not only with the PPDM algorithm, but also with the structure of the database, and with the meaning and relevance of the information stored in the database with respect to a well defined context. In the scientific literature data quality is generally considered a multi-dimensional concept that in certain contexts involves both objective and subjective parameters [3, 34]. Among the various possible parameters, the following ones are usually considered the most relevant:

- *Accuracy*: it measures the proximity of a sanitized value to the original value.

- *Completeness*: it evaluates the degree of missed data in the sanitized database.

- *Consistency*: it is related to the internal constraints, that is, the relationships that must hold among different fields of a data item or among data items in a database.

Accuracy. The accuracy is closely related to the *information loss* resulting from the hiding strategy: the less is the information loss, the better is the data quality. This measure largely depends on the specific class of PPDM algorithms. In what follows, we discuss how different approaches measure the accuracy.

As for heuristic-based techniques, we distinguish the following cases based on the modification technique that is performed for the hiding process. If the algorithm adopts a perturbation or a blocking technique to hide both raw and aggregated data, the information loss can be measured in terms of the dissimilarity between the original dataset D and the sanitized one D'. In [21], Oliveira and Zaiane propose three different methods to measure the *dissimilarity* between the original and sanitized databases. The first method is based on the difference between the frequency histograms of the original and the sanitized databases. The second method is based on computing the difference between the sizes of the sanitized database and the original one. The third method is based on a comparison between the contents of two databases. A more detailed analysis on the definition of dissimilarity is presented by Bertino et al. in [6]. They suggest to use the following formula in the case of transactional dataset perturbation:

$$Diss(D, D') = \frac{\sum_{i=1}^{n} |f_D(i) - f_{D'}(i)|}{\sum_{i=1}^{n} f_D(i)} \tag{8.5}$$

where i is a data item in the original database D and $f_D(i)$ is its frequency within the database, whereas i' is the given data item after the application of a privacy preservation technique and $f_{D'}(i)$ is its new frequency within the transformed database D'. As we can see, the information loss is defined as the ratio between the sum of the absolute errors made in computing the frequencies of the items from a sanitized database and the sum of all the frequencies of items in the original database. The formula 8.5 can also be used for the PPDM algorithms which adopt a blocking technique for inserting into the dataset uncertainty about some sensitive data items or their correlations. The frequency of the item i belonging to the sanitized dataset D' is then given by the mean value between the minimum frequency of the data item i, computed by considering all the blocking values associated with it equal to zero, and the maximum frequency, obtained by considering all the question marks equal to one.

In case of data swapping, the information loss caused by an heuristic-based algorithm can be evaluated by a parameter measuring the *data confusion* introduced by the value swappings. If there is no correlation among the different database records, the *data confusion* can be estimated by the percentage of value replacements executed in order to hide specific information.

For the multiplicative-noise-based approaches [19], the quality of the perturbed data depends on the size of the random projection matrix. In general, the error bound of the inner product matrix produce by this perturbation technique is 0 on average and the variance is bounded by the inverse of the dimensionality of the reduced space. In other words, when the dimensionality of the random projection matrix is close to that of the original data, the result of computing the inner product matrix based on the transformed or projected data is also close to the actual value. Since inner product is closely related to many distance-based metrics (e.g., Euclidean distance, cosine angle of two vectors, correlation coefficient of two vectors, etc), the analysis on error bound has direct impact on the mining results if these data mining tasks adopt certain distance-based metrics.

If the data modification consists of aggregating some data values, the information loss is given by the loss of detail in the data. Intuitively, in this case, in order to perform the hiding operation, the PPDM algorithms use some type of "Generalization or Aggregation Scheme" that can be ideally modeled as a tree scheme. Each cell modification applied during the sanitization phase using the Generalization tree introduces a data perturbation that reduces the general accuracy of the database. As in the case of the k-anonymity algorithm presented in [28], we can use the following formula. Given a database T with N_A fields and N transactions, if we identify as generalization scheme a domain generalization hierarchy GT with a depth h, it is possible to measure the *information loss* (IL) of a sanitized database T^* as:

$$IL(T^*) = \frac{\sum_{i=1}^{i=N_A} \sum_{j=1}^{i=N} \frac{h}{|GT_{Ai}|}}{|T| * |N_A|} \tag{8.6}$$

where $\frac{h}{|GT_{Ai}|}$ represent the detail loss for each cell sanitized. For hiding techniques based on sampling approach, the quality is obviously related to the size of the considered sample and, more generally, on its features.

There are some other precision metrics specifically designed for k-anonymization approaches. One of the earliest data quality metrics is based on the height of generalization hierarchies [25]. The height is the number of times the original data value has been generalized. This metric assumes that a generalization on the data represents an information loss on the original data value. Therefore, data should be generalized as fewer steps as possible to preserve maximum utility. However, this metric does not take into account that not every generalization steps are equal in the sense of information loss.

Later, Iyengar [13] proposes a general *loss metric* (LM). Suppose T is a data table with n attributes. The LM metric is thought as the average information loss of all data cells of a given dataset, defined as follows:

$$LM(T^*) = \frac{\sum_{i=1}^{n} \sum_{j=1}^{|T|} \frac{f(T^*[i][j])-1}{g(A_i)-1}}{|T| \cdot n} \tag{8.7}$$

In equation 8.7, T^* is the anonymized table of T, f is a function that given a data cell value $T^*[i][j]$, returns the number of distinct values that can be generalized to $T^*[i][j]$, and g is a function that given an attribute A_i, returns the number of distinct values of A_i.

The next metric, *classification metric* (CM), is introduced by Iyengar [13] to optimize a k-anonymous dataset for training a classifier. It is defined as the sum of the individual penalties for each row in the table normalized by the total number of rows N.

$$CM(T^*) = \frac{\sum_{all\ rows} penalty(row\ r)}{N} \tag{8.8}$$

The penalty value of row r is 1, i.e., row r is penalized, if it is suppressed or if its class label is not the majority class label of its group. Otherwise, the penalty value of row r is 0. This metric is particularly useful when we want to build a classifier over anonymous data.

Another interesting metric is the *discernibility metric*(DM) proposed by Bayado and Agrawal [4]. This discernibility metric assigns a penalty to each tuple based on how many tuples in the transformed dataset are indistinguishable from it. Let t be a tuple from the original table T, and let $G_{T^*}(t)$ be the set of tuples in an anonymized table T^* indistinguishable from t or the set of tuples in $T*$ equivalent to the anonymized value of t. Then DM is defined as follows:

$$DM(T^*) = \sum_{t \in T} |G_{T^*}(t)| \tag{8.9}$$

Note that if a tuple t has been suppressed, the size of $G_{T^*}(t)$ is the same as the size of T^*. In many situation, suppressions are considered to be most expensive in the sense of information loss. Thus, to maximize data utility, tuple suppression should be avoided whenever possible.

For any given metric M, if $M(T) > M(T')$, we say T has a higher information loss, or is less precise, than b. In other words, data quality of T is worse than that of T'. Is this true for all metrics? What is a good metric? It is not easy to answer these kinds of questions. As shown in [20], CM works better than LM in classification application. In addition, LM is better for association rule mining. It is apparent that to judge how good a particular metric is, we need to associate our judgement with specific applications (e.g., classification, mining association rules).

The CM metric and the information gain privacy loss ratio [5, 28] are more interesting measure of utility because it considers the possible application for the data. Nevertheless, it is unclear what to do if we want to build classifiers on various attributes. In addition, these two metrics only work well if the data are intended to be used for building classifiers. Is there a utility metric that works well for various applications? Having this in mind, Kifer [17] proposes a utility measure related to Kullback-Leibler divergence. In theory, using this measure, *better* anonymous datasets (for different applications) can be produced. Researchers have measured the utility of the resulting anonymous datasets. Preliminary results show that this metric works well in practical applications.

For the statistical-based perturbation techniques which aim to hide the values of a confidential attribute, the information loss is basically the lack of precision in estimating the original distribution function of the given attribute. As defined in [1], the information loss incurred during the reconstruction of estimating the density function $f_X(x)$ of the attribute X, is measured by computing the following value:

$$I(f_X, \widehat{f_X}) = \frac{1}{2} E\left[\int_{\Omega_X} \left| f_X(x) - \widehat{f_X}(x) \right| dx \right] \tag{8.10}$$

that is, half of the expected value of L_1 norm between $f_X(x)$ and $\widehat{f_X}(x)$, which are the density distributions respectively before and after the application of the privacy preserving technique.

When considering the cryptography-based techniques which are typically employed in distributed environments, we can observe that they do not use any kind of perturbation techniques for the purpose of privacy preserving. Instead, they use the cryptographic techniques to assure data privacy at each site by limiting the information shared by all the sites. Therefore, the quality of data stored at each site is not compromised at all.

Completeness and Consistency. While the accuracy is a relatively general parameter in that it can be measured without strong assumptions on the dataset analyzed, the completeness is not so general. For example, in some PPDM strategies, e.g. blocking, the completeness evaluation is not significant. On the other hand, the consistency requires to determine all the relationships that are relevant for a given dataset.

In [5], Bertino et al. propose a set of evaluation parameters including the completeness and consistency evaluation. Unlike other techniques, their approach takes into account two more important aspects: relevance of data and structure of database. They provide a formal description that can be used to magnify the aggregate information of interest for a target database and the relevance of data quality properties of each aggregate information and for each attribute involved in the aggregate information. Specifically, the completeness lack (denoted as CML) is measured as follows:

$$CML = \sum_{i=0}^{n} (DMG.N_i.CV \times DMG.N_i.CW) \quad (8.11)$$

In equation 8.11, DMG is an oriented graph where each node N_i is an attribute class. CV is the completeness value and CW is the consistency value. The consistency lack (denoted as CSL) is given by the number of constraint violations occurred in all the sanitized transaction multiplied by the weight associated with every constraints.

$$\begin{aligned} CSL &= \sum_{i=0}^{n} (DMG.SC_i.csv \times DMG.SC_i.cw) \\ &\quad + \sum_{j=0}^{m} (DMG.CC_j.csv \times DMG.CC_j.cw) \end{aligned} \quad (8.12)$$

In equation 8.11, csv indicates the number of violations, cw is the weight of the constraint, SC_i describes a simple constraint class, and CC_j describes a complex constraint class.

8.4.2 Quality of the Data Mining Results

In some situations, it can be useful and also more relevant to evaluate the quality of the data mining results after the sanitization process. This kind of metric is strictly related to the use the data are intended for. Data can be analyzed in order to mine information in terms of associations among single data items or to classify existing data with the goal of finding an accurate classification of new data items, and so on. Based on the intended data use, the information loss is measured with a specific metric, depending each time on the particular type of knowledge model one aims to extract.

If the intended data usage is data clustering, the information loss can be measured by the percentage of legitimate data points that are not well-classified after the sanitization process. As in [22], a misclassification error M_E is defined to measure the information loss.

$$M_E = \frac{1}{N} \sum_{i=1}^{k} (|Cluster_i(D)| - |Cluster_i(D')|) \tag{8.13}$$

where N represents the number of points in the original dataset, k is the number of clusters under analysis, and $|Cluster_i(D)|$ and $|Cluster_i(D')|$ represent the number of legitimate data points of the ith cluster in the original dataset D and the sanitized dataset D' respectively. Since a privacy preserving technique usually modify data for the sanitization purpose, the parameters involved in the clustering analysis is almost inevitably affected. In order to achieve high clustering quality, it is very important to keep the clustering results as consistent as possible before and after the application of a data hiding technique.

When quantifying information loss in the context of the other data usages, it is useful to distinguish between: *lost information* representing the percentage of non-sensitive patterns (i.e., association, classification rules) which are hidden as side-effect of the hiding process; and the *artifactual information* representing the percentage of artifactual patterns created by the adopted privacy preserving technique. For example, in [21], Oliveira and Zaiane define two metrics *misses cost* and *artifactual pattern* which are corresponding to *lost information* and *artifactual information* respectively. In particular, misses cost measures the percentage of non-restrictive patterns that are hidden after the sanitization process. This happens when some non-restrictive patterns lose support in the database due to the sanitization process. The misses cost (MC) is computed as follows:

$$MC = \frac{\# \sim R_P(D) - \# \sim R_P(D')}{\# \sim R_P(D)} \tag{8.14}$$

where $\# \sim R_P(D)$ and $\# \sim R_P(D')$ denote the number of non-restrictive patterns discovered from the original database D and the sanitized database D' respectively. In the best case, MC should be 0%. Notice that there is a compromise between the misses cost and the hiding failure in their approach. The more restrictive patterns they hide, the more legitimate patterns they miss. The other metric, artifactual pattern (AP), is measured in terms of the percentage of the discovered patterns that are artifacts. The formula is:

$$AP = \frac{|P'| - |P \bigcap P'|}{P'} \tag{8.15}$$

where $|X|$ denotes the cardinality of X. According to their experiments, their approach does not have any artifactual patterns, i.e., AP is always 0.

In case of association rules, the lost information can be modeled as the set of non-sensitive rules that are accidentally hidden, referred to as `lost rules`, by the privacy preservation technique, the artifactual information, instead, represents the set of new rules, also known as `ghost rules`, that can be extracted from the database after the application of a sanitization technique.

Similarly, if the aim of the mining task is data classification, e.g. by means of decision trees inductions, both the lost and artifactual information can be quantified by means of the corresponding lost and ghost association rules derived by the classification tree. These measures allow one to evaluate the high level information that are extracted from a database in form of the widely-used inference rules before and after the application of a PPDM algorithm.

It is worth noting that for most cryptography-based PPDM algorithms, the data mining results are the same as that produced from unsanitized data.

8.5 Complexity Metrics

The *complexity* metric measures the efficiency and scalability of a PPDM algorithm. Efficiency indicates whether the algorithm can be executed with good performance, which is generally assessed in terms of space and time. Space requirements are assessed according to the amount of memory that must be allocated in order to implement the given algorithm.

For the evaluation of time requirements, there are several approaches. The first approach is to evaluate the CPU time. For example, in [21], they first keep constant both the size of the database and the set of restrictive patterns, and then increase the size of the input data to measure the CPU time taken by their algorithm. An alternative approach would be to evaluate the time requirements in terms of the computational cost. In this case, it is obvious that an algorithm having a polynomial complexity is more efficient than another one with exponential complexity. Sometimes, the time requirements can even be evaluated by counting the average number of operations executed by a PPDM algorithm. As in [14], the performance is measured in terms of the number of encryption and decryption operations required by the specific algorithm. The last two measures, i.e. the computational cost and the average number of operations, do not provide an absolute measure, but they can be considered in order to perform a fast comparison among different algorithms.

In case of distributed algorithms, especially the cryptography-based algorithms (e.g. [14, 31]), the time requirements can be evaluated in terms of communication cost during the exchange of information among secure processing. Specifically, in [14], the communication cost is expressed as the number of messages exchanged among the sites, that are required by the protocol for securely counting the frequency of each rule.

Scalability is another important aspect to assess the performance of a PPDM algorithm. In particular, scalability describes the efficiency trends when data sizes increase. Such parameter concerns the increase of both performance and storage requirements as well as the costs of the communications required by a distributed technique with the increase of data sizes.

Due to the continuous advances in hardware technology, large amounts of data can now be easily stored. Databases along with data warehouses today store and manage amounts of data which are increasingly large. For this reason, a PPDM algorithm has to be designed and implemented with the capability of handling huge datasets that may still keep growing. The less fast is the decrease in the efficiency of a PPDM algorithm for increasing data dimensions, the better is its scalability. Therefore, the scalability measure is very important in determining practical PPDM techniques.

8.6 How to Select a Proper Metric

In previous section, we have discussed various types of metrics. An important question here is "which one among the presented metrics is the most relevant for a given privacy preserving technique?".

Dwork and Nissim [9] make some interesting observations about this question. In particular, according to them in the case of statistical databases *privacy* is paramount, whereas in the case of distributed databases for which the privacy is ensured by using a secure multiparty computation technique *functionality* is of primary importance. Since a real database usually contains a large number of records, the performance guaranteed by a PPDM algorithm, in terms of time and communication requirements, is a not negligible factor, as well as its trend when increasing database size. The *data quality* guaranteed by a PPDM algorithm is, on the other hand, very important when ensuring privacy protection without damaging the data usability from the authorized users.

From the above observations, we can see that a trade-off metric may help us to state a unique value measuring the effectiveness of a PPDM algorithm. In [7], the score of a masking method provides a measure of the trade-off between disclosure risk and information loss. It is defined as an average between the ranks of disclosure risk and information loss measures, giving the same importance to both metrics. In [8], a R-U confidentiality map is described that traces the impact on disclosure risk R and data utility U of changes in the parameters of a disclosure limitation method which adopts an additive noise technique. We believe that an index assigning the same importance to both the data quality and the degree of privacy ensured by a PPDM algorithm is quite restrictive, because in some contexts one of these parameters can be more relevant than the other. Moreover, in our opinion the other parameters, even less relevant ones, should be also taken into account. The efficiency and scalability measures, for instance, could be discriminating factors in choosing among a set of PPDM algorithms that ensure similar degrees of privacy and data utility. A *weighted mean* could be, thus, a good measure for evaluating by means of a unique value the quality of a PPDM algorithm.

8.7 Conclusion and Research Directions

In this chapter, we have surveyed different approaches used in evaluating the effectiveness of privacy preserving data mining algorithms. A set of criteria is identified, which are *privacy level*, *hiding failure*, *data quality* and *complexity*. As none of the existing PPDM algorithms can outperform all the others with respect to all the criteria, we discussed the importance of certain metrics for each specific type of PPDM algorithms, and also pointed out the goal of a good metric.

There are several future research directions along the way of quantifying a PPDM algorithm and its underneath application or data mining task. One is to develop a comprehensive framework according to which various PPDM algorithms can be evaluated and compared. It is also important to design good metrics that can better reflect the properties of a PPDM algorithm, and to develop benchmark databases for testing all types of PPDM algorithms.

References

[1] Agrawal, D., Aggarwal, C.C.: On the design and quantification of privacy preserving data mining algorithms. In: Proceedings of the 20th ACM SIGACT-SIGMOD-SIGART Symposium on Principle of Database System, pp. 247–255. ACM (2001)

[2] Agrawal, R., Srikant, R.: Privacy preserving data mining. In: Proceeedings of the ACM SIGMOD Conference of Management of Data, pp. 439–450. ACM (2000)

[3] Ballou, D., Pazer, H.: Modelling data and process quality in multi input, multi output information systems. Management science **31**(2), 150–162 (1985)

[4] Bayardo, R., Agrawal, R.: Data privacy through optimal k-anonymization. In: Proc. of the 21st Int'l Conf. on Data Engineering (2005)

[5] Bertino, E., Fovino, I.N.: Information driven evaluation of data hiding algorithms. In: 7th Internationa Conference on Data Warehousing and Knowledge Discovery, pp. 418–427 (2005)

[6] Bertino, E., Fovino, I.N., Provenza, L.P.: A framework for evaluating privacy preserving data mining algorithms. Data Mining and Knowledge Discovery **11**(2), 121–154 (2005)

[7] Domingo-Ferrer, J., Torra, V.: A quantitative comparison of disclosure control methods for microdata. In: L. Zayatz, P. Doyle, J. Theeuwes, J. Lane (eds.) Confidentiality, Disclosure and Data Access: Theory and Practical Applications for Statistical Agencies, pp. 113–134. North-Holland (2002)

[8] Duncan, G.T., Keller-McNulty, S.A., Stokes, S.L.: Disclosure risks vs. data utility: The R-U confidentiality map. Tech. Rep. 121, National Institute of Statistical Sciences (2001)

[9] Dwork, C., Nissim, K.: Privacy preserving data mining in vertically partitioned database. In: CRYPTO 2004, vol. 3152, pp. 528–544 (2004)

[10] Evfimievski, A.: Randomization in privacy preserving data mining. SIGKDD Explor. Newsl. **4**(2), 43–48 (2002)

[11] Evfimievski, A., Srikant, R., Agrawal, R., Gehrke, J.: Privacy preserving mining of association rules. In: 8th ACM SIGKDD International Conference on Knowledge Discovery and Data Mining, pp. 217–228. ACM-Press (2002)

[12] Fung, B.C.M., Wang, K., Yu, P.S.: Top-down specialization for information and privacy preservation. In: Proceedings of the 21st IEEE International Conference on Data Engineering (ICDE 2005). Tokyo, Japan (2005)

[13] Iyengar, V.: Transforming data to satisfy privacy constraints. In: Proc., the Eigth ACM SIGKDD Int'l Conf. on Knowledge Discovery and Data Mining, pp. 279–288 (2002)

[14] Kantarcioglu, M., Clifton, C.: Privacy preserving distributed mining of association rules on horizontally partitioned data. In: ACM SIGMOD Workshop on Research Issues in Data Mining and Knowledge Discovery, pp. 24–31 (2002)

[15] Kantarcıoğlu, M., Jin, J., Clifton, C.: When do data mining results violate privacy? In: Proceedings of the 2004 ACM SIGKDD International Conference on Knowledge Discovery and Data Mining, pp. 599–604. Seattle, WA (2004).

[16] Kargupta, H., Datta, S., Wang, Q., Sivakumar, K.: On the privacy preserving properties of random data perturbation techniques. In: Proceedings of the Third IEEE International Conference on Data Mining (ICDM'03). Melbourne, Florida (2003)

[17] Kifer, D., Gehrke, J.: Injecting utility into anonymized datasets. In: Proceedings of the 2006 ACM SIGMOD International Conference on Management of Data, pp. 217–228. ACM Press, Chicago, IL, USA (2006)

[18] Kumar Tayi, G., Ballou, D.P.: Examining data quality. Communications of the ACM **41**(2), 54–57 (1998)

[19] Liu, K., Kargupta, H., Ryan, J.: Random projection-based multiplicative data perturbation for privacy preserving distributed data mining **18**(1), 92–106 (2006)

[20] Nergiz, M.E., Clifton, C.: Thoughts on k-anonymization. In: The Second International Workshop on Privacy Data Management held in conjunction with The 22nd International Conference on Data Engineering. Atlanta, Georgia (2006)

[21] Oliveira, S.R.M., Zaiane, O.R.: Privacy preserving frequent itemset mining. In: IEEE icdm Workshop on Privacy, Security and Data Mining, vol. 14, pp. 43–54 (2002)

[22] Oliveira, S.R.M., Zaiane, O.R.: Privacy preserving clustering by data transformation. In: 18th Brazilian Symposium on Databases (SBBD 2003), pp. 304–318 (2003)

[23] Oliveira, S.R.M., Zaiane, O.R.: Toward standardization in privacy preserving data mining. In: ACM SIGKDD 3rd Workshop on Data Mining Standards, pp. 7–17 (2004)

[24] Rizvi, S., Haritsa, R.: Maintaining data privacy in association rule mining. In: 28th International Conference on Very Large Databases, pp. 682–693 (2002)

[25] Samarati, P.: Protecting respondents' identities in microdata release. IEEE Transactions on Knowledge and Data Engineering (TKDE) **13**(6), 1010–1027 (2001).

[26] Schoeman, F.D.: Philosophical Dimensions of Privacy: An Anthology. Cambridge University Press. (1984)

[27] Shannon, C.E.: A mathematical theory of communication. Bell System Technical Journal **27**, 379–423, 623–656 (1948)

[28] Sweeney, L.: Achieving *k*-anonymity privacy protection using generalization and suppression. International Journal of Uncertainty, Fuzziness and Knowledge Based Systems **10**(5), 571–588 (2002)

[29] Trottini, M.: A decision-theoretic approach to data disclosure problems. Research in Official Statistics **4**, 7–22 (2001)

[30] Trottini, M.: Decision models for data disclosure limitation. Ph.D. thesis, Carnegie Mellon University (2003).

[31] Vaidya, J., Clifton, C.: Privacy preserving association rule mining in vertically partitioned data. In: 8th ACM SIGKDD International Conference on Knowledge Discovery and Data Mining, pp. 639–644. ACM Press (2002)

[32] Verykios, V.S., Bertino, E., Nai Fovino, I., Parasiliti, L., Saygin, Y., Theodoridis, Y.: State-of-the-art in privacy preserving data mining. SIGMOD Record **33**(1), 50–57 (2004)

[33] Walters, G.J.: Human Rights in an Information Age: A Philosophical Analysis, chap. 5. University of Toronto Press. (2001)

[34] Wang, R.Y., Strong, D.M.: Beyond accuracy: what data quality means to data consumers. Journal of Management Information Systems **12**(4), 5–34 (1996)

[35] Willenborg, L., De Waal, T.: Elements of statistical disclosure control, *Lecture Notes in Statistics*, vol. 155. Springer (2001)

Chapter 9

A Survey of Utility-based Privacy-Preserving Data Transformation Methods

Ming Hua
Simon Fraser University
School of Computing Science
8888 University Drive, Burnaby, BC, Canada V5A 1S6
mhua@cs.sfu.ca

Jian Pei
Simon Fraser University
School of Computing Science
8888 University Drive, Burnaby, BC, Canada V5A 1S6
jpei@cs.sfu.ca

Abstract As a serious concern in data publishing and analysis, privacy preserving data processing has received a lot of attention. Privacy preservation often leads to information loss. Consequently, we want to minimize utility loss as long as the privacy is preserved. In this chapter, we survey the utility-based privacy preservation methods systematically. We first briefly discuss the privacy models and utility measures, and then review four recently proposed methods for utility-based privacy preservation.

We first introduce the *utility-based anonymization* method for maximizing the quality of the anonymized data in query answering and discernability. Then we introduce the *top-down specialization (TDS)* method and the *progressive disclosure algorithm (PDA)* for privacy preservation in classification problems. Last, we introduce the *anonymized marginal* method, which publishes the anonymized projection of a table to increase the utility and satisfy the privacy requirement.

Keywords: Privacy preservation, data utility, utility-based privacy preservation, k-anonymity, sensitive inference, l-diversity.

9.1 Introduction

Advanced analysis on data sets containing information about individuals poses a serious threat to individual privacy. Various methods have been proposed to tackle the privacy preservation problem in data analysis, such as anonymization and perturbation. The major goal is to protect some sensitive individual information (privacy) from being identified by the published data. For example, in k-anonymization, certain individual information is generalized or suppressed so that any individual in a released data set is indistinguishable from other $k-1$ individuals.

A natural consequence of privacy preservation is the information loss. For example, after the k-anonymization, the information describing an individual should be the same as at least other $k-1$ individuals. The loss of the specific information about certain individuals may affect the data quality. In the extreme case, the data may become totally useless.

EXAMPLE 9.1 (UTILITY LOSS IN PRIVACY PRESERVATION) *Table 9.1a is a data set used for customer analysis. Among the listed attributes,* {Age, Education, Zip Code} *can be used to uniquely identify an individual. Such a set of attributes is called a quasi-identifier.* Annual Income *is a sensitive attribute.* Target Customer *is the class label of customers.*

In order to protect the annual income information for individuals, suppose 2*-anonymity is required so that any individual is indistinguishable from another one on the quasi-identifier. Table 9.2b and 9.3c are both valid* 2*-anonymizations of 9.1a. The tuples sharing the same quasi-identifier have the same gId. However, Table 9.2b provides more accurate results than Table 9.3c in answering the following two queries.*

$Q1$: "How many customers under age 29 *are there in the data set?"*

$Q2$: "Is an individual with $age = 25$, $Education = Bachelor$, $Zip\ Code = 53712$ a target customer?"

According to Table 9.2b, the answers of $Q1$ and $Q2$ are "2" and "Y", respectively. But according to Table 9.3c, the answer to $Q1$ is an interval $[0, 4]$, because 29 *falls in the* age *range of tuple $t1$, $t2$, $t4$, and $t6$. The answer to $Q2$ is Y and N with* 50% *probability each.*

From this example, we make two observations. *First, different anonymization may lead to different information loss.* Table 9.2b and 9.3c are in the same anonymization level, but Table 9.2b provides more accurate answers to the queries. Therefore, it is crucial to minimize the information loss in privacy preservation.

Second, the data utility depends on the applications using the data. In the above example, $Q1$ is an aggregate query, thus the data is more useful if the attribute values are more accurate. $Q2$ is a classification query, so the utility

Table 9.1a. The original table

tId	*Age*	*Education*	*Zip Code*	*Annual Income*	*Target Customer*
t1	24	Bachelor	53711	40k	Y
t2	25	Bachelor	53712	50k	Y
t3	30	Master	53713	50k	N
t4	30	Master	53714	80k	N
t5	32	Master	53715	50k	N
t6	32	Doctorate	53716	100k	N

Table 9.2b. A 2-anonymized table with better utility

gId	*tId*	*Age*	*Education*	*Zip Code*	*Annual Income*	*Target Customer*
g1	t1	[24-25]	Bachelor	[53711-53712]	40k	Y
g1	t2	[24-25]	Bachelor	[53711-53712]	50k	Y
g2	t3	30	Master	[53713-53714]	50k	N
g2	t4	30	Master	[53713-53714]	80k	N
g3	t5	32	GradSchool	[53715-53716]	50k	N
g3	t6	32	GradSchool	[53715-53716]	100k	N

Table 9.3c. A 2-anonymized table with poorer utility

gId	*tId*	*Age*	*Education*	*Zip Code*	*Annual Income*	*Target Customer*
g1	t1	[24-30]	ANY	[53711-53714]	40k	Y
g2	t2	[25-32]	ANY	[53712-53716]	50k	Y
g3	t3	[30-32]	Master	[53713-53715]	50k	N
g1	t4	[24-30]	ANY	[53711-53714]	80k	N
g3	t5	[30-32]	Master	[53713-53715]	50k	N
g2	t6	[25-32]	ANY	[53712-53716]	100k	N

of data depends on how much the classification model is preserved in the anonymized data. In a word, utility is the quality of data for the intended use.

9.1.1 What is Utility-based Privacy Preservation?

The *utility-based privacy preservation* has two goals: protecting the private information and preserving the data utility as much as possible. Privacy preservation is a hard requirement, that is, it must be satisfied, and utility is the measure to be optimized. While privacy preservation has been extensively studied, the research of utility-based privacy preservation has just started. The challenges include:

Utility measure. One key issue in the utility-based privacy preservation is how to model the data utility in different applications. A good utility measure should capture the intrinsic factors that affect the quality of data for the specific application.

Balance between utility and privacy. In some situation, preserving utility and privacy are not conflicting. But more often than not, hiding the privacy information may have to sacrifice some utility. How do we trade off between the two goals?

Efficiency and scalability. The traditional privacy preservation is already computational challenging. For example, even simple restriction of optimized k-anonymity is NP-hard [3]. How do we develop efficient algorithms if utility is involved? Moreover, real data sets often contains millions of high dimensional tuples, highly scalable algorithms are needed.

Ability to deal with different types of attributes. Real life data often involve different types of attributes, such as numerical, categorical, binary or mixtures of these data types. The utility-based privacy preserving methods should be able to deal with attributes of different types.

9.2 Types of Utility-based Privacy Preservation Methods

In this section, we introduce some common privacy models and recently proposed data utility measures.

9.2.1 Privacy Models

Various privacy models have been proposed in literature. This section introduces some of the privacy models that are often used as well as the corresponding privacy preserving methods.

***K*-Anonymity.** K-anonymity is a privacy model developed for the linking attack [18]. Given a table T with attributes $(A_1, \ldots, A_n)$, a *quasi-identifier* is a minimal set of attributes $(A_{i_1}, \ldots, A_{i_l})$ $(1 \leq i_1 < \ldots < i_l \leq n)$ in T that can be joined with external information to re-identify individual records. Note that there may be more than one quasi-identifer in a table.

A table T is said *k-anonymous* given a parameter k and the quasi-identifer $QI = (A_{i_1}, \ldots, A_{i_l})$ if for each tuple $t \in T$, there exist at least another $(k-1)$ tuples $t_1, \ldots, t_{k-1}$ such that those k tuples have the same projection on the quasi-identifier. Tuple t and all other tuples indistinguishable from t on the quasi-identifier form an *equivalence class*.

Given a table T with the quasi-identifier and a parameter k, the problem of k-*anonymization* is to compute a view T' that has the same attributes as T such that T' is k-anonymous and as close to T as possible according to some quality metric.

Data suppression and value generalization are often used for anonymization. Suppression is masking the attribute value with a special value in the domain. Generalization is replacing a specific value with a more generalized one. For example, the actual age of an individual can be replaced by an interval, or the city of an individual can be replaced by the corresponding province. Certain quality measures are often used in the anonymization, such as the average equivalence class size. Theoretical analysis shows that the problem of optimal anonymization under many quality models is NP-hard [1, 14, 3]. Various k-anonymization methods are proposed [19, 20, 29, 12, 11].

One of the most important advantages of k-anonymity is that no additional noise or artificial perturbation is added into the original data. All the tuples in an anonymized data remains trustful.

l-Diversity. l-diversity [13] is based on the observation that if the sensitive values in one equivalence class lacks diversity, then no matter how large the equivalence class is, attacker may still guess the sensitive value of an individual with high probability. For example, Table 9.3c is a 2-anonymous table. Particularly, $t3$ and $t5$ are generalized into the same equivalence class. However, since their annual income is the same, an attacker can easily conclude that the annual income of $t3$ is $50k$ although the 2-anonymity is preserved. Table 9.2b has better diversity in the sensitive attribute. $t3$ and $t4$ are in the same equivalence class and their annual income is different. Therefore, the attacker only have a 50% opportunity to know the real annual income of $t3$.

l-diversity model addresses the above problem. By intuition, a table is l-diverse if each equivalence class contains at least l "well represented" sensitive values, that is, at least l most frequent values have very similar frequencies. Consider a table $T = (A_1, \ldots, A_n, S)$ and constant c and l, where $(A_1, \ldots, A_n)$ is a quasi-identifier and S is a sensitive attribute. Suppose an equivalence class EC contains value $s_1, \ldots, s_m$ with frequency $f(s_1), \ldots, f(s_m)$ (appearing in the frequency non-ascending order) on sensitive attribute S, EC satisfies (c, l)-diversity with respect to S if

$$f(s_1) < c \sum_{i=l}^{m} f(s_i)$$

l-diversity complements k-anonymity by requiring certain diversity on the sensitive attributes. It is a more practical privacy model.

Sensitive Inference. Sensitive inference [23] comes from the statistical analysis and data mining ability. The privacy inference occurs when the sensitive value can be determined from a set of other non-sensitive values with high confidence. The inference can be achieved by data mining abilities, such as association rule mining and classification.

Given a table $T = (M_1, \ldots, M_m, S_1, \ldots, S_n)$. S_i $(1 \leq i \leq n)$ is called a *sensitive attribute*. Certain values in the sensitive attributes are not accessible. M_j $(1 \leq j \leq m)$ is called a *non-sensitive attribute* and contains the non-sensitive information. M_j's and S_i's are disjoint. A *sensitive inference* is a rule $\{m_{i_1}, \ldots, m_{i_l}\} \rightarrow s_j$ with high confidence, where $m_{i_1}, \ldots, m_{i_l}$ are attribute values on non-sensitive attributes $M_{i_1}, \ldots, M_{i_l}$, respectively, and s_j is an inaccessible sensitive value on S_j. For example, suppose in Table 9.1a $AnnualIncome$ is a sensitive attribute and the values lower than $60k$ in this attribute is confidential and should not be disclosed. However, rule $Master \rightarrow 50k$ can be derived from Table 9.1a with confidence 66.7%. Therefore, although $50k$ is not disclosed, an attacker can guess the value from the non-sensitive value $Master$ with high probability.

Data suppression can be used in eliminating the sensitive inference. The intuition is to mask some non-sensitive information causing the inference, so that the confidence of the inference rule decreases to below certain threshold.

[2] deals with the sensitive inference caused by association mining. The objective is to hide a minimal set of entries so that the sensitive fields cannot be disclosed by the sensitive inferences. Other work on eliminating sensitive inference includes [15, 22].

Other Related Work. Another privacy model *Anatomy* is proposed in [26], which publishes the quasi-identifer and the sensitive information into two separate tables. Equivalence classes are formed without generalizing the values in the quasi-identifier. The advantage is that more information in the quasi-identifier is preserved. [25] proposes a privacy model *m-invariance* in a dynamic context, that is, a sequential releasing of table with any sequence of insertion and deletions. The main objective is to make each tuple indistinguishable during its lifetime in the publication. In order to prevent the attacker from linking different versions of the released tables together to obtain the sensitive information, certain "invariance" (in terms of having the similar sensitive values) in each equivalence class is required.

9.2.2 Utility Measures

In the context of privacy preservation, the data utility is both *relative* and *specific*. First, we do not consider the absolute utility of a data set, instead, we measure how much utility is preserved in the published data after privacy preservation compared to the original data. Second, different applications require different information in a data set. We cannot find a measure to quantify the amount of information contained in the data for all different applications. Therefore, the utility measure should be designed under the context of certain applications.

Query Answering Accuracy. One common use of the published data is query answering, such as the aggregate queries including *SUM*, *COUNT* and *AVERAGE*. The data quality in query answering depends on how far away each attribute value is from the original one after applying the privacy preserving methods. For example, if generalization is used in privacy preservation, then a specific value is replaced by a more general one in the published data. Intuitively, in order to maximize the query answering accuracy, the generalized value in the published data should be as close to the original value as possible. A quantitative measure is proposed in [28, 27]. It uses the normalized interval size to measure the utility loss for numeric attributes, and the normalized number of descendants in the generalization hierarchy to measure the utility loss for categorical attributes. More details will be discussed in Section 9.3.

Classification Accuracy. In classification analysis, the published data are often used to train a classifier. Thus, the data quality depends on how well the class structure is preserved. More specifically, we want to minimize the uncertainty of classification within a group of tuples indistinguishable from each other. [5, 6] propose a utility score that measures the entropy change during the anonymization. Ideally, the entropy of an equivalence class with respect to class label distribution should be minimized in the published data. Other utility measures for classification are proposed in [7, 24].

Distribution Similarity. Distribution is a fundamental characteristic of a data set. Many data analysis try to make certain conclusions about the data distribution. Therefore, how well the published data preserve the distribution of the original data is crucial for those applications. [8] develops a utility model which measures the difference between the distribution of the original data and that of the anonymized data.

Other Utility Measures. Other utility measures include *generalization height* [17], which measures the total number of generalization steps applied on the original data set. The idea behind is that, since the generalization steps cause information loss, the number of generalization steps represents the total amount of information loss. [3] considers the *discernability* of the anonymized data. It tries to minimize the average equivalence class size, because the more tuples are in the same group, the less specific information is preserved for those tuples.

9.2.3 Summary of the Utility-Based Privacy Preserving Methods

In this chapter, we introduce four utility-based privacy preservation methods. They are the *utility-based anonymization* method [28], the *top-down specialization (TDS)* method [5], the *progressive Disclosure Algorithm (PDA)* [23] and the *anonymized marginal* method [8]. The privacy models and utility measures used in the four methods are summarized in Table 9.4.

Table 9.4. Summary of utility-based privacy preserving methods

Method	*Privacy model*	*Utility measure*
Utility-based anonymization	k-Anonymity	Query answering accuracy
TDS	k-Anonymity	Classification accuracy
PDA	Sensitive Inference	Classification accuracy
Anonymized marginal	k-Anonymity & l-Diversity	Distribution similarity

The *utility-based anonymization* method will be discussed in Section 9.3. The *top-down specialization (TDS)* method and the *progressive Disclosure Algorithm (PDA)* both deal with data used in classification problems, and will be discussed in Section 9.4. Section 9.5 introduces the *anonymized marginal* method. Section 9.6 concludes this chapter.

9.3 Utility-Based Anonymization Using Local Recoding

The utility-based anonymization method proposed in [28] aims at improving the query answering accuracy on anonymized tables. The utility measure proposed captures two aspects. *First, the less generalized attribute value gives more accurate answers in query answering on the anonymized table.* For example, Table 9.2b and 9.3c are both 2-anonymous, but the *age* attribute is less generalized in Table 9.2b. If we perform the aggregate query on this attribute, Table 9.2b gives more accurate answers. *Second, different attributes may have different utility in data analysis.* For example, in Table 9.1a, suppose that the information about *annual income* is more related to *age* and *education* than the other attributes, in order to preserve the correlation among the data in anonymization, it is better to generalize other attributes which are not so related to annual income, such as "*Zip Code*".

Based on the above observations, the *weighted normalized certainty penalty* is proposed to measure the *utility* of attributes in the anonymization. For a numeric attribute value, the *normalized certainty penalty (NCP)* measures its normalized interval size after generalization; for a categorical attribute value, *NCP* measures its normalized number of descendants in the hierarchy tree after generalization. A weight is assigned to each attribute to reflect its utility in the analysis on the anonymized data. Given a table and the anonymity requirement, such as k-anonymity, the *utility-based anonymization* aims at computing a k-anonymous table that minimizes the weighted normalized certainty penalty.

In order to tackle the problem, two algorithms are proposed. The *bottom-up method* iteratively groups the tuples with similar attribute values together until each group has at least k tuples. The *top-down method* works in the opposite way. It put all tuples into one group at the beginning, and then iteratively partitions the tuples in a group into two groups, trying to maximize the difference of attribute values between the two groups. The partitioning stops when further splitting violates the k-anonymity.

To give some details, first we introduce the *local recoding* method in the anonymization. Then, we define the utility measure, *weighted normalized certainty penalty*, formally. The *bottom-up method* and *top-down method* are introduced last.

9.3.1 Global Recoding and Local Recoding

Two methods have been proposed for anonymization: *global recoding* and *local recoding*. Global recoding maps a given value in a single domain to another one globally, while local recoding maps a given tuple to some recoded tuple. Clearly, global recoding can be regarded as a specific type of local recoding.

Table 9.5a. 3-anonymous table by global recoding

gId	*tId*	*Age*	*Education*	*Zip Code*	*Annual Income*	*Target Customer*
g1	t1	[24-32]	ANY	[53711-53713]	40k	Y
g1	t2	[24-32]	ANY	[53711-53713]	50k	Y
g1	t3	[24-32]	ANY	[53711-53713]	50k	N
g2	t4	[24-32]	ANY	[53714-53716]	80k	N
g2	t5	[24-32]	ANY	[53714-53716]	50k	N
g2	t6	[24-32]	ANY	[53714-53716]	100k	N

Table 9.6b. 3-anonymous table by local recoding

gId	*tId*	*Age*	*Education*	*Zip Code*	*Annual Income*	*Target Customer*
g1	t1	[24-30]	ANY	[53711-53713]	40k	Y
g1	t2	[24-30]	ANY	[53711-53713]	50k	Y
g1	t3	[24-30]	ANY	[53711-53713]	50k	N
g2	t4	[30-32]	GradSchool	[53714-53716]	80k	N
g2	t5	[30-32]	GradSchool	[53714-53716]	50k	N
g2	t6	[30-32]	GradSchool	[53714-53716]	100k	N

EXAMPLE 9.2 (GLOBAL RECODING V.S. LOCAL RECODING) *Consider Table 9.1a. Attribute* {Age, Education, Zip Code} *is a quasi-identifer. Table 9.5a and 9.6b are 3-anonymous tables by global recoding and local recoding, respectively.* $t3$ *and* $t4$ *have the same attribute values on* Age(30) *and* Education($Master$)*. In global recoding,* 30 *is mapped to interval* $[24 - 32]$ *globally, while* $Master$ *is mapped to the most generalized value* ANY*. In local recoding, the value* 30 *and* $Master$ *in* $t3$ *is mapped to* $[24 - 30]$ *and* ANY*, respectively; but the same values* 30 *and* $Master$ *in* $t4$ *are mapped to* $[30 - 32]$ *and* $GradSchool$*, respectively. After the global recoding, the only knowledge about* Age *and* Education *in Table 9.5a is the full range* $[24 - 32]$ *in the domain and the most generalized value* ANY*, respectively; while Table 9.6b shows more specific information about* Age *and* Education.

From the above example, we can see that the local recoding may lead to less information loss than the global recoding.

9.3.2 Utility Measure

The information loss caused by the anonymization can be measured by how well the generalized tuples approximate the original ones. After the generalization, some attribute values of a tuple are generalized to an interval. The interval size reflects the accuracy loss in query answering. Therefore, we use the sum of interval size on all attributes of the generalized tuples to measure the certainty loss. The total certainty loss of the anonymized table is the sum of certainty loss of all the tuples.

Utility Measure for Numerical Attributes. Consider table T with quasi-identifier $(A_1, \ldots, A_n)$. Suppose a tuple $t = (x_1, \ldots, x_n)$ is generalized to tuple $t' = ([y_1, z_1], \ldots, [y_n, z_n])$ such that $y_i \leq x_i \leq z_i$ $(1 \leq i \leq n)$. Then, we define the *normalized certainty penalty* (NCP) of tuple t on attribute A_i as

$$NCP_{A_i}(t) = \frac{z_i - y_i}{|A_i|}, \text{ where } |A_i| = \max_{t \in T} t.A_i - \min_{t \in T} t.A_i$$

Utility Measure for Categorical Attributes. The generalization on a categorical attribute often follows a hierarchy tree, which specifies the attribute values with different granularity. Suppose a tuple t has value v on categorical attribute A_i, v is generalized to a set of values $v_1, \ldots, v_m$. We find the *common ancestor* of $v_1, \ldots, v_m$, denoted by $ancestor(v_1, \ldots, v_m)$ in the hierarchy tree, and use the size of $ancestor(v_1, \ldots, v_m)$, that is, the number of leaf nodes that are descendants of $ancestor(v_1, \ldots, v_m)$, to measure the generalization quantitatively. That is

$$NCP_{A_i}(t) = \frac{|ancestor(v_1, \ldots, v_m)|}{|A_i|}$$

where $|A_i|$ is the number of distinct values on A_i in the most specific level.

Intuitively, for a numeric attribute A_i, $NCP_{A_i}(t)$ measures *how much t is generalized on attribute A_i in terms of the generalized interval size*; for a categorical attribute A_i, $NCP_{A_i}(t)$ measures *how much t is generalized on A_i in terms of the number of distinct values the generalized value covers.* Consider both the numeric and categorical attributes, we define the *weighted normalized certainly penalty* of a tuple t as

$$NCP(t) = \sum_{i=1}^{n} (w_i \cdot NCP_{A_i}(t)), \text{ where } \sum_{i=1}^{n} w_i = 1$$

Moreover, the *weighted normalized certainly penalty* of a table T is defined as

$$NCP(T) = \sum_{t \in T} NCP(t)$$

Since in many data analysis applications, different attributes may have different utility, we assign each attribute a weight to reflect the different importance of the attribute. Therefore, $NCP(t)$ is the weighted sum of the normalized certainty penalty on all attributes. The certainty penalty on the whole table is the sum of penalty on all the tuples.

EXAMPLE 9.3 (WEIGHTED NORMALIZED CERTAINTY PENALTY) *Consider Table 9.1a and the corresponding 3-anonymous Table 9.6b. Suppose the weights of attributes* Age, Education *and* Zip Code *are* 0.5, 0.4 *and* 0.1, *respectively. Then, we have*

$$\begin{aligned}
NCP_{Age}(t1) &= \tfrac{30-24}{32-24} = \tfrac{6}{8} = 0.75 \\
NCP_{Edu}(t1) &= \tfrac{3}{3} = 1 \\
NCP_{ZipCode}(t1) &= \tfrac{53713-53711}{53716-53711} = \tfrac{2}{5} = 0.4 \\
NCP(t1) &= W_{Age} \times NCP_{Age}(t1) + W_{Edu} \times NCP_{Edu}(t1) + W_{ZipCode} \\
&\quad \times NCP_{ZipCode}(t1) = 0.5 \times 0.75 + 0.4 \times 1 + 0.1 \times 0.4 = 0.815
\end{aligned}$$

9.3.3 Anonymization Methods

As shown in [3], optimal k-anonymization under simple restrictions is NP-hard. As a generalization, the utility-based anonymization is also NP-hard. Two heuristic local recoding algorithms are proposed to solve the problem.

The Bottom-up Method. The *bottom-up method* puts a tuple in a group at the beginning, and then iteratively merges the small groups into larger ones. In each iteration, a group whose population is less than k is combined with another group such that the combined group has the minimal utility loss. The iteration terminates when each group has at least k tuples. Each group forms an equivalence class. The algorithm is illustrated using the following example.

EXAMPLE 9.4 (THE BOTTOM-UP METHOD) *Consider Table 9.1a. The 2-anonymization using the bottom-up method works as follows.*

First, each tuple forms a group. For each group whose size is smaller than 2, merge it with another group which minimizes the certainty penalty. For example, since $|g1| = |\{t1\}| < 2$, *we calculate the certainty penalty* $NCP(g1, gi)$ *for* $2 \leq i \leq 6$ *and find that merging* $g1$ *with* $g2$ *minimizes* $NCP(g1, gi)$. *Similarly,* $g3$ *is merged with* $g4$, *and* $g5$ *is merged with* $g6$. *The final anonymized table is shown in Table 9.2b.*

The Top-down Method. The *top-down method* first treats the whole table as an equivalence class (group). The utility of the table is minimal since all tuples are generalized to the same. It then recursively partitions a group into two groups if each subset contains at least k tuples and the utility is improved. The algorithm stops when further partitioning leads to the violation of the k-anonymity.

EXAMPLE 9.5 (THE TOP-DOWN METHOD) *Consider the 2-anonymization of Table 9.1a. The top-down method iteratively partitions the tuples into two groups* G_u *and* G_v, *trying to minimize the certainty penalty. It first finds a pair of tuples (as the seeds) that maximize the normalized certainty penalty.* $t1$ *and* $t6$ *are the tuples. Thus,* $t1$ *and* $t6$ *are added into* G_u *and* G_v, *respectively. Then, we assign all the other tuples to one of the two groups. For each tuple* t, *we calculate* $NCP(G_u, t)$ *and* $NCP(G_v, t)$, *and assign* t *to the group with the smaller* NCP *value. For example,* $t2$ *is assigned into* G_u *because* $NCP(G_u, t2)$ *is smaller than* $NCP(G_v, t2)$.

After all tuples are assigned, we have $G_u = \{t1, t2\}$ *and* $G_v = \{t3, t4, t5, t6\}$. *Since* $|G_u| = 2$, *we only partition* G_v *in the next iteration. Similarly, the seeds found in the next iteration are* $t3$ *and* $t6$. G_v *is partitioned into* $\{t3, t4\}$ *and* $\{t5, t6\}$. *The final anonymized table is shown in Table 9.2b.*

Finding the seeds u, v with the maximum $NCP(u, v)$ requires $O(|T|^2)$. A heuristic method can be used to accelerate the computation. It randomly picks a tuple t_1 and scans the table once to find another tuple t_2 maximizing $NCP(t_1, t_2)$. Then, by another scan, it finds the third tuple t_3 that maximizes $NCP(t_2, t_3)$. The process is repeated several times until $NCP(t_i, t_{i+1})$ does not increase significantly. Then we use t_i and t_{i+1} as the seeds.

During the top-down partition, some groups may have fewer than k tuples, we adjust such a group by combining it with another group which minimizes the certainty penalty.

The *bottom-up method* and the *top-down method* both provide satisfactory results. The *top-down method* is faster than the *bottom-up method*. This is because the *top-down method* recursively partitions the search space for local tuples. Moreover, the heuristic method for finding seeds also reduces the computational cost.

9.3.4 Summary and Discussion

The advantage of the utility-based anonymization method is two fold. First, *it increases the accuracy of query answering on anonymized tables*. The utility-based anonymization groups similar tuples together, and applies the local recoding to tuples in the same group. As a result, the generalized tuples are often in small ranges and similar to the original values. Thus, the answers to the queries are also bound in a small range around the exact answer. Second, *it naturally increases the discernability of the anonymized table*. The *weighted normalized certainty penalty* on a table is non-decreasing if the average size of equivalence classes increases. This is because all tuples in a group are generalized to the same in the anonymized table. The more tuples are in the same equivalence class, the more likely they are generalized to larger ranges. Although discernability penalty is not explicitly incorporated in the utility measure, the size of the equivalence class is kept small.

9.4 The Utility-based Privacy Preserving Methods in Classification Problems

In classification analysis, the published data are often used to train classifiers. As discussed in Section 9.2, the data quality depends on how well the class structure is preserved. In this section, we discuss two utility-based privacy preservation methods which try to preserve the privacy and retain the data utility for classification as much as possible.

The *top-down specialization* (TDS) method is based on the k-anonymity privacy model. The objective is to generalize tuples in a table, such that tuples in the same equivalence class are as pure as possible with respect to class labels. The *progressive disclosure algorithm* (PDA) is based on the sensitive inference privacy model. In order to eliminate sensitive inferences, it suppresses some attribute values so that the confidence of each inference rule is controlled lower than a user defined threshold.

The two methods share the same spirit in algorithm frameworks and data utility measures. First, both algorithms take a top-down approach. They hide all the specific information at the beginning, and progressively release the more specific information as long as the privacy requirement is not violated and the data utility in classification is increased. The operation of releasing more specific information is called *specialization* in *TDS* and *disclosure* in *PDA*. Second, *TDS* and *PDA* both measure the data utility in classification as *information gain per unit of privacy loss*. The intuition is that, in a specialization or disclosure, one goal is to maximize the information gain. At the same time, we should also satisfy the privacy requirement. If a specialization or disclosure improves the information utility greatly but sacrifices privacy fast, it is undesirable.

The critical difference between *TDS* and *PDA* is how to measure the *information gain* and the *privacy loss*. Since *TDS* is based on the k-anonymity model, in a specialization, the information gain is measured by the entropy reduction on the affected equivalence classes, and the privacy loss is defined as the reduction of the smallest equivalence class size. The reason is that, if an equivalence class size decreases, so does the anonymity (that is, how indistinguishable a tuple is from others) of tuples in the equivalence class. On the other hand, *PDA* is based on the sensitive inference model. After disclosing an attribute value, the information gain is defined as the entropy reduction on the tuples involving the disclosed value, and the privacy loss is defined as the average confidence increase of sensitive inference rules. This is because a higher confidence leads to a higher probability to guess the sensitive data successfully.

9.4.1 The Top-Down Specialization Method

The *top-down specialization* method anonymizes a table in the way that the tuples in the same equivalence class are as pure as possible in their class labels. The objective is to minimize the uncertainty of classifying tuples within an equivalence class, and thus to improve the classification accuracy. As shown in Example 9.1, Table 9.2b is a better anonymization than Table 9.3c in classification applications, because it provides better classification accuracy.

To give more details, we first introduce the specialization method, and then discuss how to evaluate a specialization. An example is given to illustrate the algorithm.

The Specialization Method. In order to specify the hierarchical structure of values with different granularity in an attribute, a user-specified *taxonomy tree* is given on each categorical attribute. A leaf node in a taxonomy tree represents a most specific value in the original table and its parent node represents a more generalized value. The root of a taxonomy tree represents the most generalized value on the corresponding attribute. For continuous attributes, a generalized value is represented as an interval. The algorithm dynamically grows a taxonomy tree for each continuous attribute at runtime. It starts from the full range of the domain on an attribute, and iteratively splits the interval into two sub intervals that maximize the information gain. Figure 9.1 shows a taxonomy tree on categorical attribute *Education* in Table 9.7a and Figure 9.2 shows a taxonomy tree on continuous attribute *Age* in the same table.

A *specialization* $v \rightarrow children(v)$ on attribute A replaces value v with one of the values in $children(v)$. $children(v)$ contains all the children values of

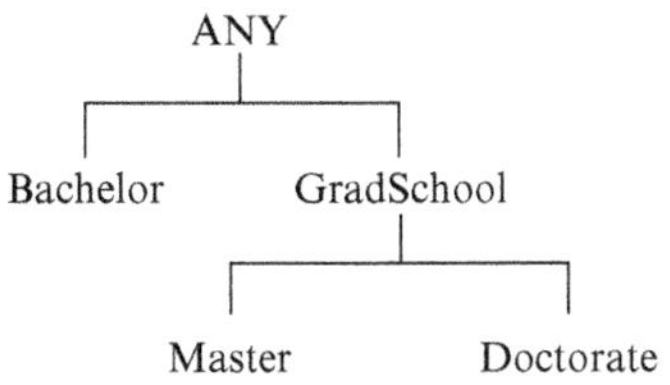

Figure 9.1. A taxonomy tree on categorical attribute *Education*

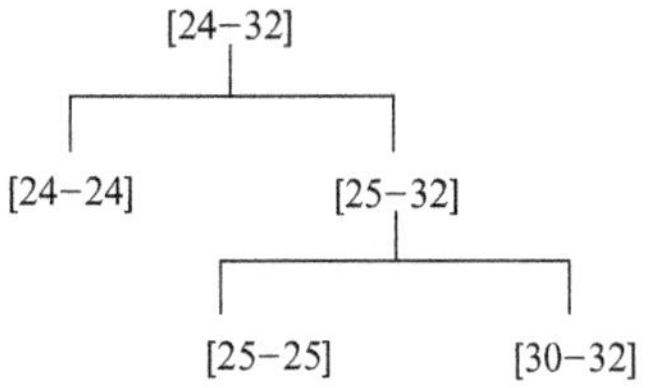

Figure 9.2. A taxonomy tree on continuous attribute *Age*

v in the taxonomy tree of A. After the specialization, the equivalence classes containing value v are partitioned into subgroups according to $children(v)$. Thus, the specialization reduces the anonymity of the tuples containing v. A specialization is *valid* (with respect to T) if the anonymity of table T after the specialization does not violate the anonymity requirement. Moreover, a specialization on v is *beneficial* (with respect to T) if more than one class is involved in the records containing v. The intuition behind is the follows. One objective of specializations is to increase the utility for classification. The information utility is achieved by partitioning a group of records with mixed classes into subgroups that have "purer" classes. Therefore, if the records containing v only involve one class, then there is no potential for the specialization on v to increase the information for classification. A specialization is performed only if it is both valid and beneficial.

Utility Score. In order to measure the benefits of a specialization quantitatively, a utility score is defined as

$$Score(v) = \begin{cases} \frac{InfoGain(v)}{AnonyLoss(v)} & \text{if } AnonyLoss(v) \neq 0 \\ InfoGain(v) & \text{otherwise} \end{cases}$$

where $InfoGain(v)$ is the information gain in the specialization and $AnonyLoss(v)$ measures the privacy loss in the specialization, which will be defined as follows.

$InfoGain(v)$**.** Given value v, all tuples containing v are denoted by R_v. The entropy of R_v with respect to the class distribution measures the randomness of classes in the tuples, which is defined as

$$H(R_v) = -\sum_{c \in cls} \frac{freq(R_v, c)}{|R_v|} \times \log_2 \frac{freq(R_v, c)}{|R_v|}$$

where cls is the set of class labels involved in R_v.

The information gain of a specialization on value v measures the reduction of entropy in the specialization, which is defined as

$$InfoGain(v) = H(R_v) - \sum_{c \in children(v)} \frac{|R_c|}{|R_v|} H(R_c)$$

$AnonyLoss(v)$. Given a table T and a quasi-identifier $QI = (A_{i_1}, \ldots, A_{i_l})$, the tuples sharing the same values on QI form *equivalence classes*. The size of an equivalence class indicates the anonymity of tuples in the class, that is, how indistinguishable the tuples are from others. The anonymity of quasi-identifier QI, denoted by $A(QI)$, is defined as the minimal equivalence class size on QI.

Then, how to measure the anonymity loss in a specialization? Suppose table T is specialized on value v in attribute A_i, since the equivalence classes involving A_i may be partitioned into subgroups in the specialization on v, the anonymity of the quasi-identifiers containing A_i may decrease. We use the average anonymity reduction of all the quasi-identifiers containing A_i to measure the anonymity loss of the specialization on v. It is defined as

$$AnonyLoss(v) = AVG_{A_i \in QI_j}\{A(QI_j) - A_v(QI_j)\}$$

where $A(QI_j)$ is the anonymity of quasi-identifier QI_j before the specialization, and $A_v(QI_j)$ is the anonymity of QI_j after the specialization.

The calculation of utility score is illustrated in the following example.

EXAMPLE 9.6 (UTILITY SCORE) *Consider Table 9.7a and the taxonomy tree on categorical attribute* Education *in Figure 9.1. After the specialization on ANY, all the tuples in Table 9.7a are partitioned into two groups, one contains value Bachelor and the other one contains GradSchool. We have*

$$\begin{aligned}
&H(R_{ANY}) = -\tfrac{2}{6} \times \log_2 \tfrac{2}{6} - \tfrac{4}{6} \times \log_2 \tfrac{4}{6} = 0.9149 \\
&H(R_{Bachelor}) = -1 \times \log_2 1 - 0 \times \log_2 0 = 0 \\
&H(R_{GradSchool}) = -0 \times \log_2 0 - 1 \times \log_2 1 = 0 \\
&InfoGain(ANY) = H(R_{ANY}) - (\tfrac{|R_{Bachelor}|}{|R_{ANY}|} \times H(R_{Bachelor}) \\
&\qquad + \tfrac{|R_{GradSchool}|}{|R_{ANY}|} \times H(R_{GradSchool})) \\
&\qquad = 0.9149 - (\tfrac{2}{6} \times 0 + \tfrac{4}{6} \times 0) = 0.9149 \\
&AnonyLoss(ANY) = 6 - 2 = 4 \\
&Score(ANY) = \frac{InfoGain(ANY)}{AnonyLoss(ANY)} = \frac{0.9149}{4} = 0.2287
\end{aligned}$$

Consider continuous attribute Age *in Table 9.7a. Since no taxonomy tree is specified on* Age, *we grow an "optimal" taxonomy tree by iteratively splitting the intervals on* Age *at the values that maximize the information gain. For example, the distinct values on* Age *include* 24, 25, 30, *and* 32, *therefore, the*

Table 9.7a. The original table

tId	*Age*	*Education*	*Zip Code*	*TargetCustomer*
1	24	Bachelor	53711	Y
2	25	Bachelor	53713	Y
3	25	Master	53715	N
4	30	Master	53712	N
5	32	Master	53714	N
6	32	Doctorate	53716	N

Table 9.8b. The anonymized table

gId	*tId*	*Age*	*Education*	*Zip Code*	*TargetCustomer*
1	1	[24-32]	Bachelor	[53711-53716]	Y
1	2	[24-32]	Bachelor	[53711-53716]	Y
2	3	[24-32]	GradSchool	[53711-53716]	N
2	4	[24-32]	GradSchool	[53711-53716]	N
2	5	[24-32]	GradSchool	[53711-53716]	N
2	6	[24-32]	GradSchool	[53711-53716]	N

possible splitting values are 24, 25, *and* 30. *The corresponding split intervals are* $[24-24]\&[25-32]$, $[24-25]\&[30-32]$, *and* $[24-30]\&[32-32]$, *respectively.* $R_{[v_1,v_2]}$ *denotes the tuples containing values in* $[v_1,v_2]$ *on* Age. *We have*

$$\begin{aligned} InfoGain(24) &= H(R_{[24-32]}) - (\tfrac{1}{6}\times H(R_{[24-24]}) + \tfrac{5}{6}\times H(R_{[25-32]})) \\ &= 0.9149 - (\tfrac{1}{6}\times 0 + \tfrac{5}{6}\times 0.7219) = 0.3133 \\ InfoGain(25) &= H(R_{[24-32]}) - (\tfrac{3}{6}\times H(R_{[24-25]}) + \tfrac{3}{6}\times H(R_{[30-32]})) \\ &= 0.9149 - (\tfrac{3}{6}\times 0.9149 + \tfrac{3}{6}\times 0) = 0.4575 \\ InfoGain(30) &= H(R_{[24-32]}) - (\tfrac{4}{6}\times H(R_{[24-30]}) + \tfrac{2}{6}\times H(R_{(32-32]})) \\ &= 0.9149 - (\tfrac{4}{6}\times 1 + \tfrac{2}{6}\times 0) = 0.2482 \end{aligned}$$

Since $InfoGain(24)$ *is the maximum, the full range* $[24-32]$ *on Age is first split into* $[24-24]$ *and* $[25-32]$. *Similarly,* $[25-32]$ *is split into* $[25-25]$ *and* $[30-32]$. *Since there is only one class involved in* $R_{[30-32]}$, *that is, no specialization on* $[30-32]$ *is beneficial, we do not split it further. The grown taxonomy tree on* Age *is shown in Figure 9.2.*

The Algorithm. The top-down specialization method starts at the root of the taxonomy tree on each attribute. That is, all tuples in a table are generalized to the same on quasi-identifiers. In each iteration, it considers all the possible specializations on the current generalized level. The scores of the valid and beneficial specializations are calculated and the specialization with the maximum score is executed. The iteration terminates when there is no valid and beneficial specialization. The following example illustrates the algorithm.

EXAMPLE 9.7 (THE TOP-DOWN SPECIALIZATION METHOD) *Consider Table 9.7a, where* {Age, Education, Zip Code} *is a quasi-identifier and* TargetCustomer *is a class label. There are two classes in the table:* $\{t|t \in T,\ t.TargetCustomer = Y\}$ *and* $\{t|t \in T,\ t.TargetCustomer = N\}$. *Suppose* 2*-anonymity is required, the* Top-Down Specialization *method works as follows.*

At first, all tuples are generalized to $([24-32], ANY, [53711-53716])$ *on the quasi-identifier. Then, the scores of all valid and beneficial specializations are calculated. Since* $Score(ANY)$ *on attribute* Education *is the maximum, the table is specialized on value* ANY. *The tuples are partitioned into two groups:* $t1$ *and* $t2$ *form an equivalence class with values* $([24-32], Bachelor, [53711-53716])$ *on the quasi-identifier;* t_3, t_4, t_5 *and* t_6 *form another equivalence class with values* $([24-32], GradSchool, [53711-53716])$ *on the quasi-identifier. Each equivalence class contains only one class. Therefore, there is no more beneficial specialization. The final* 2*-anonymized table is shown in Table 9.8b.*

9.4.2 The Progressive Disclosure Algorithm

The *progressive disclosure algorithm* suppresses some attribute values in a table so that the confidence of sensitive inferences is reduced to a user defined threshold or below, and the data utility in classification is retained as much as possible. For example, consider Table 9.9a used for classification. Suppose *TargetCustomer* is a class label and *AnnualIncome* is a sensitive attribute. Particularly, the values "$\leq 50k$" and "$> 100k$" should not be disclosed. There are two inference rules $\{[20-30], Bachelor\} \rightarrow$ "$\leq 50k$" and $\{Doctorate, Lawyer\} \rightarrow$ "$> 100k$" with high confidence. Table 9.10b is a suppressed table where the confidence of each inference rule is reduced to 50% or below. But the table remains useful for classification. That is, given a tuple t' with the same values on attribute *Age, Education*, and *Job* as any tuple t in the original table, t' receives the same class label as t with a high probability according to Table 9.10b. This is because the class label in Table 9.9a is highly related to attribute *Job*. As long as the values on *Job* are disclosed, the classification accuracy is guaranteed.

To give more details about the method, we first introduce how to define the privacy requirement using *privacy templates*, and then discuss the utility measure. Last, we use an example to illustrate the algorithm.

Privacy Template. To make a table free from sensitive inferences, it is required that the confidence of each inference rule is low.

Table 9.9a. The original table

tId	*Age*	*Education*	*Job*	*AnnualIncome*	*TargetCustomer*
1	[20-30]	Bachelor	Engineer	$\leq 50k$	Y
2	[20-30]	Bachelor	Artist	$\leq 50k$	N
3	[20-30]	Bachelor	Lawyer	$\leq 50k$	Y
4	[20-30]	Bachelor	Artist	$[50k - 100k]$	N
5	[20-30]	Master	Artist	$[50k - 100k]$	N
6	[31-40]	Master	Engineer	$[50k - 100k]$	Y
7	[20-30]	Doctorate	Lawyer	$> 100k$	N
8	[31-40]	Doctorate	Lawyer	$> 100k$	Y
9	[31-40]	Doctorate	Lawyer	$[50k - 100k]$	Y
10	[20-30]	Doctorate	Engineer	$[50k - 100k]$	N

Table 9.10b. The suppressed table

tId	*Age*	*Education*	*Job*	*AnnualIncome*	*TargetCustomer*
1	[20-30]	$\perp_{Edu}$	Engineer	$\leq 50k$	Y
2	[20-30]	$\perp_{Edu}$	Artist	$\leq 50k$	N
3	[20-30]	$\perp_{Edu}$	Lawyer	$\leq 50k$	Y
4	[20-30]	$\perp_{Edu}$	Artist	$[50k - 100k]$	N
5	[20-30]	Master	Artist	$[50k - 100k]$	N
6	$\perp_{Age}$	Master	Engineer	$[50k - 100k]$	Y
7	[20-30]	$\perp_{Edu}$	Lawyer	$> 100k$	N
8	$\perp_{Age}$	$\perp_{Edu}$	Lawyer	$> 100k$	Y
9	$\perp_{Age}$	$\perp_{Edu}$	Lawyer	$[50k - 100k]$	Y
10	[20-30]	$\perp_{Edu}$	Engineer	$[50k - 100k]$	N

Templates can be used to specify such a requirement. Consider table $T = (M_1, \ldots, M_m, \Pi_1, \ldots, \Pi_n, \Theta)$, where M_j $(1 \leq j \leq m)$ is a *non-sensitive attribute*, Π_i $(1 \leq i \leq n)$ is a *sensitive attribute*, and Θ is a class label attribute. A *template* is defined as $\langle IC \rightarrow \pi_i, h \rangle$, where π_i is a *sensitive attribute value* from sensitive attribute Π_i, IC is a set of attributes not containing Π_i and called *inference channel*, and h is a confidence threshold. An *inference* is an instance of $\langle IC \rightarrow \pi_i, h \rangle$, which has the form $ic \rightarrow \pi_i$, where ic contains values from attributes in IC. The confidence of inference $ic \rightarrow \pi_i$, denoted by $conf(ic \rightarrow \pi_i)$, is the percentage of tuples containing both ic and π_i among the tuples containing ic. That is, $conf(ic \rightarrow \pi_i) = \frac{|R_{ic,\pi_i}|}{|R_{ic}|}$, where R_v denotes the tuples containing value v. The confidence of a template is defined as the maximum confidence of all the inferences of the template. That is, $Conf(IC \rightarrow \pi_i) = \max conf(ic \rightarrow \pi_i)$. Table T *satisfies* template $\langle IC \rightarrow \pi_i, h \rangle$ if $Conf(IC \rightarrow \pi_i) \leq h$. T satisfies a set of templates if T satisfies each template in the set.

Progressive Disclosure and Utility Measure. As discussed in Section 9.2, *suppression* is an efficient method for eliminating sensitive inferences. Consider table $T = (M_1, \ldots, M_m, \Pi_1, \ldots, \Pi_N, \Theta)$. M_j $(1 \leq j \leq m)$ is a *non-sensitive attribute* and Π_i $(1 \leq i \leq n)$ is a *sensitive attribute*. Θ is a class label. The *suppression* of a value on attribute M_j is to replace all occurrences of this value by a special value $\perp_j$. For each template $\langle IC \rightarrow \pi_i, h\rangle$ not satisfied in T, some values in the inference channel IC should be suppressed so that $Conf(IC \rightarrow \pi_i)$ is reduced to not greater than h.

Disclosure is the opposite operation of suppression. Given a suppressed table T, Sup_j denotes all values suppressed on attribute M_j. A *disclosure* of value $v \in Sup_j$ replaces the special value $\perp_j$ with v in all the tuples that currently contain $\perp_j$ but originally contain v. A disclosure is *valid* if it does not lead to a template violation. Moreover, a disclosure on attribute M_j is *beneficial*, that is, it increases the information utility for classification, if more than one class is involved in the tuples containing $\perp_j$. The following utility score measures the benefit of a disclosure quantitatively.

For each suppressed attribute value v , $Score(v)$ is defined as

$$Score(v) = \frac{InfoGain(v)}{PrivLoss(v) + 1}$$

where $InfoGain(v)$ is the information gain in disclosing value v and $PrivLoss(v)$ is the privacy loss in disclosing value v, which are defined as follows.

$InfoGain(v)$**.** Given a set of tuples S and the class labels cls involved in S, the entropy is defined as

$$H(S) = -\sum_{c \in cls} \frac{freq(S,c)}{|S|} \times \log_2 \frac{freq(S,c)}{|S|}$$

where $freq(S, c)$ is the number of tuples containing class c in S.

Given value v on attribute M_j, the tuples containing v is denoted by R_v. Suppose $R_{\perp_j}$ is the set of tuples having suppressed value on M_j before disclosing v, the information gain of disclosing v is

$$InfoGain(v) = H(R_{\perp_j}) - (\frac{|R_v|}{|R_{\perp_j}|} H(R_v) + \frac{|R_{\perp_j} - R_v|}{|R_{\perp_j}|} H(R_{\perp_j} - R_v))$$

$PrivLoss(v)$**.** Given value v on attribute M_j, the privacy loss $PrivLoss(v)$ is defined as the average confidence increase of inferences.

$$PrivLoss(v) = AVG_{M_j \in IC}\{Conf'(IC \rightarrow \pi_i) - Conf(IC \rightarrow \pi_i)\}$$

where $Conf(IC \rightarrow \pi_i)$ and $Conf'(IC \rightarrow \pi_i)$ are the confidence before and after disclosing v.

EXAMPLE 9.8 (UTILITY SCORE) *Consider Table 9.9a. Suppose the privacy templates are*

$$\langle \{Age, Education\} \rightarrow \text{“} \leq 50k\text{”}, 50\% \rangle$$
$$\langle \{Education, Job\} \rightarrow \text{“} > 100k\text{”}, 50\% \rangle$$

Suppose at first all the values on attribute Job *is suppressed to* $\perp_{Job}$. *The score of disclosing value* $Engineer$ *on* Job *is calculated as follows.*

$$H(R_{\perp_{Job}}) = -\tfrac{5}{10} \times \log_2 \tfrac{5}{10} - \tfrac{5}{10} \times \log_2 \tfrac{5}{10} = 1$$
$$H(R_{Engineer}) = -\tfrac{2}{3} \times \log_2 \tfrac{2}{3} - \tfrac{1}{3} \times \log_2 \tfrac{1}{3} = 0.9149$$
$$H(R_{\perp_{Job}} - R_{Engineer}) = -\tfrac{3}{7} \times \log_2 \tfrac{3}{7} - \tfrac{4}{7} \times \log_2 \tfrac{4}{7} = 0.9852$$
$$InfoGain(Engineer) = H(R_{\perp_{Job}}) - (\tfrac{3}{10} \times H(R_{Engineer}) + \tfrac{7}{10} \times H(R_{\perp_{Job}} - R_{Engineer})) = 0.03589$$

Before disclosing Engineer:

$$Conf(\{Education, Job\} \rightarrow > 100k)$$
$$= conf(\{\perp_{Education}, \perp_{Job}\} \rightarrow > 100k) = 0.2$$

After disclosing Engineer:

$$conf(\{\perp_{Education}, Engineer\} \rightarrow > 100k) = 0$$
$$conf(\{\perp_{Education}, \perp_{Job}\} \rightarrow > 100k) = 0.286$$
$$Conf'(\{Education, Job\} \rightarrow > 100k) = \max\{0, 0.286\} = 0.286$$
$$PrivLoss(Engineer) = 0.286 - 0.2 = 0.086$$
$$Score(Engineer) = \frac{0.03589}{0.086+1} = 0.033$$

The Algorithm. The *Progressive Disclosure Algorithm* first suppresses all non-sensitive attribute values in a table and then iteratively discloses the attribute values that are helpful for classification without violating privacy templates. In each iteration, the score of each suppressed value is calculated and the one with the maximum score is disclosed. The iteration terminates when there is no valid and beneficial disclosure. The algorithm is illustrated using the following example.

EXAMPLE 9.9 (THE PROGRESSIVE DISCLOSURE ALGORITHM) *Consider the following templates on Table 9.9a.*

$$(1) \quad \langle \{Age, Education\} \rightarrow \text{“} \leq 50k\text{”}, 50\% \rangle$$
$$(2) \quad \langle \{Education, Job\} \rightarrow \text{“} > 100k\text{”}, 50\% \rangle$$

At first, the values on attribute Age, Education, *and* Job *are suppressed to* $\perp_{Age}$, $\perp_{Education}$, *and* $\perp_{Job}$, *respectively. The candidate disclosing values include* $[20 - 30]$, $[31 - 40]$, $Bachelor$, $Master$, $Doctorate$, $Engineer$,

$Artist$, and $Lawyer$. In order to find the most beneficial disclosure, the score of each value is calculated. Since $Artist$ has the maximum score, it is disclosed in this iteration. At the next iteration, the scores of the rest candidates are updated, and the one with the maximum score, $[20 - 30]$, is disclosed. All the valid and beneficial disclosures are executed similarly in the rest iterations. The finally published table is shown in Table 9.10b. Note that in the finally published table, $Bachelor$ and $Doctorate$ are suppressed because disclosing them violates the privacy templates; $[31 - 40]$ is suppressed because disclosing it is not beneficial.

9.4.3 Summary and Discussion

The top-down specialization (TDS) method and the progressive disclosure algorithm (PDA) are based on the observation that the goals of privacy preservation and classification modeling may not be always conflicting. Privacy preservation is to hide the sensitive individual (specific) information, while classification modeling draws the general structure of the data. *TDS* and *PDA* try to achieve the "win-win" goal that the specific information hidden for privacy preservation is the information misleading or not useful for classification. Therefore, the quality of the classification model built on the table after using *TDS* or *PDA* may be even better than that built on the original table.

9.5 Anonymized Marginal: Injecting Utility into Anonymized Data Sets

One drawback of the anonymization method is that after the generalization on quasi-identifiers, the distribution of the more specific data is lost. For example, consider Table 9.11a and the corresponding 2-anonymous Table 9.12b. After the anonymization, all the values on attribute *Age* are generalized to the full range in the domain without any specific distribution information. However, if we publish Table 9.13a in addition to Table 9.12b, more information about *Age* is published and the 2-anonymity is still guaranteed. Table 9.13a is called a *marginal* on *Age*.

On the other hand, not all marginals preserve privacy. For example, Table 9.14b satisfies 2-anonymity itself, but if an attacker knows an individual living in 53715 with $Doctorate$ degree is in the original table, he/she may link the information from Table 9.12b and 9.14b together and conclude that the annual income of the individual is $80k$.

Based on the above observation, [8] models the utility of anonymized tables as how much they preserve the *distribution* of the original table. It then proposes to publish more than one anonymized tables to better approximate the original distribution.

Table 9.11a. The original table

tId	*Age*	*Education*	*Zip Code*	*AnnualIncome*
1	27	Bachelor	53711	40k
2	28	Bachelor	53713	50k
3	27	Master	53715	50k
4	28	Doctorate	53716	80k
5	30	Master	53714	50k
6	30	Doctorate	53712	100k

Table 9.12b. The anonymized table

gId	*tId*	*Age*	*Education*	*Zip Code*	*AnnualIncome*
1	1	[27-30]	Bachelor	[53711-53713]	40k
1	2	[27-30]	Bachelor	[53711-53713]	50k
2	3	[27-30]	GradSchool	[53715-53716]	40k
2	4	[27-30]	GradSchool	[53715-53716]	80k
3	5	[27-30]	GradSchool	[53712-53714]	50k
3	6	[27-30]	GradSchool	[53712-53714]	100k

Table 9.13a. *Age* Marginal

Age	*Count*
27	2
28	2
30	2

Table 9.14b. (*Education, AnnualIncome*) Marginal

Education	*AnnualIncome*	*Count*
Bachelor	40k	1
Bachelor	50k	1
Master	50k	2
Doctorate	80k	1
Doctorate	100k	1

Now the problem becomes, *which additional anonymized tables should be published and how to check the privacy if more than one anonymized table are released.* First of all, we introduce the concept of *anonymized marginal* and the utility measure to evaluate the quality of a set of anonymized marginals.

9.5.1 Anonymized Marginal

Consider a table $T = (A_1, \ldots, A_n)$. $\{A_{i_1}, \ldots, A_{i_m}\}$ $(1 \leq i_1 < \ldots < i_m \leq n)$ is a set of attributes in T. A marginal table $T_{A_{i_1},\ldots,A_{i_m}}$ can be created by the following SQL statement. (Attribute *Count* is the number of tuples in $T_{A_{i_1},\ldots,A_{i_m}}$ sharing the same values on $A_{i_1}, \ldots, A_{i_m}$).

$$\begin{array}{lll} \texttt{CREATE} & \texttt{TABLE} & T_{A_{i_1},\ldots,A_{i_m}} \\ \texttt{AS} & \texttt{(SELECT} & A_{i_1},\ldots,A_{i_m}, \texttt{COUNT(*)AS}\ Count \\ & \texttt{FROM} & T \\ & \texttt{GROUP BY} & A_{i_1},\ldots,A_{i_m}) \end{array}$$

The marginal table indicates the distribution of the tuples from T in domain $D(A_{i_1}) \times \ldots \times D(A_{i_m})$, where $D(A_i)$ is the domain of attribute A_i. A marginal is *anonymized* if some of its attribute values are generalized.

9.5.2 Utility Measure

Distribution is an intrinsic characteristic of a data set. Many data analysis discover the patterns from data distribution, such as classification which discovers the class distribution in a data set. Therefore, whether the distribution of a data set is preserved after anonymization is crucial for the utility of data. In this spirit, a utility measure is defined as the difference between the distribution of the original data and that of the anonymized data.

Empirical distribution of the original table. Consider a table $T = (A_1, \ldots, A_m)$. In the probabilistic view, the tuples in T can be considered as an i.i.d. (identically and independently distributed) sample generated from an underlying distribution F. Reversely, F can be estimated using the *empirical distribution* $\widehat{F}_T$ of T. Given any instance $x = (x_1, \ldots, x_m)$ in the domain of T, the empirical probability $\widehat{p}_T(x)$ is the posteriori probability of x in table T. In other words, $\widehat{p}_T(x)$ is the proportion of tuples in T having the same attribute values as x, that is, $\widehat{p}_T(x) = \frac{|\{t | t \in T, t.A_i = x_i,\ 1 \leq i \leq m\}|}{|T|}$.

Maximum entropy probability distribution of anonymized marginals. Similarly, the anonymized marginals of T can be viewed as a set of constraints on the underlying distribution. For example, *Age Marginal* in Table 9.13a indicates that 33.3% of the tuples in Table 9.11a have age 27, 28 and 30, respectively.

Given a set of constraints, the *maximum entropy probability distribution* is the distribution that maximizes the entropy among all the probability distributions satisfying the constraints. It is often used to estimate the underlying distribution given some constraints. The intuition is that, by maximizing the entropy, the prior knowledge about the distribution is minimized.

Consider a table $T = (A_1, \ldots, A_m)$ and a set of marginals $M = \{M_1, \ldots, M_n\}$, each marginal $M_i = (A_{i_1}, \ldots, A_{i_k}, Count)$ $(1 \leq i_1 < \ldots < i_k \leq m)$ contains the projection of T on attribute $\{A_{i_1}, \ldots, A_{i_k}\}$ and the count of tuples. A distribution F satisfies M_i if for any instance t in M_i, the probability in F satisfies

$$\sum_{\Pi_{A_{i_1},\ldots,A_{i_k}} x=t} p(x) = \frac{t.Count}{|T|}$$

where x is an instance from the domain of T and $\Pi_{A_{i_1},\ldots,A_{i_k}} x = t$ means that the projection of x on $A_{i_1}, \ldots, A_{i_k}$ is the same as t. The above equation means that the projection of distribution F on $A_{i_1}, \ldots, A_{i_k}$ is the same as the empirical distribution of M_i. F satisfies a set of marginals M if F satisfies each marginal M_i in M. The maximum entropy probability distribution $\widehat{F}_M$ is the distribution with the maximum entropy in all the distributions satisfying M.

Kullback-Leibler divergence (KL-divergence). Suppose the empirical distribution of a table T is $\widehat{F}_1$ and the maximum entropy probability distribution of the anonymized marginals M is $\widehat{F}_2$, the *Kullback-Leibler divergence (KL-divergence)* [9] is used to measure the difference between the two distributions. (Note that KL-divergence is not a metric.)

$$D_{KL}(\widehat{F}_1, \widehat{F}_2) = \sum_i p_i^{(1)} \log \frac{p_i^{(1)}}{p_i^{(2)}} = H(\widehat{F}_1, \widehat{F}_2) - H(\widehat{F}_1)$$

where $p_i^{(1)}$ and $p_i^{(2)}$ are the probabilities of an instance from distribution $\widehat{F}_1$ and $\widehat{F}_2$, respectively. $H(\widehat{F}_1)$ is the entropy of $\widehat{F}_1$, which measures how much effort it needs to identify an instance from distribution $\widehat{F}_1$. $H(\widehat{F}_1, \widehat{F}_2)$ is the cross-entropy of $\widehat{F}_1$ and $\widehat{F}_2$, which measures the effort needed to identify an instance from distribution $\widehat{F}_1$ and $\widehat{F}_2$. A smaller KL-divergence indicates that the two distributions are more similar. KL-divergence is non-negative and it is minimized when $\widehat{F}_1 = \widehat{F}_2$. Given a table T, the entropy $H(\widehat{F}_1)$ is constant. Therefore, minimizing $D_{KL}(\widehat{F}_1, \widehat{F}_2)$ is mathematically equivalent to minimizing $H(\widehat{F}_1, \widehat{F}_2)$.

Therefore, the utility of a set of anonymized marginals $M = \{M_1, \ldots, M_n\}$ can be measured by the KL-divergence between $\widehat{F}_M$ and $\widehat{F}_T$. A smaller KL-divergence value indicates better utility of M.

9.5.3 Injecting Utility Using Anonymized Marginals

Based on the above utility measure, ideally, we want to search all the possible sets of anonymized marginals and find the one with the minimum KL-divergence. There are two challenges.

Calculating the KL-divergence is computational challenging. First, generating all the possible sets of marginals needs exhaustive search. Second, finding the optimal k-anonymization for a single marginal is already NP-hard [3]. Third, given a set of constraints, calculating the maximum entropy probability distribution requires iterative algorithms [4, 16], which may be time-consuming.

Since there is a close-form algorithm [10] to compute the maximum entropy probability distribution on decomposable tables, the anonymized marginal method restricts the search to only including *decomposable marginals.* The concept of decomposable marginals is derived from the *decomposable graphical model* [10]. If a set of marginals are decomposable, then they are *conditionally independent.* Instead of giving the formal definition, we use the following example to illustrate the decomposable marginals and how to calculate the maximum entropy probability on decomposable marginals.

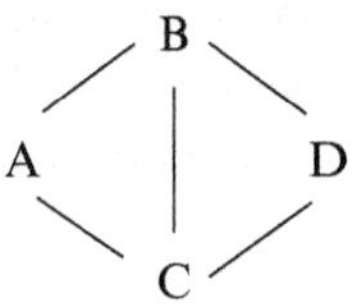

Figure 9.3. Interactive graph

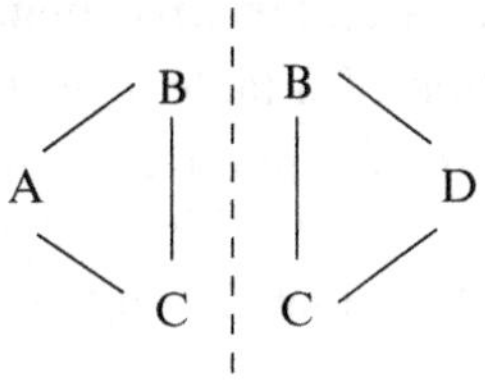

Figure 9.4. A decomposition

EXAMPLE 9.10 (DECOMPOSABLE MARGINAL) *Consider a set of marginals* $M_1 = (A, B, C, Count)$ *and* $M_2 = (B, C, D, Count)$. *We create an* interactive graph *(Figure 9.3) for them by generating a vertex for each attribute. An edge between two vertices is created if the corresponding attributes are in the same marginal.* M_1 *and* M_2 *are decomposable because they satisfies the following two conditions:*
(1) in the corresponding interactive graph, clique BC *separates* A *and* D *(the two components after the decomposition are shown in Figure 9.4);*
(2) each maximal clique in the interactive graph is covered by a marginal.

An example of non-decomposable marginals is $M_1 = (A, B, C, Count)$, $M_2 = (B, D, Count)$ *and* $M_3 = (C, D, Count)$. *They have the same interactive graph as shown in Figure 9.3, but the maximal clique* BCD *is not covered by any marginal. Therefore, they are not decomposable marginals.*

A set of decomposable marginals can be viewed as a set of conditionally independent relations. For example, attributes A and D in marginals $M_1 = (A, B, C, Count)$ and $M_2 = (B, C, D, Count)$ are independent given attributes BC. The calculation of the maximum entropy probability distribution for decomposable marginals is illustrated in the following example.

EXAMPLE 9.11 (MAXIMUM ENTROPY PROBABILITY) *Consider marginals* $M_1 = (A, B, Count)$, *and* $M_2 = (B, C, Count)$ *of table*

$T = (A, B, C)$. M_1 and M_2 are decomposable and B separates A and C. Therefore, attribute A and C are independent given B.

If M_1 and M_2 are ordinary marginals: *The attribute values in ordinary marginals are not generalized. For any instance $x = (a, b, c)$ in the domain of T, the maximum entropy probability of x is*

$$\begin{aligned} p(x) &= p(a, b, c) \\ &= p(a, c|b) \cdot p(b) \\ &= p(a|b) \cdot p(c|b) \cdot p(b) \\ &= \frac{p(a,b) \cdot p(b,c)}{p(b)} \end{aligned}$$

where $p(a, b)$ is the proportion of tuples in M_1 having value a and b on attribute A and B, respectively.

If M_1 and M_2 are anonymized marginals: *Some attribute values in anonymized marginals are generalized. For any instance $x = (a, b, c)$ in the domain of T, suppose a', b', c' are the corresponding generalized values in M_1 and M_2. The maximum entropy probability of x is:*

$$p(x) = p(a, b, c) = \frac{p(a', b') \cdot p(b', c')}{p(b')} \cdot \frac{1}{|R_{a'}| \cdot |R_{b'}| \cdot |R_{c'}|}$$

where $p(a', b')$ is the fraction of tuples having value a' and b' on attribute A and B in M_1, respectively. $R_{a'}$ is the set of tuples having value a' on A in M_1.

Since finding all the possible decomposable marginals requires exhaustive search, a search algorithm like genetic algorithm or random walk is needed.

Guarantee the privacy. Another challenge is that given a set of marginals $\{M_1, \ldots, M_n\}$, how to check whether the information obtained from combining $\{M_1, \ldots, M_n\}$ satisfies k-anonymity and l-diversity?

The theoretical results in [8] show that in order to check k-anonymity of a set of decomposable marginals $\{M_1, \ldots, M_n\}$, we only need to check whether each marginal M_i satisfies k-anonymity. But checking whether $\{M_1, \ldots, M_n\}$ satisfies l-diversity is more difficult. We have to join all the marginals together and test whether the joined table satisfies l-diversity.

Several propositions help reduce the computation. First, if there is one marginal that violates l-diversity, then the whole set of marginals violate l-diversity. Second, only the marginals containing sensitive attributes need to be joined together to check for l-diversity. Third, if a subset of marginals do not satisfy l-diversity, then the whole set of marginals do not satisfy l-diversity.

9.5.4 Summary and Discussion

Anonymized marginal is very effective in improving the utility of the anonymized data. However, searching all the possible decomposable marginals for the optimal solution requires a lot of computation. A simpler yet effective method is, given table T, first compute an traditional k-anonymous table T', and then create a set of anonymous marginals M containing single attribute from T. Experimental results [8] show that publishing T' together with M still dramatically decreases the KL-divergence.

9.6 Summary

Utility-based privacy preserving methods are attracting more and more attention. However, the concept of utility is not new in privacy preservation problems. *Utility* is often used as one of the criteria for the privacy preserving methods [21] and measures the information loss after using the privacy preservation technique on data sets.

Then, *what makes the utility-based privacy preservation methods special?* Traditional privacy preserving methods often do not make explicit assumptions about the applications where the data are used. Therefore, the utility measure is often very general and thus not so effective. For example, traditionally, in the sensitive inference privacy model, the utility is often considered maximal if the number of suppressed entries is minimized. It is true only for certain applications. As a comparison, the utility-based privacy preservation methods target at a class of applications based on the same data utility. Therefore, the developed methods are effective in reducing the information loss for the intended applications while preserving privacy as well.

In addition to the four methods discussed in this chapter, there are many applications which utilize some special functions of data. How to extend the utility-based privacy preserving methods to various applications is highly interesting. For example, in the data set where ranking queries are usually issued, the utility of data should be measured as how much the dominance relationship among tuples is preserved. None of the existing models can handle this problem. Moreover, the utility-based privacy preserving methods can also be extended to other types of data, such as stream data where the temporal characteristics are considered more important in analysis.

Acknowledgements

This work is supported in part by the NSERC Grants 312194-05 and 614067, and an IBM Faculty Award. All opinions, findings, conclusions and recommendations in this paper are those of the authors and do not necessarily reflect the views of the funding agencies.

References

[1] Charu C. Aggarwal. On k-anonymity and the curse of dimensionality. In *Proceedings of the 31st International Conference on Very Large Data Bases*, pages 901–909, August 2005.

[2] Charu C. Aggarwal, Jian Pei, and Bo Zhang. On privacy preservation against adversarial data mining. In *Proceedings of the 12th ACM SIGKDD international conference on Knowledge discovery and data mining*, pages 510 – 516. ACM Press, 2006.

[3] Roberto J. Bayardo and Rakesh Agrawal. Data privacy through optimal k-anonymization. In *Proceedings of the 21st International Conference on Data Engineering (ICDE'05)*, pages 217 – 228. IEEE Computer Society, 2005.

[4] A.L. Berger, S.A. Della-Pietra, and V.J. Della-Pietra. A maximum entropy approach to natural language processing. *Computational Linguistics*, 22(1):39–71, 1996.

[5] Benjamin C. M. Fung, Ke Wang, and Philip S. Yu. Top-down specialization for information and privacy preservation. In *Proceedings of the 21st International Conference on Data Engineering (ICDE'05)*, volume 00, pages 205 – 216. IEEE Computer Society, 2005.

[6] Benjamin C. M. Fung, Ke Wang, and Philip S. Yu. Anonymizing classification data for privacy preservation. *IEEE Transactions on Knowledge and Data Engineering*, 19(5):711–725, May 2007.

[7] Vijay S. Iyengar. Transforming data to satisfy privacy constraints. In *Proceedings of the eighth ACM SIGKDD international conference on Knowledge discovery and data mining*, pages 279 – 288. ACM Press, 2002.

[8] Daniel Kifer and Johannes Gehrke. Injecting utility into anonymized datasets. In *Proceedings of the 2006 ACM SIGMOD international conference on Management of data*, pages 217 – 228. ACM Press, 2006.

[9] S. Kullback and R. Leibler. On information and sufficiency. *Annals of Mathematical Statistics*, 22:79–87, 1951.

[10] Steffen L. Lauritzen. *Graphical Models.* Oxford Science Publicatins, 1996.

[11] F. Giannotti M. Atzori, F. Bonchi and D. Pedreschi. Blocking anonymity threats raised by frequent itemset mining. In *Proceedings of the Fifth IEEE International Conference on Data Mining (ICDM'05)*, November 2005.

[12] F. Giannotti M. Atzori, F. Bonchi and D. Pedreschi. k-anonymous patterns. In *Proceedings of the Ninth European Conference on Principles and Practice of Knowledge Discovery in Databases (PKDD'05)*, volume 3721 of *Lecture Notes in Computer Science, Springer*, Porto, Portugal, October 2005.

[13] Ashwin Machanavajjhala, Johannes Gehrke, Daniel Kifer, and Muthuramakrishnan Venkitasubramaniam. l-diversity: Privacy beyond k-anonymity. In *Proceedings of the 22nd International Conference on Data Engineering (ICDE'06)*, page 24, 2006.

[14] Adam Meyerson and Ryan Williams. On the complexity of optimal k-anonymity. In *Proceedings of the Twenty-third ACM SIGACT-SIGMOD-SIGART Symposium on Principles of Database Systems*, pages 223–228, June 2004.

[15] Stanley R. M. Oliveira and Osmar R. Zaïane. Privacy preserving frequent itemset mining. In *CRPITS'14: Proceedings of the IEEE international conference on Privacy, security and data mining*, pages 43–54, Darlinghurst, Australia, Australia, 2002. Australian Computer Society, Inc.

[16] Adwait Ratnaparkhi. A maximum entropy part-of-speech tagger. In *Proceedings of the Conference on Empirical Methods in Natural Language Processing*, pages 133–142, University of Pennsylvania, May 1996. ACL.

[17] P. Samarati. Protecting respondents' identities in microdata release. *IEEE Transactions on Knowledge and Data Engineering*, 13(6): 1010 – 1027, November 2001.

[18] Pierangela Samarati and Latanya Sweeney. Generalizing data to provide anonymity when disclosing information. Technical report, March 1998.

[19] Latanya Sweeney. Achieving k-Anonymity Privacy Protection Using Generalization and Suppression. *International Journal on Uncertainty, Fuzziness and Knowledge-based Systems*, 10(5):571–588, 2002.

[20] Latanya Sweeney. k-anonymity: a model for protecting privacy. *Int. J. Uncertain. Fuzziness Knowl.-Based Syst.*, 10(5):557–570, 2002.

[21] Vassilios S. Verykios, Elisa Bertino, Igor Nai Fovino, Loredana Parasiliti Provenza, Yucel Saygin, and Yannis Theodoridis. State-of-the-art in privacy preserving data mining. *ACM SIGMOD Record*, 33(1):50 – 57, 2004.

[22] Vassilios S. Verykios, Ahmed K. Elmagarmid, Elisa Bertino, Yucel Saygin, and Elena Dasseni. Association rule hiding. *IEEE Transactions on Knowledge and Data Engineering*, 16(4):434–447, 2004.

[23] Ke Wang, Benjamin C. M. Fung, and Philip S. Yu. Template-based privacy preservation in classification problems. In *Proceedings of the Fifth IEEE International Conference on Data Mining*, pages 466 – 473. IEEE Computer Society, 2005.

[24] Ke Wang, Philip S. Yu, and Sourav Chakraborty. Bottom-up generalization: A data mining solution to privacy protection. In *Proceedings of the Fourth IEEE International Conference on Data Mining (ICDM'04)*, volume 00, pages 249 – 256. IEEE Computer Society, 2004.

[25] Xiaokui Xiao and Yufei Tao. m-invariance: Towards privacy preserving re-publication of dynamic datasets. In *To appear in ACM Conference on Management of Data (SIGMOD), 2007.*

[26] Xiaokui Xiao and Yufei Tao. Anatomy: simple and effective privacy preservation. In *Proceedings of the 32nd international conference on Very large data bases*, volume 32, pages 139 – 150. VLDB Endowment, 2006.

[27] Jian Xu, Wei Wang, Jian Pei, Xiaoyuan Wang, Baile Shi, and Ada Wai-Chee Fu. Utility-based anonymization for privacy preservation with less information loss. *ACM SIGKDD Explorations Newsletter*, 8(2):21–30, December 2006.

[28] Jian Xu, Wei Wang, Jian Pei, Xiaoyuan Wang, Baile Shi, and Ada Wai-Chee Fu. Utility-based anonymization using local recoding. In *Proceedings of the 12th ACM SIGKDD international conference on Knowledge discovery and data mining*, pages 785 – 790. ACM Press, 2006.

[29] Sheng Zhong, Zhiqiang Yang, and Rebecca N. Wright. Privacy-enhancing k-anonymization of customer data. In *Proceedings of the twenty-fourth ACM SIGMOD-SIGACT-SIGART symposium on Principles of database systems(PODS '05)*, pages 139–147, New York, NY, USA, 2005. ACM Press.

Chapter 10

Mining Association Rules under Privacy Constraints

Jayant R. Haritsa
Database Systems Lab
Indian Institute of Science, Bangalore 560012, INDIA
haritsa@dsl.serc.iisc.ernet.in

Abstract Data mining services require accurate input data for their results to be meaningful, but privacy concerns may impel users to provide spurious information. In this chapter, we study whether users can be encouraged to provide correct information by ensuring that the mining process cannot, with any reasonable degree of certainty, violate their privacy. Our analysis is in the context of extracting association rules from large historical databases, a popular mining process that identifies interesting correlations between database attributes. We analyze the various schemes that have been proposed for this purpose with regard to a variety of parameters including the degree of trust, privacy metric, model accuracy and mining efficiency.

Keywords: Privacy, data Mining, association rules.

10.1 Introduction

The knowledge models produced through data mining techniques are only as good as the accuracy of their input data. One source of data inaccuracy is when users deliberately provide wrong information. This is especially common with regard to customers who are asked to provide personal information on Web forms to e-commerce service providers. The compulsion for doing so may be the (perhaps well-founded) worry that the requested information may be misused by the service provider to harass the customer. As a case in point, consider a pharmaceutical company that asks clients to disclose the diseases they have suffered from in order to investigate the correlations in their occurrences – for example, "Adult females with malarial infections are also prone to contract tuberculosis". While the company may be acquiring the data solely for genuine data mining purposes that would eventually reflect itself in better service to the client, at the same time the client might worry that if her medical records are either inadvertently or deliberately disclosed, it may adversely affect her future employment opportunities.

In this chapter, we study whether customers can be encouraged to provide correct information by ensuring that the mining process cannot, with any reasonable degree of certainty, violate their privacy, but at the same time produce sufficiently accurate mining results. The difficulty in achieving these goals is that privacy and accuracy are typically contradictory in nature, with the consequence that improving one usually incurs a cost in the other [3]. A related issue is the degree of trust that needs to be placed by the users in third-party intermediaries. And finally, from a practical viability perspective, the time and resource overheads imposed on the data mining process due to supporting the privacy requirements.

Our study is carried out in the context of extracting *association rules* from large historical databases [7], an extremely popular mining process that identifies interesting correlations between database attributes, such as the one described in the pharmaceutical example. By the end of the chapter, we will attempt to show that the state-of-the-art is such that it is indeed possible to simultaneously achieve all the desirable objectives (i.e. privacy, accuracy, and efficiency) for association rule mining.

In the above discussion, and for the most part in this chapter, the focus is on maintaining the confidentiality of the *input* user data. However, it is also conceivable to think of the complementary aspect of maintaining *output secrecy*, that is, the privacy of sensitive association rules that are an outcome of the mining process – a summary discussion on these techniques is included in our coverage of the literature.

10.2 Problem Framework

In this section, we describe the framework of the privacy mining problem in the context of association rules.

10.2.1 Database Model

We assume that the original (true) database U consists of N records, with each record having M categorical attributes. Note that boolean data is a special case of this class, and further, that continuous-valued attributes can be converted into categorical attributes by partitioning the domain of the attribute into fixed length intervals.

The domain of attribute j is denoted by S_U^j, resulting in the domain S_U of a record in U being given by $S_U = \prod_{j=1}^{M} S_U^j$. We map the domain S_U to the index set $I_U = \{1, \ldots, |S_U|\}$, thereby modeling the database as a set of N

values from I_U. If we denote the ith record of U as U_i, then $U = \{U_i\}_{i=1}^N, U_i \in I_U$.

To make this concrete, consider a database U with 3 categorical attributes *Age*, *Sex* and *Education* having the following category values:

Age	Child, Adult, Senior
Sex	Male, Female
Education	Elementary, Graduate

For this schema, $M = 3$, S_U^1={Child, Adult, Senior}, S_U^2={Male, Female}, S_U^3={Elementary, Graduate}, $S_U = S_U^1 \times S_U^2 \times S_U^3$, $|S_U| = 12$. The domain S_U is indexed by the index set $I_U = \{1, ..., 12\}$, and hence the set of records

U				U
Child	Male	Elementary		1
Child	Male	Graduate	maps	2
Child	Female	Graduate	to	4
Senior	Male	Elementary		9

10.2.2 Mining Objective

The goal of the data-miner is to compute *association rules* on the above database. Denoting the set of attributes in the U database by C, an association rule is a (statistical) implication of the form $C_x \Longrightarrow C_y$, where $C_x, C_y \subset C$ and $C_x \cap C_y = \phi$. A rule $C_x \Longrightarrow C_y$ is said to have a *support* (or frequency) factor *s* iff at least $s\%$ of the transactions in U satisfy $C_x \cup C_y$. A rule $C_x \Longrightarrow C_y$ is satisfied in U with a *confidence* factor *c* iff at least $c\%$ of the transactions in U that satisfy C_x also satisfy C_y. Both support and confidence are fractions in the interval [0,1]. The support is a measure of statistical significance, whereas confidence is a measure of the strength of the rule.

A rule is said to be "interesting" if its support and confidence are greater than user-defined thresholds sup_{min} and con_{min}, respectively, and the objective of the mining process is to find all such interesting rules. It has been shown in [7] that achieving this goal is effectively equivalent to generating all subsets of C that have support greater than sup_{min} – these subsets are called *frequent* itemsets. Therefore, the mining objective is, in essence, to efficiently discover all frequent itemsets that are present in the database.

10.2.3 Privacy Mechanisms

We now move on to considering the various mechanisms through which privacy of the user data could be provided. One approach to address this problem is for the service providers to assure the users that the databases obtained from their information would be anonymized through the variety of techniques proposed in the statistical database literature [1, 38], before being supplied to the data miners. For example, the swapping of values between different customer records, as proposed in [17]. Depending on the service provider to guarantee privacy can be referred to as a “B2B (business-to-business)” privacy environment.

However, in today’s world, most users are (perhaps justifiably) cynical about such assurances, and it is therefore imperative to demonstrably provide privacy at the point of data collection itself, that is, *at the user site*. This is referred to as the “B2C (business-to-customer)” privacy environment [47]. Note that in this environment, any technique that requires knowledge of other user records becomes infeasible, and therefore the B2B approaches cannot be applied here.

The bulk of the work in privacy-preserving data mining of association rules has addressed the B2C environment (e.g. [2, 9, 19, 34]), where the user’s true data has to be anonymized at the source itself. Note that the anonymization process has to be implemented by a program which could be supplied either by the service provider or, more likely, by an independent trusted third-party vendor. Further, this program has to be verifiably secure – therefore, it must be simple in construction, eliminating the possibility of the true data being surreptitiously supplied to the service provider. In a nutshell, the goal of these techniques is to ensure the privacy of the raw local data at the source, but, at the same time, to support accurate reconstruction of the global data mining models at the destination.

Within the above framework, the general approach has been to adopt a *data perturbation* strategy, wherein each individual user’s true data is altered in some manner before forwarding to the service provider. Here, there are two possibilities: *statistical distortion*, which has been the predominant technique, and *algebraic distortion*, proposed in [47]. In the statistical approach, a common randomizing algorithm is employed at all user sites, and this algorithm is disclosed to the eventual data miner. For example, in the MASK technique [34], targeted towards “market-basket” type of sparse boolean databases, each bit in the true user transaction vector is independently flipped with a parametrized probability.

While there is only one-way communication from users to the service provider in the statistical approach, the algebraic scheme, in marked contrast, requires *two-way communication* between the data miner and the user. Here, the data miner supplies a user-specific perturbation vector, and the user then returns the perturbed data after applying this vector on the true data, discretizing the output and adding some noise. The vector is dependent on the *current contents* of the perturbed database available with the miner and, for large enterprises, the data collection process itself could become a bottleneck in the efficient running of the system.

Within the statistical approach, there are two further possibilities: (a) A simple *independent attribute perturbation*, wherein the value of each attribute in the user record is perturbed independently of the rest; or (b) A more generalized *dependent attribute perturbation*, where the perturbation of each attribute may be affected by the perturbations of the other attributes in the record. Most of the statistical perturbation techniques in the literature, including [18, 19, 34], fall into the independent attribute perturbation category. Notice, however, that this is in a sense antithetical to the original goal of association rule mining, which is to identify *correlations across attributes*. This limitation is addressed in [10], which employs a dependent attribute perturbation model, with each attribute in the user's data vector being perturbed based on its own value as well as the perturbed values of the earlier attributes.

Another model of privacy-preserving data mining is the k-anonymity model [35, 2], where each record value is replaced with a corresponding generalized value. Specifically, each perturbed record cannot be distinguished from at least k other records in the data. However, this falls into the B2C model since the intermediate database-forming-server can learn or recover precise records.

10.2.4 Privacy Metric

Independent of the specific scheme used to achieve privacy, the end result is that the miner receives as input the perturbed database V and the perturbation technique T used to produce this database. From these inputs, the miner attempts to reconstruct the original *distribution* of the true database U, and mine this reconstructed database to obtain the association rules. Given this framework, the general notion of privacy in the association rule mining literature is the level of certainity with which the data miner can reconstruct the true data values of users. The certainity can be evaluated at various levels:

Average Privacy. This metric measures the reconstruction probability of a random value in the database.

Worst-case Privacy. This metric measures the maximum reconstruction probability across all the values in the database.

Re-interrogated Privacy. A common system environment is where the miner does not have access to the perturbed database after the completion of the mining process. But it is also possible to have situations wherein the miner can use the mining output (i.e. the association rules) to subsequently *re-interrogate* the perturbed database, possibly resulting in reduced privacy.

Amplification Privacy. A particularly strong notion of privacy, called "amplification", was presented in [18], which guarantees strict limits on privacy breaches of individual user information, *independent of the distribution of the true data*. Here, the property of a data record U_i is denoted by $Q(U_i)$. For example, consider the following record from the example dataset U discussed earlier:

Age	*Sex*	*Education*
Child	Male	Elementary

Sample properties of the record include
$Q_1(U_i) \equiv$ "$Age = Child$ **and** $Sex = Male$", and
$Q_2(U_i) \equiv$ "$Age = Child$ **or** $Adult$".

In this context, the *prior probability* of a property of a customer's private information is the likelihood of the property in the absence of any knowledge about the customer's private information. On the other hand, the *posterior probability* is the likelihood of the property given the perturbed information from the customer and the knowledge of the prior probabilities through reconstruction from the perturbed database. In order to preserve the privacy of some property of a customer's private information, the posterior probability of that property should not be *unduly different* to that of the prior probability of the property for the customer. This notion of privacy is quantified in [18] through the following results, where ρ_1 and ρ_2 denote the prior and posterior probabilities, respectively:

Privacy Breach: An upward ρ_1-to-ρ_2 privacy breach exists with respect to property Q if $\exists v \in S_V$ such that

$$P[Q(U_i)] \leq \rho_1 \quad \text{and} \quad P[Q(U_i)|R(U_i) = v] \geq \rho_2.$$

Conversely, a downward ρ_2-to-ρ_1 privacy breach exists with respect to property Q if $\exists v \in S_V$ such that

$$P[Q(U_i)] \geq \rho_2 \quad \text{and} \quad P[Q(U_i)|R(U_i) = v] \leq \rho_1.$$

Amplification: Let the perturbed database be $V = \{V_1, \ldots, V_N\}$, with domain S_V, and corresponding index set I_V. For example, given the sample database U discussed above, and assuming that each attribute is distorted to produce a value within its original domain, the distortion may result in

V
5
7
2
12

which maps to

V		
Adult	Male	Elementary
Adult	Female	Elementary
Child	Male	Graduate
Senior	Female	Graduate

Let the probability of an original customer record $U_i = u, u \in I_U$ being perturbed to a record $V_i = v, v \in I_V$ be $p(u \rightarrow v)$, and let A denote the matrix of these transition probabilities, with $A_{vu} = p(u \rightarrow v)$.

With the above notation,a randomization operator $R(u)$

$$\forall u_1, u_2 \in S_U : \frac{p[u_1 \to v]}{p[u_2 \to v]} \leq \gamma$$

where $\gamma \geq 1$ and $\exists u : p[u \to v] > 0$. Operator $R(u)$ is at most γ-amplifying if it is at most γ-amplifying for all qualifying $v \in S_V$.

Breach Prevention: Let R be a randomization operator, $v \in S_V$ be a randomized value such that $\exists u : p[u \to v] > 0$, and ρ_1, ρ_2 $(0 < \rho_1 < \rho_2 < 1)$ be two probabilities as per the above privacy breach definition. Then, if R is at most γ-amplifying for v, revealing "$R(u) = v$" will cause neither upward (ρ_1-to-ρ_2) nor downward (ρ_2-to-ρ_1) privacy breaches with respect to any property if the following condition is satisfied:

$$\frac{\rho_2(1-\rho_1)}{\rho_1(1-\rho_2)} > \gamma$$

If this situation holds, R is said to support (ρ_1, ρ_2) privacy guarantees.

10.2.5 Accuracy Metric

For association rule mining on a perturbed database, two kinds of errors can occur: Firstly, there may be *support* errors, where a correctly-identified frequent itemset may be associated with an incorrect support value. Secondly, there may be *identity* errors, wherein either a genuine frequent itemset is mistakenly classified as rare, or the converse, where a rare itemset is claimed to be frequent.

The **Support Error** (μ) metric reflects the average relative error (in percent) of the reconstructed support values for those itemsets that are correctly identified to be frequent. Denoting the number of frequent itemsets by $|F|$, the reconstructed support by $\widehat{sup}$ and the actual support by sup, the support error is computed over all frequent itemsets as

$$\mu = \frac{1}{\mid F \mid} \Sigma_{f \in F} \frac{\mid \widehat{sup}_f - sup_f \mid}{sup_f} * 100$$

The **Identity Error** (σ) metric, on the other hand, reflects the percentage error in identifying frequent itemsets and has two components: σ^+, indicating the percentage of false positives, and σ^- indicating the percentage of false negatives. Denoting the reconstructed set of frequent itemsets with R and the correct set of frequent itemsets with F, these metrics are computed as

$$\sigma^+ = \frac{|R-F|}{|F|} * 100 \qquad \sigma^- = \frac{|F-R|}{|F|} * 100$$

Note that in some papers (e.g. [47]), the accuracy metrics are taken to be the worst-case, rather than average-case, versions of the above errors.

10.3 Evolution of the Literature

From the database perspective, the field of privacy-preserving data mining was catalyzed by the pioneering work of [9]. In that work, developing privacy-preserving *data classifiers* by adding noise to the record values was proposed and analyzed. This approach was extended in [3] and [26] to address a variety of subtle privacy loopholes.

Concurrently, the research community also began to look into extending privacy-preserving techniques to alternative mining patterns such as association rules, clustering, etc. For association rules, two streams of literature emerged, as mentioned earlier, one looking at providing input data privacy, and the other considering the protection of sensitive output rules. An important point to note here is that unlike the privacy-preserving classifier approaches that were based on adding a noise component to continuous-valued data, the privacy-preserving techniques in association-rule mining are based on *probabilistic mapping* from the domain space to the range space, over categorical atttributes.

With regard to input data privacy, the early papers include [34, 19], which proposed the MASK algorithm and the Cut-and-Paste operators, respectively.

MASK. In MASK [34], a simple probabilistic distortion of user data, employing random numbers generated from a pre-defined distribution function, was proposed and evaluated in the context of sparse boolean databases, such as those found in "market-baskets". The distortion technique was simply to flip each 0 or 1 bit with a parametrized probability p, or to retain as is with the complementary probability $1-p$, and the privacy metric used was average privacy. Through a theoretical and empirical analysis, it was shown that the p parameter could be carefully tuned to simultaneously achieve acceptable average privacy and good accuracy.

However, it was also found that mining the distorted database could be orders of magnitude more time-consuming as compared to mining the original database. This issue was addressed in a followup work [12] which showed that by generalizing the distortion process to perform symbol-specific distortion (i.e. different flipping probabilities for different values), appropriately chooosing these distortion parameters, and applying a variety of set-theoretic optimizations in the reconstruction process, runtime efficiencies that are well within an order of magnitude of undistorted mining can be achieved.

Cut-and-Paste Operator. The notion of a privacy breach was introduced in [19] as the following: The presence of an itemset I in the randomized transaction causes a privacy breach of level ρ if it is possible to infer, for some transaction in the true database, that the probability of some item i occuring in it exceeds rho.

With regard to this worst-case privacy metric, a set of randomizing privacy operators were presented and analyzed in [19]. The starting point was *Uniform Randomization*, where each existing item in the true transaction is, with probability p, replaced with a new item not present in the original transaction. (Note that this means that the number of items in the randomized transaction is always equal to the number in the original transaction, and is therefore different from MASK where the number of items in the randomized transaction is usually significantly more than its source since the flipping is done on both the 1's *and the 0's* in the transaction bit vector.) It was then pointed out that a basic deficiency of the uniform randomization approach is that while it might, with a suitable choice of p, be capable of providing acceptable average privacy, its worst case privacy could be significantly weaker.

To address this issue, an alternative *select-a-size* **(SaS)** randomization operator was proposed, which is composed of the following steps, employed on a per-transaction basis:

Step 1: For customer transaction t_i of length m, a random integer j from $[1, m]$ is first chosen with probability $p_m[j]$.

Step 2: Then, j items are uniformly and randomly selected from the true transaction and inserted into the randomized transaction.

Step 3: Finally, a uniformly and randomly chosen fraction ρ_m of the remaining items in the database that are not present in the true transaction (i.e. $C-$ items in t_i), are inserted into the randomized transaction.

In short, the final randomized transaction is composed of a subset of true items from the original transaction and additional false items from the complementary set of items in the database.

A variant of the SaS operator studied in detail in [19] is the *cut-and-paste* **(C&P)** operator. Here, an additional parameter is a cutoff integer, K_m, with the integer j being chosen from $[1, K_m]$, rather than from $[1, m]$. If it turns out that $j > m$, then j is set to m (which means that the entire original transaction is copied to the randomized transaction). Apart from the cutoff threshold, another difference between $C\&P$ and SaS is that the subsequent ρ_m randomized insertion (Step 3 above) is carried out on (a) the items that are not present in the true transaction (as in SaS), and (b) additionally, *on the remaining items in the true transaction* that were not selected for inclusion in Step 2.

An issue in the C&P operator is the optimal selection of the ρ_m and K_m parameters, and combinatorial formulae for determining their values are given in [19]. Through a detailed set of experiments on real-life datasets, it was shown that even with a challenging privacy requirement of not permitting any breaches with $\rho > 50\%$, mining a C&P-randomized database was able to correctly identify around 80 to 90% of the "short" frequent itemsets, that is frequent itemsets of lengths upto 3. The issue of how to safely randomize and mine long transactions was left as an open problem, since directly using C&P in such environments could result in unacceptably poor accuracy.

The above work was significantly extended in [18] through, as discussed in Section 10.2.4, the formulation of strict amplification-based privacy metrics and delineating a methodology for limiting the associated privacy breaches.

Distributed Databases. Maintaining input data privacy was also considered in [41, 25] in the context of databases that are *distributed* across a number of sites with each site only willing to share data mining results, but not the source data. While [41] considered data that is vertically partitioned (i.e., each site hosts a disjoint subset of the matrix columns), the complementary situation where the data is horizontally partitioned (i.e., each site hosts a disjoint subset of the matrix rows) is addressed in [25]. The solution technique in [41] requires generating and computing a large set of independent linear equations – in fact, the number of equations and the number of terms in each equation is proportional to the *cardinality* of the database. It may therefore prove to be expensive for market-basket databases which typically contain millions of customer transactions. In [25], on the other hand, the problem is modeled as a secure multi-party computation [23] and an algorithm that minimizes the information shared without incurring much overhead on the mining process is presented. Note that in these formulations, a pre-existing true database at each site is assumed, i.e. a B2B model.

Algebraic Distortion. Then, in [47], an algebraic-distortion mechanism was presented that unlike the statistical approach of the prior literature, requires *two-way communication* between the miner and the users. If V_c is the current perturbed database, then E_k is computed by the miner, which corresponds to the eigenvectors corresponding to the largest k eigenvalues of ${V_c}^T V_c$, where ${V_c}^T$ is the transpose of V_c. The choice of k makes a tradeoff between privacy and accuracy – large values of k give more accuracy and less privacy, while small values provide higher privacy and less accuracy. E_k is supplied to the user, who then uses it on her true transaction vector, discretizes the output, and then adds a noise component.

Their privacy metric is rather different, in that they evaluate the level of privacy by measuring the probability of an "unwanted" item to be included in the perturbed transaction. The definition of unwanted here is that it is an item that does not contribute to association rule mining in the sense that it does not appear in any frequent itemset. An implication is that privacy estimates can be *conditional* on the choices of association rule mining parameters

(sup_{min}, con_{min}). This may encourage the miner to experiment with a variety of values in order to maximize the breach of privacy.

Output Rule Privacy. We now turn our attention to the issue of maintaining the privacy of output rules. That is, we would like to alter the original database in a manner such that only the association rules deemed to be sensitive by the owner of the data source cannot be identified through the mining process. The proposed solutions involve either falsifying some of the entries in the true database or replacing them with null values. Note that, by definition, these techniques require a completely materialized true database as the starting point, in contrast to the B2C techniques for input data privacy.

In [13], the process of transforming the database to hide sensitive rules is termed as "sanitization", and in practical terms, this requires reducing either the support or the confidence of the sensitive rules to below the sup_{min} or con_{min} thresholds. Specifically, using R to refer to the set of all rules, and S to refer to the set of sensitive rules, the goal is to hide all the S rules by reducing the supports or confidences, and simultaneously minimize the number of rules in $R - S$ that may also become hidden as a side-effect of the sanitization process. (Note that the objective is only to maintain the visibility of rules in $R - S$, allowing the specific supports or confidences obtained by the miner for the $R - S$ rules to be altered if required. That is, it would be perfectly acceptable for the database to be sanitized such that a rule with high support or confidence in $R - S$ became a rule that was just above the threshold in the sanitized database.)

The sanitization can be achieved in different ways: 1) By changing the values of individual entries in the database; or, 2) By removing entire transactions from the database. It was shown in the initial work of [13], which only considere the lowering of support values, that, irrespective of the sanitization approach, finding the optimal (w.r.t. minimizing the impact on $R-S$) sanitization is an NP-Hard problem (through reduction from the Hitting Set problem [21]). A greedy heuristic technique was suggested, where the S set is ordered in decreasing order of support, and then each element is hidden in the ordered set is hidden in an iterative fashion. The hiding is done by performing a greedy search through the ancestors of the itemset, selecting at each level the parent with the maximum support and setting the selected parent as the new itemset that needs to be hidden. At the end of the process, a frequent item has been selected. The algorithm searches through the common list of transactions that support both the selected item and the initial frequent itemset to be hidden in order to identify the transaction that affects the minimum number of 2-itemsets. After this transaction is identified, then the selected frequent item is removed from the identified transaction. The effects of this database alteration are propagated to the other itemset elements, and the process repeats until the itemset is hidden.

The above work was extended in [15] to achieve hiding by also using the the confidence criterion. Unlike the purely support-based hiding approach where only 1's are converted to 0's, hiding through the confidence criterion can be achieved by converting 0's into 1's. However, an associated danger is that there can now be false positives, that is, infrequent rules may be incorrectly promoted into the frequent category. A detailed treatment of this issue is presented in [44].

An alternative approach for output rule privacy proposed in [37, 36] is to use the concept of "data blocking", wherein some values in the database are replaced with NULLs signifying unknowns. In this framework, the notions of itemset support and confidence are converted into *intervals*, with the actual support and confidence lying within these intervals. For example, the minimum support of itemset C_x is the percentage of transactions that have 1's for this itemset, while the maximum possible support is the percentage of transactions that contain either 1 or NULL for this itemset. Greedy algorithms for implementing the hiding are presented, and a discussion of their effectiveness is provided in [36]. More recently, decision-theoretic approaches based on data blocking are presented in [30, 22], which also utilize the "border theory" of frequent itemsets [40] – however, these approaches can be computationally demanding.

The rule-hiding techniques have limitations in that (a) they crucially depend on the data miner processing the database only with the specified supports and confidence levels – this may be hard to ensure in practice; (b) they may introduce significant false positives and false negatives in the non-sensitive set of rules; (c) they may introduce significant changes in the supports and confidences of the non-sensitive set of rules; and (c) in the case of data blocking, it may be sometimes possible to infer the hidden rules by assigning values to the null attributes.

Frameworks. A common trend in the input data privacy literature was to propose *specific* perturbation techniques, which are then analyzed for their privacy and accuracy properties. Recently, in [10], the problem was approached from a different perspective, wherein a generalized matrix-theoretic *framework* that facilitates a systematic approach to the *design* of random perturbation schemes for privacy-preserving mining was proposed. This framework supports amplification-based privacy, and its execution and memory overheads are comparable to that of classical mining on the true database. The distinguishing feature of FRAPP is its quantitative characterization of the *sources of error* in the random data perturbation and model reconstruction processes.

In fact, although it uses dependent attribute perturbation, it is fully decomposable into the perturbation of individual attributes, and hence has the *same run-time complexity* as any independent perturbation method. Through the framework, many of the earlier techniques are cast as special instances of the FRAPP perturbation matrix. More importantly, it was shown that through appropriate choices of matrix elements, new perturbation techniques can be constructed that provide highly accurate mining results even under strict amplification-based [18] privacy guarantees. In fact, a perturbation matrix with provably *minimal condition number*[1], was identified, substantially improving the accuracy under the given constraints. Finally, an efficient integration of this optimal matrix with the association mining process was outlined.

10.4 The FRAPP Framework

In the remainder of this chapter, we present, as a representative example, the salient details of FRAPP and discuss how it simultaneously provides strong privacy, high accuracy and good efficiency, in a B2C privacy-preserving environment of mining association rules.

As mentioned earlier, let the probability of an original customer record $U_i = u, u \in I_U$ being perturbed to a record $V_i = v, v \in I_V$ be $p(u \rightarrow v)$, and let A denote the matrix of these transition probabilities, with $A_{vu} = p(u \rightarrow v)$. This random process maps to a Markov process, and the perturbation matrix A should therefore satisfy the following properties [39]:

$$A_{vu} \geq 0 \quad \text{and} \quad \sum_{v \in I_V} A_{vu} = 1 \quad \forall u \in I_U, v \in I_V \tag{10.1}$$

Due to the constraints imposed by Equation 10.1, the domain of A is a *subset* of $\mathbf{R}^{|S_V| \times |S_U|}$. This domain is further restricted by the choice of perturbation method. For example, for the MASK technique [34], all the entries of matrix A are decided by the choice of a single parameter, namely, the flipping probability.

We now explore the *preferred choices* of A to simultaneously achieve privacy guarantees and high accuracy, without restricting *ab initio* to a particular perturbation method.

From the previously-mentioned results of [18], the following condition on the perturbation matrix A in order to support (ρ_1, ρ_2) privacy can be derived:

$$\frac{A_{vu_1}}{A_{vu_2}} \leq \gamma < \frac{\rho_2(1-\rho_1)}{\rho_1(1-\rho_2)} \qquad \forall u_1, u_2 \in I_U, \forall v \in I_V \tag{10.2}$$

[1] In the class of symmetric positive-definite matrices (refer Section 10.4.2.1).

That is, the choice of perturbation matrix A should follow the restriction that the *ratio of any two matrix entries (in a row) should not be more than* γ.

10.4.1 Reconstruction Model

We now analyze how the distribution of the original database is reconstructed from the perturbed database. As per the perturbation model, a client C_i with data record $U_i = u, u \in I_U$ generates record $V_i = v, v \in I_V$ with probability $p[u \rightarrow v]$. This event of generation of v can be viewed as a Bernoulli trial with success probability $p[u \rightarrow v]$. If the outcome of the ith Bernoulli trial is denoted by the random variable Y_v^i, the total number of successes Y_v in N trials is given by the sum of the N Bernoulli random variables:

$$Y_v = \sum_{i=1}^{N} Y_v^i \tag{10.3}$$

That is, the total number of records with value v in the perturbed database is given by Y_v.

Note that Y_v is the sum of N independent *but non-identical* Bernoulli trials. The trials are non-identical because the probability of success varies from trial i to trial j, depending on the values of U_i and U_j, respectively. The distribution of such a random variable Y_v is known as the Poisson-Binomial distribution [45].

From Equation 10.3, the expectation of Y_v is given by

$$E(Y_v) = \sum_{i=1}^{N} E(Y_v^i) = \sum_{i=1}^{N} P(Y_v^i = 1) \tag{10.4}$$

Using X_u to denote the number of records with value u in the original database, and noting that $P(Y_v^i = 1) = p[u \rightarrow v] = A_{vu}$ for $U_i = u$, results in

$$E(Y_v) = \sum_{u \in I_U} A_{vu} X_u \tag{10.5}$$

Let $X = [X_1 X_2 \cdots X_{|S_U|}]^T$, $Y = [Y_1 Y_2 \cdots Y_{|S_V|}]^T$. Then, the following expression is obtained from Equation 10.5:

$$E(Y) = AX \tag{10.6}$$

At first glance, it may appear that X, the distribution of records in the original database (and the objective of the reconstruction exercise), can be directly obtained from the above equation. However, an immediate difficulty is that that the data miner does not possess $E(Y)$, but only *a specific instance* of Y, with which she has to approximate $E(Y)$.[2] Therefore, the following approximation to Equation 10.6 is resorted to:

$$Y = A\widehat{X} \tag{10.7}$$

where X is estimated as $\widehat{X}$. This is a system of $|S_V|$ equations in $|S_U|$ unknowns, and for the system to be uniquely solvable, a necessary condition is that the space of the perturbed database is a superset of the original database (i.e. $|S_V| \geq |S_U|$). Further, if the inverse of matrix A exists, the solution of this system of equations is given by

$$\widehat{X} = A^{-1}Y \tag{10.8}$$

providing the desired estimate of the distribution of records in the original database. Note that this estimation is *unbiased* because $E(\widehat{X}) = A^{-1}E(Y) = X$.

10.4.2 Estimation Error

To analyze the error in the above estimation process, the following well-known theorem from linear algebra applies [39]:

THEOREM 10.1 *Given an equation of the form $Ax = b$ and that the measurement of b is in-exact, the relative error in the solution $x = A^{-1}b$ satisfies*

$$\frac{\| \delta x \|}{\| x \|} \leq c \frac{\| \delta b \|}{\| b \|}$$

where c is the condition number of matrix A.

For a positive-definite matrix, $c = \lambda_{max}/\lambda_{min}$, where λ_{max} and λ_{min} are the maximum and minimum eigen-values of matrix A, respectively. Informally, the condition number is a measure of the sensitivity of a matrix to numerical operations. Matrices with condition numbers near one are said to be *well-conditioned*, i.e. stable, whereas those with condition numbers much greater than one (e.g. 10^5 for a $5 * 5$ Hilbert matrix [39]) are said to be *ill-conditioned*, i.e. highly sensitive.

Equations 10.6 and 10.8, coupled with Theorem 10.1, result in

$$\frac{\| \widehat{X} - X \|}{\| X \|} \leq c \frac{\| Y - E(Y) \|}{\| E(Y) \|} \tag{10.9}$$

---

[2] If multiple distorted versions are provided, then $E(Y)$ is approximated by the observed average of these versions.

which means that the error in estimation arises from two sources: First, the sensitivity of the problem, indicated by the condition number of matrix A; and second, the deviation of Y from its mean, i.e. the deviation of perturbed database counts from their expected values, indicated by the variance of Y. In the remainder of this sub-section, we determine how to reduce this error by (a) appropriate choice of perturbation matrix to minimize the condition number, and (b) identifying the minimum size of the database required to (probabilistically) bound the deviation within a desired threshold.

10.4.2.1 Minimizing the Condition Number. The perturbation techniques proposed in the literature primarily differ in their choices for perturbation matrix A. For example:

- MASK [34] uses a matrix A with

$$A_{vu} = p^k(1-p)^{M_b-k} \tag{10.10}$$

  where M_b is the number of *boolean* attributes when each categorical attribute j is converted into $\mid S_U^j \mid$ boolean attributes, $(1-p)$ is the bit flipping probability for each boolean attribute, and k is the number of attributes with matching bits between the perturbed value v and the original value u.

- The *cut-and-paste* (C&P) randomization operator [19] employs a matrix A with

$$\begin{aligned} A_{vu} &= \sum_{z=0}^{M} p_M[z] \\ &\cdot \sum_{q=max\{0,z+l_u-M,l_u+l_v-M_b\}}^{min\{z,l_u,l_v\}} \frac{{}^{l_u}C_q \; {}^{M-l_u}C_{z-q}}{{}^{M}C_z} \\ &\cdot {}^{M_b-l_u}C_{l_v-q}\rho^{(l_v-q)}(1-\rho)^{(M_b-l_u-l_v+q)} \end{aligned} \tag{10.11}$$

  where

$$\begin{aligned} p_M[z] &= \sum_{w=0}^{min\{K,z\}} {}^{M-w}C_{z-w}\rho^{(z-w)}(1-\rho)^{(M-z)} \\ &\cdot \begin{cases} 1 - M/(K+1) & \text{if } w = M \;\&\; w < K \\ 1/(K+1) & \text{o.w.} \end{cases} \end{aligned}$$

  Here l_u and l_v are the number of 1 bits in the original record u and its corresponding perturbed record v, respectively, while K and ρ are operator parameters.

To enforce strict privacy guarantees, the choice of listed parameters for the above methods are bounded by the constraints, given in Equations 10.1 and 10.2, on the values of the elements of the perturbation matrix A. It turns out that for practical values of privacy requirements, the resulting matrix A for these previous schemes is extremely *ill-conditioned* – in fact, the condition numbers in our experiments were of the order of 10^5 and 10^7 for MASK and C&P, respectively.

Such ill-conditioned matrices make the reconstruction very sensitive to the variance in the distribution of the perturbed database. Thus, it is important to carefully choose the matrix A such that it is well-conditioned (i.e has a low condition number). If a distortion method is decided *ab initio*, as in the earlier techniques, then there is little room for making specific choices of perturbation matrix A. Therefore, the opposite approach of *first designing matrices of the required type*, and then devising perturbation methods that are compatible with these matrices, is taken.

Choosing a suitable matrix starts from the intuition that for $\gamma = \infty$, the obvious matrix choice is the *unity matrix*, which both satisfies the constraints on matrix A (Equations 10.1 and 10.2), and has the lowest possible condition number, namely, 1. Hence, for a given γ, the following matrix can be chosen:

$$A_{ij} = \begin{cases} \gamma x & \text{if } i = j \\ x & \text{o.w.} \end{cases} \quad \text{where } x = \frac{1}{\gamma + (|S_U| - 1)} \tag{10.12}$$

which is of the form

$$x \begin{bmatrix} \gamma & 1 & 1 & \dots \\ 1 & \gamma & 1 & \dots \\ 1 & 1 & \gamma & \dots \\ \vdots & \vdots & \vdots & \ddots \end{bmatrix}$$

It is easy to see that the above matrix, which incidentally is symmetric and Toeplitz [39], also satisfies the conditions given by Equations 10.1 and 10.2. Further, its condition number can be algebraically computed to be $1 + \frac{|S_U|}{\gamma - 1}$. At an intuitive level, this matrix implies that the probability of a record u remaining as u after perturbation is γ times the probability of its being distorted to some $v \neq u$. This matrix is termed as the "Gamma-Diagonal matrix" in [10].

At this point, an obvious question is whether it is possible to design matrices that have even lower condition number than the gamma-diagonal matrix. In [11], it is proven that the gamma-diagonal matrix has the *lowest* possible condition number among the class of symmetric perturbation matrices satisfying the constraints of the problem, that is, it is an *optimal choice* (albeit non-unique).

10.4.2.2 Database Size and Mining Accuracy. An analysis of the dependence of deviations of itemset counts in the perturbed database from their expected values, with respect to the size of the database, is carried out in [11]. Based on the analysis, which is based on an application of Hoeffding's General Bound [31], the following threshold on the database size, N, required for obtaining a desired accuracy, Δ, with a confidence of at least ϵ, is derived:

$$\Rightarrow N \geq \ln(2/(1-\epsilon))/(2\Delta^2) \tag{10.13}$$

That is, the miner must collect data from at least the number of customers given by the above bound. For example, with $\Delta = 0.001$ and $\epsilon = 0.95$, this turns out to be $N \geq 2 \times 10^6$, which is well within the norm for typical e-commerce environments. Further, note that these acceptable values are obtained with the comparatively loose Hoeffding Bound, and that in practice the minimum data requirements could be still lower.

10.4.3 Randomizing the Perturbation Matrix

The estimation models discussed thus far implicitly assumed the perturbation matrix A to be *deterministic*. However, it appears intuitive that if the perturbation matrix parameters were themselves *randomized*, so that each client uses a perturbation matrix not specifically known to the miner, the privacy of the client will be further increased. Of course, it may also happen that the reconstruction accuracy suffers in this process.

This trade-off is evaluated in [10] by replacing the deterministic matrix A with a randomized matrix $\tilde{A}$, where each entry $\tilde{A}_{vu}$ is a random variable with $E(\tilde{A}_{vu}) = A_{vu}$. The values taken by the random variables for a client C_i provide the specific parameter settings for her perturbation matrix.

The experimental results in [10] indicate that the trade-off turns out such that the two opposing effects almost cancel each other out, making the error only *marginally worse than the deterministic case.*

10.4.4 Efficient Perturbation

Having discussed the privacy and accuracy issues of the FRAPP approach, we now turn our attention to the efficient *implementation* of the perturbation algorithm described in Section 10.4. This requires generating, for each $U_i = u$, a discrete distribution with PMF $P(v) = A_{vu}$ and CDF $F(v) = \sum_{i \leq v} A_{iu}$, defined over $v = 1, \ldots, \mid S_V \mid$. To achieve this, the following algorithm whose complexity is proportional to the *sum* of the cardinalities of the attribute domains, is presented in [11]:

Specifically, the perturbation of record $U_i = u$ can be written as $P(V_i; U_i = u)$
$= P(V_{i1}, \ldots, V_{iM}; u)$

$= P(V_{i1}; u) \cdot P(V_{i2}|V_{i1}; u) \cdots P(V_{iM}|V_{i1}, \ldots, V_{i(M-1)}; u)$
where V_{ij} denotes the jth attribute of record V_i. For the perturbation matrix A, this works out to be

$$\begin{aligned} P(V_{i1} = a; u) &= \sum_{\{v|v(1)=a\}} A_{vu} \\ P(V_{i2} = b|V_{i1} = a; u) &= \frac{P(V_{i2} = b, V_{i1} = a; u)}{P(V_{i1} = a; u)} \\ &= \frac{\sum_{\{v|v(1)=a \text{ and } v(2)=b\}} A_{vu}}{P(V_{i1} = a; u)} \\ &\ldots \text{and so on} \end{aligned}$$

where $v(i)$ denotes the value of the ith attribute for the record with value v.

When A is chosen to be the gamma-diagonal matrix, and n_j is used to represent $\prod_{k=1}^{j} | S_U^k |$, the following expressions for the above probabilities are obtained after some simple algebraic manipulations:

$$\begin{aligned} P(V_{i1} = b; U_{i1} = b) &= (\gamma + \frac{n_M}{n_1} - 1)x \\ P(V_{i1} = b; U_{i1} \neq b) &= \frac{n_M}{n_1}x \end{aligned} \tag{10.14}$$

and for the jth attribute

$$P(V_{ij} = b|V_{i1}, \ldots, V_{i(j-1)}; U_{ij} = b)$$
$$= \begin{cases} \frac{(\gamma + \frac{n_M}{n_j} - 1)x}{\prod_{k=1}^{j-1} p_k} & \text{if } \forall k < j, V_{ik} = U_{ik} \\ \frac{(\frac{n_M}{n_j})x}{\prod_{k=1}^{j-1} p_k} & \text{o.w.} \end{cases} \tag{10.15}$$

$$P(V_{ij} = b|V_{i1}, \ldots, V_{i(j-1)}; U_{ij} \neq b) = \frac{(\frac{n_M}{n_j})x}{\prod_{k=1}^{j-1} p_k}$$

where p_k is the probability that V_{ik} takes value a, given that a is the outcome of the random process performed for the kth attribute, i.e. $p_k = P(V_{ik} = a|V_{i1}, \ldots, V_{i(k-1)}; U_i)$.

The above perturbation algorithm takes M steps, one for each attribute. For the first attribute, the probability distribution of the perturbed value depends only on the original value for the attribute and is given by Equation 10.13. For any subsequent column j, to achieve the desired random perturbation, both its original value and the *perturbed values* of the previous $j - 1$ columns are used as inputs, and the perturbed value for j is then generated as per the discrete distribution given in Equation 10.15. This is an example of *dependent column perturbation*, in contrast to the independent column perturbations used in most of the literature.

Finally, to assess the complexity of the algorithm, it is easy to see that the maximum number of iterations for generating the jth discrete distribution is $|S_U^j|$, and hence the maximum number of iterations for generating a perturbed record is $\sum_j |S_U^j|$.

10.4.5 Integration with Association Rule Mining

The core computation in association rule mining is to identify "frequent itemsets", that is, all those itemsets whose support (i.e. frequency) in the database is in excess of a user-specified threshold sup_{min}. Equation 10.8 can be *directly used* to estimate the support of itemsets containing all M categorical attributes. However, in order to incorporate the reconstruction procedure into bottom-up association rule mining algorithms such as *Apriori* [8], we need to also be able to estimate the supports of itemsets consisting of only a *subset* of attributes – this procedure is described next.

Let C denote the set of all attributes in the database, and C_s be a subset of these attributes. Each of the attributes $j \in C_s$ can assume one of the $|S_U^j|$ values. Thus, the number of itemsets over attributes in C_s is given by $I_{C_s} = \prod_{j \in C_s} |S_U^j|$. Let $\mathcal{L}, \mathcal{H}$ denote itemsets over this subset of attributes.

A user record *supports* an itemset $\mathcal{L}$ if the attributes in C_s take the values given by the itemset $\mathcal{L}$. Let the support cardinality of any itemset $\mathcal{L}$ in the original and distorted databases be denoted by $sup_{\mathcal{L}}^U$ and $sup_{\mathcal{L}}^V$, respectively. Then,

$$sup_{\mathcal{L}}^V = \frac{1}{N} \sum_{v \text{ supports } \mathcal{L}} Y_v$$

where Y_v denotes the number of records in V with value v (refer Section 10.4.1). From Equation 10.7, it is known that

$$Y_v = \sum_{u \in I_U} A_{vu} \widehat{X}_u$$

and therefore, using the fact that A is symmetric,

$$\begin{aligned} sup_{\mathcal{L}}^V &= \frac{1}{N} \sum_{v \text{ supports } \mathcal{L}} \sum_{u} A_{vu} \widehat{X}_u \\ &= \frac{1}{N} \sum_{u} \widehat{X}_u \sum_{v \text{ supports } \mathcal{L}} A_{vu} \\ &= \frac{1}{N} \sum_{\mathcal{H}} \sum_{u \text{ supports } \mathcal{H}} \widehat{X}_u \sum_{v \text{ supports } \mathcal{L}} A_{vu} \end{aligned}$$

If for all u which support a given itemset $\mathcal{H}$, $\sum_{v \text{ supports } \mathcal{L}} A_{vu} = \mathcal{A}_{\mathcal{HL}}$, then the above equation can be written as:

$$
\begin{aligned}
sup_{\mathcal{L}}^{V} &= \frac{1}{N} \sum_{\mathcal{H}} \mathcal{A}_{\mathcal{HL}} \sum_{u \text{ supports } \mathcal{H}} \widehat{X}_u \\
&= \sum_{\mathcal{H}} \mathcal{A}_{\mathcal{HL}} \quad \widehat{sup^{U}}_{\mathcal{H}}
\end{aligned}
$$

The next step is to identify the matrix $\mathcal{A}$ for the gamma-diagonal matrix. Using the above formula for $\mathcal{A}_{\mathcal{HL}}$, the $\mathcal{A}$ corresponding to itemsets over subset C_s is obtained as

$$
\mathcal{A}_{\mathcal{HL}} = \begin{cases} \gamma x + (\frac{I_C}{I_{C_s}} - 1)x & \text{if } \mathcal{H} = \mathcal{L} \\ \frac{I_C}{I_{C_s}} x & \text{o.w.} \end{cases} \tag{10.16}
$$

i.e. the probability of an itemset remaining the same after perturbation is $\dfrac{\gamma + I_C/I_{C_s} - 1}{I_C/I_{C_s}}$ times the probability of its being distorted to any other itemset.

Using the above $I_{C_s} \times I_{C_s}$ matrix, the supports of itemsets over any subset C_s of attributes can be estimated. A legitimate concern here might be that the matrix inversion could be time-consuming if I_{C_s} is large. Fortunately, the inverse for this matrix has a simple closed-form expression, as explained in [11], that can be directly used in the reconstruction process, greatly reducing both space and time resources.

Thus, FRAPP can efficiently reconstruct the counts of itemsets over any subset of attributes without requiring to construct all the counts, and the scheme can be implemented efficiently on bottom-up association rule mining algorithms such as Apriori [8]. Further, it is trivially easy to incorporate FRAPP even in *incremental* association rule mining algorithms such as DELTA [32] which operate periodically on changing historical databases, and use the results of previous mining operations to minimize the amount of work done during each new mining operation.

10.5 Sample Results

We move on, in this section, to presenting sample quantitative results on the privacy, accuracy and efficiency levels that can be supported for association rule mining.

The results are obtained on the **CENSUS** dataset, derived from a real-world census database available at the UCI repository [51]. Three categorical (`native-country, sex, race`) attributes and three continuous (`age, fnlwgt, hours-per-week`) attributes from the census database are used in the experiment, with the continuous attributes partitioned into discrete intervals to convert them into categorical attributes. The specific categories used for these six attributes are listed in Table 10.1. (The reason only a subset of the attributes in the original database is considered is that it has been established in several sociological studies[14, 46] that users typically expect privacy on only a few of the database fields – usually sensitive attributes such as health, income, etc.)

Table 10.1. CENSUS Dataset

Attribute	*Categories*
race	White, Asian-Pac-Islander, Amer-Indian-Eskimo, Other, Black
sex	Female, Male
native-country	United-States, Other
age	$[15-35), [35-55), [55-75), \geq 75$
fnlwgt	$[0-1e5], [1e5-2e5), [1e5-3e5), [3e5-4e5), \geq 4e5$
hours-per-week	$[0-20), [20-40), [40-60), [60-80), \geq 80$

The association rule mining accuracy on the CENSUS dataset is evaluated for a user-specified minimum support of $sup_{min} = 2\%$. Table 10.2 gives the number of frequent itemsets in the dataset for this support threshold, as a function of the itemset length.

Performance Metrics. The performance of the system is measured with regard to the accuracy that can be provided for a given privacy requirement specified by the user. The (ρ_1, ρ_2) amplification-based strict privacy measure from [18] is the privacy metric, and the results are presented for a $(5, 50)$ setting. $\rho_1 = 5$ is representative of the fact that users typically want to hide uncommon values which set them apart from the rest, while $\rho_2 = 50$ indicates that the user can still plausibly deny any value attributed to him or her since it is equivalent to a random coin-toss attribution.

To quantify data mining accuracy, the Support Error and Identity Error metrics presented earlier in Section 10.2.5 are utilized.

Table 10.2. Frequent Itemsets for $sup_{min} = 0.02$

Data Set	Itemset Length 1	2	3	4	5	6	7
CENSUS	19	102	203	165	64	10	–

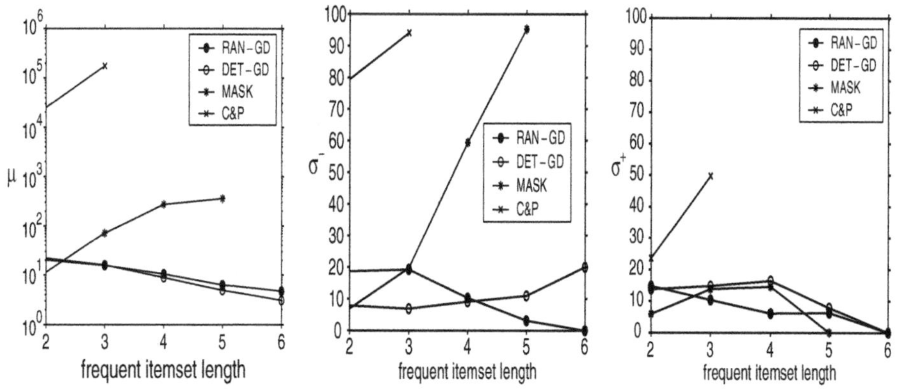

(a) Support error μ (b) False negatives σ^- (c) False positives σ^+

Figure 10.1. CENSUS ($\gamma = 19$)

Perturbation Algorithms. The experimental results are presented for FRAPP and representative prior techniques. For all the perturbation mechanisms, mining on the distorted database was accomplished using the *Apriori* [8] algorithm, with an additional support reconstruction phase at the end of each pass to recover the original supports from the perturbed database supports computed during the pass [12, 34].

Specifically, the perturbation mechanisms evaluated here are the following: (a) **DET-GD**: A deterministic gamma-diagonal perturbation matrix A (Section 10.4.2.1) is used for perturbation and reconstruction; (b) **RAN-GD**: A uniformly-distributed randomized gamma-diagonal perturbation matrix $\tilde{A}$ (Section 10.4.3) is used for perturbation and reconstruction; (c) **MASK**: This is the MASK scheme [34] with flipping parameter p set to 0.439, and the categorical attributes mapped to boolean attributes by making each value of the category an attribute; and (d) **C&P**: This is the Cut-and-Paste perturbation scheme [19], with algorithmic parameters K and ξ set to 3 and 0.494, respectively.

Results. For the CENSUS dataset, the support (μ) and identity (σ^-, σ^+) errors of the four perturbation mechanisms (DET-GD, RAN-GD, MASK, C&P) for $\gamma = 19$ are shown in Figure 10.1, as a function of the length of the frequent itemsets. Note that the support error (μ) graphs are plotted on a *log-scale*.

In these figures, we first note that DET-GD performs, on an absolute scale, extremely well, the error being of the order of 10 percent for the longer itemsets. Further, its performance is visibly better than that of MASK and C&P. In fact, as the length of the frequent itemset increases, the performance of both

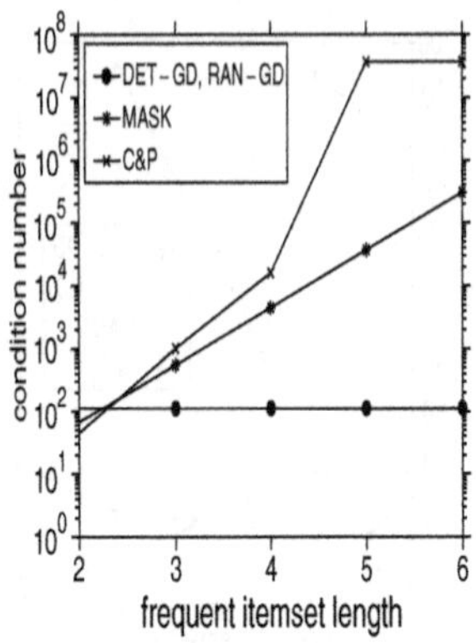

Figure 10.2. Perturbation Matrix Condition Numbers ($\gamma = 19$)

MASK and C&P degrade drastically. Specifically, MASK is not able to find any itemsets of length above 4 while C&P cannot identify itemsets beyond length 3.

The second point to note is that the accuracy of RAN-GD, although employing a randomized matrix, is only marginally lower than that of DET-GD. In return, it provides a substantial increase in the privacy – its worst case (determinable) privacy breach is only 33% as compared to 50% with DET-GD [11].

The primary reason for DET-GD and RAN-GD's good performance is the low *condition numbers* of their perturbation matrices. This is quantitatively shown in Figure 10.2, which plots these condition numbers on a *log-scale* (the condition numbers of DET-GD and RAN-GD are identical in this graph because $E(\tilde{A}) = A$). Note that the condition numbers are not only low but also *independent* of the frequent itemset length.

In marked contrast, the condition numbers for MASK and C&P increase *exponentially* with increasing itemset length, resulting in drastic degradation in accuracy. Thus, the choice of a gamma-diagonal matrix indicates highly promising results for discovery of long patterns.

Finally, with regard to actual mining response times also, FRAPP takes about *the same time* as Apriori for the complete mining process on the original and perturbed databases, respectively. This is because, as mentioned before, the reconstruction component shows up only *in between* mining passes and involves very simple computations. Further, the initial pre-processing step of perturbation of the large datasets took only a very modest amount of time even on vanilla PC hardware. Specifically, on a P-IV 2.0GHz PC with 1 GB RAM and 40 GB hard disk, perturbing 2.5 million records of CENSUS took only about a minute.

10.6 Closing Remarks

We studied in this chapter supporting privacy concerns in the association rule mining process. These concerns can arise with regard to the input data provided by users or with regard to the association rules arising from the mining process. We described the various kinds of privacy metrics and outlined an evolution of the rich body of related literature. Finally, using the FRAPP framework as a representative example, we quantitatively demonstrated how the state-of-the-art has reached a level whereby it is indeed feasible to simultaneously achieve the conflicting goals of strong privacy, high accuracy, and good efficiency in association rule mining.

Looking into the future, with regard to input data privacy, the ideal transformation technique would be one that allowed the randomized database to be directly mined to obtain the rules without involving any explicit reconstruction process. On the other hand, for output rule privacy, the ideal technique would be one that while hiding the sensitive rules, would minimize not just the number of affected rules in the non-sensitive set, but also the impact on their support and confidence values. As the ultimate holy grail, we look forward to the development of database transformation techniques that will be uniformly robust towards any kind of data mining activity.

Acknowledgements

This work was supported in part by a Swarnajayanti Fellowship from the Dept. of Science & Technology, Govt. of India.

References

[1] N. Adam and J. Wortman. Security control methods for statistical databases. *ACM Computing Surveys*, 21(4), 1989.

[2] C. Aggarwal and P. Yu. A condensation approach to privacy preserving data mining. *Proc. of 9th Intl. Conf. on Extending Database Technology (EDBT)*, March 2004.

[3] D. Agrawal and C. Aggarwal. On the design and quantification of privacy preserving data mining algorithms. *Proc. of ACM Symp. on Principles of Database Systems (PODS)*, May 2001.

[4] R. Agrawal, R. Bayardo, C. Faloutsos, J. Kiernan, R. Rantzau and R. Srikant. Auditing compliance with a hippocratic database. *Proc. of 30th Intl. Conf. on Very Large Data Bases (VLDB)*, August 2004.

[5] R. Agrawal, J. Kiernan, R. Srikant and Y. Xu. Hippocratic databases. *Proc. of 28th Intl. Conf. on Very Large Data Bases (VLDB)*, August 2002.

[6] R. Agrawal, A. Kini, K. LeFevre, A. Wang, Y. Xu and D. Zhou. Managing healthcare data hippocratically. *Proc. of ACM SIGMOD Intl. Conf. on Management of Data*, June 2004.

[7] R. Agrawal, T. Imielinski and A. Swami. Mining association rules between sets of items in large databases. *Proc. of ACM SIGMOD Intl. Conf. on Management of Data*, May 1993.

[8] R. Agrawal and R. Srikant. Fast algorithms for mining association rules. *Proc. of 20th Intl. Conf. on Very Large Data Bases (VLDB)*, September 1994.

[9] R. Agrawal and R. Srikant. Privacy-preserving data mining. *Proc. of ACM SIGMOD Intl. Conf. on Management of Data*, May 2000.

[10] S. Agrawal and J. Haritsa. A Framework for High-Accuracy Privacy-Preserving Mining. *Proc. of 21st IEEE Intl. Conf. on Data Engineering (ICDE)*, April 2005.

[11] S. Agrawal and J. Haritsa. A Framework for High-Accuracy Privacy-Preserving Mining. Tech. Rep. TR-2004-02, DSL/SERC, Indian Institute of Science, 2004. `http://dsl.serc.iisc.ernet.in/pub/TR/TR-2004-02.pdf`

[12] S. Agrawal, V. Krishnan and J. Haritsa. On addressing efficiency concerns in privacy-preserving mining. *Proc. of 9th Intl. Conf. on Database Systems for Advanced Applications (DASFAA)*, March 2004.

[13] M. Atallah, E. Bertino, A. Elmagarmid, M. Ibrahim and V. Verykios. Disclosure limitation of sensitive rules. *Proc. of IEEE Knowledge and Data Engineering Exchange Workshop (KDEX)*, November 1999.

[14] L. Cranor, J. Reagle and M. Ackerman. Beyond concern: Understanding net users' attitudes about online privacy. AT&T Tech. Rep. 99.4.3, April 1999.

[15] E. Dasseni, V. Verykios, A. Elmagarmid and E. Bertino. Hiding association rules by using confidence and support. *Proc. of 4th Intl. Information Hiding Workshop (IHW)*, April 2001.

[16] P. de Wolf, J. Gouweleeuw, P. Kooiman, and L. Willenborg. Reflections on PRAM. *Proc. of Statistical Data Protection Conf.*, March 1998.

[17] D. Denning. *Cryptography and Data Security*. Addison-Wesley, 1982.

[18] A. Evfimievski, J. Gehrke and R. Srikant. Limiting privacy breaches in privacy preserving data mining. *Proc. of ACM Symp. on Principles of Database Systems (PODS)*, June 2003.

[19] A. Evfimievski, R. Srikant, R. Agrawal and J. Gehrke. Privacy preserving mining of association rules. *Proc. of 8th ACM Intl. Conf. on Knowledge Discovery and Data Mining (KDD)*, July 2002.

[20] W. Feller. *An Introduction to Probability Theory and its Applications (Vol. I).* Wiley, 1988.

[21] M. Garey and D. Johnson. *Computers and Intractability: A Guide to the Theory of NP-Completeness.* W. H. Freeman, 1979.

[22] A. Gkoulalas-Divanis and V. Verykios. An integer programming approach for frequent itemset hiding. *Proc. of 15th ACM Conf. on Information and Knowledge Management (CIKM)*, November 2006.

[23] O. Goldreich. Secure Multi-party Computation. *www.wisdom.weizmann.ac.il/˜oded/pp.html*, 1998.

[24] J. Gouweleeuw, P. Kooiman, L. Willenborg and P. de Wolf. Post randomisation for statistical disclosure control: Theory and implementation. *Journal of Official Statistics*, 14(4), 1998.

[25] M. Kantarcioglu and C. Clifton. Privacy-preserving distributed mining of association rules on horizontally partitioned data. *Proc. of ACM SIGMOD Workshop on Research Issues in Data Mining and Knowledge Discovery (DMKD)*, June 2002.

[26] H. Kargupta, S. Datta, Q. Wang and K. Sivakumar. On the privacy preserving properties of random data perturbation techniques. *Proc. of the 3rd IEEE Intl. Conf. on Data Mining (ICDM)*, December 2003.

[27] K. LeFevre, R. Agrawal, V. Ercegovac, R. Ramakrishnan, Y. Xu and D. DeWitt. Limiting disclosure in hippocratic databases. *Proc. of 30th Intl. Conf. on Very Large Data Bases (VLDB)*, 2004.

[28] N. Mishra and M. Sandler. Privacy via pseudorandom sketches. *Proc. of 25th ACM Symp. on Principles of Database Systems (PODS)*, 2006.

[29] T. Mitchell. *Machine Learning.* McGraw Hill, 1997.

[30] G. Moustakides and V. Verykios. A Max-Min Approach for Hiding Frequent Itemsets. *Proc. of 6th IEEE Intl. Conf. on Data Mining - Workshops*, December 2006.

[31] R. Motwani and P. Raghavan. *Randomized Algorithms.* Cambridge University Press, 1995.

[32] V. Pudi and J. Haritsa. Quantifying the Utility of the Past in Mining Large Databases. *Information Systems*, Elsevier Science Publishers, vol. 25, no. 5, July 2000, pgs. 323-344

[33] J. R. Quinlan. *C4.5: Programs for Machine Learning.* Morgan Kaufmann, 1993.

[34] S. Rizvi and J. Haritsa. Maintaining data privacy in association rule mining. *Proc. of 28th Intl. Conf. on Very Large Databases (VLDB)*, August 2002.

[35] P. Samarati and L. Sweeney. Generalizing data to provide anonymity when disclosing information. *Proc. of 17th ACM Symp. on Principles of Database Systems (PODS)*, June 1998.

[36] Y. Saygin, V. Verykios and C. Clifton. Using unknowns to prevent discovery of association rules. *ACM SIGMOD Record*, vol. 30, no. 4, 2001.

[37] Y. Saygin, V. Verykios and A. Elmagarmid. Privacy preserving association rule mining. *Proc. of 12th Intl. Workshop on Research Issues in Data Engineering (RIDE)*, February 2002.

[38] A. Shoshani. Statistical databases: Characteristics, problems and some solutions. *Proc. of 8th Intl. Conf. on Very Large Databases (VLDB)*, September 1982.

[39] G. Strang. *Linear Algebra and its Applications*. Thomson Learning Inc., 1988.

[40] H. Toivonen. Sampling large databases for association rules. *Proc. of 22nd Intl. Conf. on Very Large Databases (VLDB)*, August 1996.

[41] J. Vaidya and C. Clifton. Privacy preserving association rule mining in vertically partitioned data. *Proc. of 8th ACM Intl. Conference on Knowledge Discovery and Data Mining (KDD)*, July 2002.

[42] J. Vaidya and C. Clifton. Privacy-preserving k-means clustering over vertically partitioned data. *Proc. of 9th ACM Intl. Conf. on Knowledge Discovery and Data Mining (KDD)*, August 2003.

[43] J. Vaidya and C. Clifton. Privacy preserving naive bayes classifier for vertically partitioned data. *Proc. of SIAM Intl. Conf. on Data Mining*, April 2004.

[44] V. Verykios, A. Elmagarmid, E. Bertino, Y. Saygin and E. Dasseni. Association Rule Hiding. *IEEE Trans. on Knowledge and Data Engineering*, 16(4), 2004.

[45] Y. Wang. On the number of successes in independent trials. *Statistica Silica 3*, 1993.

[46] A. Westin. Freebies and privacy: What net users think. Tech. Rep., Opinion Research Corporation, 1999.

[47] N. Zhang, S. Wang and W. Zhao. A new scheme on privacy-preserving association rule mining. *Proc. of 8th European Conference on Principles and Practice of Knowledge Discovery in Databases (PKDD)*, September 2004.

[48] Data from US Census beaurau : National Health Interview Survey : Person, 1993. http://dataferrett.census.gov.

[49] http://en.wikibooks.org/wiki/cookbook:frapp%c3%a9_coffee.

[50] http://www.cs.waikato.ac.nz/ml/weka.

[51] http://www.ics.uci.edu/ mlearn/mlsummary.html.

Chapter 11

A Survey of Association Rule Hiding Methods for Privacy

Vassilios S. Verykios
Dept. of Computer and Communication Engineering
University of Thessaly, Volos, GREECE
verykios@inf.uth.gr

Aris Gkoulalas-Divanis
Dept. of Computer and Communication Engineering
University of Thessaly, Volos, GREECE
arisgd@inf.uth.gr

Abstract Data and knowledge hiding are two research directions that investigate how the privacy of raw data, or information, can be maintained either before or after the course of mining the data. By focusing on the knowledge hiding thread, we present a taxonomy and a survey of recent approaches that have been applied to the association rule hiding problem. Association rule hiding refers to the process of modifying the original database in such a way that certain sensitive association rules disappear without seriously affecting the data and the non-sensitive rules. We also provide a thorough comparison of the presented approaches, and we touch upon hiding approaches used for other data mining tasks. A detailed presentation of metrics used to evaluate the performance of those approaches is also given. Finally, we conclude our study by enumerating interesting future directions in this research body.

Keywords: Privacy preserving data mining, knowledge hiding, frequent itemset hiding, association rule hiding.

11.1 Introduction

Privacy preserving data mining is a new research area that investigates the side-effects of data mining methods that originate from the penetration into the privacy of individuals and organizations. From a general point of view, we may classify privacy issues related to the application of data mining, into two broad categories. The first is related to the data per se and is known as *data hiding*, while the second concerns the information, or else the knowledge, that a data mining method may discover after having analyzed the data, and is known as *knowledge hiding*. Data hiding tries to remove confidential or private information from the data before its disclosure. Knowledge hiding, on the other hand, is concerned with the sanitization of confidential knowledge from the data.

In this expository study we present an overview in a specific class of methods in the knowledge hiding area, known as *frequent itemset* and *association rule hiding*. Other classes of methods, under the same area, include *classification rule hiding*, *clustering model hiding*, *sequence hiding* and so on and so forth. "Association rule hiding" (a term used for brevity instead of the longer title "frequent itemset and association rule hiding") has been mentioned for the first time in 1999 in a workshop paper by Atallah et al. [5]. The authors in [5], tried to apply general ideas regarding the implications of data mining in security and privacy of information – first presented by Clifton and Marks in [9] – to the association rule mining [3] framework. Clifton and Marks following the suggestions of D.E. O'Leary [24] – who was the very first to point out the security and privacy breaches that originate from data mining algorithms – indicated the need to consider different data mining approaches under the prism of preserving the privacy of information. Along these lines, they proposed a number of solutions like fuzzification of the source database, limiting access to the source database, as well as releasing of samples instead of the entire database.

The following scenario exemplifies the necessity of applying association rule hiding algorithms to protect sensitive knowledge. Let us suppose that we are negotiating with Dedtrees Paper Company, as purchasing directors of BigMart, a large supermarket chain. They offer their products in reduced prices, provided that we agree to give them access to our database of customer purchases. We accept the deal and Dedtrees starts mining our data. By using an association rule mining tool, they find that people who purchase skim milk also purchase Green Paper. Dedtrees now runs a coupon marketing campaign offering a 50 cents discount on skim milk with every purchase of a Dedtrees product. The campaign cuts heavily into the sales of Green Paper, which increases the prices to us, based on the lower sales. During our next negotiation with Dedtrees, we find out that with reduced competition they are unwilling to offer to us a low price. Finally, we start losing business to our competitors, who were able to negotiate a better deal with Green Paper. In other words, the aforementioned scenario indicates that BigMart should sanitize competitive information (and other important corporate secrets of course) before delivering their database to Dedtrees, so that Dedtrees does not monopolize the paper market.

We should emphasize here that the association rule hiding problem can be considered as a variation of the well known *database inference control* [12] problem in statistical and multilevel databases. The primary goal, in the database inference control, is to protect access to sensitive information that can be obtained through non-sensitive data and inference rules. In association rule hiding we consider that it is not the data but the sensitive rules that create a breach to privacy. Given a set of sensitive association rules, which are specified by the security administrator, the task of the association rule hiding algorithms is to sanitize the data so that the association rule mining algorithms, that will be applied to this data, (i) will be incapable of discovering the sensitive rules under certain parameter settings, and (ii) will be able to mine all the non-sensitive rules. Another problem has been investigated recently, which even though it is not targeted to addressing privacy issues per se, it does give a special solution to the association rule hiding problem. The problem is known as *inverse frequent itemset mining* [20].

11.2 Terminology and Preliminaries

Association rule mining is the process involving the discovery of sets of items (a.k.a. *itemsets*) that frequently co-occur in a transactional database so as to produce association rules that hold for the data. Each association rule is defined as an implication of the form $A \Rightarrow B$, where A, B are frequent itemsets in the transactional database, such that $A \cap B = \varnothing$. The itemset $A \cup B$ that lead to the generation of an association rule is known as the *generating* itemset and consists of two parts, the *Left Hand Side* (LHS), which is the part on the left of the arrow of the rule (here A), and the *Right Hand Side* (RHS), which is the part on the right of the arrow of the rule (here B). Two metrics, *support* and *confidence*, are incorporated in the task of association rule mining to drive the generation of association rules and expose only the ones that are expected to be interesting to the owner of the data. The reader can refer to the work of Agrawal et al. [2, 3] for a detailed overview of the association rule mining process and a set of computationally efficient algorithms for the production of the association rules. The theory of border is also important in our discussion. For a better understanding of the concepts underlying the border theory, the reader is encouraged to refer to the work of Mannila and Toivonen [18]. The need to revise the border so as to hide certain sensitive association rules is paramount to a subset of the presented algorithms. A presentation of the theory of border revision is given in the work of Moustakides and Verykios [21], while efficient algorithms for the computation of the borders can be found in [18, 13].

Knowledge hiding, in the context of association rule mining, aims at sanitizing the original dataset in a way that at least one of the following goals is accomplished: (i) no rule that is considered as sensitive from the owner's perspective, can be revealed from the sanitized dataset when this is mined at pre-specified thresholds of confidence and support (or at any value higher than these thresholds), (ii) all the non-sensitive rules can be successfully mined in the sanitized database at pre-specified thresholds of confidence and support (or higher), and (iii) no rule that was not found in the original dataset can be found at the sanitized database when mining this database at pre-specified thresholds of confidence and support (or higher). The first goal requires sensitive rules to disappear. The second goal simply states that there should be no *lost rules* in the sanitized dataset. The third goal says that no *false rules* should be produced as a side-effect of the sanitization process. Generally speaking, in the typical case hiding scenario, the sanitization process has to be accomplished in a way that minimally affects the original dataset, preserves the general patterns and trends, and achieves to conceal the sensitive knowledge.

11.3 Taxonomy of Association Rule Hiding Algorithms

In this section, we present a taxonomy of frequent itemset and association rule hiding algorithms after having reviewed a large collection of independent works in the area. In order to be able to classify the various algorithms, we propose a set of orthogonal dimensions based on which we will present the existing approaches. As a first dimension, we consider whether the hiding algorithm uses the support or the confidence of the rule to drive the hiding process. In this way we separate the hiding algorithms into *support*-based and *confidence*-based. The second dimension in the classification is related to the modification in the raw data that is caused by the hiding algorithm. The two forms of the modification comprise the *distortion* and the *blocking* of the original values. Distortion is the process of replacing 1's by 0's and 0's by 1's, while blocking refers to replacing original values by question marks. The third dimension, refers to whether a single rule or a set of rules can be hidden during an iteration of the hiding algorithms. Based on this criterion we differentiate hiding algorithms into *single rule* and *multiple rule* schemes. The fourth dimension has to do with the nature of the hiding algorithm, which can be either *heuristic* or *exact*. Heuristic techniques rely on optimizing certain sub-goals in the hiding process, while they do not guarantee optimality. The formulation of the association rule hiding problem presented in Section 11.2 implies that there are two specific sub-goals that need to be attained by every association rule hiding algorithm. The first sub-goal, which is basically the most important, is to try to hide as many sensitive rules as possible. The second sub-goal is to manage to hide the sensitive rules by minimizing the possible side-effects. As side-effects in the hiding process, we consider (a) the number of data items affected by the hiding process, (b) the number of non-sensitive rules which were accidentally hidden during the hiding process, and (c) the number of rules which were created by the hiding process. Different hiding algorithms give different priorities to the satisfaction of the sub-goals presented, producing in this way a list of hiding primitives. Exact techniques, on the other hand, rely on formulating the association rule hiding problem in such a way, that a solution can be found that satisfies all the sub-goals. Of course, there is a possibility that an exact approach fails to give a solution, and for this reason, some of the sub-goals need to be relaxed. However, this relaxation process is still part of the exact approach, which makes it different from the heuristic approaches. The fifth and final dimension determines whether a hiding algorithm preprocesses the user specified sensitive rules so that a minimal set of sensitive rules are given as input to the hiding technique. For the time being, there is only one technique which has been proposed to serve this task. This technique makes use of the border of the frequent itemsets, and it modifies it appropriately by recomputing it, in such a way, that a minimal set of sensitive rules joins the newly computed negative border. The algorithm is then driven by the negative and positive border for hiding the rules.

11.4 Classes of Association Rule Algorithms

Association rule hiding algorithms can be divided into three distinct classes, namely *heuristic* approaches, *border-based* approaches and *exact* approaches. The first class of approaches involves efficient, fast algorithms that selectively sanitize a set of transactions from the database to hide the sensitive knowledge. Due to their efficiency and scalability, the heuristic approaches have been the focus of attention for the vast majority of researchers in the knowledge hiding field. However, there are several circumstances in which they suffer from undesirable side-effects that lead them to suboptimal solutions.

The second set of approaches considers the task of sensitive rule hiding through modification of the original borders in the lattice of the frequent and the infrequent patterns in the dataset. In these schemes, the sensitive knowledge is hidden by enforcing the revised borders (which accommodate the hiding of the sensitive itemsets) in the sanitized database. The algorithms in this class differ both in the borders that they track and use for the hiding strategy, and in the methodology that they follow to enforce the revised borders in the sanitized dataset.

Finally, the third class of approaches contains non-heuristic algorithms which conceive the hiding process as a constraint satisfaction problem that they solve by using integer or linear programming. The main difference of these approaches, compared to the previous ones, is the fact that the sanitization process guarantees optimality in the hiding solution, provided that an optimal solution exists. On the other hand, these approaches are usually several orders of magnitude slower than the heuristic ones, especially due to the runtime of the integer/linear programming solver.

11.4.1 Heuristic Approaches

In this section, we review support-based and confidence-based heuristic approaches, which are based on either distortion or blocking of the original values. Between these two categories of approaches, the distortion-based are the ones commonly adopted by the overwhelming majority of researchers.

Support-based and Confidence-based Distortion Schemes. Atallah et al. [5] were the first to propose an algorithm for the hiding of sensitive association rules through the reduction in the support of their generating itemsets. The authors propose the construction of a lattice-like graph in the database. Through this graph, the hiding of a large itemset, related to the existence of a sensitive rule, is achieved by a greedy iterative traversal of its immediate subsets, selection of the subset that has the maximum support among all candidates (therefore is less probable to be hidden) and setting of this itemset as the new candidate to be hidden. By iteratively following these steps, the algorithm identifies the 1-itemset ancestor of the initial sensitive itemset, having the highest support. Then, by identifying the supporting transactions for both the initial candidate and the currently identified 1-itemset, the algorithm removes the 1-itemset from the supporting transaction which affects the least number of 2-itemsets. In sequel, the algorithm propagates the results of this action to the affected itemsets in the graph. When hiding a set of sensitive rules, the algorithm first sorts the corresponding large itemsets based on their support and then proceeds to hide them in a one-by-one fashion, using the methodology presented above. One of the most significant contributions of this work is the proof regarding the *NP-hardness* of finding an optimal sanitization of a dataset. On the negative side, the proposed approach is not interested in the extent of the loss of support for a large itemset, as long as it remains frequent in the sanitized outcome.

Dasseni et al. [11] generalize the problem in the sense that they consider the hiding of both sensitive frequent itemsets and sensitive rules. The authors propose three single rule heuristic hiding approaches that are based on the reduction of either the support or the confidence of the sensitive rules, but not both. In all three approaches, the goal is to hide the sensitive rules while minimally affecting the support of the non-sensitive itemsets. The first two strategies reduce the confidence of the sensitive rule either (i) by increasing the support of the rule antecedent, through transactions that partially support it, until the rule confidence decreases below the minimum confidence threshold, or (ii) by decreasing the frequency of the rule consequent through transactions that support the rule, until the rule confidence is below the minimum threshold. The third strategy decreases the frequency of a sensitive rule, by decreasing the support of either the antecedent or the rule consequent, until either the confidence or the support lies below the minimum threshold. A basic drawback of the proposed schemes is the strong assumption that all the items appearing in a sensitive rule do not appear in any other sensitive rule. Under this assumption, hiding of the rules one at a time or altogether makes no difference. Moreover, since this work aims at hiding all the sensitive knowledge appearing in the dataset, it fails to avoid undesired side-effects such as lost and false rules.

Verykios et al. [37] extend the work of Dasseni et al. [11] by improving and evaluating the algorithms for their performance under different sizes of input datasets and different sets of sensitive rules. Moreover, the authors propose two heuristic algorithms that incorporate the third strategy presented earlier. The first of these algorithms protects the sensitive knowledge by hiding the item having the maximum support from the minimum length transaction. The hiding of the generating itemsets of the sensitive rules is performed in a decreasing order of size and support and in a one-by-one fashion. Similarly to the first algorithm, the second algorithm first sorts the generating itemsets with respect to their size and support, and then hides them in a round-robin fashion as follows. First, for each generating itemset, a random ordering of its items and of its supporting transactions is attained. Then, the algorithm proceeds to remove the items from the corresponding transactions in a round-robin fashion, until the support of the sensitive itemset drops below the minimum support threshold. The intuition behind hiding in a round-robin fashion is fairness and the proposed algorithm (although rather naïve) serves as a baseline for conducting a series of experiments.

Oliveira and Zaïane [25] were the first to introduce multiple rule hiding approaches. The proposed algorithms are efficient and require two scans of the database, regardless of the number of sensitive itemsets to hide. During the first scan, an index file is created to speed up the process of finding the sensitive transactions and to allow for an efficient retrieval of the data. In the second scan, the algorithms sanitize the database by selectively removing the least amount of individual items that accommodate the hiding of the sensitive knowledge. An interesting novelty of this work is the fact that the proposed methodology takes into account not only the impact of the sanitization on hiding the sensitive patterns, but also the impact related to the hiding of non-sensitive knowledge. Three item restriction-based (MinFIA, MaxFIA, and IGA) algorithms are proposed that selectively remove items from sensitive transactions. The first algorithm, MinFIA, proceeds as follows. For each restrictive pattern it identifies the supporting transactions and the item having the smallest support in the pattern (called *victim item*). Then, by using a user-supplied disclosure threshold, it first sorts the identified transactions in ascending order of degree of conflict and then selects the number of transactions (among them) that need to be sanitized. Finally, from each selected transaction the algorithm removes the victim item. The MaxFIA algorithm proceeds exactly as the MinFIA with the only difference of selecting as the victim item the one that has the maximum support in the sensitive rule. Finally, IGA aims at clustering the restricted patterns into groups that share the same itemsets. By identifying overlapping clusters, the algorithm proceeds to hide the corresponding sensitive patterns at once (based on the sensitive itemsets they share) and consequently reduces the impact on the released dataset.

A more efficient approach than the one in [25] and the works of [11, 33, 34] was proposed by Oliveira and Zaïane [26]. The proposed algorithm, called SWA, is an efficient, scalable, one-scan heuristic which aims at providing a balance between the needs for privacy and knowledge discovery in association rule hiding. It achieves to hide multiple rules in only one pass through the dataset, regardless of its size or the number of sensitive rules that need to be protected. The algorithm proceeds in five steps that are applied to every group of K transactions (thus formulating a window of size K) read from the original database. Firstly, the non-sensitive transactions are separated from the sensitive ones and copied directly to the sanitized database. For each sensitive rule, the item having the highest frequency is selected and the supporting transactions are identified. Then, a disclosure threshold, potentially different for each sensitive rule, is used to capture the severity characterizing the release of the rule. Based on this threshold, SWA computes the number of supporting transactions that need to be sanitized for each rule and then sorts them in ascending order of size. For each selected transaction, the corresponding item is removed and then the transaction is copied to the sanitized dataset. The authors present a set of computational tests to demonstrate that SWA outperforms state-of-the-art approaches in terms of concealing all the sensitive rules, while maintaining high data utility of the released dataset.

Amiri [4] proposes three effective, multiple rule hiding heuristics that outperform SWA by offering higher data utility and lower distortion, at the expense of computational cost. Although similar in the philosophy to the previous approaches, the proposed schemes do a better job in modelling the overall objective of a rule hiding algorithm. The first approach, called *Aggregate* approach, computes the union of the supporting transactions for all sensitive itemsets. Among them, the transaction that supports the most sensitive and the least non-sensitive itemsets is selected and expelled from the database. The same process is repeated until all the sensitive itemsets are hidden. Similarly to this approach, the *Disaggregate* approach aims at removing individual items from transactions, rather than removing the entire transaction. It achieves that by computing the union of all transactions supporting sensitive itemsets and then, for each transaction and supporting item, by calculating the number of sensitive and non-sensitive itemsets that will be affected if this item is removed from the transaction. Finally, it chooses to remove the item from the transaction that will affect the most sensitive and the least non-sensitive itemsets. The third approach, called *Hybrid*, is a combination of the previous two, since it uses the *Aggregate* approach to identify the sensitive transactions and the *Disaggregate* approach to selectively delete items of these transactions, until the sensitive knowledge is hidden.

Wu et al. [41] propose a sophisticated methodology that removes the assumption of [11], regarding the disjoint relation among the items of the various sensitive rules. By using set theory, the authors formalize a set of constraints related to the possible side-effects of the hiding process and allow item modifications to enforce these constraints. However, the existing correlation among the rules can make impossible the hiding of the sensitive knowledge, without the violation of any constraints. For this reason, the user is permitted to specify which of the constraints she considers more significant and relaxes the rest. A drawback of the approach is the simultaneous relaxation (without the users' consent) of the constraint regarding the hiding of all the sensitive itemsets. To accommodate for rule hiding, the new scheme defines a class of allowable modifications that are represented as templates and are selected in a one-by-one fashion. A template contains the item to be modified, the applied operation, the items to be preserved or removed from the transaction and coverage information regarding the number of rules that are affected. Based on this information the algorithm can select and apply only the templates that are considered as beneficial based on the fact that they involve the least side-effects.

Pontikakis et al. [31] propose two distortion-based heuristics to selectively hide the sensitive rules. On the positive side, the proposed schemes use effective data structures for the representation of the rules and effectively prioritize the selection of transactions for sanitization. However, in both algorithms the proposed hiding process may introduce a number of side-effects, either by generating rules which were previously unknown, or by eliminating existing non-sensitive rules. The first algorithm, called *Priority-based Distortion Algorithm* (PDA), reduces the confidence of a rule by reversing 1's to 0's in items belonging in its consequent. On the other hand, the second algorithm, called *Weight-based Sorting Distortion Algorithm* (WDA), concentrates on the optimization of the hiding process in an attempt to achieve the least side-effects and the minimum complexity. This is achieved through the use of priority values assigned to transactions based on weights. Regarding performance, the proposed schemes tend to produce hiding solutions of comparable or slightly higher quality than the algorithms in [34] by generally introducing less side-effects. However, both algorithms are computationally demanding, with PDA requiring typically twice the time of the schemes in [34] to perform the hiding process.

Support-based and Confidence-based Blocking Schemes. Saygin et al. [33, 34] are the first to propose the use of *unknowns* (represented as question marks in the database) instead of transforming 1's to 0's and the opposite, for the hiding of sensitive association rules. As demonstrated in [33], the use of unknowns provides a safer alternative especially in critical real life applications where the distinction between "false" and "unknown" is vital. In their work, the authors introduce three simple heuristic approaches. The first approach, relies on the reduction in the support of the generating itemsets of the rule, while the other two rely on the reduction of the rule confidence of the rule, below the minimum thresholds. The definitions of both the support and the confidence measures are extended to capture the notion of an interval instead of being crisp values, while the algorithms consider both 0 and 1 values to use for hiding (in some proportion) so that it is difficult for an adversary to conclude upon the value hidden behind a question mark. A universal *safety margin* is applied to capture how much below the minimum thresholds should the new support and confidence of a sensitive rule lie, in order to consider that the rule is safely hidden. An important contribution of this work, apart from the methodology itself, is a discussion regarding the effect of the algorithms towards hiding of the sensitive knowledge, the possibility of reconstruction of the hidden patterns by an adversary and the importance of choosing an adequate safety margin when hiding the sensitive rules.

Wang and Jafari [39] propose two modification schemes that incorporate unknowns and aim at the hiding of predictive association rules, i.e. rules containing the sensitive items on their LHS. Both algorithms rely on the distortion of a portion of the database transactions to lower the confidence of the association rules. Compared to the work of Saygin et al. [33, 34], the algorithms presented in [39] require a reduced number of database scans and exhibit an efficient pruning strategy. However, by construction, they are assigned the task of hiding *all* the rules containing the sensitive items on their LHS, while the algorithms in the work of Saygin et al. can hide any specific rule. The first strategy, called ISL, decreases the confidence of a rule by increasing the support of the itemset in its LHS. The second approach, called DSR, reduces the confidence of the rule by decreasing the support of the itemset in its RHS. Both algorithms experience the *item ordering effect* under which, based on the order that the sensitive items are hidden, the produced sanitized databases are different. Moreover, the DSR algorithm seems to be more effective when the sensitive items have high support.

Pontikakis et al. [30] argue that the main disadvantage of a blocking algorithm is the fact that the dataset, apart from the blocked values (a.k.a.

unknowns), is not distorted. Thus, an adversary can disclose the hidden rules by identifying those generating itemsets that contain question marks and lead to rules with a maximum confidence that lies above the minimum confidence threshold. If the number of these rules is small then the probability of identifying the sensitive ones among them becomes high. To avoid this issue, the authors propose a blocking algorithm that purposely creates rules that were not existent in the original dataset (a.k.a. *ghost* rules) and their generating itemsets contain unknowns. Thus, the identification of the sensitive rules becomes harder since the adversary is unable to tell which of the rules that have a maximum confidence above the minimum threshold are the sensitive and which are the ghost ones. However, the introduction of ghost rules leads to a decrement in the data quality of the sanitized outcome. To balance the trade-off between privacy and data loss the proposed algorithm incorporates a safety margin that corresponds to the extend of sanitization that is performed in the dataset. The higher the safety margin the better the protection of the sensitive rules and the worse the data quality of the resulting dataset.

11.4.2 Border-based Approaches

In this section, we review two border-based approaches for the hiding of sensitive rules. The work of Sun and Yu [35] was the first to introduce the process of *border revision* for the hiding of the sensitive association rules. In their work, the authors propose a heuristic approach that uses the notion of the *border* (further analyzed in [18]) of the non-sensitive frequent itemsets to track the impact of altering transactions in the database. The proposed scheme, first computes the positive and the negative borders in the lattice of all itemsets and then focuses on preserving the quality of the computed borders during the hiding process. The quality of the borders directly affects the quality of the sanitized database that is produced, which can be maintained by greedily selecting those modifications that lead to minimal side-effects. In the proposed heuristic, a weight is assigned to each element of the expected positive border (which is the original positive border after it has been shaped up with the removal of the sensitive itemsets) in an attempt to quantify its vulnerability of being affected by item deletion. These weights are dynamically computed (during the sanitization process) as a function of the current support of the corresponding itemsets in the database. To reduce the support of a sensitive itemset from the negative border, the algorithm calculates the impact of the possible item deletions by computing the sum of the weights of the positive border elements that will be affected. Then, it proceeds to delete the candidate item that will have the minimal impact on the positive border.

Moustakides and Verykios [21] follow a similar approach to [35] by proposing two heuristics that use the revised positive and negative borders, produced by the removal of the sensitive itemsets and their supersets from the old frequent itemset lattice. The proposed algorithms try to remove from the database all the sensitive itemsets that belong to the revised negative border, while maintaining frequent all the itemsets of the revised positive border. For every item of a sensitive itemset, the algorithms list the set of positive border itemsets which depend on it. Then, from among all minimum border itemsets, the one with the highest support is selected as it is the one with the maximum distance away from the border. This itemset, called the *max-min* itemset, determines the item through which the hiding of the sensitive itemset will incur. The proposed algorithms try to modify this item in such a way that the support of the max-min itemset is minimally affected. When hiding multiple itemsets, the algorithms perform the sanitization in a one-by-one fashion, starting from the itemsets that have lower supports. Finally, the second algorithm improves the first one and, through experimental evaluation, is shown to provide better hiding solutions than [35], in the majority of the tested settings.

11.4.3 Exact Approaches

In this section, we review two exact approaches for the hiding of sensitive association rules. Exact approaches are typically capable of providing superior solutions compared to the ones of the heuristic schemes, at a high computational cost. They achieve this by formulating the sanitization process as a constraint satisfaction problem and by solving it using an integer/linear programming solver. Thus, the sanitization of the dataset is performed as an atomic operation which avoids the local minima experienced by the heuristic approaches.

Menon et al. [19] propose a scheme that consists of an exact and a heuristic part for the hiding of sensitive frequent patterns. The exact part formulates a *Constraint Satisfaction Problem* (CSP) with the objective of identifying the minimum number of transactions that need to be sanitized for the proper hiding of all the sensitive knowledge. To avoid the *NP-hardness* issue, the authors reduce the problem size considering only the sensitive itemsets, requesting that their support remains below the minimum support threshold. The optimization process is driven by a criterion function that is inspired by the measure of accuracy [17]. Moreover, the constraints imposed in the CSP formulation capture the number of supporting transactions that need to be sanitized for the hiding of each sensitive itemset. An integer programming solver is then applied to identify the best solution of the CSP and to derive the objective. In turn, this objective is provided as input to a heuristic sanitization algorithm that is assigned the task of identifying the actual transactions within the database and performing their sanitization. An important contribution of the authors, apart from the algorithm itself, is a discussion over the possibility of parallelization of the exact part. As demonstrated, based on the underlying properties of the dataset to be sanitized, it is possible for the produced CSP to be decomposed into parts that are solved independently. Bearing in mind the exponential complexity involving the solution of a CSP, this process can drastically reduce the required computational time for the hiding of the sensitive knowledge.

Gkoulalas and Verykios [13] propose an exact approach for the hiding of sensitive rules that uses the itemsets belonging in the revised positive and the revised negative borders to identify the candidate itemsets for sanitization. Through a set of theorems, involving existing relations among itemsets, the authors achieve to further reduce the set of candidates to a small fraction of its original size. The hiding process is then performed by formulating a CSP in which the status (frequent vs infrequent) of each of the itemsets in the reduced set is controlled through a set of constraints. By using a process of constraints degree reduction, all the participating constraints in the CSP become linear and have no coefficients. Moreover, all the variables involved in the CSP are of binary nature. These facts allow for an efficient solution of the CSP, by using binary integer programming, which is typically faster compared to the use of integer or linear programming (w.r.t. same problem sizes). The provided solution is proved to lead to an exact (without any side-effects) hiding of the sensitive patterns. A heuristic approach that relaxes the initial CSP to allow for the identification of a good solution, is applied only when the initial CSP is infeasible and therefore an exact solution cannot be attained. An important property of the proposed scheme is the fact that although the problem formulation leads to a CSP with a size that is typically larger than the one of [19], the hiding algorithm achieves good efficiency. We feel that this can be attributed both to the binary nature of the variables and to the linear (and without coefficients) constraints involved in the formulation of the CSP.

11.5 Other Hiding Approaches

Association rule hiding algorithms aim at protecting sensitive knowledge depicted in the form of frequent patterns and the related association rules. However, sensitive knowledge may appear in various forms directly related to the applied data mining algorithm that achieved to expose it. As a consequence, a set of hiding approaches have been devised recently to allow for the safeguarding of sensitive knowledge exposed by data mining tasks such as *clustering*, *classification* and *sequence* mining. In what follows, we briefly cover some state-of-the-art research work involving the hiding of sensitive knowledge depicted in one of the aforementioned formats.

Classification rule hiding algorithms consider a set of classification rules as sensitive and proceed to protect them by using either *suppression*-based or *reconstruction*-based techniques. The suppression-based techniques aim at reducing the *confidence* of a sensitive classification rule (measured in terms of the owner's belief regarding the rule's holding given the data), by distorting some attributes in the dataset belonging to transactions related to its existence. Chang and Moskowitz [7], were the first to address the inference problem caused by the downgrading of the data in the context of decision rules. Through a blocking technique, called *parsimonious downgrading*, the authors block the inference channels that lead to the identification of the sensitive rules by selectively sanitizing transactions so that missing values appear in the released dataset. This has as an immediate consequence the lowering of the confidence for the holding of the sensitive rules. Wang et al. [38] propose a heuristic approach that achieves to fully eliminate all the sensitive inferences, while effectively handling overlapping rules. The algorithm identifies the set of attributes that influence the existence of each sensitive rule the most and removes them from those supporting transactions that affect the non-sensitive rules the least. On the other hand, reconstruction-based approaches, inspired by the work in [32, 8] and introduced by Natwichai et al. [22], target at reconstructing the dataset by using only supporting transactions of the non-sensitive rules. These approaches are advantageous over the heuristic data modification approaches, since they hardly introduce any side-effects [36]. They first perform a rule-based classification of the original dataset to enable the owner of the data to identify the sensitive rules. Then, they proceed to construct a decision tree that is constituted only on non-sensitive rules approved by the data owner. The constructed dataset remains similar to the original one, except from the sensitive part, while the difference between the two datasets is proved to reduce as the number of rules increases. In [23] the authors achieve to further improve the quality of the reconstructed dataset. This is accomplished by extracting additional characteristic information from the original dataset with regard to the classification issue and by improving the decision tree building process. Furthermore, with the aid of information gain, the usability of the released dataset is substantially ameliorated even in the case of hiding many sensitive rules with high discernability in records classification.

The field of privacy preserving clustering collects techniques that aim at protecting the underlying attribute values and thus assure the privacy of individuals when sharing data for clustering. Achieving privacy preservation when sharing data for clustering is a challenging problem since the privacy requirements should be met, while the clustering results remain valid. The various methodologies can be separated into two broad categories; the *transformation*-based approaches and the *protocol*-based approaches. The transformation-based approaches are directly related to the distortion-based approaches of association rule hiding. They operate by performing a data transformation of the original dataset that maintains the similarity among the various pairs of attributes and they are usually independent of the clustering algorithm that is used. In the transformed space, the similarity between the distorted attribute pairs can still provide accurate results that allow for the correct clustering of the various objects. Some interesting approaches in this category involve the work of Oliveira and Zaïane [27, 28]. On the other hand, protocol-based approaches assume a distributed scenario where a set of data owners want to share their data for clustering, without compromising the privacy of their data by revealing any secrets. The algorithms of this category make an assumption regarding the partitioning of the data among the interested parties and are typically the privacy-aware versions of commonly used clustering algorithms, such as K-means. The proposed protocols control the information that is communicated among the data owners and guarantee that no sensitive knowledge can be learned from the model. Approaches in this category include the work of Jha et al. [16] and the work of Jagannathan et al. [15]. A somewhat different kind of approach that targets on density-based clustering is presented in [10]. The authors propose a kernel-based distributed clustering algorithm that uses an approximation of density estimation in an attempt to harden the reconstruction process for the original dataset. Each site computes a local density estimate for the data it holds and transmits it to a trusted third party. In sequel, the trusted party builds a global density estimate and returns it to the peers. By making use of this estimate, the sites can locally execute density-based clustering. Finally, the work of İnan and Saygin [14] extends the protocol-based approaches to capture the clustering of spatio-temporal data. The proposed protocol is in compliance with a series of trajectory comparison functions and allows for secure similarity computations through the use of a trusted third party.

The hiding of sensitive sequences is one of the most recent research directions in privacy preserving data mining, particularly due to the close relation that exists between sequences and trajectories. The given problem has the same underlying properties as the rule hiding one in the sense that a set of sensitive sequential patterns need to be hidden from a database while causing the least side-effects on their non-sensitive counterparts. The work of Abul et al. [1] is the first to concentrate on the NP-hardness issue involving the optimal hiding of sequences and to provide a heuristic, polynomial time algorithm that carries out the sanitization task. The proposed algorithm enables the incorporation of a disclosure threshold and can effectively hide sequences based on the minimum gap, the maximum gap and the maximum window of their events.

11.6 Metrics and Performance Analysis

In this section, we present two categories of measures related to the performance of a hiding algorithm. The first category consists of measures that can either be optimized by a hiding scheme in the course of its execution, or be adopted to allow for a fair comparison among different hiding schemes under a unified framework. The measures belonging in this category are called *internal* and were proposed by Oliveira et al. [29]. They are classified as either *data sharing*-based or *pattern sharing*-based. The data sharing-based measures quantify the extend of side-effects regarding sensitive association rules that failed to be hidden, legitimate rules that were accidentally missed, and artifactual association rules that were created by the sanitization process. On the other hand, the pattern sharing-based measures quantify the extend of side-effects regarding non-sensitive association rules that were lost or sensitive rules that were improperly hidden and can be easily be recovered through the use of inference channels. Furthermore, we proceed to present another set of metrics, which measure external parameters such as the behavior of the algorithm when applied to large datasets, its computational speed, and so on and so forth. The measures of this category are called *external* and were proposed by Bertino et al. [6].

The proposed data-sharing based measures are the following:

Hiding Failure (HF) This measure quantifies the percentage of the sensitive patterns that remain exposed in the sanitized dataset. It is defined as the fraction of the restrictive association rules that appear in the sanitized database divided by the ones that appeared in the original dataset. Formally,

$$\text{HF} = \frac{|R_P(D')|}{|R_P(D)|}$$

where $R_P(D')$ corresponds to the sensitive rules discovered in the sanitized dataset D', $R_P(D)$ to the sensitive rules appearing in the original dataset D and $|X|$ is the size of set X. Ideally, the hiding failure should be 0%.

Misses Cost (MC) This measure quantifies the percentage of the non-restrictive patterns that are hidden as a side-effect of the sanitization process. It is computed as follows:

$$\text{MC} = \frac{|\tilde{R}_P(D)| - |\tilde{R}_P(D')|}{|\tilde{R}_P(D)|}$$

where $\tilde{R}_P(D)$ is the set of all non-sensitive rules in the original database D and $\tilde{R}_P(D')$ is the set of all non-sensitive rules in the sanitized database D'. As one can notice, there exists a compromise between the misses cost and the hiding failure, since the more sensitive association rules one needs to hide, the more legitimate association rules is expected to miss.

Artifactual Patterns (AF) This measure quantifies the percentage of the discovered patterns that are artifacts. It is computed as follows:

$$\text{AP} = \frac{|P'| - |P \cap P'|}{|P'|}$$

where P is the set of association rules discovered in the original database D and P' is the set of association rules discovered in D'.

Dissimilarity (Diss) The measure of dissimilarity quantifies the difference between the original and the sanitized datasets by comparing their histograms, where the horizontal axis contains the items in the dataset and the vertical axis corresponds to their frequencies. It is calculated as follows:

$$\text{Diss}(D, D') = \frac{1}{\sum_{i=1}^{n} f_D(i)} \times \sum_{i=1}^{n} [f_D(i) - f_{D'}(i)]$$

where $f_X(i)$ represents the frequency of the i−th item in the dataset X, and n is the number of distinct items in the original dataset D.

The proposed pattern-sharing based metrics are the following:

Side-Effect Factor (SEF) Similarly to the measure of misses cost, the side-effect factor is used to quantify the amount of non-sensitive association rules that are removed as an effect of the sanitization process. It is defined as follows:

$$\text{SEF} = \frac{|P| - (|P'| + |R_P(D)|)}{|P| - |R_P|}$$

Recovery Factor (RF) This measure expresses the possibility of an adversary to recover a sensitive rule based on the non-sensitive ones. The recovery factor of a pattern takes into account the existence of its subsets. If *all* the subsets of a sensitive rule can be recovered from the sanitized dataset, then the recovery of the rule itself is possible, thus it is assigned an RF value of 1; otherwise RF = 0. However, this measure is not certain since, for instance, an adversary may not learn an itemset despite knowing its subsets.

Bertino et al. [6] propose a set of measures that are directly related to the performance of a hiding algorithm as far as external parameters are concerned. These "process performance" measures are clustered into four categories, as follows:

Efficiency This category consists of measures that quantify the ability of a privacy preserving algorithm to efficiently use the available resources and execute with good performance. Efficiency is measured in terms of CPU-time, space requirements (related to the memory usage and the required storage capacity) and communication requirements.

Scalability This category consists of measures that evaluate how effectively the privacy preserving technique handles increasing sizes of the data from which information needs to be mined and privacy needs to be ensured. Scalability is measured based on the decrease in the performance of the algorithm or the increase of the storage requirements along with the communications cost (if in a distributed setting), when the algorithm is provided with larger datasets.

Data Quality The data quality of a privacy preservation algorithm depends on two parameters. There are the quality of the dataset after the sanitization process, and the quality of the data mining results when applied to this dataset, compared to the ones attained when using the original dataset. Among the various possible measures for the quantification of the data quality, the most preferable are: (i) *accuracy*, which measures the proximity of a sanitized value to the original one and is closely related to the information loss resulting from the hiding strategy, (ii) *completeness*, which is used to evaluate the degree of missed data in the sanitized database and (iii) *consistency*, which is related to the relationships that must continue to hold among the different fields of a data item or among data items in a sanitized database.

Privacy Level This category consists of measures that estimate the degree of uncertainty according to which, the protected information can still be predicted. Measures, such as the information entropy, the level of privacy and the J-measure [6], are some among the possible metrics that one can apply to quantify the privacy level attained by a hiding scheme.

11.7 Discussion and Future Trends

There is a plethora of open issues related to the association rule hiding problem that are still under investigation. First of all, the emergence of sophisticated exact hiding approaches of very high complexity, especially for very large databases, causes the consideration of efficient parallel approaches to be employed for the solution of this problem. The parallel approaches will allow for a decomposition of the constraints satisfaction problem into numerous components that can be solved independently. The overall solution is then attained as a function of the objectives of the individual solutions.

Regarding the use of unknowns in blocking algorithms, a lot more research work is in need to provide hiding solutions that take advantage of the capabilities offered by their use. We feel that the use of unknowns in several real life scenarios is much more preferable than the use of conventional distortion techniques. This is true because distortion techniques fail to provide a distinction between the real values in the dataset and the ones that were distorted by the hiding algorithm in order to allow for its proper sanitization. Therefore, it is our belief that research in association rule hiding should target towards providing sophisticated and efficient solutions that make use of unknowns.

A different research direction concerns the use of database reconstruction approaches. Prominent research efforts towards this direction include the work of several researchers in the field of inverse frequent itemset mining [20, 40]. On going work considers yet another solution which is to append to the original database a synthetically generated database part so that the sensitive knowledge is hidden in the combined database which is disclosed to the public.

Other interesting future trends include, but are certainly not limited to, (i) the extension of the border revision idea to cover the direct hiding of association rules, instead of their indirect hiding through their generating itemsets, (ii) the introduction of techniques for correlation rule hiding, which is a more general problem than the one of association rule hiding, (iii) the provision and unification of more advanced measures for the comparison of the different hiding strategies, and (iv) the inception of spatio-temporal privacy preserving rule hiding methodologies that will prohibit the leakage of sensitive rules related to sensitive spatial and/or temporal information. The hiding of spatio-temporal patterns is currently a hot research issue since it imposes much greater challenges than the traditional knowledge hiding approaches.

11.8 Conclusions

Privacy preserving data mining is a new body of research focusing on the implications originating from the application of data mining algorithms to large public databases. In this study, we have delved into the deep waters of knowledge hiding, which is primarily concerned with the privacy of knowledge that is hidden in large databases. More specifically, we have surveyed a research direction that investigates how sensitive association rules can escape the scrutiny of malevolent data miners by modifying certain values in the database. We have also presented a thorough analysis and comparison of the surveyed approaches, as well as a classification of association rule hiding algorithms to facilitate the organization in our presentation. Before we conclude our study we have provided a synopsis of other related hiding approaches and we have introduced a set of metrics for the evaluation of the association rule hiding algorithms. Our study indicates that the state-of-the-art has greatly advanced from the inception of this research area till now. There is also a fertile ground of on-going work that guarantees even more substantial achievements in the days to come. There is definitely a lot more work that is needed before this area can be considered as mature. Moreover, we strongly believe that the emergence in the association rule hiding area will come into play in the evolution of other related fields in data mining and will cause new waves of research study. At that point, we will be certain that our expectations regarding the destiny of this field will have been fulfilled.

References

[1] O. Abul, M. Atzori, F. Bonchi, and F. Giannotti. Hiding sequences. Technical report, Pisa KDD Laboratory, ISTI-CNR, Area della Ricerca di Pisa, Nov. 2006.

[2] R. Agrawal, T. Imielinski, and A. Swami. Mining association rules between sets of items in large databases. In *Proceedings of the 1993 ACM SIGMOD International Conference on Management of Data (SIGMOD'93)*, pages 207–216, 1993.

[3] R. Agrawal and R. Srikant. Fast algorithms for mining association rules in large databases. In *Proceedings of the 20th International Conference on Very Large Databases (VLDB'94)*, pages 487–499, 1994.

[4] A. Amiri. Dare to share: Protecting sensitive knowledge with data sanitization. *Decision Support Systems*, 43(1):181–191, 2007.

[5] M. Atallah, E. Bertino, A. Elmagarmid, M. Ibrahim, and V. S. Verykios. Disclosure limitation of sensitive rules. In *Proceedings of the 1999 IEEE Knowledge and Data Engineering Exchange Workshop (KDEX'99)*, pages 45–52, 1999.

[6] E. Bertino, I. N. Fovino, and L. P. Povenza. A framework for evaluating privacy preserving data mining algorithms. *Data Mining and Knowledge Discovery*, 11(2):121–154, 2005.

[7] L. Chang and I. S. Moskowitz. Parsimonious downgrading and decision trees applied to the inference problem. In *Workshop on New Security Paradigms*, 1998.

[8] X. Chen, M. Orlowska, and X. Li. A new framework of privacy preserving data sharing. In *Proceedings of the 4th IEEE International Workshop on Privacy and Security Aspects of Data Mining*, pages 47–56, 2004.

[9] C. Clifton and D. Marks. Security and privacy implications of data mining. In *Proceedings of the 1996 ACM SIGMOD International Conference on Management of Data (SIGMOD'96)*, pages 15–19, Feb. 1996.

[10] J. C. da Silva and M. Klusch. Inference on distributed data clustering. In *Proceedings of the 4th International Conference on Machine Learning*

and Data Mining in Pattern Recognition (MLDM 2005), pages 610–619, 2005.

[11] E. Dasseni, V. S. Verykios, A. K. Elmagarmid, and E. Bertino. Hiding association rules by using confidence and support. In *Proceedings of the 4th International Workshop on Information Hiding*, pages 369–383, 2001.

[12] C. Farkas and S. Jajodia. The inference problem: A survey. *ACM SIGKDD Exploration Newsletter*, 4(2):6–11, 2002.

[13] A. Gkoulalas-Divanis and V. S. Verykios. An integer programming approach for frequent itemset hiding. In *Proceedings of the 2006 ACM Conference on Information and Knowledge Management (CIKM 2006)*, pages 748–757, 2006.

[14] A. Ìnan and Y. Saygin. Privacy preserving spatio-temporal clustering on horizontally partitioned data. In *Proceedings of the 8th International Conference on Data Warehousing and Knowledge Discovery (DaWaK 2006)*, pages 459–468, 2006.

[15] G. Jagannathan, K. Pillaipakkamnatt, and R. N. Wright. A new privacy-preserving distributed k-clustering algorithm. In *Proceedings of the 2006 SIAM International Conference on Data Mining (SDM 2006)*, 2006.

[16] S. Jha, L. Kruger, and P. McDaniel. Privacy preserving clustering. In *Proceedings of the 10th European Symposium on Research in Computer Security (ESORICS 2005)*, pages 397–417, 2005.

[17] G. Lee, C.-Y. Chang, and A. L. P. Chen. Hiding sensitive patterns in association rules mining. In *28th Annual International Computer Software and Applications Conference (COMPSAC 2004)*, pages 424–429, 2004.

[18] H. Mannila and H. Toivonen. Levelwise search and borders of theories in knowledge discovery. *Data Mining and Knowledge Discovery*, 1(3):241–258, 1997.

[19] S. Menon, S. Sarkar, and S. Mukherjee. Maximizing accuracy of shared databases when concealing sensitive patterns. *Information Systems Research*, 16(3):256–270, 2005.

[20] T. Mielikainen. On inverse frequent set mining. In W. Du and C. W. Clifton, editors, *Proceedings of the 2nd Workshop on Privacy Preserving Data Mining*, pages 18–23, 2003.

[21] G. V. Moustakides and V. S. Verykios. A max-min approach for hiding frequent itemsets. In *Workshops Proceedings of the 6th IEEE International Conference on Data Mining (ICDM 2006)*, pages 502–506, 2006.

[22] J. Natwichai, X. Li, and M. Orlowska. Hiding classification rules for data sharing with privacy preservation. In *Proceedings of the 7th International*

Conference on Data Warehousing and Knowledge Discovery (DaWaK 2005), pages 468–477, 2005.

[23] J. Natwichai, X. Li, and M. Orlowska. A reconstruction-based algorithm for classiciation rules hiding. In *Proceedings of the 17th Australasian Database Conference (ADC 2006)*, pages 49–58, 2006.

[24] D. E. O'Leary. Knowledge discovery as a threat to database security. In *Proceedings of the 1st International Conference on Knowledge Discovery in Databases*, pages 507–516, 1991.

[25] S. R. M. Oliveira and O. R. Zaïane. Privacy preserving frequent itemset mining. In *Proceedings of the 2002 IEEE International Conference on Privacy, Security and Data Mining (CRPITS 2002)*, pages 43–54, 2002.

[26] S. R. M. Oliveira and O. R. Zaïane. Protecting sensitive knowledge by data sanitization. In *Proceedings of the Third IEEE International Conference on Data Mining (ICDM 2003)*, pages 211–218, 2003.

[27] S. R. M. Oliveira and O. R. Zaïane. Achieving privacy preservation when sharing data for clustering. In *Proceedings of the 2004 SIAM International Conference on Data Mining (SDM 2004)*, 2004.

[28] S. R. M. Oliveira and O. R. Zaïane. Privacy-preserving clustering by object similarity-based representation and dimensionality reduction transformation. In *Proceedings of the Second IEEE International Conference on Data Mining (ICDM 2004)*, pages 21–30, 2004.

[29] S. R. M. Oliveira and O. R. Zaiane. A unified framework for protecting sensitive association rules in business collaboration. *International Journal of Business Intelligence and Data Mining*, 1(3):247–287, 2006.

[30] E. Pontikakis, Y. Theodoridis, A. Tsitsonis, L. Chang, and V. S. Verykios. A quantitative and qualitative analysis of blocking in association rule hiding. In *Proceedings of the 2004 ACM Workshop on Privacy in the Electronic Society (WPES 2004)*, pages 29–30, 2004.

[31] E. D. Pontikakis, A. A. Tsitsonis, and V. S. Verykios. An experimental study of distortion-based techniques for association rule hiding. In *Proceedings of the 18th Conference on Database Security (DBSEC 2004)*, pages 325–339, 2004.

[32] S. Rizvi and J. R. Haritsa. Maintaining data privacy in association rule mining. In *Proceedings of the 28th International Conference on Very Large Databases (VLDB 2002)*, 2002.

[33] Y. Saygin, V. S. Verykios, and C. Clifton. Using unknowns to prevent discovery of association rules. *ACM SIGMOD Record*, 30(4):45–54, 2001.

[34] Y. Saygin, V. S. Verykios, and A. K. Elmagarmid. Privacy preserving association rule mining. In *Proceedings of the 2002 International*

Workshop on Research Issues in Data Engineering: Engineering E-Commerce/E-Business Systems (RIDE 2002), pages 151–163, 2002.

[35] X. Sun and P. S. Yu. A border-based approach for hiding sensitive frequent itemsets. In *Proceedings of the Fifth IEEE International Conference on Data Mining (ICDM 2005)*, pages 426–433, 2005.

[36] V. S. Verykios, E. Bertino, I. N. Fovino, L. P. Provenza, Y. Saygin, and Y. Theodoridis. State-of-the-art in privacy preserving data mining. *ACM SIGMOD Record*, 33(1):50–57, 2004.

[37] V. S. Verykios, A. K. Emagarmid, E. Bertino, Y. Saygin, and E. Dasseni. Association rule hiding. *IEEE Transactions on Knowledge and Data Engineering*, 16(4):434–447, 2004.

[38] K. Wang, B. C. M. Fung, and P. S. Yu. Template-based privacy preservation in classification problems. In *Proceedings of the Fifth IEEE International Conference on Data Mining (ICDM 2005)*, pages 466–473, 2005.

[39] S.-L. Wang and A. Jafari. Using unknowns for hiding sensitive predictive association rules. In *Proceedings of the 2005 IEEE International Conference on Information Reuse and Integration (IRI 2005)*, pages 223–228, 2005.

[40] X. Wu, Y. Wu, Y. Wang, and Y. Li. Privacy aware market basket data set generation: A feasible approach for inverse frequent set mining. In *Proceedings of the 2005 SIAM International Conference on Data Mining (SDM 2005)*, 2005.

[41] Y.-H. Wu, C.-M. Chiang, and A. L. P. Chen. Hiding sensitive association rules with limited side effects. *IEEE Transactions on Knowledge and Data Engineering*, 19(1):29–42, 2007.

Chapter 12

A Survey of Statistical Approaches to Preserving Confidentiality of Contingency Table Entries

Stephen E. Fienberg
Department of Statistics, Machine Learning Department, and Cylab,
Carnegie Mellon University
Pittsburgh PA 15213-3890, U.S.A.
fienberg@stat.cmu.edu

Aleksandra B. Slavkovic
Department of Statistics
Pennsylvania State University
University Park PA 16802, U.S.A.
sesa@stat.psu.edu

Abstract In the statistical literature, there has been considerable development of methods of data releases for multivariate categorical data sets, where the releases come in the form of marginal and conditional tables corresponding to subsets of the categorical variables. In this chapter we provide an overview of this methodology and we relate it to the literature on the release of association rules which can be viewed as conditional tables. We illustrate this with two examples. A related problem, "association rule hiding" is often independently studied in the database community.

Keywords: Algebraic geometry, association rules, conditional tables, contingency tables, disclosure limitation, marginal tables, privacy preservation.

12.1 Introduction

The cross-classification of individuals or other units according to multiple categorical variables produces multi-way tables of counts, better known as contingency tables. There is an extensive statistical literature on the analysis of such tables, e.g., see [1], [4], [15], and [25]. When the number of variables is large, the cells of the resulting contingency tables often contain a substantial number of small counts. These pose potential problems of disclosure risk. One strategy for protecting the confidentiality of the entries in such circumstances has been the release of subsets of the data in the form of marginal and conditional tables. In this chapter we provide a survey of the literature that explains the effectiveness of this strategy both for the protection of confidentiality and utility in connection with log-linear and logit model methods.

The search for association rules in datamining focuses on the detection of relationships or "associations" between specific values of categorical variables in large data sets, i.e., multi-way contingency table. This search requires working with observed conditional distributions for an outcome variable or feature given one or more explanatory variables. Thus the search for association rules requires the construction of marginal and then conditional tables from the full contingency table, i.e., datamining for association rules in effect involve the efficient construction and storage of marginal and conditional tables, e.g., see [2] and [27]. Different datamining methods use these marginal and conditional tables in different ways. Some approach the problem by focusing solely on low-dimensional marginal tables while others utilize the full power of log-linear and logit models and use higher-dimensional marginal tables. The methods we describe here are relevant to both approaches.

Our methods described here relate to "association rule hiding" problem studied by the privacy-preserving data mining and database community. In this volume, Verykios et al. [39] give a survey of association rule hiding methods. They do not describe any related statistical disclosure limitation methods. What they refer to as "data hiding" in SDL literature is labeled usually as data masking. They point out that in general the sensitivity of the rules is determined by security administrator, while the focus is on efficiency and algorithmic approaches for hiding of the rules rather than the usability. Our methodology offers a way for detecting a sensitivity of a rule based on the data utility relevant for valid statistical analysis.

12.2 The Statistical Approach Privacy Protection

Statisticians have approached this search problem in the following fashion. Supose we have k-way cross-classification of counts arising from a sample of size n from a large population of size N, e.g., the size of the US adult population, or that from California. We want to report as much information from this table as possible without releasing data that would allow an intruder to identify one or more individuals with substantial probability. For the release to be useful, an analyst needs to be able to use what is released to reach some statistical conclusions that she would have tried to reach with the full k-way array.

Statisticians often define usefulness in this case in terms of fitting and interpreting the parameters in a log-linear model. The relevant quantities for doing this are marginal totals that correspond to the highest order interaction terms—these are the "best" data summaries, or *minimal sufficient statistics* for the mdel. The difficulty is: which log-linear model? To understand this we must do some form of model search, e.g., based on a search of model space and using some criterion like the Bayesian Information Criterion (BIC), e.g., see the paper by [26]. Releasing just those minimal sufficient margins that correspond to the model which minimizes BIC does not let the analyst check the fit of the model relative to others so we may wish to release even more i.e., higher-order margins that include these. When we fit the model we begin with the presence of certain interaction terms and we estimate their value along with asymptotic standard errors. The latter typical involve functions that are sums of inverses of the values in the minimal sufficient margins. This is extremely important since BIC and other criteria pick models where the asymptotic variance of the discarded terms are the same order of magnitude as the estimates. The implication is that for "good" log-linear models the minimal sufficient margins tend to have substantial sized counts typically on the order of 10 or more, and sometimes 100 or more! They will almost never have zeros in them, because that yields special estimability and fit problems and they will rarely include very small counts.

To check on privacy protection, we ask whether the information in the marginal and conditional tables used in the construction of association rules discloses confidential data about individuals or units represented in the full multi-way contingency table. Much of the statistical focus has tended to be on identification of small cell counts, e.g., "1" and "2." The first order of business is to assess the contribution from sampling. Roughly speaking, the probability that an individual record that is unique in the sample is also unique in the population from which the sample was drawn equals the sampling fraction, n/N, e.g., see [18]. Thus for a sample of size 2,000 drawn from a population of 200,000,000 adults the sampling fraction is 2,000/200,000,000 or 0.00001. The bottom line therefore is that sampling protects, just not absolutely or even in the formal sense that computer scientists have suggested, e.g., see [13]. Thus we go further and look directly at the table and compute several quantities, such as upper and lower bounds for the cell counts in the k-way table, or the number of possible tables satisfying the marginal or possibly marginal and conditional constraints, or we might look at the distribution over these possible tables to assure themselves that the probabilities don't lump up on just a few of the values between the bounds, e.g., see [9]. We provide some details in the remainder of the chapter.

12.3 Datamining Algorithms, Association Rules, and Disclosure Limitation

Association rules are often described using a market-basket metaphor that assumes that there are a large number of products that can be purchased by the customer, either in a single transaction, or over time in a sequence of transactions. Customers fill their basket with only a fraction of what is on display—i.e., with a sample. Association rules can be extracted from a database of transactions, to determine which products are frequently purchased together. For example, one might find that A ="purchases of diapers" typically coincide with B ="purchases of dog food" in the same basket. We then evaluate the usefulness of the rule using some form of statistical summary such as "support" and "confidence". For example,

Rule form: A ⇒ B [support, confidence]

Example: buys(x, "diapers") ⇒ buys(x, "dog food") [0.55%, 68%]

More generally, we have k-tuples based on k possible product types and the transactions or market baskets produce counts for a k-way contingency table with attributes corresponding to the presence or absence of the product types. Our new goal is to discover association rules involving the variables that make up this contingency table. For an association rule of the form: $\{A, B, C, \dots\} \Rightarrow \{E, F, G, \dots\}$, we define:

Confidence (accuracy) of $A \Rightarrow B$: $P(B|A)$ = (# of transactions containing both A and B) / (# of transactions containing A).

Support (coverage) of $A \Rightarrow B$: $P(A, B)$ = (# of transactions containing both A and B) / (total # of transactions)

There are many other possible criteria for assessing the usefulness of rules, e.g., [38] uses a variation on support and confidence while [29] and [30] use chi-square statistics for independence and conditional independence computed on the marginal tables.

Machine learning approaches often attempt to treat every possible combination of attribute values as a separate class, learn rules using the rest of attributes as input and then evaluate them for "support" and "confidence". This essentially involves examining all possible marginal tables corresponding to the attributes. The problem is that this approach tends to be computationally intractable, i.e., there are too many classes and consequently, too many rules. Alternatively criteria involve looking for rules that exceed pre-defined support (minimum support) and have high confidence. If we include among the objects of interest the negations of the items, or in statistical terms all of the categories of the variables, then in fact we are simply relying on full marginal and conditional tables for empirical evaluation and rule search. We reiterate this key point: *Support is a marginal table, and confidence is a conditional table, both corresponding to a subset of variables making up the full table.*

There is a major issue about what we mean by "the release of association rules." Many of the authors in the datamining literature have taken this notion to simply mean announcing or releasing the form of the rule, i.e., the variables involved. We believe that this is essentially a vacuous approach, since using the association rule requires the data that allow one to make predictions. To us, releasing a rule means releasing the data on which it is based, i.e., the corresponding conditional and/or marginal table. The more complex the rules and the more rules the greater the risk of disclosure of individual information and thus the violation of confidentiality promised to and the privacy of those whose data are represented in the table. The real differences between between the machine learning literature on association rules and the statistical literature on contingency tables is how they deal with the marginal and conditional tables, and what is reported or shared with others. We address the latter point in the next section.

Fienberg and Slavkovic [20] describe results based on release of exact marginals and conditionals that can help us determine which rules to hide in order to preserve privacy but to allow sufficient information for statistical inference; in this paper we highlight some of those results. In the computer science literature there are a number of alternative approaches, e.g., perturbing the full data array as proposed by [14], [28], and [23].

12.4 Estimation and Disclosure Limitation for Multi-way Contingency Tables

There is a separate literature on privacy and confidentiality in categorical statistical data bases that approaches a number of the issues raised directly or indirectly in the datamining literature but with a different and heavier emphasis on the tradeoff between preserving confidentiality and assuring utility of the released data in the sense of allowing for proper statistical inferences.

For the present purposes we can group the approaches in the statistical literature into perturbational and aggregation or collapsing. For continuous data, aggregation methods go under names such as micro-aggregation and k-anonymity. For categorical data, aggregation typically involve combining categories of variables with more than two values, but a special example of collapsing involves summing over variables to produce marginal tables. Thus instead of reporting the full muti-way contingency table we might report multiple collapsed versions of it. The release of multiple sets of marginal totals has the virtue of allowing statistical inferences about the relationships among the variables in the original table using log-linear model methods. Barak et al. [3]

present a novel approach to contingency tables using perturbation and aggregation ideas.

Notation and Definitions. Let $X = (X_1, X_2, ..., X_k)$ be a discrete random vector with probability function

$$p(x) = P(X = x) = P(X_1 = x_1, ..., X_k = x_k)$$

where $x = (x_1, ..., x_k)$. Each X_i is defined on a finite set of integers $[d_i] = \{1, 2, ..., d_i\}, d_i \geq 1, i = 1, ..., k$, with $\mathcal{D} = [d_1] \times ... \times [d_k]$. A k-way contingency table of counts, $\mathbf{n} = \mathbf{n}(i), i \in \mathcal{D}$, is a k-way dimensional array of non-negative integers such that each cell entry $\mathbf{n}(i) = \#\{X = i\}$ represents the number of times the configuration i is observed in a series of independent realizations of $X_1, ..., X_k$. The data of interest are counts in a k-way contingency table, $d_1 \times d_2 \times \cdots \times d_k$. Defined in this way, a table of counts is a point in a simplex of dimension equal to $\mathcal{D} - 1$, i.e., the number of cells-1. The values of X_i are lattice points in a convex polytope. Parameter sets lie in a related simplex. This sets up a link between contingency tables and algebraic geometry and allows us to use tools from algebraic geometry to describe the space of tables all satisfying some constraints or a model.

Consider disjoint subsets A and B of $K = \{1, ..., k\}$. The marginal table X_A with probabilities is defined as $p(x_A) = \sum_{K \setminus A} p(x_K)$, or equivalently $x_A = (x_j : j \in A)$. For example, if $A = \{1, 4\}$, then $x_A = (x_1, x_4)$. We define a conditional table $X_{A|B}$ with conditional probability values as a multi-conditional array $p(x_A|x_B) = \frac{p(x_{AB})}{p(x_B)}$ (e.g., Table 12.1).

Suppose that that we observe an arbitrary set of conditional and marginal tables, $\mathcal{T}$. We define the *fiber* $\mathcal{F}_\mathbf{t}$ as a set of all k-way non-negative integer tables that satisfy the constraints $\mathcal{T} = \mathbf{t}$. Consider a sublattice $\mathcal{L}_\mathbf{t}$ of $\mathbb{Z}^\mathcal{D}$ that depends on a collection $\mathcal{T}$ and a finite subset $\mathcal{B}_\mathbf{t}$ (e.g., a Markov basis is the smallest such subset) of $\mathcal{L}_\mathbf{t}$.

Each element of $\mathcal{B}_\mathbf{t}$, $\mathbf{z}$, can be thought of as a contingency table with values in $\mathbb{Z}^\mathcal{D}$, and each is called a *move* that satisfies $A_\mathbf{t}(\mathbf{n}+\mathbf{z}) = A_\mathbf{t}\mathbf{n}$, where $A_\mathbf{t}$ is a matrix that defines the constraints $\mathcal{T} = \mathbf{t}$ imposed on table $\mathbf{n}$. The most important property of Markov bases, for our purposes, is that they *connect* all tables satisfying the same set of constraints; thus they can be used for data swaps and for building a connected Markov chain. Helpful references for tools on algebraic statistics, including the calculation and use of Markov and Gröbner bases, are [6], [34], and [31].

Log-linear Models. Consider an $I \times J \times K$ table of observed counts $\{n_{ijk}\}$, with corresponding estimated expected values, $\{m_{ijk}\}$ under a multinomial sampling model. The saturated log-linear model for $\{m_{ijk}\}$ takes the form

$$\log m_{ijk} = u + u_{1(i)} + u_{2(j)} + u_{3(k)} + u_{12(ij)} + u_{13(ik)} + u_{23(jk)} + u_{123(ijk)}, \tag{12.1}$$

where each subscripted u-term sums to zero over any subscript, e.g.,

$$\sum_i u_{123(ijk)} = \sum_j u_{123(ijk)} = \sum_k u_{123(ijk)} = 0.$$

We get *unsaturated* models from (12.1) by setting sets of u-terms equal to zero, e.g., if we set

$$u_{123} = 0 \text{ for all } i, j, k, \tag{12.2}$$

we have the model of no second-order interaction. A logit model involves conditioning on a marginal total and for all practical purposes can be thought of as equivalent for the present purposes to the corresponding log-linear model which includes the u-terms that correspond to the marginal conditioned upon. These ideas and the definition of log-linear models generalize naturally from 3 to k dimensions.

Estimation and Assessing Goodness-of-Fit. We have the following key features associated with inference for log-linear models:

- The relevant statistical models focus on simultaneous interactions among sets of variables that define the contingency table.
- Special subsets of these models include the family of conditional independence models and the family of graphical models, which involve simultaneous occurrence of conditional independencies. For more details on graphical models in statistics see [25], and in machine learning see [22].
- The minimal sufficient statistics (i.e., sufficient data summaries) for a log-linear model are the marginal totals corresponding to the highest-order interaction terms in the model. For example, for the no second-order interaction model for three-way tables in equation (12.2) above, the minimal sufficient statistics are the three sets of two-way marginal totals, $\{n_{ij+}\}$, $\{n_{i+k}\}$ and $\{n_{+jk}\}$ corresponding to $\{u_{12(ij)}\}$, $\{u_{13(ik)}\}$, and $\{u_{23(jk)}\}$, respectively.
- The maximum likelihood estimates for the expected cell values are found by setting the minimal sufficient statistics equal to their expectations. For example, for the no-second-order interaction model for

three-way tables in equation (12.2) above:

$$\begin{aligned}\hat{m}_{ij+} &= n_{ij+} \text{ for all } i, j,\\ \hat{m}_{i+k} &= n_{i+k} \text{ for all } i, k,\\ \hat{m}_{+jk} &= n_{+jk} \text{ for all } j, k.\end{aligned}$$

- Maximum likelihood estimates for expected cell values under logit models are the same as corresponding log-linear models which include terms associated with the fixed margins that the logit model conditions upon, e.g., see the discussion in [4] and [15].

- Decomposable log-linear models are graphical models for which the maximum likelihood estimates have an explicit closed-form expression. They correspond to triangulated graphs. See [25].

- Standard methods of goodness-of-fit allow the user to assess how well the model and its minimal sufficient statistical margins can explain or reconstruct the original cell counts. These include goodness-of-fit criteria such as likelihood ratio statistics for separate models or for comparing nested models, and penalized criteria such as the BIC, e.g., see Madigan and Raftery [26]. In particular, the likelihood ratio test for comparing a pair of nested log-linear models is expressible in terms of the minimal sufficient marginals of the more complex model, a result implicit in formulae in [4] and [25], and made explicit in [16].

Disclosure Limitation and Bounds on Cell Counts. To check on the disclosure limitation provided by releasing only a subset of marginal totals one can consider the information in the margins for the construction of bounds for the individual cell entries. Consider an $I \times J$ table with entries $\{n_{ij}\}$ and row margins $\{n_{i+}\}$ and column margins $\{n_{+j}\}$. Then it is well-known that

$$\min\{n_{i+}, n_{+j}\} \geq n_{ij} \geq \max\{0, n_{i+} + n_{+j} - n_{++}\}, \tag{12.3}$$

and that these bounds, also known as Fréchet bounds, are sharp. Now consider the situation where instead of releasing a full k-way contingency table, we release a set of lower-dimensional marginal totals from it. Any contingency table with non-negative integer entries and fixed marginal totals is a lattice point in the convex polytope defined by the linear system of equations induced by the released marginals. The constraints given by the values in the released marginals induce upper and lower bounds on the interior cells of the initial table. In principle, we can obtain these bounds by solving the corresponding linear programming (LP) problem, but in general this is an NP-hard problem. Dobra and Fienberg [7, 8] have derived explicit formulas for several interesting sets of margins corresponding to special subsets of graphical log-linear models and they have proposed strategies for using these methods to find sets of margins that would not allow an intruder to make sharp inferences about the entries in the original table. In particular, [7] provide simple and explicit bounds formulas that are generalizations of equation (12.3) when the margins correspond to the minimal sufficient statistics of decomposable log-linear models.

It is important to recognize that as the number and size of the released margins grow, we tighten the bounds on the cells in the table (based on in increasing amount of information available) and the tightening takes on subtly complex forms because of the interlocking structure of the margins. Slavkovic [31] explored the form of linear and integer programming (IP) bounds for given conditionals. We illustrate the bounds approach in the present paper and describe some extensions to it involving combinations of margins and conditionals.

A major theme in the literature on disclosure limitation deals with the trade off between disclosure risk and data utility. See especially [36], and selected papers in [10]. Duncan with a variety of coauthors has stressed a graphical representation for this trade-off which they call the R-U map, e.g., see [12] for a discussion in the context of categorical data. Trottini and Fienberg [35] take the trade-off formalism several steps further and embeds it in a fully Bayesian decision-theoretic framework. Following [16] we adopt a somewhat more informal assessment process by considering maximal releases of marginal and conditional tables subject to limited disclosure risk in terms of bounds on cell entries in the table.

Releasing Marginal and Conditional Tables. Because data from both marginal and conditional tables are potentially of interest in assessing and reporting association rules, we need to understand how they differ in terms of the information they convey about the entries in multi-way contingency tables. For example, we want to do is check to see whether or not sets of marginal and conditional distributions for a contingency table are sufficient to uniquely identify the existing joint distribution. If so, we might as well release the full table!

The joint distribution for any two-way table is uniquely identified by any of the following sets of distributions: (1) $P(X_1|X_2)$ and $P(X_2|X_1)$, (2) $P(X_1|X_2)$ and $P(X_2)$, or (3) $P(X_2|X_1)$ and $P(X_1)$. Cell entries are allowed to be zero as long as we do not condition on an event of zero probability. Sometimes the sets $P(X_1|X_2), P(X_1)$ and $P(X_2|X_1), P(X_2)$ uniquely identify the joint distribution. The following result, due to [33] and [31], describes this situation and a generalization for a k-way table.

THEOREM 12.1 *(Slavkovic(2004)) Consider a k-way table and a collection $T = \{p_{A|B}, p_A\}$, where $A, B \subset K$. If given matrices with conditional* *probability values have a full rank, and $d_A \geq d_B$, then T uniquely identifies marginal table p_{AB}.*

Trivially, for bivariate tables, the joint probability distribution is the *support*, and thus along with the knowledge of sample size n, an association rule will reveal all cell counts. The above results also imply that releasing the *confidence* of a rule along with some marginal information, again will identify all entries in a table, although we are concerned primarily with the identification of cells with small counts.

Often, there are multiple realizations of the joint distribution for X, i.e., there is more than one table that satisfies the constraints imposed by them. Slavkovic [33], and [31] describe the calculation of bounds given an arbitrary collection of marginals and conditionals. They use LP and IP and discuss potential inadequacies in treating conditional constraints via LP. These results rely on the fact that any k-way table satisfying a set of compatible marginals and/or conditionals is a point in a convex polytope defined by a system of linear equations induced by released conditionals and marginals.

If a cell count is small and the upper bound is close to the lower bound, the intruder knows with a high degree of certainty that there is only a small number of individuals possessing the characteristics corresponding to the cell. This may pose a risk of disclosure of the identity of these individuals. For example, equation (12.3) gives the bounds when all that is released are the two one-way marginals in a two-way table. When a single marginal or a single conditional is given, the cell's probability is bounded below by zero and above by a corresponding marginal or a conditional value. This translates into bounds for cell counts as long as we have the knowledge of sample size n which is implicitly given by releasing the observed margins, while it must be provided as an additional piece of information for the released conditional probabilites.

When the conditions of Theorem 12.1 are not satisfied, we can obtain bounds for cell entries, and in some two-way cases there are closed form solutions. These bounds are sharp for a set of low dimensional tables with nicely rounded conditional probability values. For higher dimensions linear approximations of the bounds could be very far off from the true solution for the table of counts, and thus these bounds may mask the true disclosure risk. To calculate sharp IP bounds, we need either nicely rounded conditional probability values, which rarely occur in practice, or we need the observed cell counts. The latter implies that in practice the database owner is the only one which can produce the "true" bounds in the case of the conditionals; see [32].

Using the tools of computational commutative algebra such as Gröbner and Markov bases in statistics, we can find feasible solutions to the constrained maximization/minimization problem. Some advantages of this approach are that (1) we obtain sharp bounds when the linear or integer program approach fails, and (2) we can use it to describe all possible tables satisfying given costraints. In particular, a set of minimal Markov bases (moves) allows us to build a connected Markov chain and perform a random walk over the space of tables of counts that have the same fixed marginals and/or conditionals. This will allow us to either enumerate or sample from the space of tables via Sequential Importance Sampling (SIS) or Markov Chain Monte Carlo sampling. Some disadvantages of algebraic approach are that (1) calculation of Markov bases can be computationally infeasible for k-way tables, and (2) for conditionals, Markov bases are extremely sensitive to rounding of cell probabilities. A technical description of calculation and structure of Markov bases given fixed conditionals for two-way tables can be found in [31]. The reported results in the examples below rely on use of this methodology.

In a two-way case, we only deal with so called *full* conditionals because they involve all variables in the data base. Theorem 1.1 also describes the relationship between a conditional and a marginal table that involves a subset of variables from the data base. In other words, it describes a relationship between confidence and support for a rule that involves a subset of characteristics from a data base. Related theorems, their heuristics and constructions are illustrated in [31], and [20] who also further elaborate on relationships between a Markov basis set and the confidence and support, and implications for privacy. Here we focus on some of the consequences of these theorems relevant to establishing bounds on cells for evaluating potential disclosure.

One result implies that given the full conditional and the sample size n, the value of the moves can be used to determine if we have a unique solution. Other results imply that, for the same sample size n, the number of solutions for a fixed small conditional, $p_{A|B}$, is greater than or equal to the number of solutions we obtain by fixing the margin X_{AB}. This in turn should lead to wider bounds on some of the cell entries. We can study a specific subsets of Markov basis and determine if we are in the situation where the bounds given the small conditionals are the same as given its corresponding marginal. In a number of examples that we have examined to date, however, we have obtained the exact same bounds. This observation has led us to consider a set of conditions and heuristics that we can use in practice to determine when the bounds on cells given these two sets of released information are the same.

To evaluate the effect of releasing an association rule has on disclosure, we want to evaluate both confidence and support of the rule. The results of this section imply that it is sufficient to evaluate the support.

12.5 Two Illustrative Examples

12.5.1 Example 1: Data from a Randomized Clinical Trial

Koch et al. [24] report the data in Table 12.1 on the results of a randomized clinical trial on the effectiveness of an analgesic drug for patients of two different statuses and from two different centers. We use a shorthand notation to describe variables and marginals from the full tables, denoting Status as [S], Center as [C], Treatment as [T] with levels Active = 1 and Placebo = 2, and Response as [R] with levels Poor = 1, $|hboxModerate = 2$, Excellent = 3. Given that individuals in the clinical trial form a "population," confidentiality questions focus on the potential harm associated with the release of information on the four cells with counts of "3" in this table, corresponding to two sets of three individuals in 'Center 1,' and two sets of three individuals in 'Center 2.' In [19, 20] we analyzed these data with a focus on the links between the uniqueness and bounds results to association rules. Here we add to these earlier analyses and findings.

We are interested in the effect of the treatment on the response, controlling for the other two variables. More specifically, we are interested in answering: Which association rules are safe to release and provide enough information for an analyst to make proper inferences about the question of interest. We could be interested in evaluating the following association rules: $T \Rightarrow R$, $CS \Rightarrow R$, $CST \Rightarrow R$, and $CS \Rightarrow T$. In particular, the analyst needs the margins, or support, to go with a "good" log-linear model that fits the data well.

First, consider an association rule, $CST \Rightarrow R$. Support is the joint marginal distribution of $[CRST]$ and confidence $[R|CST]$ is a table with conditional probability values (see Table 12.1). It is trivial to see that release of the support of this rule results in full disclosure since it is the full four-way table. These probabilities along with the sample size n uniquely identify all cell counts.

If we just release the confidence associated with this rule we can explore an important inferential question of treatment effect by using the empirical conditional probability values from a full conditional distribution of $[R|CST]$. If we also have the 3-way margin $[CST]$, we can clearly reconstruct the full 4-way table! Given $[R|CST]$ with sample size n, there are 7,703,002 tables all having

Table 12.1. Results of clinical trial for the effectiveness of an analgesic drug. Source: Koch et al. [24]. The second panel contains observed counts, and the third panel has corresponding observed conditional probability values for $[R|CST]$.

		R	1	2	3	1	2	3
C	S	T						
1	1	1	3	20	5	0.107	0.714	0.179
1	1	2	11	14	8	0.333	0.424	0.242
1	2	1	3	14	12	0.103	0.483	0.414
1	2	2	6	13	5	0.250	0.542	0.208
2	1	1	12	12	0	0.500	0.500	0
2	1	2	11	10	0	0.524	0.476	0
2	2	1	3	9	4	0.188	0.563	0.250
2	2	2	6	9	3	0.333	0.500	0.167

the same conditional probability values. We give LP relaxation bounds in Table 12.2. The tightest bound for the count of "3" is [1, 16.48] in cell (1,2,1,1). We supplement these bounds by sharp integer bounds which in this case can be calculated only by using observed counts (see [32]). These bounds are much sharper than the LP bounds, with some cell counts being uniquely identified such as the above mentioned cell (1,2,1,1). Thus both the LP bounds and the number of possible tables can be misleading in evaluating the disclosure risk. More generally, [31] shows that with knowledge of the sample size n full conditionals are too risky to be released, and clearly in this example the release of confidence $[R|CST]$ is not safe! Fienberg and Slavkovic [20] demonstrate that we could potentially approximate "safely" the knowledge of the release of this association rule by treating the data in Table 12.1 as if they come from a two-way 8×3 table and compute the Fréchet bounds for margins $[CST]$ and $[R]$ (c.f., Table 1.5 in [20]).

We note that this single conditional release reveals the zero counts in the table unlike the release of margins, where we needed 3 3-way margins to learn the position of zeros. While the disclosure of zero in this example does not have much impact on an overall confidentiality risk, for larger and sparser k-way tables the presence of a large fraction of 0 cells that are identified as such may substantially increase the risk of disclosure of sensitive non-zero cells by constraining them even more than the constraints that come directly from the marginals.

Because this is a randomized clinical trial, in order to perform meaningful statistical analysis, we need to include the three-way margin for the three explanatory variables, i.e., $[CST]$. Most model search procedures would narrow the focus to two models, Model 1: $[CST]$ $[CSR]$, or Model 2: $[CST][CSR][RT]$, both of which fit the data well. Model 1 is a special case of Model 2 and the likelihood ratio test for the difference between them takes the value $\Delta G^2 = 5.4$ with 2 degrees of freedom, a value that is not significant

Table 12.2. Second panel has LP relaxation bounds, and third panel has sharp IP bounds for cell entries in Table 1.1 given $[R|CST]$ conditional probability values

		R	1	2	3	1	2	3
C	S	T						
1	1	1	[1,17.03]	[6.67,113.55]	[1.7,28.4]	[3,6]	[20,40]	[5,10]
1	1	2	[1.4,51.26]	[1.75,65.23]	[1,37.28]	[11,11]	[14,14]	[8,8]
1	2	1	[1,16.48]	[4.67,76.91]	[4,65.92]	[3, 3]	[14,14]	[12,12]
1	2	2	[1.2, 38.61]	[2.60,83.66]	[1,32.18]	[6,12]	[13,26]	[5,10]
2	1	1	[1.10,79.44]	[1,72.26]	0	[1,18]	[1,18]	[0]
2	1	2	[1.10,79.48]	[1,72.26]	0	[11,11]	[10,10]	[0]
2	2	1	[1,29.06]	[3,87.17]	[1,38.74]	[3,9]	[9,27]	[4,12]
2	2	2	[2,51.89]	[3,77.83]	[1,25.94]	[2,12]	[3,18]	[1,6]

at the 0.10 level when compared with a chi-squared distribution with 2 degrees of freedom. Thus one might reasonably conclude that the effect of the treatment on the response is explained through the interactive effect of Center and Status.

Note that we need three sets of marginal totals to make this inference: $[CST]$, $[CSR]$, and $[RT]$. We can think of these marginal tables as supports of the following association rules: $CS \Rightarrow T$, $CS \Rightarrow R$, and $T \Rightarrow R$. Thus we want to evaluate the release of these marginals in combination with appropriate confidences, that is conditional tables such as $[T|CS]$, $[R|CS]$ and $[R|T]$. By applying theorems mentioned in Section 3, we can draw a number of interesting conclusions. For example, bounds on cells given only the confidence $[R|T]$ will be as wide or wider than given only the rule's support $[RT]$. The same observation holds for the other association rules we are considering in this example. This result implies that for each rule it should be sufficient to evaluate only its support to determine if the release is safe.

Sometimes, however, we only have partial information on a rule, such as its confidence, and want to evaluate those along with other data summaries. For example, if we release $[R|T]$ and $[R]$, Theorem 1.1, tells us that we have $[RT]$. On the other hand, theoretically, $[R|CS]$ and $[R]$ will not uniquely identify $[CRS]$ because the number of levels in $[R]$ is not greater than in $[CS]$ which is four. The number of tables for $[CRS]$ is 31,081,397,760,000, and for $[R|CS]$ is 31,081,579,235,840. The LP relaxation bounds for releasing the conditional $[R|CS]$ instead of the margin $[CRS]$ are much wider, see Table 12.3. For example, the upper LP bound for (1,1,1,1) cell for $[R|CS]$ is 37.42 while for $[CRS]$ is 14. Based on these bounds, we could mistakenly conclude that it is safer to release the conditional, i.e., the confidence of the rule. The sharp bounds for $[R|CS]$ in place of $[CRS]$ are the same even though they produce a larger space of possible tables; however, the latter can have potential implications for estimating distributions over the space of solutions.

Table 12.3. Sharp upper and lower bounds for cell entries in Table 12.1 given the $[CSR]$ margin, and LP relaxation bounds given $[R|CS]$ conditional probability values

		R	1	2	3	1	2	3
C	S	T						
1	1	1	[0,14]	[0,34]	[0,13]	[1,37.42]	[1,92.31]	[1,34.68]
1	1	2	[0,14]	[0,34]	[0,13]	[1,37.42]	[1,74.73]	[1,34.68]
1	2	1	[0,9]	[0,27]	[1,17]	[1,27.84]	[0,57.10]	[0,53.47]
1	2	2	[0,9]	[0,27]	[0,17]	[1,27.84]	[1,85.51]	[1,53.48]
2	1	1	[0,23]	[0,22]	[0,0]	[1,32.22]	[1,78.36]	0
2	1	2	[0,23]	[0,22]	[0,0]	[1,75.04]	[1,11.23]	0
2	2	1	[0,9]	[0,18]	[0,7]	[1,43.40]	[1,87.81]	[1,33.54]
2	2	2	[0,9]	[2,18]	[0,7]	[1,43.40]	[1,87.81]	[1,33.54]

In our example, releasing the three association rules turns out to be safe based on an examination of the bounds given the rule's supports (c.f., [20], Table 1.7). As before, all of the upper bounds are reasonably far from the lower bounds except for the (2,1,2,3) cell where the upper and lower bounds are now 0, and perhaps the (2,2,1,3) and (2,2,2,3) cells where the bounds are [0,7]. If we released the $[CST]$, $[CSR]$, and $[RT]$ margins an intruder would be far from certain what entries belonged in the four cells that actually contain the count of "3."

12.5.2 Example 2: Data from the 1993 U.S. Current Population Survey

Table 12.4 describes data extracted from the 1993 Current Population Survey. Versions of these data have been used previously to illustrate several other approaches to confidentiality protection. The resulting 8-way table contains 2880 cells and is based on 48,842 cases; 1185 cells approximately 41%, contain 0 count cells. This is an example of a sparse table, too often present in practice, which poses significant problems in the model fitting and estimation. Almost all lower level margins (e.g., 2-way margins) contain 0 counts. Thus the existence of maximum likelihood estimates is an issue. These zeros propagate into the corresponding conditional tables.

Table 12.4. Description of variables in CPS data extract

Variable	Label	Categories
Age (in years)	A	$< 25, 25 - 55, > 55$
Employer Type (*Empolyment*)	B	Gov, Pvt, SE, Other
Education	C	<HS, HS, Bach, Bach+, Coll
Marital status (*Marital*)	D	Married, Other
Race	E	White, Non-White
Sex	F	Male, Female
Hours Worked (*HrsWorked*)	G	$< 40, 40, > 40$
Annual Salary (*Salary*)	H	$< \$50K, \$50K+$

From disclosure risk perspective we are interested in protecting cells with small counts such as "1" and "2". There are 361 cells with count of 1 and 186 with count of 2. Our task is to reduce a potential disclosure risk for at least 19% of our sample, while still providing sufficient information for a "valid" statistical analysis.

To alleviate estimation problems, we recoded variables B and G from 5 and 2 categories respectively to 2 categories each yielding a reduced 8-way table with 768 cells. This table is still sparse. There are 193 zero count cells, or about 25% of the cells. About 16% of cells have high potential disclosure risk; there are 73 cells with counts of 1 and 53 with counts of 2. For this table we find two reasonable log-liner models

Model 1: $[ABCFG][ACDFG][ACDGH][ADEFG]$,

Model 2: $[ACDGH][ABFG][ABCG][ADFG][BEFG][DEFG]$,

with goodness-of-fit statistics $G^2 = 1870.64$ with 600 degrees of freedom and $G^2 = 2058.91$ with 634 degrees of freedom, respectively.

Model 1 is a decomposable graphical log-linear model whose minimal sufficient statistics are the released margins. We first evaluate if these five-way marginal tables are safe to release by analyzing number of cells with small counts. Most of the cell counts are large and do not seem to present an immediate disclosure risk. Two of the margins are potentially problematic. Marginal table $[ABCFG]$ has 1 cell with count of "5" in (1,4,2,1,2) cell, while the margin $[ACDGH]$ has a low count of "4" and two cells with count of "8"; e.g., see Table 12.5. Even without out any further analysis, most agencies would not release such margins. Because we are fitting a decomposable models this initial exploratory analysis reveals that there will be at least one cell with a tight sharp upper bound of size "4". Bellow we investigate if these margins are indeed safe to release accounting for the log-linear model we can fit and the estimates they provide for the reduced and full eight-way tables.

Table 12.5. Marginal table $[ACDGH]$ from 8-way CPS table

		A	1		2		3	
		C	1	2	1	2	1	2
D	G	H						
1	1	1	198	139	943	567	2357	2225
		2	11	19	240	715	1009	3781
	2	1	246	144	765	294	3092	2018
		2	8	14	274	480	1040	2465
2	1	1	2327	2558	835	524	2794	3735
		2	8	14	51	105	114	770
	2	1	1411	1316	617	359	3738	3953
		2	4	15	32	68	78	372

Model 1 is easy to fit and evaluate: it is decomposable and there are closed-form solutions for bounds given the margins. Almost all lower bounds are 0. As expected from the analysis above, the smallest upper bound is 4 counts. There are 16 such cells, of which 4 contain counts of "1" and rest contain "0". The next smallest upper bound is 5, for 7 "0" cell counts and for 1 cell with a count of "5". The 5 cells with counts of "1" have the highest risk of disclosure. The next set of cells with a considerably high disclosure risk are cells with an upper bound of size 8. There are 32 such cells (23 contain counts of "0", 4 contain counts of "1", 3 contain counts of "2", and 2 contain counts of "3"). If we focus on count cells of "1" and "2", with the release of this model we directly identified 12 out of 126 sensitive cells.

Table 12.6. Summary of difference between upper and lower bounds for small cell counts in the full 8-way CPS table under Model 1 and under Model 2

	Model 1						Model 2					
Bound diff. / Cell count	0	1	2	3	4	5	0	1	2	3	4	5
0	226	112	66	52	69	62	192	94	58	40	36	26
1	-	12	15	14	13	20	-	10	8	6	2	10
2	-	-	1	3	8	4	-	-	2	2	4	4
3	-	-	-	1	4	2	-	-	-	0	0	0

If we fit the same model to the full 8-way table with 2,880 cells, there are 660 cells with difference in bounds less than equal to 5, with all lower bounds being 0. Most of these are "0" cell counts; however, a high disclosure risk exists for 74 cells with count of "1", 16 cells with cell count equal "2", and 7 cells with counts of "3"; see the summary in Table 12.6. Thus releasing the margins corresponding to Model 1 poses a substantial risk of disclosure.

Model 2 is non-decomposable log-linear model and it requires an iterative algorithm for parameter estimation and extensive calculation for bounds. This model has 5 marginals as sufficient statistics. The 5-way margin $[ACDGH]$ is still problematic; however, the 4 4-way margins all appear to be safe to release with the smallest count of size "46" appearing in cell (1,4,1,1) of the margin $[ABFG]$.

We focus our discussion only on cells with small counts, as we did for the Model 1. Since Model 2 is non-decomposable, no closed-form solutions exist for cell bounds, and we must rely on LP and IP which sometimes may not produce sharp bounds. In this case this was not an issue. For the reduced 8-way table, all lower bounds are 0 and the minimum upper bound again is 4. There are 16 cells with upper bound of 4, of which four cells have count "1", and the rest are "0". The next smallest upper bound is 8, and there are 5 such cells with counts of "1", 4 cells with counts of "2", and 3 cells with counts of "3". With these margins, in comparison to the released margins under Model 1, we have eliminated the effect of the margin $[ABCFG]$, and reduced a disclosure risk for a subset of small cell counts; however, we did not reduced the disclosure risk for the small cell counts with the highest disclosure risk. For the full 8-way table, we compare the distribution of small cell bounds for the small cell counts under the two models; see Table 12.6. There are no cells with counts of "3" that have very tight bounds. For the cells with counts of "2", the number of tight bounds have not substantially decreased (e.g., 16 under Model 1 vs. 12 under Model 2), but there has been a significant decrease in the number of tight bounds for the cells with count of "1" (e.g., from 74 under Model 1 to 36 under Model 2).

In theory we could enumerate the number of possible tables utilizing algebraic techniques and software such as LattE [5], MCMC, or SIS. Due to large dimension of the solution polytope for this example, however, LattE is currently unable the execute the computation because the space of possible tables is extremely large. We have also been unable to fine-tune the SIS procedure to obtain a reasonable estimate except "infinity". While it is possible to find a Markov basis corresponding to the second log-linear model, utilizing those for calculating bounds and or sampling from the space of tables is also currently computationally infeasible. But the practicality of such calculations is likely to change with increased computer power and memory.

Based on Model 1, variables B and H are conditionally independent given the remaining 6 variables. Thus we can collapse the 8-way table to a 6-way table and carry out a disclosure risk analysis on it. The collapsed table has only 96 cells, and there is only one small cell count of size "2" that would raise an immediate privacy concern. Furthermore, we have collapsed over the two "most" sensitive and most interesting variables for statistical analysis: Type of Employer and Income. We do not pursue this analysis here but, if other variables are of interest, we could again focus on search for the best decomposable model. With various search algorithms and criteria, out of 32,768 possible decomposable models all searches converge to $[ACFG][ADEFG]$, a model with a likelihood ratio chi-square of $G^2 = 144.036$ and 36 degrees of freedom.

In this case, we could simply provide the margins of the above model to the user to construct association rules provided that they do not provide precise information on three sensitive cells. Numerous association rules can be derived from the given margins. Some interesting rules, for example could be $AFG \Rightarrow C$, and $AFG \Rightarrow DE$. As we did in in the clinical trial example, we can evaluate how safe the release of these rules are by determining the bounds on the cells given the marginal and conditional constraints, that is the rules' support and confidence.

12.6 Conclusions

The literature on datamining for association rules has focused on extracting rules with high predictive utility, measured by criteria such as support and confidence. For categorical data bases, coming in the form of multi-way contingency tables, these rules and criteria essentially are extracting marginal tables and linked conditionals. Some authors have recognized the relevance of log-linear and related models for this type of datamining activity, e.g., see [11], and [37], but few have addressed the issue of preserving the privacy of individuals represented in the data base being mined, with no links to date to ideas from log-linear and related models. In this chapter we have provided an overview of the totally separate statistical literature focused on protecting against disclosure limitation in contingency tables, while providing marginal and conditional tables for analysis and reporting.

From the perspective of privacy preservation the methods described in this chapter for bounds on cell counts provide an alternative approach to that found in most of the machine learning literature. These methods stress the link between the ensemble of data to be released, i.e., margins and conditionals, and their ability to characterize the data base through the use of log-linear and related statistical models and assessments of goodness-of-fit. Measures of privacy preservation based on bounds and other statistically related quantities may suggest that "the best association rules" may not be releasable without possibly compromising confidentiality.

New to this enterprise, and especially new to datamining are the tools from computational algebraic geometry. We have attempted to illustrate their applicability here largely through the examples. For more details we refer the interested reader to [6], [17], [31], and papers in a special 2006 issue of the *Journal of Symbolic Computation* devoted to problems at the interface of statistics and algebraic geometry.

Machine learning has made major progress in the efficient extraction of association rules from large data bases. The statistical literature has focused more heavily on understanding the utility of the the extracted information and on related methodologies for assessing disclosure limitation or privacy preservation. Our goal in reviewing the points of convergence in these two literatures has been to stimulate a fusion of the different methodologies and computational tools. Barak et al. [3] adds the element of perturbation to our toolkit and we hope to compare their methods with those described in this paper in the near future.

Acknowledgements

We owe special thanks to Alan Karr for pointing out the close correspondence between the contingency tables and association rule mining and to Cynthia Dwork for getting us to explain the sense in which our approach addresses confidentiality protection. This research was supported in part by NSF Grants EIA-98-76619 and IIS-01-31884 to the National Institute of Statistical Sciences, by Army Contract DAAD19-02-1-3-0389 to CyLab and by NSF Grant DMS-0631589 to the Department of Statistics, both at Carnegie Mellon University, by NSF Grant SES-0532407 to the Department of Statistics, Pennsylvania State University, and by NSF Grant DMS-0439734 to the Institute for Mathematics and Its Application, University of Minnesota.

References

[1] Agresti, A. (2002). *Categorical Data Analysis.* 2nd Edition. New York: Wiley.

[2] Anderson, B. and Moore, A. (1998). AD-trees for Fast Counting and for Fast Learning of Association Rules, *Knowledge Discovery from Databases Conference.*

[3] Barak, B., Chaudhuri, K., Dwork, C., Kale, S., McSherry, M., and Talwar, K. (2007). Privacy, accuracy, and consistency too: a holistic solution to contingency table release, *PODS '07: Proceedings of 26th ACM SIGMOD-SIGACT-SIGART Symposium on Principles of Database Systems*, New York: ACM Press, 273–282.

[4] Bishop, Y. M. M., Fienberg, S. E., and Holland, P. W. (1975). *Discrete Multivariate Analysis: Theory and Practice.* Cambridge, MA: MIT Press.

[5] De Loera, J., Haws, D., Hemmecke, R., Huggins, P., Tauzer, J., and Yoshida, R. (2003). *A User's Guide for LattE v1.1.* University of California, Davis.

[6] Diaconis, P. and Sturmfels, B. (1998). Algebraic Algorithms for Sampling From Conditional Distributions, *Annals of Statistics*, 26, 363–397.

[7] Dobra, A. and Fienberg, S. E. (2000). Bounds for Cell Entries in Contingency Tables Given Marginal Totals and Decomposable Graphs, *Proceedings of the National Academy of Sciences*, 97, 11885–11892.

[8] Dobra, A. and Fienberg, S. E. (2001). Bounds for Cell Entries in Contingency Tables Induced by Fixed Marginal Totals, *Statistical Journal of the United Nations ECE,* 18, 363–371.

[9] Dobra, A., Fienberg, S. E., and Trottini, M. (2003). Assessing the Risk of Disclosure of Confidential Categorical Data (with discussion), In J. Bernardo et al. eds., *Bayesian Statistics 7,* Clarendon: Oxford University Press, 125–144.

[10] Domingo-Ferrer, J. and Torra, V. (eds.) (2004). *Privacy in Statistical Databases, Lecture Notes in Computer Science No. 3050,* New York: Springer-Verlag.

[11] DuMouchel, W. and Pregibon, D. (2001). Empirical Bayes Screening for Multi-Item Associations, *Proceedings of the ACM SIGKDD Intentional Conference on Knowledge Discovery in Databases & Data Mining (KDD01),* ACM Press, 67–76.

[12] Duncan, G. T., Fienberg, S. E., Krishnan, R., Padman, R., and Roehrig, S. F. (2001). Disclosure Limitation Methods and Information Loss for Tabular Data, In P. Doyle, J. Lane, J. Theeuwes, and L. Zayatz (eds.) *Confidentiality, Disclosure and Data Access: Theory and Practical Applications for Statistical Agencies,* Amsterdam: Elsevier, 135–166.

[13] Dwork, C., McSherry, F., Nissim, K. and Smith, A. (2006). Calibrating Noise to Sensitivity of Functions in Private Data Analysis, *3rd Theory of Cryptography Conference (TCC) 2006*, 265–284.

[14] Evfimievski, A., Srikant, R., Agrawal, R., and Gehrke, J. (2002). Privacy Preserving Mining of Association Rules, *Proceedings of the 8th ACM*

SIGKDD International Conference on Knowledge Discovery in Databases and Data Mining, Edmonton, Canada, July 2002.

[15] Fienberg, S. E. (1980). *The Analysis of Cross-Classified Categorical Data.* 2nd edition. Cambridge, MA: MIT Press.

[16] Fienberg, S. E. (2004). Datamining and Disclosure Limitation for Categorical Statistical Databases, *Proceedings of Workshop on Privacy and Security Aspects of Data Mining, Fourth IEEE International Conference on Data Mining (ICDM 2004)*, Brighton, UK, November 2004.

[17] Fienberg, S. E., Makov, U. E., Meyer, M. M., and Steele, R. J. (2001). Computing the Exact Distribution for a Multi-way Contingency Table Conditional on its Marginals Totals, In A. K. M. E. Saleh, ed. *Data Analysis from Statistical Foundations: Papers in Honor of D. A. S. Fraser*, Huntington, NY: Nova Science Publishing, 145–165.

[18] Fienberg, S. E. and Makov, U. E. (1998). Confidentiality, Uniqueness, and Disclosure Limitation for Categorical Data, *Journal of Official Statistics,* 14, 385–397.

[19] Fienberg, S. E. and Slavkovic, A. B. (2004). Making the Release of Confidential Data from Multi-Way Tables Count, *Chance*, 17(3), 5–10.

[20] Fienberg, S. E. and Slavkovic, A. B. (2005). Preserving the Confidentiality of Categorical Statistical Data Bases When Releasing Information for Association Rules, *Data Mining and Knowledge Discovery.* 11, 155–180.

[21] Hemmecke, R. and Hemmecke, R. (2003). 4ti2 Version 1.1—Computation of Hilbert bases, Graver bases, toric Gröbner bases, and more.
`http://www.4ti2.de.`

[22] Jordan, M. I. (ed.) (1998). *Learning in Graphical Models.* Cambridge MA: MIT Press.

[23] Kargupta, H., Datta, S., Wang, Q., and Sivakumar, K. (2003). Random Data Perturbation Techniques and Privacy Preserving Data Mining, *Proceedings of the 3rd IEEE International Conference on Data Mining (ICDM 2003),* Melbourn, Florida, USA, December 2003.

[24] Koch, G., Amara, J., Atkinson, S., and Stanish, W. (1983). Overview of categorical analysis methods, *SAS-SUGI*, 8, 785–795.

[25] Lauritzen, S. L. (1996). *Graphical Models.* Oxford: Oxford University Press.

[26] Madigan, D. and Raftery, A. E. (1994). Model Selection and Accounting for Model Uncertainty in Graphical Models Using Occams Window, *Journal of the American Statistical Association,* 89: 1535–1546.

[27] Moore, A. and Schneider, J. (2002). Real-valued All-Dimensions Search: Low-overhead Rapid Searching Over Subsets of Attributes, *Proceedings of the 18th Conference on Uncertainty in Artificial Intelligence, July, 2002*, San Francisco: Morgan Kaufmann Publishers, 360–369.

[28] Rizvi, S. and Haritsa, J. (2002). Maintaining Data Privacy in Association Rule Mining, *Proceedings of the 28th Conference on Very Large Data Base (VLDB'02).*

[29] Silverstein, C., Brin, S., and Motwani, R. (1998). Beyond Market Baskets: Generalizing Association Rules to Dependence Rules, *Data Mining and Knowledge Discovery,* 2,39–68.

[30] Silverstein, C., Brin, S., Motwani, R. and Ullman, J. (2000). Scalable Techniques for Mining Causal Structures, *Data Mining and Knowledge Discovery*, 4, 163–192.

[31] Slavkovic, A. B. (2004). *Statistical Disclosure Limitation Beyond the Margins.* Ph.D. Thesis, Department of Statistics, Carnegie Mellon University.

[32] Slavkovic, A. B. and Smucker, B. (2007). *Calculating Cell Bounds in Contingency Tables Based on Conditional Frequencies.* Technical Report, Department of Statistics, Penn State University.

[33] Slavkovic, A. B. and Fienberg, S. E. (2004). Bounds for Cell Entries in Two-way Tables Given Conditional Relative Frequencies, In Domingo-Ferrer, J. and Torra, V. (eds.), *Privacy in Statistical Databases, Lecture Notes in Computer Science No. 3050*, 30–43. New York: Springer-Verlag.

[34] Sturmfels, B. (2003). *Algebra and Geometery of Statistical Models.* John von Neumann Lectures at Munich University.

[35] Trottini, M. and Fienberg, S. E. (2002). Modelling User Uncertainty for Disclosure Risk and Data Utility, *International Journal of Uncertainty, Fuzziness and Knowledge Based Systems,* 10, 511–528.

[36] Willenborg, L. C. R. J. and de Waal, T. (2000). *Elements of Statistical Disclosure Control.* Lecture Notes in Statistics, Volume 155, New York: Springer-Verlag.

[37] Wu, X., Barbará, D. and Ye, Y. (2003). Screening and Interpreting Multi-item Associations Based on Log-linear modeling, *Proceedings of the ACM SIGKDD Intentional Conference on Knowledge Discovery in Databases & Data Mining (KDD03)*, ACM Press, 276–285.

[38] Zaki M. J. (2004). *Mining Non-Redundant Association Rules, Data Mining and Knowledge Discovery*, 9, 223–248.

[39] Verykios S. Vassilios and Gkoulalas-Divani A.(2007) *A Survey of Association Rule Hiding Methods for Privacy,* in this volume .

Chapter 13

A Survey of Privacy-Preserving Methods Across Horizontally Partitioned Data

Murat Kantarcioglu
Computer Science Department
University of Texas at Dallas
muratk@utdallas.edu

Abstract

Data mining can extract important knowledge from large data collections, but sometimes these collections are split among various parties. Data warehousing, bringing data from multiple sources under a single authority, increases risk of privacy violations. Furthermore, privacy concerns may prevent the parties from directly sharing even some meta-data.

Distributed data mining and processing provide a means to address this issue, particularly if queries are processed in a way that avoids the disclosure of any information beyond the final result. This chapter describes methods to mine horizontally partitioned data without violating privacy and discusses how to use the data mining results in a privacy-preserving way. The methods described here incorporate cryptographic techniques to minimize the information shared, while adding as little as possible overhead to the mining and processing task.

Keywords: Privacy, distributed data mining, horizontally partitioned data and homomorphic encryption.

13.1 Introduction

Data mining technology has emerged as a means of identifying patterns and trends from large quantities of data. Recently, there has been growing concern over the privacy implications of data mining. Some of this is public perception: The "Data Mining Moratorium Act of 2003" introduced in the U.S. Senate [8] was based on a fear of government searches of private data for individual information, rather than what the technical community views as Data Mining. However, concerns remain. While data mining is generally aimed at producing general models rather than learning about specific individuals, the *process* of data mining creates integrated data warehouses that pose real privacy issues. Data that is of limited sensitivity by itself becomes highly sensitive when integrated, and gathering the data under a single roof greatly increases the opportunity for misuse. Even though some of the distributed data mining tasks protect individual data privacy, they still require that each site reveals some partial information about the local data. What if even this information is sensitive?

For example, suppose the Centers for Disease Control (CDC), a public agency, would like to mine health records to try to find ways to reduce the proliferation of antibiotic resistant bacteria. Insurance companies have data on patient diseases and prescriptions. CDC may try to mine association rules of the form $X \Rightarrow Y$ such that the $Pr(X\&Y)$ and $Pr(Y|X)$ are above some certain thresholds. Mining this data for association rules would allow the discovery of rules such as $Augmentin\&Summer \Rightarrow Infection\&Fall$, i.e., people taking Augmentin in the summer seem to have recurring infections.

The problem is that insurance companies will be concerned about sharing this data. Not only must the privacy of patient records be maintained, but insurers will be unwilling to release rules pertaining only to them. Imagine a rule indicating a high rate of complications with a particular medical procedure. If this rule doesn't hold globally, the insurer would like to know this; they can then try to pinpoint the problem with their policies and improve patient care. If the fact that the insurer's data supports this rule is revealed (say, under a Freedom of Information Act request to the CDC), the insurer could be exposed to significant public relations or liability problems. This potential risk could exceed their own perception of the benefit of participating in the CDC study.

One solution to this problem is to avoid disclosing data beyond its source, while still constructing data mining models equivalent to those that would have been learned on an integrated data set. Since we prove that data is not disclosed beyond its original source, the opportunity for misuse is not increased by the process of data mining.

The definition of privacy followed in this line of research is conceptually simple: no site should learn anything new from the *process* of data mining. Specifically, anything learned during the data mining process must be derivable given one's own data and the final result. In other words, nothing is learned about any other site's data that isn't inherently obvious from the data mining result. The approach followed in this research has been to select a type of data mining model to be learned and develop a protocol to learn the model while meeting this definition of privacy.

In addition to the type of data mining model to be learned, the different types of data distribution result in a need for different protocols. For example, the first paper in this area proposed a solution for learning decision trees on horizontally partitioned data: each site has complete information on a distinct set of entities, and an integrated dataset consists of the union of these datasets.

In contrast, vertically partitioned data has different types of information at each site; each has partial information on the same set of entities. In this case an integrated dataset would be produced by *joining* the data from the sites. While [25] showed how to generate ID3 decision trees on horizontally partitioned data, a completely new method was needed for vertically partitioned data [6]. (We will not further discuss the vertically partitioned data case in this chapter. Please see Vaidya's chapter in this book for the discussion of vertically partitioned data case) This chapter presents solutions such that the parties learn (almost) nothing beyond the global results. We assume homogeneous databases and horizontally partitioned data: All sites have the same schema, but each site has information on different entities. Given solutions are relatively efficient and proved to preserve privacy under some reasonable assumptions. Specifically, in section 13.2, we briefly discuss the necessary cryptographic definitions and tools. In section 13.3, we summarize how the basic cryptographic tools could be used to create privacy-preserving sub-protocols. Later on, in section 13.4, we outline how the privacy-preserving distributed data mining protocols are created using these few sub-protocols. In section 13.6, we discuss how to extend current algorithms to withstand different adversarial models. In section 13.8, we give an overview of other privacy issues related to data mining results. Finally, in section 13.9, we conclude with possible future research directions.

13.2 Basic Cryptographic Techniques for Privacy-Preserving Distributed Data Mining

Privacy-preserving distributed data mining algorithms require collaboration between parties to compute the results, while provably preventing the disclosure of any information except the data mining results. To achieve this goal, we will use tools from secure multiparty computation (SMC) domain. The concept of privacy in this approach is based on a solid body of theoretical work. First, we briefly discuss the basic ideas from SMC domain. Then, we describe a useful variant of public-key cryptography system called homomorphic encryption.

Privacy Definitions and Proof Techniques

Secure Multiparty Computation (SMC) originated with Yao's Millionaires' problem [33]. The basic problem is that two millionaires would like to know who is richer, with neither revealing their net worth. Abstractly, the problem is to simply compare two numbers, each held by one party, without either party revealing its number to the other. Yao[33] presented a generic circuit evaluation based solution for this problem as well as generalizing it to any efficiently computable function restricted to two parties.

The SMC literature defines two basic adversarial models:

Semi-Honest: Semi-honest (or Honest but Curious) adversaries follow the protocol faithfully, but can try to infer the secret information of the other parties from the data they see during the execution of the protocol.

Malicious: Malicious adversaries may do anything to infer secret information. They can abort the protocol at any time, send spurious messages, spoof messages, collude with other (malicious) parties, etc.

While the semi-honest model may seem questionable for privacy (if a party can be trusted to follow the protocol, why don't we trust them with the data?), we believe that it meets several practical needs for early adoption of the technology. Consider the case where credit card companies jointly build data mining models for credit card fraud detection. In many cases the parties involved already have authorization to see the data (e.g., the theft of credit card information from CardSystems [30] involved data that CardSystems was expected to see during processing). The problem is that *storing* the data brings with it a responsibility (and cost) of protecting that data; CardSystems was supposed to delete the information once the processing was complete. If parties could develop the desired models without seeing the data, then they are saved the responsibility (and cost) of protecting it. Also the simplicity and efficiency possible with semi-honest protocols will help speed adoption so that trusted parties are saved the expense of protecting data other than their own. As the technology gains acceptance, malicious protocols will become viable for uses where the parties are not mutually trusted. (Please see section 13.6 for the discussion of malicious parties)

In either adversarial model, there exist formal definitions of privacy [13]. Informally, the definition of privacy is based on equivalence to having a trusted third party perform the computation. This is the gold standard of secure multiparty computation. Imagine that each of the data sources gives their input to a (hypothetical) trusted third party. This party, acting in complete isolation, computes the results and reveals them. After revealing the results, the trusted party forgets everything it has seen. A secure multiparty computation approximates this standard: no party learns more than it would in the trusted third party approach.

One fact is immediately obvious: no matter how secure the computation, some information about the inputs may be revealed. This is a result of the computed function itself. For example, if one party's net worth is \$100,000, and the other party is richer, one has a lower bound on their net worth. This is captured in the formal SMC definitions: any information that can be inferred from one's own data and the result can be revealed by the protocol. Thus, there are two kinds of information leaks; the information leak from the function computed irrespective of the process used to compute the function and the information leak from the specific process of computing the function. Whatever is leaked from the function itself is unavoidable as long as the function has to be computed (We discuss the privacy issues related to data mining results in section 13.8). In secure computation the second kind of leak is provably prevented. There is *no* information leak whatsoever due to the process. Some algorithms improve efficiency by trading off some security (leak a small amount of information). Even if this is allowed, the SMC style of proof provides a tight bound on the information leaked; allowing one to determine if the algorithm satisfies a privacy policy.

This leads to the primary proof technique used to demonstrate the security of privacy-preserving distributed data mining: a simulation argument. Given only its own input and the result, a party must be able to simulate what it sees during execution of the protocol.

One key point is the restriction of the simulator to polynomial time algorithms, and that the views only need to be *computationally* indistinguishable. Algorithms meeting this definition need not be proven against an adversary capable of trying an exponential number of possibilities in a reasonable time frame. While some protocols do not require this restriction, most make use of cryptographic techniques that are only secure against polynomial time adversaries. This is adequate in practice (as with cryptography); security parameters can be set to ensure that the computing resources to break the protocol in any reasonable time do not exist.

While the Yao's generic circuit evaluation method has been proven secure by the above definition, it poses significant computational problems. Given the size and computational cost of data mining problems, representing algorithms as a boolean circuit results in unrealistically large circuits. The challenge of privacy-preserving distributed data mining is to develop algorithms that have reasonable computation and communication costs on real-world problems, and prove their security with respect to the above definition.

The composition theorem [13] is another very useful theorem from the SMC literature.

THEOREM 13.2.1 *Composition Theorem for the semi-honest model.*

Suppose that g is privately reducible to f and that there exists a protocol for privately computing f. Then there exists a protocol for privately computing g.

Informally, the theorem states that if a protocol is shown to be secure except for several invocations of sub-protocols, and if the sub-protocols themselves are proven to be secure, then the entire protocol is secure. The immediate consequence is that, with care, we can combine secure sub-protocols to produce new secure protocols. Also, if many algorithms depend on a few common sub-protocols, efficient implementation of these sub-protocols significantly improves the overall efficiency. The following section shows that many privacy preserving data mining algorithms can be developed using few sub-protocols.

Homomorphic Encryption

As mentioned above, most of the protocols devised for privacy-preserving distributed data mining could be implemented using few sub-protocols. For the ease of exposition, we describe those sub-protocols using homomorphic encryption techniques.

In a nutshell, we can describe the homomorphic encryption as follows: Let $E_{pk}(.)$ denote the encryption function with public key pk and $D_{pr}(.)$ denote the decryption function with private key pr. A secure public key cryptosystem is called homomorphic if it satisfies the following requirements: (1) Given the encryption of m_1 and m_2, $E_{pk}(m_1)$ and $E_{pk}(m_2)$, there exists an efficient algorithm to compute the public key encryption of m_1+m_2, denoted $E_{pk}(m_1+m_2) := E_{pk}(m_1) +_h E_{pk}(m_2)$. (2) Given a constant k and the encryption of m_1, $E_{pk}(m_1)$, there exists an efficient algorithm to compute the public key encryption of $k \cdot m_1$, denoted $E_{pk}(k \cdot m_1) := k \times_h E_{pk}(m_1)$. Please refer to [29] for more details.

13.3 Common Secure Sub-protocols Used in Privacy-Preserving Distributed Data Mining

We will briefly describe the common secure sub-protocols used in Privacy-preserving Distributed Data Mining. For each sub-protocol, if possible, we describe a version that only uses homomorphic encryption. Unless otherwise stated, all the sub-protocols are secure in the semi-honest model with no collusion, and all the arithmetic operations are defined in some large enough finite field.

In later sections, we will show how different algorithms could be implemented using these secure sub-protocols. Since these common building blocks are quite general, using theorem 13.2.1, they can be combined to create new privacy preserving algorithms in the future.

Secure Sum

Secure Sum securely calculates the sum of values from individual sites. Assume that each site i has some value v_i and all sites want to securely compute $v = \sum_{l=1}^{m} v_l$ where v is known to be in the range $[0..n]$. Homomorphic encryption could be used to calculate secure sum as follows:

1: Site 1 creates a homomorphic encryption public and private key pair, and sends the public key to all sites

2: Site 1 sets $s_1 = E_{pk}(v_1)$

3: Each site i where $m \geq i > 1$, gets s_{i-1} from site $i-1$ and computes $s_i = s_{i-1} +_h E_{pk}(v_i)$ using additive property of the homomorphic encryption
4: Site m sends s_m to site 1
5: Site 1 sends $D_{pr}(s_m)$ to all parties

The above protocol is secure because any party other than site 1 cannot decrypt the s_i values. It also correctly calculates the summation because $s_m = s_{m-1} +_h E_{pk}(v_m) = E_{pk}(\sum_{l=1}^{m} v_l)$ and $D_{pr}(s_m) = v$.

Assuming three or more parties and no collusion, a more efficient method can be found in [19].

Secure Comparison / Yao's Millionaire Problem

Assume that two sites, each having one value, want to compare the two values without revealing anything else other than the comparison result. Secure Comparison methods can be used to solve the above problem. To the best of our knowledge, secure circuit evaluation based approaches still provide the best performance [33].

Dot Product Protocol

Securely computing the dot product of two vectors is another important sub-protocol required in many privacy-preserving data mining tasks. Many secure dot product protocols have been proposed in the past [5, 31, 14, 12]. Among those proposed techniques, the method of Goethals et al. [12] is quite simple and provably secure. We now briefly describe it here.

The problem is defined as follows: Alice has a n-dimensional vector $\vec{X} = (x_1, \ldots, x_n)$ while Bob has a n-dimensional vector $\vec{Y} = (y_1, \ldots, y_n)$. At the end of the protocol, Alice should get $r_a = \vec{X} \cdot \vec{Y} + r_b$ where r_b is a random number chosen from uniform distribution that is known only to Bob, and $\vec{X} \cdot \vec{Y} = \sum_{i=1}^{n} x_i \cdot y_i$. The key idea behind the protocol is to use a homomorphic encryption system described in section 13.2. Using such a system, it is quite simple to build a dot product protocol. If Alice encrypts her vector and sends in encrypted form to Bob, using the additive homomorphic property, Bob can compute the dot product. The specific details are given below:

Require: Alice has input vector $\vec{X} = \{x_1, \ldots, x_n\}$
Require: Bob has input vector $\vec{Y} = \{y_1, \ldots, y_n\}$
Require: Alice and Bob get outputs r_A, r_B respectively such that $r_A + r_B = \vec{X} \cdot \vec{Y}$

1: Alice generates a homomorphic private and public key pair.
2: Alice sends public key to Bob.
3: **for** $i = 1 \ldots n$ **do**
4:     Alice sends to Bob $c_i = E_{pk}(x_i)$.
5: **end for**

6: Bob computes $w_i = (c_i \times_h y_i)$
7: Bob computes $w = w_1 +_h w_2 +_h \dots +_h w_n$
8: Bob generates a random plaintext r_B.
9: Bob sends to Alice $w' = w +_h E_{pk}(-r_B)$.
10: Alice computes $r_A = D_{pr}(w') = \vec{X} \cdot \vec{Y} - r_B$.

Oblivious Evaluation of Polynomials

Another important sub-protocol required in privacy-preserving data mining is the secure polynomial evaluation protocol. Consider the case where Alice has a polynomial P of degree k over some finite field $\mathcal{F}$. Bob has an element $x \in \mathcal{F}$ and also knows k. Alice would like to let Bob compute the value $P(x)$ in such a way that Alice does not learn x and Bob does not gain any additional information about P (except $P(x)$). This problem was first investigated by [28]. Subsequently, there have been more protocols improving the communication and computation efficiency [2] as well as extending the problem to floating point numbers [1].

We now briefly describe the protocol used for oblivious polynomial evaluation that uses the secure dot product above. Given a dot product protocol, we can easily create a protocol for polynomial evaluation as follows: Let $P(y) = \sum_{i=0}^{k} a_i y^i$ be Alice's input and x be Bob's input, using secure dot product, Bob can evaluate the $P(x)$ as follows

Alice forms Bob forms

$$\vec{U} = \begin{bmatrix} a_0 \\ a_1 \\ \vdots \\ a_k \end{bmatrix} \quad \vec{V} = \begin{bmatrix} 1 \\ x \\ \vdots \\ x^k \end{bmatrix}$$

Alice and Bob engage in secure dot product so that (only) Bob gets $r = \vec{U}.\vec{V}$

Clearly $r = \sum_{i=0}^{k} a_i x^i = P(x)$. Using theorem 13.2.1, it can be shown that if the dot product protocol is secure, then the above protocol is also secure.

Privately computing $\ln x$

For entropy measures used in data mining, we need to be able to privately compute $\ln x$, where $x = x_1 + x_2$ with x_1 known to Alice and x_2 known to Bob. Thus, Alice should get y_1 and Bob should get y_2 such that $y_1 + y_2 = \ln x = \ln(x_1 + x_2)$. One of the key results presented in [26] was a cryptographic protocol for this computation. We now describe the protocol in brief: Note that $\ln x$ is *Real* while general cryptographic tools work over finite fields. We multiply the $\ln x$ with a known constant to make it integral.

The basic idea behind computing random shares of $\ln(x_1 + x_2)$ is to use the Taylor approximation for $\ln x$. Remember that the Taylor approximation gives us:

$$\begin{aligned}\ln(1+\epsilon) &= \sum_{i=1}^{\infty} \frac{(-1)^{i-1}\epsilon^i}{i} \\ &= \epsilon - \frac{\epsilon^2}{2} + \frac{\epsilon^3}{3} - \frac{\epsilon^4}{4} + \ldots \text{ for } -1 < \epsilon < 1\end{aligned}$$

For an input x, let $n = \lfloor \log_2 x \rfloor$. Then 2^n represents the closest power of 2 less than x. Therefore, $x = x_1 + x_2 = 2^n(1+\epsilon)$ where $-1/2 \le \epsilon \le 1/2$. Consequently,

$$\begin{aligned}\ln(x) &= \ln(2^n(1+\epsilon)) \\ &= \ln 2^n + \ln(1+\epsilon) \\ &\approx \ln 2^n + \sum_{i=1\ldots k} (-1)^{i-1}\epsilon^i/i \\ &= \ln 2^n + T(\epsilon)\end{aligned}$$

where $T(\epsilon)$ is a polynomial of degree k. This error is exponentially small in k.

There are two phases to the protocol. Phase 1 finds an appropriate n and ϵ. Let N be a predetermined (public) upper-bound on the value of n. First, Yao's circuit evaluation is applied to the following small circuit which takes x_1 and x_2 as input and outputs random shares of $\epsilon 2^N$ and $2^N n \ln 2$. Note that $\epsilon 2^n = x - 2^n$, where n can be determined by simply looking at the two most significant bits of x, and $\epsilon 2^N$ is obtained simply by shifting the result by $N-n$ bits to the left. Thus, the circuit outputs random α_1 and α_2 such that $\alpha_1 + \alpha_2 = \epsilon 2^N$, and also outputs random β_1 and β_2 such that $\beta_1 + \beta_2 = 2^N n \ln 2$. This circuit can be easily constructed. Random shares are obtained by having one of the parties input random values $\alpha_1, \beta_1 \in \mathcal{F}$ into the circuit and having the circuit output $\alpha_2 = \epsilon 2^N - \alpha_1$ and $\beta_2 = 2^N n \ln 2 - \beta_1$ to the other party.

Phase 2 of the protocol involves computing shares of the Taylor series approximation, $T(\epsilon)$. This is done as follows: Alice chooses a random $w_1 \in \mathcal{F}$ and defines a polynomial $Q(x)$ such that $w_1 + Q(\alpha_2) = T(\epsilon)$. Thus $Q(\cdot)$ is defined as

$$Q(x) = \text{lcm}(2, \ldots, k) \sum_{i=1}^{k} \frac{(-1)^{i-1}}{2^{N(i-1)}} \frac{(\alpha_1 + x)^i}{i} - w_1$$

Alice and Bob then execute a secure polynomial evaluation defined above with Alice inputting $Q(\cdot)$ and Bob inputting α_2, in which Bob obtains $w_2 = Q(\alpha_2)$. Alice and Bob define $u_1 = lcm(2, \ldots, k)\beta_1 + w_1$ and $u_2 = lcm(2, \ldots, k)\beta_2 + w_2$. We have that $u_1 + u_2 \approx 2^N lcm(2, \ldots, k) \ln x$. Further details on the protocol, as well as the proof of security, can be found in [26].

Secure Intersection

Secure Intersection methods are useful in data mining to find common rules, frequent itemsets etc., without revealing the owner of the item. Many algorithms have been developed for calculating Secure Set Intersection. For example, [32] provides an efficient solution.

Here we describe a secure set intersection protocol that uses secure polynomial evaluation. [23, 9] Let us assume that Alice has set $X = \{x_1, \ldots, x_n\}$ and Bob has set $Y = \{y_1, \ldots, y_n\}$. Our goal is to securely calculate $X \cap Y$. By representing set X as a polynomial and using polynomial evaluation, Alice and Bob can calculate $X \cap Y$ securely. The specific details are given below:

Require: Alice has input set $X = \{x_1, \ldots, x_n\}$
Require: Bob has input set $Y = \{y_1, \ldots, y_n\}$
Require: Alice and Bob learn $X \cap Y$
Alice generates a homomorphic private and public key pair
Alice sends public key to Bob
Alice creates a polynomial $P(z) = \sum_{i=0}^{n} a_i z^i$ such that $P(x_i) = 0$ for all x_i (This is possible using interpolation)
for $i = 1 \ldots n$ **do**
    Alice sends to Bob $c_i = E_{pk}(a_i)$
end for
for $i = 1 \ldots n$ **do**
    Using c_i values, and random non-zero r_i, Bob computes $w_i = E_{pk}(r_i \cdot P(y_i) + y_i)$ (This is possible due to homomorphic encryption)
end for
Bob permutes w_i values and send it to Alice
Alice decrypts all w_i values and outputs $D_{pr}(w_i)$ as an element of $X \cap Y$ if $D_{pr}(w_i) \in X$

Note that above protocol works, because if $y_i \in X \cap Y$ then $P(y_i) = 0$, and if $P(y_i) = 0$, then $D_{pr}(w_i) = y_i$. On the other hand, if $y_i \notin X \cap Y$ then $P(y_i) \neq 0$, and then $D_{pr}(w_i)$ will be some random number based on r_i. See [9] for further details.

Secure Set Union

Secure union methods are useful in data mining to allow each party to give its rules,decision trees etc. without revealing the owner of the item. Union of items can be easily evaluated using SMC methods if the domain of the items is small. Each party creates a binary vector (where the ith entry is 1 if the ith item is present locally). At this point, a simple circuit that *or's* the corresponding vectors can be built and securely evaluated using general secure multi-party circuit evaluation protocols. However, in data mining, the domain of the items are usually very large, potentially infinite. This problem can be overcome using approaches based on commutative encryption [20].

13.4 Privacy-preserving Distributed Data Mining on Horizontally Partitioned Data

In this section, we will give an overview of how different sub-protocols described in section 13.3 could be used to create various privacy-preserving distributed data mining algorithms on horizontally partitioned data (PPDDM). In each of the discussed PPDDM algorithms general data mining functionality is reduced to a computation of secure sub-protocols. Figure 13.1 shows the correspondence between algorithms and constituent secure sub-protocols.

In all the algorithms described below, we assume that the data is horizontally partitioned. This assumption implies that different sites collect the same set of information about different entities. For example, different credit card companies may collect credit card transactions of different individuals. In relational terms, with horizontal partitioning, the relation to be mined is the union of the relations at the sites. Also, at the end of this section, we briefly discuss the relationship between the privacy-preserving algorithms developed for horizontally and vertically partitioned data.

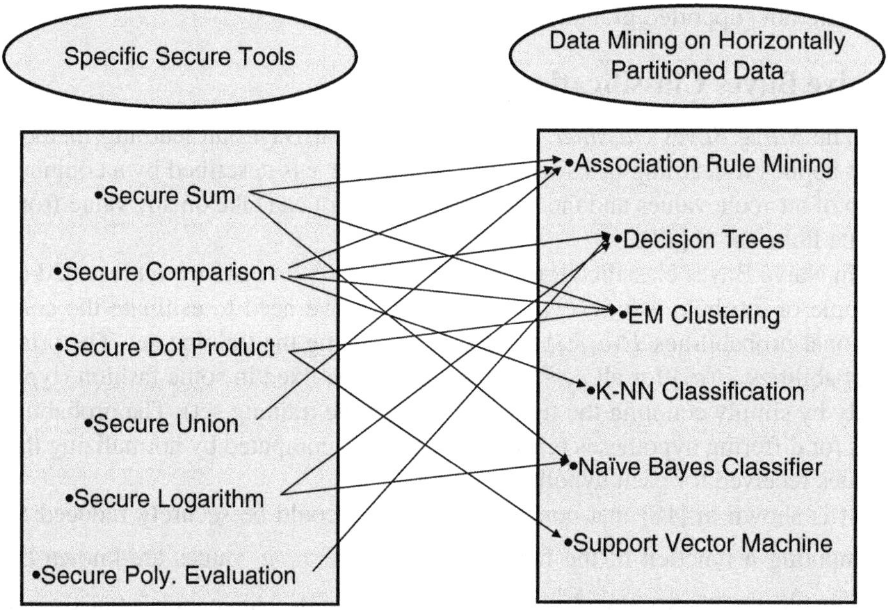

Figure 13.1. Relationship between Secure Sub-protocols and Privacy Preserving Distributed Data Mining on Horizontally Partitioned Data

ID3 Decision Tree Mining

In the first work on privacy-preserving distributed data mining on horizontally partitioned data [25], the goal is to securely build an ID3 decision tree where the training set is horizontally distributed between two parties. The basic idea is that finding the attribute that maximizes information gain is equivalent to finding the attribute that minimizes the conditional entropy. The conditional entropy for an attribute for two parties can be written as a sum of the expression of the form $(v_1 + v_2) \times \log(v_1 + v_2)$. The authors use the secure log algorithm, secure polynomial evaluation, and secure comparison sub-protocols to securely calculate the expression $(v_1 + v_2) \times \log(v_1 + v_2)$ and show how to use this function for building the ID3 securely.

Association Rule Mining

The goal of privacy-preserving association rule mining is to compute rules of the form $X \Rightarrow Y$ (e.g Diaper implies Beer) that has a global support and confidence over some certain threshold. It is proven in [20] that this could be achieved using secure set union, secure summation and secure comparison sub-protocols.

The algorithm described in [20] has two phases. The first phase uses secure set union to get the union of candidate association rules. In the second phase, secure summation and secure comparison are used to filter the candidate items that are not supported globally.

Naive Bayes Classification

The *Naive Bayes classifier* is a highly practical Bayesian learning method that applies to learning tasks where each instance x is described by a conjunction of attribute values and the target function $f(x)$ can take on any value from some finite set C [27].

In Naive Bayes classification, in order to classify an instance represented as a tuple of attribute values $< a_1, a_2, \ldots, a_n >$, we need to estimate the conditional probabilities $P(a_i|c_j)$ for all $c_j \in C$ using the training set. The prior probabilities $P(c_j)$ for all $c_j \in C$ also need to be fixed in some fashion (typically by simply counting the frequencies from the training set). The probabilities for differing hypotheses (classes) can also be computed by normalizing the values received for each hypothesis (class).

It is shown in [18] that computing $P(a_i|c_j)$ could be securely reduced to computing a function of the form $\frac{\sum_{i=1}^{n} x_i}{\sum_{i=1}^{n} y_i}$ where x_i, y_i values are known by site i. At the same time, $\frac{\sum_{i=1}^{n} x_i}{\sum_{i=1}^{n} y_i}$ could be securely calculated using secure summation and secure $\ln(x)$ protocol.

k-NN Classification

k-NN classification predicts the class value of an instance using the k nearest examples based in the training data. Various distance metrics are used to determine the k nearest examples. [11]

In [21], a privacy preserving k-nn algorithm is suggested under the assumption that the instance that needs to be classified is public. The approach given in [21] makes use of an untrusted, non-colluding party that is not allowed to learn anything about any of the data, but is trusted not to collude with other parties to reveal private information.

The basic idea is that each site finds its own k-nearest neighbors, (this is possible since the instance that needs to be classified is public and the data is horizontally partitioned) and encrypts the class with the public key of the site that sent the instance for classification (querying site). The parties securely compare their k-nearest neighbors with those of all other sites – except that the comparison gives each site a random share of the result, so no party learns the result of the comparison. The results from all sites are combined, scrambled, and given to the untrusted, non-colluding site. This site combines the random shares to get the comparison result for each pair, enabling it to sort and select the global k-nearest neighbors (but without learning the source or values of the items). The querying site and the untrusted, non-colluding site then engage in a protocol to find the class value.

Support Vector Machine Classification

Support Vector Machine (SVM) classification is an another important classification technique. In [34], a privacy-preserving solution for horizontally partitioned case is given using secure dot product sub-protocol. The solution given in [34] uses the observation that to build the SVM, only the kernel matrix K is needed. To calculate the Kernel matrix K, the gram matrix G where $G_{ij} = x_i \cdot x_j$ is needed to be computed securely for all training instance pairs x_i, x_j. Clearly G_{ij} could be calculated using the secure dot-product protocol.

k-means and EM Clustering

Clustering is a well studied data mining technique that tries to group similar instances in a given data set into clusters to minimize some objective function. In k-means clustering, the goal is to partition data into k clusters. Usually, k initial cluster centers are chosen, and then the cluster centroids are updated using an iterative method. In [15], it is shown that k-means clustering could be achieved on arbitrarily partitioned data using secure dot product, secure summation and secure comparison. Similarly, in [24], secure clustering using the expectation maximization method is given for horizontally partitioned data using secure summation protocol.

13.5 Comparison to Vertically Partitioned Data Model

The privacy preserving algorithms developed for vertically partitioned data also uses the common sub-protocols discussed above. To illustrate the difference between the vertically partitioned and horizontally partitioned data model, let us revisit the association rule mining on both data models. In both data models, to mine association rules, we need to check whether the global support of an itemset X (e.g. the global support of an itemset that contains beer and diaper) is bigger than some certain threshold. In the horizontally partitioned data model, a transaction database DB is assumed to be partitioned among n sites (namely $S_1, S_2, \ldots, S_n$) where $DB = DB_1 \cup DB_2 \cup \cdots \cup DB_n$ and DB_i resides at site S_i ($1 \leq i \leq n$). The itemset X has *local* support count of $X.sup_i$ at site S_i if $X.sup_i$ of the transactions contains X. The *global* support count of X is given as $X.sup = \sum_{i=1}^{n} X.sup_i$. An itemset X is *globally supported* if $X.sup \geq s \times (\sum_{i=1}^{n} |DB_i|)$. To check whether an itemset X is globally supported or not, we can check the following equivalent condition:

$$
\begin{aligned}
X.sup &\geq s * (\sum_{i=1}^{n} |DB_i|) \\
\sum_{i=1}^{n} X.sup_i &\geq s * (\sum_{i=1}^{n} |DB_i|) \\
\sum_{i=1}^{n} (X.sup_i - s * |DB_i|) &\geq 0
\end{aligned}
$$

Clearly, in the horizontally partitioned data case, we can check whether an itemset X is globally supported or not by using a secure sum protocol that involves at most n values (i.e. the sum of the values $(X.sup_i - s * |DB_i|)$ for $1 \leq i \leq n$ where n is the number of sites) and a secure comparison protocol.

In the case of vertically partitioned data [31], a transaction database DB is assumed to be partitioned among n sites where $DB = DB_1 \bowtie DB_2 \bowtie \cdots \bowtie DB_n$. In other words, information about each transaction is distributed among multiple sites. In [31], it is shown that to compute whether an itemset X is globally supported or not, we need to compute a dot product that involves all the transactions. This means that if the original DB has m transactions, we need to run secure dot product algorithm with vector sizes m to compute a single global support. In practice, the total number of transactions (i.e. m) is much larger than the total number of possible sites (i.e. n). Due to these reasons, privately mining association rules over vertically partitioned data is much more expensive then privately mining association rules over horizontally partitioned data.

Similar phenomenon emerges in other types of privacy preserving distributed data mining algorithms. Usually, privacy-preserving algorithms running on the vertically partitioned data require secure dot product protocol executions over large vectors. On the other hand, for the horizontally partitioned data, it may be possible to aggregate local information (e.g. local support count of an itemset in the association rule mining) for efficient distributed processing. The above observation implies that for horizontally partitioned data, we may need to restrict the number of sites participating in the protocol execution for efficiency purposes. However, for the vertically partitioned data, we may need to control both the number of sites participating in the protocol and the total size of the data.

13.6 Extension to Malicious Parties

Most of the work described in the previous sections deals only with semi-honest adversaries, which are assumed to follow the prescribed protocol but try to infer private information using the messages they receive during the protocol. Although the semi-honest model is reasonable in some cases, it is unrealistic to assume that adversaries will always follow the protocols exactly. In particular, malicious adversaries could deviate arbitrarily from their prescribed protocols. Secure protocols that are developed against malicious adversaries require utilization of expensive techniques. Clearly, protocols that can withstand malicious adversaries provide more security. However, there is an obvious trade-off: protocols that are secure against malicious adversaries are generally more expensive than those secure only against semi-honest adversaries. In this section, we give a brief overview of how to make commonly used sub-protocols secure against malicious adversaries. Again, our exposition is based on the homomorphic encryption. First, we discuss few additional cryptographic tools needed to devise protocols secure against malicious parties. Later on, we discuss how these tools could be used to improve secure dot product protocol.

Threshold Homomorphic Encryption

From SMC literature, we know that any semi-honest protocol could be transformed into a protocol that is secure against malicious adversaries [13]. Zero Knowledge proofs are the key ingredients in such transformations. Using zero knowledge proofs, each party could prove that it follows the prescribed protocol without revealing any information. For the sake of completeness, here we describe the zero knowledge proofs needed to extend homomorphic encryption based semi-honest sub-protocols. The implementation details of those protocols for Paillier encryption can be found in [4, 3].

- **Threshold Decryption (two-party case):** Given the common public key pk, the private key pr corresponding to pk is divided into two pieces pr_0 and pr_1. There exists an efficient, secure protocol $D_{pr_i}(E_{pk}(a))$ that outputs the random share of the decryption result s_i along with the non-interactive zero knowledge proof $POD(pr_i, E_{pk}(a), s_i)$ showing that pr_i is used correctly. Those shares can be combined to calculate the decryption result. Also any single share of the private key pr_i cannot be used to decrypt the ciphertext alone. In other words s_i does not reveal anything about the final decryption result. We also need a special version of a threshold decryption such that only one party learns the decryption result. Such a protocol could be easily implemented exploiting the fact that for any given $E_{pk}(a)$, the party that needs to learn the decryption result could generate $E_{pk}(r1)$ and then both parties could jointly decrypt $E_{pk}(a) +_h E_{pk}(r1)$. Since only one party knows $r1$, only that party can learn the correct decryption result.

- **Proving that you know a plaintext:** A party P_i can compute the zero knowledge proof $POK(e_a)$ if he knows an element a in the domain of valid plaintexts such that $D_{pr}(e_a) = a$.

- **Proving that multiplication is correct:** Assume that party P_i is given an encryption $E_{pk}(a)$, chooses a constant c, and calculates $E_{pk}(a.c)$. Later on, P_i can give zero knowledge proof $POMC(e_a, e_c, e_{a.c})$ such that $D_{pr}(e_a) = D_{pr}(a)$ and $D_{pr}(e_{a.c}) = D_{pr}(e_c).D_{pr}(e_a)$.

Converting Secure Dot protocol in the Semi-Honest Model to Malicious Model. If we look at the dot product protocol in the semi-honest model carefully, we need to make sure that the Bob does the multiplications correctly and all the encryptions sent to Alice are valid. These could be easily achieved using the zero knowledge protocols described above. Alice sends the encrypted values along with the associated proofs of correct encryption to Bob. For each multiplication, Bob generates the zero knowledge proof of correct multiplication and sends those to Alice. Later on, Alice can check those proofs to make sure that the dot product was calculated correctly. Such a generic transformation (i.e. using zero knowledge proofs) could be applied for other sub-protocols as well.

In some cases, generic transformation can further be improved in terms of efficiency by specializing them in the malicious model. As an example,in [17], the authors provide a more efficient algorithm for secure dot product in the malicious model.

13.7 Limitations of the Cryptographic Techniques Used in Privacy-Preserving Distributed Data Mining

Privacy is not free. Especially, in the case of privacy preserving distributed data mining, we need to use expensive cryptographic operations. Further more, protocols that are secure against malicious parties are even more expensive. These results indicate that we need to carefully set the parameters used in privacy preserving distributed data mining protocols. [1] For example, if we set the support threshold for association rules too low, this may cause an explosion in the number of locally supported itemsets, which in return, require many expensive cryptographic operations during secure set union phase. Similarly, for building Naive Bayes models, we need calculate the occurrence probability of each attribute value given the class attribute. Therefore, using attributes with large number of discrete values may require much higher computation times.

Although privacy preserving distributed data mining algorithms are developed to reveal nothing other than the final result, not revealing anything could be an overkill in some situations. For example, in the privacy-preserving association rule mining protocol, we need to run one secure summation and one secure comparison to securely check whether an itemset is globally supported or not. If revealing the total support count of an itemset is not a privacy threat, then we may not need to execute the secure comparison protocol. Therefore, the privacy requirements should be considered carefully before executing the privacy preserving distributed data mining protocols.

Compared to noise addition methods used in privacy-preserving data mining, cryptographic techniques for privacy-preserving distributed data mining do not allow easy trade-off between privacy and accuracy. For instance, in the noise addition techniques, variance of the noise could be adjusted to increase privacy while potentially lowering the result accuracy. In contrast, by adjusting the key sizes used in the cryptographic protocols, we can trade off between privacy and efficiency. As a result, new approaches are needed for privacy-preserving distributed data mining to trade off between privacy and accuracy systematically. One way to satisfy this goal is to introduce new "approximate privacy-preserving distributed data mining" protocols that can cheaply approximate the required data mining result, and allow trades-off between accuracy of the approximation versus efficiency. We believe that the work of Feigenbau et al. [7] can provide a good starting point in that direction.

Another limitation with current privacy-preserving data mining protocols is that each party is only assumed to be either honest, semi-honest or malicious. We believe that there are many real-world scenarios where parties participating

---

[1] As discussed in Section 13.5, for the vertically partitioned data, we need to also carefully choose the total data used for privacy preserving data mining

in the protocols are "rational". In other words, the parties are willing to share their data to achieve some certain gain and they will cheat only if cheating increases their gain. Such rational adversary assumption could potentially affect the resulting privacy-preserving distributed data mining protocols. For example, in [16], it is shown that if the participating parties are rational, we can achieve significant cost reductions in the malicious model. Clearly, further research is needed to explore the effect of rational behavior in privacy preserving distributed data mining.

Finally, all the tools and techniques discussed until this point do not consider the privacy effect of the data mining results. In the next section, we will explore this issue in more details.

13.8 Privacy Issues Related to Data Mining Results

In the previous sections, we discussed provably secure distributed data mining protocols that reveal nothing but the resulting data mining model. This work still leaves a privacy question open: Do the resulting data mining models inherently violate privacy? This question is important because the full impact of privacy-preserving data mining will only be realized when we can guarantee that the resulting models do not violate privacy as well.

Here, in this section, we give an overview of the model developed in [22] that presents a start on methods and metrics for evaluating the privacy impact of data mining models. Although the methods discussed in [22] provide results only for classification, these results give a good cross-section of what needs to be done, and a demonstration of techniques to analyze the privacy impact.

To make the privacy implications of data mining results clear, consider the following "medical diagnosis" scenario. Suppose we want to create a "medical diagnosis" model for public use: a classifier that predicts the likelihood of an individual getting a terminal illness. Most individuals would consider the classifier output to be sensitive – for example, when applying for life insurance. The classifier takes some public information (age, address, cause of death of ancestors), together with some private information (eating habits, lifestyle), and gives a probability that the individual will contract the disease at a young age. Since the classifier requires some information that the insurer is presumed not to know, can we state that the classifier does not violate privacy?

The answer is not as simple as it seems. Since the classifier uses some public information as input, it would appear that the insurer could *improve* an estimate of the disease probability by repeatedly probing the classifier with the known public information and "guesses" the unknown information. At first glance, this appears to be a privacy violation. Surprisingly, given reasonable assumptions on the external knowledge available to an adversary, it can be *proven* that the adversary learns nothing new [22]. To analyze similar cases, in [22], the authors categorize the data by into three classes:

- **Public Data:**(P) This data is accessible to everyone, including the adversary.
- **Private/Sensitive Data:**(S) It is assumed that this kind of data must be protected: The values should remain unknown to the adversary.
- **Unknown Data:**(U) This is the data that is not known to the adversary, and is not *inherently* sensitive. However, before disclosing this data to an adversary (or enabling an adversary to estimate it, such as by publishing a data mining model) we must show that it does not help the adversary to discover sensitive data.

Later on, the authors analyze the cases where giving a classifier to an adversary could violate privacy. The most obvious way a classifier can compromise privacy is by taking Public data and predicting Sensitive values. However, it turns out that there are many other ways a classifier can be misused to violate privacy. In [22], the authors have analyzed the following cases:

1 $P \rightarrow S$: Classifier that produces sensitive data given public data.

2 $PU \rightarrow S$: Classifier taking public and unknown data into sensitive data.

3 $PS \rightarrow P$: Classifier taking public and sensitive data into public data.

4 Assuming that the adversary has access to Sensitive data for some individuals, what is the effect on privacy of giving the following classifiers to an adversary?

   (a) $P \rightarrow S$: Can the adversary do better with such a classifier because of his/her background knowledge?

   (b) $P \rightarrow U$: Can giving the adversary a predictor for Unknown data improve its ability to build a classifier for Sensitive data?

The long list of possible privacy violations due to data mining results given above indicates that we need to be really careful in revealing data mining results. Recently, in [10], the authors gave a new decision tree learning algorithm which guarantees that the data mining result does not violate the k-anonymity of the individuals represented in the training data.Although, current work in this area resulted in some interesting results, we believe that more research is needed to understand the privacy implications of data mining results.

13.9 Conclusion

This chapter presents a survey of efficient solutions for many privacy preserving data mining tasks on horizontally partitioned data. We show that many privacy preserving distributed data mining protocols on horizontally partitioned data can be efficiently implemented by securely reducing them to few basic secure building blocks. Also we give an overview of some of the initial solutions on how to use the data mining results without violating privacy.

We believe that the need for mining of data where access is restricted due to privacy concerns will increase. Examples include knowledge discovery among intelligence services of different countries and collaboration among corporations without revealing trade secrets. Even within a single multi-national company, privacy laws in different jurisdictions may prevent sharing individual data. This increasing need for privacy preserving data mining techniques will require flexible and efficient solutions that could be tailored for individual privacy needs for different distributed data mining tasks. Current solutions do not allow users to trade off between efficiency, accuracy, and privacy easily. We believe that more flexible and more efficient solutions are needed for future wide-scale adoption of the privacy preserving data mining techniques.

References

[1] Chang, Yan-Cheng and Lu, Chi-Jen (2001). Oblivious polynomial evaluation and oblivious neural learning. *Lecture Notes in Computer Science*, 2248:369+.

[2] Cramer, R., Gilboa, Niv, Naor, Moni, Pinkas, Benny, and Poupard, G. (2000). Oblivious Polynomial Evaluation. Can be found in the Privacy Preserving Data Mining paper by Naor and Pinkas.

[3] Cramer, Ronald, Damgård, Ivan, and Nielsen, Jesper B. (2001). Multiparty computation from threshold homomorphic encryption. *Lecture Notes in Computer Science*, 2045:280+.

[4] Damgard, I., Jurik, M., and Nielsen, J. (2003). A generalization of paillier's public-key system with applications to electronic voting.

[5] Du, Wenliang and Atallah, Mikhail J. (2001). Privacy-preserving statistical analysis. In *Proceeding of the 17th Annual Computer Security Applications Conference*, New Orleans, Louisiana, USA.

[6] Du, Wenliang and Zhan, Zhijun (2002). Building decision tree classifier on private data. In Clifton, Chris and Estivill-Castro, Vladimir, editors, *IEEE International Conference on Data Mining Workshop on Privacy, Security, and Data Mining*, volume 14, pages 1–8, Maebashi City, Japan. Australian Computer Society.

[7] Feigenbaum, Joan, Ishai, Yuval, Malkin, Tal, Nissim, Kobbi, Strauss, Martin J., and Wright, Rebecca N. (2006). Secure multiparty computation of approximations. *ACM Trans. Algorithms*, 2(3):435–472.

[8] Feingold, Mr., Corzine, Mr., Wyden, Mr., and Nelson, Mr. (2003). Data Mining Moratorium Act of 2003. U.S. Senate Bill (proposed).

[9] Freedman, Michael J., Nissim, Kobbi, and Pinkas, Benny (2004). Efficient private matching and set intersection. In *Eurocrypt 2004*, Interlaken, Switzerland. International Association for Cryptologic Research (IACR).

[10] Friedman, Arik, Wolff, Ran, and Schuster, Assaf (to appear). Providing k-anonymity in data mining. *VLDB Journal*.

[11] Fukunaga, Keinosuke (1990). *Introduction to Statistical Pattern Recognition*. Academic Press, San Diego, CA.

[12] Goethals, Bart, Laur, Sven, Lipmaa, Helger, and Mielikäinen, Taneli (2004). On Secure Scalar Product Computation for Privacy-Preserving Data Mining. In Park, Choonsik and Chee, Seongtaek, editors, *The 7th Annual International Conference in Information Security and Cryptology (ICISC 2004)*, volume 3506, pages 104–120.

[13] Goldreich, Oded (2004). *The Foundations of Cryptography*, volume 2, chapter General Cryptographic Protocols. Cambridge University Press.

[14] Ioannidis, Ioannis, Grama, Ananth, and Atallah, Mikhail (2002). A secure protocol for computing dot-products in clustered and distributed environments. In *The 2002 International Conference on Parallel Processing*, Vancouver, British Columbia.

[15] Jagannathan, Geetha and Wright, Rebecca N. (2005). Privacy-preserving distributed k-means clustering over arbitrarily partitioned data. In *Proceedings of the 2005 ACM SIGKDD International Conference on Knowledge Discovery and Data Mining*, pages 593–599, Chicago, IL.

[16] Jiang, Wei, Clifton, Chris, and Kantarcioglu, Murat (To appear.). Transforming semi-honest protocols to ensure accountability. *Data and Knowledge Engineering*.

[17] Kantarcioglu, Murat and Kardes, Onur (2006). Privacy-preserving data mining in malicious model. Technical Report CS-2006-06, Stevens Institute of Technology.

[18] Kantarcioglu, Murat and Vaidya, Jaideep (2003). Privacy preserving naive bayes classifier for horizontally partitioned data. In *the Workshop on Privacy Preserving Data Mining held in association with The Third IEEE International Conference on Data Mining*, Melbourne, FL.

[19] Kantarcıoğlu, Murat and Clifton, Chris (2002). Privacy-preserving distributed mining of association rules on horizontally partitioned data. In

The ACM SIGMOD Workshop on Research Issues on Data Mining and Knowledge Discovery (DMKD'02), pages 24–31, Madison, Wisconsin.

[20] Kantarcıoğlu, Murat and Clifton, Chris (2004a). Privacy-preserving distributed mining of association rules on horizontally partitioned data. *IEEE TKDE*, 16(9):1026–1037.

[21] Kantarcıoğlu, Murat and Clifton, Chris (2004b). Privately computing a distributed k-nn classifier. In Boulicaut, Jean-Franois, Esposito, Floriana, Giannotti, Fosca, and Pedreschi, Dino, editors, *PKDD2004: 8th European Conference on Principles and Practice of Knowledge Discovery in Databases*, pages 279–290, Pisa, Italy.

[22] Kantarcıoğlu, Murat, Jin, Jiashun, and Clifton, Chris (2004). When do data mining results violate privacy? In *Proceedings of the 2004 ACM SIGKDD International Conference on Knowledge Discovery and Data Mining*, pages 599–604, Seattle, WA.

[23] Kissner, L. and Song, D. (2005). Privacy-preserving set operations. In *Advances in Cryptology — CRYPTO 2005.*

[24] Lin, Xiaodong, Clifton, Chris, and Zhu, Michael (2005). Privacy preserving clustering with distributed EM mixture modeling. *Knowledge and Information Systems*, 8(1):68–81.

[25] Lindell, Yehuda and Pinkas, Benny (2000). Privacy preserving data mining. In *Advances in Cryptology – CRYPTO 2000*, pages 36–54. Springer-Verlag.

[26] Lindell, Yehuda and Pinkas, Benny (2002). Privacy preserving data mining. *Journal of Cryptology*, 15(3):177–206.

[27] Mitchell, Tom (1997). *Machine Learning*. McGraw-Hill Science/Engineering/Math, 1st edition.

[28] Naor, Moni and Pinkas, Benny (1999). Oblivious transfer and polynomial evaluation. In *Proceedings of the Thirty-first Annual ACM Symposium on Theory of Computing*, pages 245–254, Atlanta, Georgia, United States. ACM Press.

[29] Paillier, P. (1999). Public key cryptosystems based on composite degree residuosity classes. In *Advances in Cryptology - Eurocrypt '99 Proceedings, LNCS 1592*, pages 223–238. Springer-Verlag.

[30] Perry, John M. (2005). Statement of john m. perry, president and ceo, cardsystems solutions, inc. before the united states house of representatives subcommittee on oversight and investigations of the committee on financial services. `http://financialservices.house.gov/hearings.asp?formmode=detail&hearing=407&comm=4`.

[31] Vaidya, Jaideep and Clifton, Chris (2002). Privacy preserving association rule mining in vertically partitioned data. In *The Eighth ACM*

SIGKDD International Conference on Knowledge Discovery and Data Mining, pages 639–644, Edmonton, Alberta, Canada.

[32] Vaidya, Jaideep and Clifton, Chris (2005). Secure set intersection cardinality with application to association rule mining. *Journal of Computer Security*, 13(4).

[33] Yao, Andrew C. (1986). How to generate and exchange secrets. In *Proceedings of the 27th IEEE Symposium on Foundations of Computer Science*, pages 162–167. IEEE.

[34] Yu, Hwanjo, Jiang, Xiaoqian, and Vaidya, Jaideep (2006). Privacy-preserving svm using nonlinear kernels on horizontally partitioned data. In *SAC '06: Proceedings of the 2006 ACM symposium on Applied computing*, pages 603–610, New York, NY, USA. ACM Press.

Chapter 14

A Survey of Privacy-Preserving Methods Across Vertically Partitioned Data

Jaideep Vaidya
MSIS Department and CIMIC
Rutgers University
jsvaidya@rbs.rutgers.edu

Abstract The goal of data mining is to extract or "mine" knowledge from large amounts of data. However, data is often collected by several different sites. Privacy, legal and commercial concerns restrict centralized access to this data, thus derailing data mining projects. Recently, there has been growing focus on finding solutions to this problem. Several algorithms have been proposed that do distributed knowledge discovery, while providing guarantees on the non-disclosure of data.

Vertical partitioning of data is an important data distribution model often found in real life. Vertical partitioning or heterogeneous distribution implies that different features of the same set of data are collected by different sites. In this chapter we survey some of the methods developed in the literature to mine vertically partitioned data without violating privacy and discuss challenges and complexities specific to vertical partitioning.

Keywords: Vertically partitioned data, privacy-preserving data mining.

14.1 Introduction

Today, the collection of data is ubiquitous. With the rapid increase in computing, storage and networking resources, data is not only collected and stored but also analyzed. Indeed, data is often anonymized and released for public use. However, this brings the problem of privacy into sharp focus. Our personal data is *supposed* to be private. However, as several high profile infractions have shown, this is not really the case.

This creates a serious problem since it means that data really cannot be shared without appropriate security. One possibility is to only use local data and not worry about integrating or using global data. While this would be

Table 14.1. The Weather Dataset

outlook	temperature	humidity	windy	play
sunny	hot	high	false	no
sunny	hot	high	true	no
overcast	hot	high	false	yes
rainy	mild	high	false	yes
rainy	cool	normal	false	yes
rainy	cool	normal	true	no
overcast	cool	normal	true	yes
sunny	mild	high	false	no
sunny	cool	normal	false	yes
rainy	mild	normal	false	yes
sunny	mild	normal	true	yes
overcast	mild	high	true	yes
overcast	hot	normal	false	yes
rainy	mild	high	true	no

perfect from the security standpoint, it would not be very useful. Therefore, the key challenge is how to use data without really having complete access to it? While this may sound counter-intuitive, advances in cryptography show that it is possible. The challenge is to do this in an efficient manner.

In general data can be distributed in an arbitrary fashion. This means that different parties may own partial information about different sets of entities. Table 14.1 shows the famous weather dataset consisting of 14 items and 5 features. Tables 14.2(a)-14.2(b) show an arbitrary partitioning of the dataset between 2 parties. While this is possible in general, in practice, such arbitrary partitioning rarely happens. Two special cases of arbitrary partitioning – horizontal partitioning of data and vertical partitioning of data are a lot likelier. Horizontal partitioning of data means that different sites collect the same features of information for different entities. We have already seen in [1] how privacy-preserving data mining is done over horizontally partitioned data.

Vertically partitioned data means that different sites collect different features of data for the same set of entities. Integrating the local datasets gives the global dataset. Tables 14.3(a) and 14.3(b) show a vertical partitioning of the dataset between 2 parties. This happens in many real life situations. For example, consider a medical research study which wants to compare medical outcomes of different treatment methods of a particular disease. (E.g., to answer the question "will this treatment for this patient be successful or not?") The insurance companies must not disclose individual patient data without permission [13], and details of patient treatment plans are similarly protected data held by hospitals. Similar constraints arise in many applications; European Community legal restrictions apply to disclosure of any individual data[9].

Table 14.2. Arbitrary partitioning of data between 2 sites

(a) Site 1

outlook	temperature	humidity	windy	play
sunny	–	–	false	no
-	hot	-	true	no
overcast	hot	-	-	-
-	mild	high	false	-
rainy	-	normal	-	yes
rainy	-	-	true	-
-	-	normal	-	yes
-	mild	-	-	no
sunny	cool	-	-	-
rainy	-	-	false	-
-	-	normal	true	-
overcast	-	-	true	yes
-	hot	normal	-	yes
rainy	-	high	true	no

(b) Site 2

outlook	temperature	humidity	windy	play
-	hot	high	-	-
sunny	-	high	-	-
-	hot	-	false	yes
rainy	-	-	-	yes
-	cool	-	false	-
-	cool	normal	-	no
overcast	cool	-	true	-
sunny	-	high	false	-
-	-	normal	false	yes
-	mild	normal	-	yes
sunny	mild	-	-	yes
-	mild	high	-	-
overcast	-	-	false	-
-	mild	-	-	-

In general, with vertically partitioned data, more data significantly improves the quality of the models built from the dataset. Overall, the data analysis results are significantly more real and useful. While this is also the case with horizontally partitioned data (more data is always good), but it has a more critical impact with vertically partitioned data. This is because data from different parties give significantly different additional information about the entities. For example, consider Figure 14.1 that shows points plotted in a two dimensional space along with their projections on the X and Y axis. Assume that the data is vertically partitioned between two parties (one having the X-coordinate for each point, while the other has the Y-coordinate for each point). Suppose we

Table 14.3. Vertical partitioning of data between 2 sites

(a) Site 1

outlook	temperature
sunny	hot
sunny	hot
overcast	hot
rainy	mild
rainy	cool
rainy	cool
overcast	cool
sunny	mild
sunny	cool
rainy	mild
sunny	mild
overcast	mild
overcast	hot
rainy	mild

(b) Site 2

humidity	windy	play
high	false	no
high	true	no
high	false	yes
high	false	yes
normal	false	yes
normal	true	no
normal	true	yes
high	false	no
normal	false	yes
normal	false	yes
normal	true	yes
high	true	yes
normal	false	yes
high	true	no

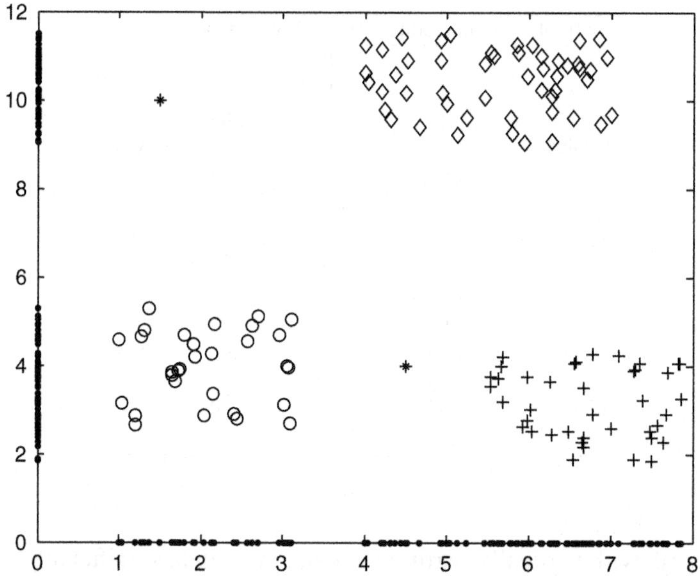

Figure 14.1. Two dimensional problem that cannot be decomposed into two one-dimensional problems

wanted to cluster the points. From the two dimensional plot it is obvious that there are at least three distinct clusters approximately centered around (2,4), (7,2.5), and (5.5,10.2). However neither site can figure this out on their own. From the Y-axis, it looks like two clusters centered at approximately 3.8, and 10.5. From the X-axis, the two clusters would be centered around 2, and 6. In fact, it is unclear if there should be only two clusters or several. The situation is equally bad if we want to identify outliers or anomalies. Again, looking at the two dimensional plot, it is obvious that the points at (1.5,10) and (4.5,4) are outliers. However, on the basis of the one-dimensional projections, neither point is identified as an outlier on either the X-axis or the Y-axis. Thus, we clearly get incorrect results with partial data. The situation only worsens with higher dimensional data.

The complexity of privacy-preserving data mining is significantly increased due to the vertical partitioning of data. In contrast to horizontal partitioning of data, vertical partitioning of data raises several unique questions with respect to the way data is processed, results are obtained and shared. We now survey different types of privacy-preserving data mining algorithms following the main data mining tasks of association rule mining, classification, clustering and outlier detection. We briefly survey the first three while going into more detail on the fourth. Thus, outlier detection serves as the expository example of privacy-preserving data mining over vertically partitioned data. In each case, we also examine some of the complications specific to vertical partitioning of data and some of the inherent challenges.

14.2 Classification

Classification refers to the problem of categorizing observations into classes. Predictive modeling uses samples of data for which the class is known to generate a model for classifying new observations. One issue with classification for vertically partitioned data is whether the class attribute is shared by all of the parties or is local to only one of them. Having the class attribute known to all of the parties simplifies the problem. However, that may not always be the case. If the class attribute is known to only one party, any process that needs to count the number of entities having a particular value for an attribute and a particular class will have to be secure. This means that computing the information gain, etc. needs to be completely secure.

Another issue with classification is how is the classification model shared between parties? One possibility is to let all of the parties know the developed model – but often this may reveal too much information. The completely secure alternative is to keep the created model completely split between the parties. However, this may have a significant impact on the classification time. Other alternatives are also possible with differing tradeoffs between security and cost. We will now see how these have affected the proposed solutions for classification.

14.2.1 Naïve Bayes Classification

Naïve Bayes is a simple but highly effective classifier. This combination of simplicity and effectiveness has lead to its use as a baseline standard by which other classifiers are measured. Vaidya and Clifton [27] present a privacy-preserving solution for vertically partitioned data. The Naïve Bayes classifier applies to learning tasks where each instance x is described by a conjunction of attribute values and the target function $f(x)$ can take on any value from some finite set C.

The Bayesian approach to classifying an new instance is to assign the most probable target value, c_{MAP}, given the attribute values $< a_1, a_2, \ldots, a_n >$ that describe the instance.

$$c_{MAP} = \underset{c_j \in C}{argmax} \left(P(c_j | a_1, a_2, \ldots, a_n) \right)$$

The Naïve Bayes classifier makes the simplifying assumption that all attributes are independent. Therefore,

$$c_{NB} = \underset{c_j \in C}{argmax} \left(P(c_j) \prod_i P(a_i | c_j) \right) \tag{14.1}$$

where c_{NB} denotes the target value output by the Naïve Bayes classifier.

Therefore, the key problem is to compute these conditional probabilities. When considering a secure solution, an important question is the location of the class attribute. There are two possibilities: the class may be known to all parties or it may be private to some party. This impacts the way the model is built and the way evaluation of a new instance is done. Both cases are realistic and model different situations. In the first case, each party can easily estimate all the required counts for nominal attributes and means and variances for numeric attributes locally, causing no privacy breaches. Prediction is also simple – each party can independently estimate the probabilities. All parties then securely multiply the probabilities and compare to obtain the predicted class. As such, we do not further discuss this. The other case is more challenging and is discussed below.

The method in [27] is fully secure in the sense that even the model built is split between the participants. Thus, none of the participants knows the actual model parameters. The only information revealed is when a new instance is classified – the class of the instance. The downside to this, of course, the performance drop. A secure protocol has to be run for every classification. If this performance penalty has to be avoided, the global model must be made available to all of the parties.

The way to compute the model parameters is somewhat different for nominal and numeric attributes. For a nominal attribute, the conditional probability is given simply by the ratio of number of instances having that attribute value and that class to the total number of instances having that class. If we encode presence of the attribute value (viz. class value) in a instance as 1, and absence as 0, we create boolean vectors such that the scalar product of the vectors gives the correct result. The scalar product of [11] also randomly splits the results between the parties. Thus, the numerator and denominator of the ratio are split between the parties. Now a secure division protocol must be run to compute splits of the conditional probability. More details can be found in [27].

For a numeric attribute, the process is more complicated. The necessary parameters are the mean μ and variance σ^2 for each class. Again, the necessary information is split between all of the parties. To compute the mean, each party needs to sum the attribute values for all appropriate instances having the same class value. These local sums are added together and the global sum is divided by the total number of instances having that same class to get the mean for that class value. This can be done, once again, by carefully constructing the vectors for each class and using the secure scalar product protocol. The party owning the class attribute builds a vector of $1/n_i$ and 0 depending on whether the training entity is in the class or not. The mean for the class is the scalar product of this vector with the projection of the data onto the attribute. The scalar product will give shares of the mean. Computing the variance is more complicated as it requires summing the square of the distances between values and the mean, without revealing values to the party owning the class attribute or the classes to the party owning the data attribute or the means to either. Thus, to compute the variance σ_y^2, it is necessary to subtract the appropriate mean from each value, square the difference and sum all such values together. Finally the global sum needs to be divided by the global number of instances having the same class y to give the required variance σ_y^2. Homomorphic encryption is used to get the differences, and secure square computation protocol is used to get shares of the square. Finally the scalar product is used as earlier to get the variance.

To evaluate a new instance, the secure ln protocol of [20] is used to get shares of the conditional probability for each attribute. Finally a secure addition and comparison circuit is used to determine the class label of the maximal class. More details can be found in [27].

14.2.2 Bayesian Network Structure Learning

Bayesian Networks relax the attribute independence assumption of the Naïve Bayes classifier, capturing situations where dependencies between attributes affect the class. A Bayesian Network is a graphical model; the vertices correspond to attributes, and the edges to probabilistic relationships between the attributes (Naïve Bayes is thus a Bayesian Network with no edges.) The probability of a given class is similar to Equation 14.1, except that the probabilities associated with an attribute are conditional on the parents of that attribute in the network.

Wright and Yang [31] propose a privacy-preserving protocol for learning the Bayesian network structure for vertically partitioned data. This protocol is limited two parties. The basic approach is to emulate the $K2$ algorithm [6], which starts with a graph with no edges, then chooses a node and greedily adds the "parent" edge to that node that most improves a score for the network, stopping when a threshold for number of parents is reached.

Since the structure of the final network is presumed to be part of the outcome (and thus not a privacy concern), the only issue is to determine which attribute most improves the score. This is similar to the decision tree induction protocol presented below ; the difference is in the score function. Instead of information gain $K2$ algorithm uses:

$$f(i, \pi_i) = \prod_{j=1}^{q_i} \frac{d_i - 1)!}{\alpha_{ij} + d_i - 1)!} \prod_{k=1}^{d_i} \alpha_{ijk}! \tag{14.2}$$

(For full details including the notation, please see [31]; our purpose here is not to give the full algorithm but to show the novel ideas with respect to privacy-preserving data mining.)

The privacy-preserving solution works by first modifying the scoring function (taking the natural log of $f(i, \pi_i)$). While this changes the output, it doesn't affect the *order*; since all that matters is determining which attribute gives the highest score, the actual value is unimportant and the resulting network is unchanged. This same technique – transforming scoring functions in ways that do not alter the final result – has proven beneficial in designing other privacy-preserving data mining algorithms.

Note that by pushing the logarithm into Equation 14.2, the products turn into summations. Moreover, taking a page from [21] they approximate a difficult to compute value (in this case, Stirling's approximation for factorial.) Ignoring small factors in the approximation, the formula reduces to a sum of factors, where each factor is of the form $\ln x$ or $x \ln x$ (except for a final factor based on the number of possible values for each attribute, which they consider public knowledge.) This now reduces to secure summation and the $\ln x$ and $x \ln x$ protocols of [21].

14.2.3 Decision Tree Classification

A solution for constructing ID3 on vertically partitioned data was proposed by Du and Zhan[8]. Their work assumes that the data is vertically partitioned between two parties. The class of the training data is assumed to be shared, but some the attributes are private. Thus most of the steps of the ID3 algorithm can be evaluated locally. The main problem is computing which site has the best attribute to split on – each can compute the gain of their own attributes without reference to the other site. [29] propose a solution that solves a more general problem – constructing an ID3 decision tree when the training data is vertically partitioned between many parties (≥ 2) and the class attribute is known to only a single party. Since each party has knowledge of only some of the attributes, knowing the structure of the tree (especially, knowledge of an unknown attribute and its breakpoints for testing) constitutes a violation of the privacy of the individual parties.

Ideally, to ensure zero leakage of extra information, even the structure of the tree should be hidden, with an oblivious protocol for classifying a new instance. However, the cost associated with this is typically unacceptable. A compromise is to hide the attribute tests used in the tree while still revealing the basic structure of the tree. A distributed protocol can then be run to to evaluate a new instance. As in Naïve Bayes, the drawback of this is that all parties have to be online in order to classify any new instance.

Before we go ahead, we briefly review the ID3 algorithm. The ID3 algorithm is a recursive partitioning algorithm. At start, all the training examples are at the root. Examples are then partitioned recursively based on selected attributes. ID3 is a greedy algorithm – in each case the attribute with the highest information gain is selected as the partitioning attribute. Partitioning stops either when all samples for a given node belong to the same class, there are no remaining attributes, or there are no samples left. In order to construct a cloaked decision tree, the parties must together figure out all how to solve all of these problems in a privacy-preserving way.

Determination of the majority class for a node requires a secure protocol since only one site knows the class. First, each site determines which of its transactions *might* reach that node of the tree. The intersection of these sets with the transactions in a particular class gives the number of transactions that reach that point in the tree having that particular class. Once this is done for all classes, the class site can now determine the distribution and majority class, and return a (leaf) node identifier. The identifier is used to map to the distribution at the time of classification.

The intersection process itself needs to be secure – this can be done by using a protocol for securely determining the cardinality of set intersection. Many protocols for doing so are known [30, 10, 2]. To formalize the whole process, the notion of a *Constraint Set* is introduced. As the tree is being built, each party i keeps track of the values of its attributes used to reach that point in the tree in a filter $Constraints_i$. Initially, this is composed of all don't care values ('?'). However, when an attribute A_{ij} at site i is used to partition, entry j in $Constraints_i$ is set to the appropriate value before recursing to build the subtree. Now, the majority class (and class distributions) are determined by computing for each class $\bigcap_{i=1..k} Y_i$, where Y_k includes a constraint on the class value.

Determining if all transactions have the same class can use the same distribution count idea described above to get the distribution counts and then check if all transactions in that node do belong to the same class. Figuring out if all attributes are used up or all transactions are done is easily done using the secure sum protocol. The main challenge lies in finding the partitioning attribute – i.e., the attribute with the largest information gain. [29] show that finding the information gain of a attribute can be done simply by counting transactions. If the number of transactions reaching a node can be determined, the number in each class c, and the same two after partitioning with each possible attribute value $a \in A$, the gain due to A can be computed. The constraint set is once again used to apply appropriate filters to get the correct count of transactions.

Finding the best attribute is a simple matter of finding out the information gain due to each attribute and selecting the best one. A naïve efficient implementation would leak the information gain due to each attribute. If even this minimal information should not be leaked, the information gain can be split between the parties, and a sequence of secure comparisons carried out to determine the best attribute. Thus, the entire ID3 tree can be built in a secure manner using these sub-protocols.

Classifying a new instance again requires a distributed protocol. Given that the structure of the tree is known, the root site first makes a decision based on its data. It then looks at the node this decision leads to and tells the site responsible for that node the node and the instance to be classified. This continues until a leaf is reached, and which point the site that originally held the class value knows the predicted class of the new instance. While this does lead to some disclosure of information (knowing the path followed, a site can say if instances have the same values for data not known to that site), specific values need not be disclosed.

14.3 Clustering

One question with clustering is how are the clusters shared? Specifically, is only cluster membership shared or is more information about the clusters shared, and if so, how? Based on cluster membership, each party can locally compute its share of the cluster means. However, are the complete cluster means shared with all of the parties? In this case other parties could easily learn a lot of information about the other attributes.

[26] proposed the first method for clustering over vertically partitioned data – a privacy-preserving protocol perform do k-means clustering. Though all parties know the final assignment of data points to clusters, they retain only partial information for each cluster. The cluster centers μ_i are assumed to be

semiprivate information, i.e., each site can learn only the components of μ that correspond to the attributes it holds. Thus, all information about a site's attributes (not just individual values) is kept private; if sharing the μ is desired, an evaluation of privacy/secrecy concerns can be performed after the values are known.

The basic protocol proposed closely follows the original K-means protocol. There are two major challenges – figuring out how to assign points to clusters in each iteration, and figuring out when to stop. Since the means at each iteration are not considered private information, figuring out when to stop is quite simple. Each party can locally compute the difference between their shares of the mean, and finally check if the total difference is less than the threshold. Since all arithmetic takes place in a field, the threshold evaluation at the end is somewhat non-obvious. *Intervals* are compared rather than the actual numbers. Further details can be found in [26]. The assignment of points to clusters in each iteration is carried out through a secure protocol utilizing three key ideas:

1 Disguise the site components of the distance with random values that cancel out when combined.

2 Compare distances so only the comparison result is learned; no party knows the distances being compared.

3 Permute the order of clusters so the real meaning of the comparison results is unknown.

One drawback of the Vaidya and Clifton protocol is that it is not completely secure since intermediate results are revealed. Essentially, the intermediate cluster assignment of data points is known to every party for each iteration, though the final result only specifies the *final* clusters. However, this compromise is required for efficiency. [15] propose a completely secure protocol for arbitrarily partitioned data. Their protocol is very similar to the Vaidya and Clifton protocol with the added complexity of splitting the intermediate cluster centers. Thus, no information is leaked whatsoever.

14.4 Association Rule Mining

[25] first showed how secure association rule mining can be done for vertically partitioned data by extending the apriori algorithm. Vertical partitioning implies that an itemset could be split between multiple sites. Most steps of the apriori algorithm can be done locally at each of the sites. The crucial step involves finding the support count of an itemset. If the support count of an itemset can be securely computed, one can check if the support is greater than threshold, and decide whether the itemset is frequent. Using this, association rules can be easily mined securely.

The key insight of [25] is that computing the support of an itemset is exactly the scalar product of the vectors representing the sub-itemsets with different parties. Thus, the entire secure association rule mining problem can be reduced to computing the scalar product of two vectors in a privacy-preserving way. [25] also proposed an algebraic method to compute the scalar product. While this method is not provably secure, it is quite efficient. A strong point of the secure association rule mining protocol is that it is not tied to any specific scalar product protocol. Indeed, there have been a number of secure scalar product protocols proposed[7, 14, 11, 33, 24], out of which at least two are provably secure. All of them have differing tradeoffs of security, efficiency, and utility (some are limited to scalar products over boolean data). Any of these could be used. [1] shows one possible secure protocol to compute the scalar product using homomorphic encryption.

While there are now several solutions using scalar product computation, one alternative solution needs to be mentioned. [30] provide an innovative alternative solution to the association rule mining problem. There are two key insights provided in this solution. First, if we encode the vectors as sets (with position numbers as elements), the scalar product is the same as the size of the intersection set. For example, assume we have vector $\vec{X} = (1,0,0,1,1)$ and $\vec{Y} = (0,1,0,1,0)$. Then the scalar product $\vec{X} \cdot \vec{Y} = \sum_{i=1}^{5} x_i * y_i$. Now, the corresponding set encodings are $XS = (1,4,5)$ and $YS = (2,4)$. Once can see that the size of the intersection set $|XS \bigcap YS| = 1$ is exactly the same as the scalar product. This idea is used to compute the scalar product.

The basic idea is to use commutative encryption to encrypt all of the items in each party's set. Commutative encryption is an important tool used in many cryptographic protocols. An encryption algorithm is commutative if the order of encryption does not matter. Thus, for any two encryption keys E1 and E2, and any message m, E1(E2(m)) = E2(E1(m)). The same property applies to decryption as well – thus to decrypt a message encrypted by two keys, it is sufficient to decrypt it one key at a time. The basic idea is for each source to encrypt its data set with its keys and pass the encrypted data set to the next source. This source again encrypts the received data using its encryption keys and passes the encrypted data to the next source until all sources have encrypted the data. Since we are using commutative encryption, the encrypted values of the set items across different data sets will be equal if and only if their original values are equal. Thus, all the intersection of the encrypted values gives the logical AND of the vectors, and counting the size of the intersection set gives the total number of 1s (i.e., the scalar product). The encryption prevents any party from knowing the actual value of any local item. This scalar product method only works for boolean vectors, but it will still work for the association rule mining problem. This idea is also used by [2] to compute Set Union, Set Intersection, Size of Set Union, and Size of Set Intersection. However, their work is limited to two parties. [10] also propose techniques using homomorphic encryption to do private matching and set intersection for two parties which can guard against malicious adversaries in the random oracle model as well.

While this is a good alternative, the real innovativeness lies in realizing the fact that once all of the items are encrypted by the keys of all of the parties, all parties can locally compute all of the frequent itemsets. This implies that the overall cost of secure association rule mining is simply the cost of completely encrypting all of the items. If there are k parties, n items and m transactions, the total cost of association rule mining is $O(nmk)$ since these will be the total number of encryptions required (the encryption time dominates all other costs). Note that this is independent of the number of frequent itemsets which can easily be in the tens of thousands. Thus, the protocol in [30] is extremely efficient in the global sense and makes privacy-preserving association rule mining really feasible.

Most of the protocols developed typically assume a semi-honest model, where the parties involved will honestly follow the protocol but can later try to infer additional information from whatever data they receive through the protocol. One result of this is that parties are not allowed to give spurious input to the protocol. If a party is allowed to give spurious input, they can probe to determine the value of a specific item at other parties. For example, if a party gives the input $(0, \ldots, 0, 1, 0, \ldots, 0)$, the result of the scalar product (1 or 0) tells the malicious party if the other party the transaction corresponding to the 1. Attacks of this type can be termed probing attacks and need to be protected against. The protocol in [30] can partially protect against such attacks.

14.5 Outlier detection

Outlier / anomaly detection is one of the most common data mining tasks carried out in practice. Hawkins [12] defines an outlier as an observation which deviates so much from other observations so as to arouse suspicions that it was generated by a different mechanism. Outlier detection has been used to find uncommon sequences in gene data, to find fradulent transactions in credit card records, fraud discovery in mobile phones, to find intrusions from network traffic data[3, 19], etc. Indeed even the search for terrorism involves outlier detection – detecting previously unknown suspicious behavior is a clear outlier detection problem. Many of these applications also have privacy concerns, and organizations must be careful to avoid overstepping the bounds of privacy legislation[9].

So what does it mean to protect privacy in this context? By definition, outlier detection means finding outliers. Thus, the output of outlier detection would be a list of detected outliers. This is highly specific information – anomalous entities/transactions are highlighted. There is no summarization carried out. Thus, implicitly, no information about a true outlier should be protected/concealed. However, no information about the other entities should be revealed. Indeed, the process of finding outliers should not reveal any extra information. Privacy-preserving outlier detection will ensure these concerns are balanced, allowing us to get the benefits of outlier detection without worrying about legal or privacy concerns. However, what about false positives? i.e., what about entities identified as outliers without really being so. While this seems problematic, a couple of caveats exist. First, no detection technique is fool-proof and false positives always exist. We merely reduce the privacy leakage and problems. Secondly, technical solutions exist. All the identifiers can be eliminated to begin with. The outliers detected are hand examined and if sufficient cause exists, the anonymization is taken away and the real identity is revealed (just as it occurs in real life with a court order).

While there are numerous different definitions of outliers as well as techniques to find them, the first privacy-preserving outlier detection technique developed was for distance-based outliers. The method developed by Vaidya and Clifton[28] finds distance-based outliers without any party gaining knowledge beyond learning which items are outliers. Ensuring that data is not disclosed maintains privacy, i.e., no privacy is lost beyond that inherently revealed in knowing the outliers. This is the absolute minimum information that must be revealed for privacy-preserving outlier detection over vertically partitioned data.

Before going into specifics, we first briefly review the notion of distance-based outliers. Knorr and Ng [17] define the notion of a Distance Based outlier as follows: *An object O in a dataset T is a DB(p,dt)-outlier if at least fraction p of the objects in T lie at distance greater than dt from O.* Other distance based outlier techniques also exist[18, 22]. The advantages of distance based outliers are that no explicit distribution needs to be defined to determine unusualness, and that it can be applied to any feature space for which we can define a distance measure. Euclidean distance is the standard, although the algorithms are easily extended to general Minkowski distances. There are other non distance based techniques for finding outliers as well as significant work in statistics [4], but there is little work on finding them in a privacy-preserving fashion – thus, this is a rich area for future work.

For Euclidean distance, for vertically partitioned data, the distance dt is fixed by the local parties deciding on the local distances dt_i (i.e., $dt = \sum_{i=1}^{k} dt_i$), since no site globally knows all of the attributes. An object X is an outlier if at least $p\%$ of the other objects lie at a distance greater than dt.

The approach of [28] duplicates the results of the outlier detection algorithm of [17]. The idea is that an object O is an outlier if more than a percentage p of the objects in the data set are farther than distance dt from O. The basic idea is that parties compute the portion of the answer they know, then engage in a secure sum to compute the total distance. The key is that this total is (randomly) split between sites, so nobody knows the actual distance. A secure protocol is used to determine if the actual distance between any two points exceeds the threshold; again the comparison results are randomly split such that summing the splits (over a closed field) results in a 1 if the distance exceeds the threshold, or a 0 otherwise.

For a given object O, each site can now sum all of its shares of comparison results (again over the closed field). When added to the sum of shares from other sites, the result is the correct count; all that remains is to compare it with the percentage threshold p. This addition/comparison is also done with a secure protocol, revealing only the result: if O is an outlier. The pairwise comparison of all points may seem excessive, but early termination could disclose information about relative positions of points. The asymptotic complexity still equals that of [17].

Note that a secure solution requires that all operations are carried out modulo some field. For the algorithms, the field D is used for distances, and F is used for counts of the number of entities. The field F must be over twice the number of objects. Limits on D are based on maximum distances; details on the size are given with each algorithm.

We now present the actual algorithm, followed by the complete proof of security for the algorithm. This is especially instructive for readers wishing to develop their own algorithms since the proof of security forms a significantly important component necessary for trust in the overall solution. A discussion of the computational and communication complexity of the algorithm rounds off this section, and affords the opportunity to discuss avenues for future work in this area.

14.5.1 Algorithm

For each object i, the protocol iterates over every other object j. Since each party owns some of the attributes, each party can compute the distance between two objects for those attributes. Thus, each party can compute a *share* of the pairwise distance locally; the sum of these shares is the total distance. However, revealing the distance still reveals too much information, therefore a secure protocol is used to get shares of the pairwise comparison of distance and threshold. The key to this protocol is that the 1 or 0 is actually two shares r'_q and r'_s returned to the two parties, such that $r'_q + r'_s = 1 \text{ (or 0)} \pmod F$. Looking at only one share, neither party can learn anything.

Once all points have been compared, the parties individually sum their shares. Since the shares add to 1 for distances exceeding the distance threshold, and 0 otherwise, the total sum $\pmod F$ gives the number of points for which the distance exceeds the threshold. Explicit computation of this sum would still reveal the actual number of points distant. So the parties do not actually compute this sum; instead all parties pass their (random) shares to a designate to add, and the designated party and the party holding the point engage in a secure protocol that reveals only if the sum of the shares exceeds $p\%$. Thus, the only result of the protocol is to reveal whether the point is an outlier or not.

An interesting side effect of this algorithm is that the parties need not reveal any information about the attributes they hold, or even the number of attributes. Each party locally determines the distance threshold for its attributes (or more precisely, the share of the overall threshold for its attributes). Instead of computing the local pairwise distance, each party computes the difference between the local pairwise distance and the local threshold. If the sum of these differences is greater than 0, the pairwise distance exceeds the threshold.

Algorithm 2 gives the full details. In steps 6-10, the sites sum their local distances (actually the difference between the local distance and the local threshold). The random x added by P_1 masks the distance from each party. In steps 11-13, Parties P_1 and P_k get shares of the pairwise comparison result. The comparison is a test if the sum is greater than 0 (since the threshold has already been subtracted.) These two parties keep a running sum of their shares. At the end, in step 15 these shares are added and compared with the percentage threshold.

At several stages in the algorithm, a protocol is required to securely compare the sum of two numbers, with the output split between the parties holding those numbers. This can be accomplished using the generic circuit evaluation technique first proposed by Yao[32].

14.5.2 Security Analysis

The protocol described above can be proven to be secure using the proof techniques of Secure Multiparty Computation. The idea is that since what a party sees during the protocol (its shares) are randomly chosen from a uniform distribution over a field, it learns nothing in isolation. (Of course, collusion with other parties could reveal information, since the *joint* distribution of the shares is not random). The idea of the proof is based on a simulation argument: If we can define a simulator that uses the algorithm output and a party's own data to simulate the messages seen by a party during a real execution of the protocol, then the real execution isn't giving away any new information (as long as the simulator runs in polynomial time).

Since all parties know the number (and identity) of objects in O, they can set up the loops; the simulator just runs the algorithm to generate most of the simulation. The only communication is at lines 8, 11, 15, and 16.

Protocol 2 Finding DB(p,D)-outliers

Require: k parties, $P_1, \ldots, P_k$; each holding a subset of the attributes for all objects O.
Require: dt_r : local distance threshold for P_r (e.g., $dt^2 + m_r/m$).
Require: Fields D larger than twice the maximum distance value (e.g., for Euclidean this is actually $Distance^2$), F larger than $|O|$

1: **for all** objects $o_i \in O$ **do**
2: $m'_1 \leftarrow m'_k \leftarrow 0 \pmod F$
3: **for all** objects $o_j \in O, o_j \neq o_i$ **do**
4: P_1: Randomly choose a number x from a uniform distribution over the field D
5: P_1: $x' \leftarrow x$
6: **for** $r \leftarrow 1, \ldots, k-1$ **do**
7: At P_r: $x' \leftarrow x' + Distance_r(o_i, o_j) - dt_r \pmod D$ $\{Distance_r$ is local distance at $P_r\}$
8: P_r sends x' to P_{r+1}
9: **end for**
10: At P_k: $x' \leftarrow x' + Distance_k(o_i, o_j) - dt_k \pmod D$
11: P_1 and P_k engage in the secure comparison protocol to get m_1 and m_k respectively such that the following condition holds: if $0 < x' + (-x) \pmod D < |D|/2$, then $m_1 + m_k = 1 \pmod F$, otherwise $m_1 + m_k = 0 \pmod F$
12: At P_1: $m'_1 \leftarrow m'_1 + m_1 \pmod F$
13: At P_k: $m'_k \leftarrow m'_k + m_k \pmod F$
14: **end for**
15: P_1 and P_k engage in the secure comparison protocol to get $temp_1$ and $temp_k$ respectively such that the following condition holds: if $m'_1 + m'_k \pmod F > |O| * p\%$, then $temp_1 + temp_k \leftarrow 1$ (o_i is an outlier), otherwise $temp_1 + temp_k \leftarrow 0$
16: P_1 and P_k send $temp_1$ and $temp_k$ to the party authorized to learn the result; if $temp_1 + temp_k = 1$ then o_i is an outlier.
17: **end for**

Step 8: Each party P_s sees $x' = x + \sum_{r=1}^{s-1} Distance_r(o_i, o_j)$, where x is the random value chosen by P_1. $Pr(x' = y) = Pr(x + \sum_{r=1}^{s-1} Distance_r(o_i, o_j) = y) = Pr(x = y - \sum_{r=0}^{s-1} Distance_r(o_i, o_j)) = \frac{1}{|D|}$. Thus we can simulate the value received by choosing a random value from a uniform distribution over D.

Steps 11 and 15: Each step is a secure comparison. Assuming this is secure, the messages in this step can be easily simulated.

Step 16: This is the final result, and can be easily simulated. $temp_1$ is simulated by choosing a random value, $temp_k = result - temp_1$. By the same argument on random shares used above, the distribution of simulated values is indistinguishable from the distribution of the shares.

The simulator clearly runs in polynomial time (the same as the algorithm). Since each party is able to simulate the view of its execution (i.e., the probability of any particular value is the same as in a real execution with the same inputs/results) in polynomial time, the algorithm is secure with respect to the semi-honest SMC definitions.

Without collusion and assuming a malicious-model secure comparison, a malicious party is unable to learn anything it could not learn from altering its input. Step 8 is particularly sensitive to collusion, but can be improved (at cost) by splitting the sum into shares and performing several such sums (see [16] for more discussion of collusion-resistant secure sum).

14.5.3 Computation and Communication Analysis

In general we do not discuss the computational/communicational complexity of any of the algorithms in detail. However, in this case the algorithmic complexity raises interesting issues vis-a-vis security. Therefore we discuss it below in detail.

Algorithm 2 suffers the drawback of having quadratic computation complexity due to the nested iteration over all objects. Due to the nested iteration, Algorithm 2 also requires $O(n^2)$ secure comparisons (step 11), where n is the total number of objects. While operation parallelism can be used to reduce the round complexity of communication, the key practical issue is the computational complexity of the encryption required for the secure comparison and scalar product protocols.

This quadratic complexity is troubling since the major focus of new algorithms for outlier detection has been to reduce the complexity, since n^2 is assumed to be inordinately large. However, achieving lower than quadratic complexity is challenging – at least with the basic algorithm. Failing to compare all pairs of points is likely to reveal information about the relative distances of the points that *are* compared. Developing protocols where such revelation can be proven not to disclose information beyond that revealed by simply knowing the outliers is a challenge. Otherwise, completely novel techniques must be developed which do not require *any* pairwise comparison. When there are three or more parties, assuming no collusion, much more efficient solutions that reveal some information can be developed. Essentially a much more efficient secure comparison can be used [5] that still reveals nothing to the third party. While not completely secure, the privacy versus cost tradeoff may be acceptable in some situations. An alternative (and another approach to future work) is demonstrating lower bounds on the complexity of fully secure outlier detection. However, significant work is required to make any of this happen – thus opening a rich area for future work.

[23] use very similar techniques to perform privacy-preserving nearest neighbor search. They further show how this can be used to perform privacy-preserving LOF outlier detection, SNN clustering and kNN classification.

14.6 Challenges and Research Directions

This chapter presents a survey of efficient solutions for many privacy preserving data mining tasks on vertically partitioned data. Like horizontally partitioned data, it can be seen that even for vertically partitioned data, many privacy-preserving algorithms can be efficiently implemented by combining specific basic secure building blocks. However, inherently, the main challenge with techniques dealing with vertically partitioned data lies with efficiency. Unlike, horizontally partitioned data, it is very difficult to carry out much local aggregation beforehand. For example, in a lot of the protocols seen above, the secure scalar product is a critical component. Utilizing the [11] protocol, a single scalar product of two vectors of length n will require n encryptions, n modular exponentiations, n modular multiplications and 1 decryption. The cost for the encryptions and exponentiations dominate. With the current speed of encryption/exponentiation, it still takes a significant amount of time to carry out a single scalar product. For example, the scalar product of two vectors of length 1000 takes approximately 40s with 512 bit encryption and 270s with 1024 bit encryption. Since data mining is typically done over millions of transactions, this cost significantly balloons up. Therefore we clearly need more efficient protocols. Indeed, very few of the protocols are actually implemented. This definitely needs to change to ensure deployment of these algorithms into real life. The other technical challenge lies with the adversarial model of the protocols. Almost all of the protocols seen above assume semi-honest participants – i.e., participants that will follow the protocol exactly but may later try to find additional information. While this is a good starting model, eventually we need protocols that would work in the presence of malicious adversaries.

Overall, we believe that the trend towards usage of privacy-preserving algorithms is on the rise. Due to increasing privacy and security concerns as well as the need to leverage commercial assets, there is a clear need for flexible and efficient privacy-preserving solutions that could be tailored for individual privacy needs. Development of such flexible and efficient solutions will be instrumental in wide-scale adoption of this technology.

References

[1] Murat Kantarcioglu. A survey of Privacy-Preserving Methods across Horizontall Partitioned Data. *Privacy-Preserving Data Mining: Models and Algorithms. Ed. Charu Aggarwal, Philip Yu, Springer, 2008.*

[2] Rakesh Agrawal, Alexandre Evfimievski, and Ramakrishnan Srikant. Information sharing across private databases. In *Proceedings of ACM SIGMOD International Conference on Management of Data*, San Diego, California, June 9-12 2003.

[3] Daniel Barbará, Ningning Wu, and Sushil Jajodia. Detecting novel network intrusions using bayes estimators. In *First SIAM International Conference on Data Mining*, Chicago, Illinois, April 5-7 2001.

[4] Vic Barnett and Toby Lewis. *Outliers in Statistical Data*. John Wiley and Sons, 3rd edition, 1994.

[5] Christian Cachin. Efficient private bidding and auctions with an oblivious third party. In *Proceedings of the 6th ACM conference on Computer and communications security*, pages 120–127. ACM Press, 1999.

[6] Gregory F. Cooper and Edward Herskovits. A bayesian method for the induction of probabilistic networks from data. *Mach. Learn.*, 9(4):309–347, 1992.

[7] Wenliang Du and Mikhail J. Atallah. Privacy-preserving statistical analysis. In *Proceeding of the 17th Annual Computer Security Applications Conference*, New Orleans, Louisiana, USA, December 10-14 2001.

[8] Wenliang Du and Zhijun Zhan. Building decision tree classifier on private data. In Chris Clifton and Vladimir Estivill-Castro, editors, *IEEE International Conference on Data Mining Workshop on Privacy, Security, and Data Mining*, volume 14, pages 1–8, Maebashi City, Japan, December 9 2002. Australian Computer Society.

[9] Directive 95/46/EC of the european parliament and of the council of 24 october 1995 on the protection of individuals with regard to the processing of personal data and on the free movement of such data. *Official Journal of the European Communities*, No I.(281):31–50, October 24 1995.

[10] Michael J. Freedman, Kobbi Nissim, and Benny Pinkas. Efficient private matching and set intersection. In *Eurocrypt 2004*, Interlaken, Switzerland, May 2-6 2004. International Association for Cryptologic Research (IACR).

[11] Bart Goethals, Sven Laur, Helger Lipmaa, and Taneli Mielikäinen. On Secure Scalar Product Computation for Privacy-Preserving Data Mining. In Choonsik Park and Seongtaek Chee, editors, *The 7th Annual International Conference in Information Security and Cryptology (ICISC 2004)*, volume 3506, pages 104–120, December 2–3, 2004.

[12] D. M. Hawkins. *Identification of Outliers*. Chapman and Hall, 1st edition, 1980.

[13] Standard for privacy of individually identifiable health information. *Federal Register*, 66(40), February 28 2001.

[14] Ioannis Ioannidis, Ananth Grama, and Mikhail Atallah. A secure protocol for computing dot-products in clustered and distributed environments. In *The 2002 International Conference on Parallel Processing*, Vancouver, British Columbia, August 18-21 2002.

[15] Geetha Jagannathan and Rebecca N. Wright. Privacy-preserving distributed k-means clustering over arbitrarily partitioned data. In *Proceedings of the 2005 ACM SIGKDD International Conference on Knowledge Discovery and Data Mining*, pages 593–599, Chicago, IL, August 21-24 2005.

[16] Murat Kantarcıoğlu and Chris Clifton. Privacy-preserving distributed mining of association rules on horizontally partitioned data. *IEEE Transactions on Knowledge and Data Engineering*, 16(9):1026–1037, September 2004.

[17] Edwin M. Knorr and Raymond T. Ng. Algorithms for mining distance-based outliers in large datasets. In *Proceedings of 24th International Conference on Very Large Data Bases (VLDB 1998)*, pages 392–403, New York City, NY, USA, August24-27 1998.

[18] Edwin M. Knorr, Raymond T. Ng, and Vladimir Tucakov. Distance-based outliers: algorithms and applications. *The VLDB Journal*, 8(3-4):237–253, 2000.

[19] Aleksandar Lazarevic, Aysel Ozgur, Levent Ertoz, Jaideep Srivastava, and Vipin Kumar. A comparative study of anomaly detection schemes in network intrusion detection. In *SIAM International Conference on Data Mining (2003)*, San Francisco, California, May 1-3 2003.

[20] Yehuda Lindell and Benny Pinkas. Privacy preserving data mining. In *Advances in Cryptology – CRYPTO 2000*, pages 36–54. Springer-Verlag, August 20-24 2000.

[21] Yehuda Lindell and Benny Pinkas. Privacy preserving data mining. *Journal of Cryptology*, 15(3):177–206, 2002.

[22] Sridhar Ramaswamy, Rajeev Rastogi, and Kyuseok Shim. Efficient algorithms for mining outliers from large data sets. In *Proceedings of the 2000 ACM SIGMOD international conference on Management of data*, pages 427–438. ACM Press, 2000.

[23] Mark Shaneck, Yongdae Kim, and Vipin Kumar. Privacy preserving nearest neighbor search. In *ICDM Workshops*, pages 541–545. IEEE Computer Society, 2006.

[24] Dragos Trinca and Sanguthevar Rajasekaran. Towards a collusion-resistant algebraic multi-party protocol for privacy-preserving association rule mining in vertically partitioned data. In *3rd International Workshop on Information Assurance*, April11–13 2007.

[25] Jaideep Vaidya and Chris Clifton. Privacy preserving association rule mining in vertically partitioned data. In *The Eighth ACM SIGKDD International Conference on Knowledge Discovery and Data Mining*, pages 639–644, Edmonton, Alberta, Canada, July 23-26 2002.

[26] Jaideep Vaidya and Chris Clifton. Privacy-preserving k-means clustering over vertically partitioned data. In *The Ninth ACM SIGKDD International Conference on Knowledge Discovery and Data Mining*, pages 206–215, Washington, DC, August 24-27 2003.

[27] Jaideep Vaidya and Chris Clifton. Privacy preserving naïve bayes classifier for vertically partitioned data. In *2004 SIAM International Conference on Data Mining*, pages 522–526, Lake Buena Vista, Florida, April 22–24 2004.

[28] Jaideep Vaidya and Chris Clifton. Privacy-preserving outlier detection. In *Proceedings of the Fourth IEEE International Conference on Data Mining (ICDM'04)*, pages 233–240, Los Alamitos, CA, November 1 – 4 2004. IEEE Computer Society Press.

[29] Jaideep Vaidya and Chris Clifton. Privacy-preserving decision trees over vertically partitioned data. In *The 19th Annual IFIP WG 11.3 Working Conference on Data and Applications Security*, Storrs, Connecticut, August 7-10 2005. Springer.

[30] Jaideep Vaidya and Chris Clifton. Secure set intersection cardinality with application to association rule mining. *Journal of Computer Security*, 13(4):593–622, November 2005.

[31] Rebecca Wright and Zhiqiang Yang. Privacy-preserving bayesian network structure computation on distributed heterogeneous data. In *Proceedings of the 10th ACM SIGKDD International Conference on Knowledge Discovery and Data Mining*, Seattle, WA, August22-25 2004.

[32] Andrew C. Yao. How to generate and exchange secrets. In *Proceedings of the 27th IEEE Symposium on Foundations of Computer Science*, pages 162–167. IEEE, 1986.

[33] Sheng Zhong. Privacy-preserving algorithms for distributed mining of frequent itemsets. *Information Sciences*, 177(2):490–503, 2007.

Chapter 15

A Survey of Attack Techniques on Privacy-Preserving Data Perturbation Methods

Kun Liu[1], Chris Giannella[2], and Hillol Kargupta[3]

[1] *IBM Almaden Research Center*
650 Harry Road, San Jose, CA 95120
kun@us.ibm.com

[2] *Department of Computer Science*
Loyola College in Maryland
4501 N. Charles Street, Baltimore, MD. 21210
cgiannel@acm.org

[3] *Department of Computer Science and Electrical Engineering*
University of Maryland, Baltimore County
1000 Hilltop Circle, Baltimore, MD 21250
Also affiliated with AGNIK, LLC
8840 Stanford Blvd. Suite 1300, Columbia, MD 21045
hillol@cs.umbc.edu

Abstract We focus primarily on the use of additive and matrix multiplicative data perturbation techniques in privacy preserving data mining (PPDM). We survey a recent body of research aimed at better understanding the vulnerabilities of these techniques. These researchers assumed the role of an attacker and developed methods for estimating the original data from the perturbed data and any available prior knowledge. Finally, we briefly discuss research aimed at attacking k-anonymization, another data perturbation technique in PPDM.

Keywords: Data perturbation, additive noise, matrix multiplicative noise, attack techniques, k-anonymity.

15.1 Introduction

Data perturbation represents one common approach in privacy preserving data mining (PPDM). It builds on a longer history in the areas of statistical disclosure control and statistical databases [1] where the original (private) dataset is perturbed and the result is released for data analysis. Typically, a "privacy/accuracy" trade-off is faced. On the one hand, perturbation must not allow the original data records to be adequately recovered. On the other, it must allow "patterns" in the original data to be mined. Data perturbation includes a wide variety of techniques including (but not limited to): additive, multiplicative [24], matrix multiplicative, k-anonymization [38, 41], micro-aggregation [3, 26], categorical data perturbation [10, 45], data swapping [11], resampling [27], data shuffling [34] (see [1, 28] for a more complete survey).

In this chapter we mostly focus on two types of data perturbation that apply to continuous data: additive and matrix multiplicative. Additive data perturbation was originally introduced in statistical disclosure control more that twenty years ago and was further studied in the PPDM community in the last eight years. Matrix multiplicative data perturbation were introduced only five years ago in the PPDM community and is in its early stages of study. In order to better understand the privacy offered by these techniques, some PPDM researchers have assumed the role of an attacker and developed techniques for breaching privacy by estimating the original data from the perturbed data and any available additional prior knowledge. Their work offers insight into vulnerabilities of this type of data perturbation. We provide a detailed survey of their work in an effort to allow the reader to observe common themes and future directions. Moreover, due to its rapidly growing study, we also provide a brief overview of attacks on k-anonymization.

This chapter is organized as follows. Section 15.2 describes definitions and notation used throughout. Section 15.3 discusses additive data perturbation, its uses and several attack techniques in detail. Section 15.4 describes matrix multiplicative data perturbation, its uses and several attack techniques in detail. Section 15.5 discusses k-anonymization and recent literature addressing vulnerabilities of this data perturbation model. Finally, Section 15.6 concludes the paper with a summary.

15.2 Definitions and Notation

Throughout this chapter, the original dataset is represented as an $n \times m$, real-valued matrix X, with each column a data record. The data owner perturbs X to produce an $n' \times m$ data matrix Y, which is then released to the public or another party for analysis. The attacker uses Y and any other available information to produce an estimation of X, denoted by $\hat{X}$. Unless otherwise stated, we will assume that each record of the original dataset arose as an independent sample from an n-dimensional random vector $\mathcal{X}$ with unknown probability density function (*p.d.f.*) (and this assumption is public knowledge). Let $\Sigma_{\mathcal{X}}$ denote the covariance matrix of $\mathcal{X}$. We will also assume that $\Sigma_{\mathcal{X}}$ has all distinct and non-zero eigenvalues (more details later) since, as argued in [20, pg. 27], this assumption holds in most practical situations.

Unless otherwise stated, all vectors are column-vectors. Given a matrix A, A^T denotes its transpose and A^{-1} denotes its inverse (provided one exists). I denotes the identity matrix with dimensions specified by context. Given vector x, $||x||$ denotes the Euclidean distance of x to the origin *i.e.* the Euclidean norm.

15.3 Attacking Additive Data Perturbation

The data owner replaces the original dataset X with

$$Y = X + R, \tag{15.1}$$

where R is a noise matrix with each column generated independently from a n-dimensional random vector $\mathcal{R}$ with mean vector zero. As is commonly done, we assume throughout that $\Sigma_{\mathcal{R}}$ equals $\sigma^2 I$, *i.e.*, the entries of R were generated independently from some distribution with mean zero and variance σ^2 (typical choices for this distribution include Gaussian and uniform). In this case, R is sometimes referred to as *additive white noise.*

While having a long history in the statistical disclosure control and statistical database fields (see [6] for a comprehensive survey), additive data perturbation was first revisited to address PPDM problems by Agrawal and Srikant [5]. They assumed the *p.d.f.* of $\mathcal{R}$ is public. They developed a technique for estimating the *p.d.f.* of $\mathcal{X}$ from Y and show how a decision tree classifier can then be constructed. Their distribution recovery technique is further developed in [4, 9].

We describe five different attack techniques against additive perturbation. The first three attacks filter off the random noise by analyzing the eigenstates of the data: spectral filtering [22], singular value decomposition (SVD) filtering [17], and principal component analysis (PCA) filtering [18]. They all use *eigen-analysis* for filtering out the protected data. The fourth attack is a Bayes approach based on maximum a posteriori probability (MAP) estimation [18]. The fifth attack shows that if the *p.d.f.* of $\mathcal{X}$ is reconstructed, in some cases, it can lead to disclosure. We refer to this attack as *distribution analysis.* Note that in all five we assume that the attacker knows the *p.d.f.* of $\mathcal{R}$, and attacker implicitly knows that the perturbed data records arose as independent samples from random vector $\mathcal{Y} = \mathcal{X} + \mathcal{R}$. Next, we describe each of these attacks in detail.

15.3.1 Eigen-Analysis and PCA Preliminaries

Before describing eigen-analysis based attacks, we first provide a brief background of eigen-analysis and PCA. Let $\mathcal{X}$ be an n-dimensional random vector. Generally speaking the eigenvalues of covariance $\Sigma_{\mathcal{X}}$ are the n roots (possible including repeats) of the degree n polynomial $|\Sigma_{\mathcal{X}} - I\lambda|$ where $|.|$ denotes the matrix determinant. Since $\Sigma_{\mathcal{X}}$ is positive semi-definite, all its eigenvalues are non-negative and real [13, pg. 295]. If we assume that they are also all distinct and non-zero, they can be denoted as $\lambda^1_{\mathcal{X}} > \ldots > \lambda^n_{\mathcal{X}} > 0$. Associated with $\lambda^j_{\mathcal{X}}$ is its *normalized eigenspace*, $\mathbb{V}^j_{\mathcal{X}} = \{v \in \mathbb{R}^n : \Sigma_{\mathcal{X}} v = v\lambda^j_{\mathcal{X}} \text{ and } ||v|| = 1\}$. These normalized eigenspaces are pair-wise orthogonal and have dimension one [13, pg. 295]. Hence each can be written as $\{v^j_{\mathcal{X}}, -v^j_{\mathcal{X}}\}$ where $v^j_{\mathcal{X}}$ is lexicographically larger than $-v^j_{\mathcal{X}}$. Let $V_{\mathcal{X}}$ denote the normalized eigenvector matrix $[v^1_{\mathcal{X}} \cdots v^n_{\mathcal{X}}]$ (which is orthogonal).

As is standard practice in PCA, we assume that $\mathcal{X}$ has mean vector zero (if not, it is replaced by $\mathcal{X} - E[\mathcal{X}]$). The jth *principal component (PC)* of $\mathcal{X}$ is ${v^j_{\mathcal{X}}}^T \mathcal{X}$ (or $-{v^j_{\mathcal{X}}}^T \mathcal{X}$). It can be shown that the PCs are pair-wise uncorrelated and capture the maximum possible variance in the following sense. For each $1 \leq j \leq n$, there does not exist $v \in \mathbb{R}^n$ orthogonal to v_ℓ for all $1 \leq \ell < j$ such that $Var(v^T\mathcal{X}) > Var({v^j_{\mathcal{X}}}^T \mathcal{X})$. It can further be shown that $Var({v^j_{\mathcal{X}}}^T \mathcal{X}) = \lambda^j_{\mathcal{X}}$. Therefore, the dimensionality of $\mathcal{X}$ can be reduced by choosing $1 \leq k \leq n$ and transforming $\mathcal{X}$ to $\tilde{\mathcal{X}} = \tilde{V}^T_{\mathcal{X}} \mathcal{X}$ where $\tilde{V}_{\mathcal{X}}$ denotes the leftmost k columns of $V_{\mathcal{X}}$. The amount of "information" preserved is typically quantified by

$$100\frac{\sum_{\ell=1}^{k} \lambda^\ell_{\mathcal{X}}}{\sum_{\ell=1}^{n} \lambda^\ell_{\mathcal{X}}}.$$

This is commonly referred to as the percentage of variance captured by $\tilde{\mathcal{X}}$. If this percentage is large, most of the information is preserved in the sense that $\tilde{V}_{\mathcal{X}}\tilde{\mathcal{X}}$ is a good approximation to $\mathcal{X}$. Indeed, if the percentage is 100, *i.e.*, $k = n$, then $\tilde{V}_{\mathcal{X}}\tilde{\mathcal{X}} = \tilde{V}_{\mathcal{X}}\tilde{V}^T_{\mathcal{X}}\mathcal{X} = \mathcal{X}$. The properties of left multiplication to $\mathcal{X}$ by $\tilde{V}_{\mathcal{X}}\tilde{V}^T_{\mathcal{X}}$ have special significance in the eigen-analysis based attacks. We call this transformation, a *projection through* the first k PCs.

In practice, one has a collection of data tuples on which dimensionality reduction via PCA is desired. If the tuples can all be regarded as independent samples from $\mathcal{X}$, PCA can be fruitfully carried out on their standard sample covariance matrix (after subtracting from each the row-mean vector of the dataset). The eigen-analysis based attacks will make critical use of the projection of the dataset through its first k PCs.

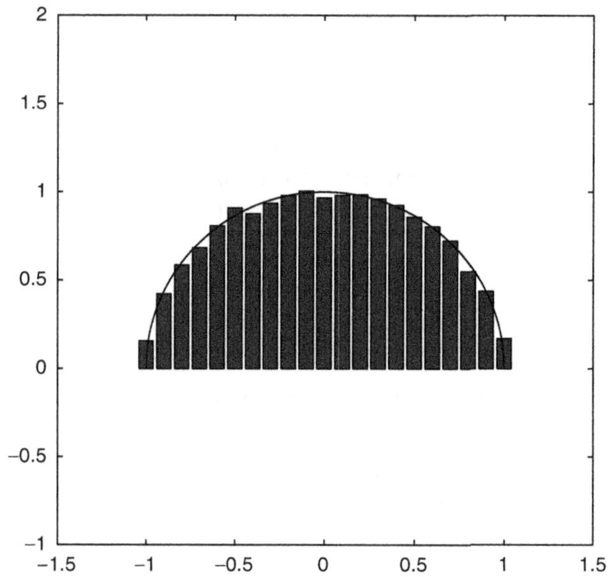

Figure 15.1. Wigner's semi-circle law: a histogram of the eigenvalues of $\frac{A+A'}{2\sqrt{2p}}$ for a large, randomly generated A

15.3.2 Spectral Filtering

This technique, developed by Kargupta *et al.* [22], utilizes the fact that the eigenvalues of a random matrix are distributed in a fairly predictable manner. For example, Wigner's semi-circle law [47] says that if A is a $p \times p$ matrix whose entries were generated independently from a distribution with zero mean and unit variance, then, for large p, the distribution of the eigenvalues of $\frac{A+A'}{2\sqrt{2p}}$ has *p.d.f.* depicted in Figure 15.1; it takes the shape of a semi-circle. As another example, consider $n \times m$ matrix R whose entries were generated independently from a distribution with mean zero and variance σ^2. For large m and n, the distribution of the eigenvalues of the sample covariance matrix of R is similar to the semi-circle law. And, key to the spectral filtering technique, this result allows bounds on these eigenvalues to be computed.

Kargupta *et al.* observe that if the jth eigenvalue arising from Y is "large", it is a good approximation to the jth eigenvalue arising from X. Therefore, the projection of Y through its PCs corresponding to these large eigenvalues (say the first k) is a good approximation to the projection of X through its first k PCs. As such $\hat{X}$ is set to the projection of Y through its first k PCs. Results from matrix perturbation theory and spectral analysis of large random matrices provide the basis for this observation.

LEMMA 15.1 *[40, Corollary 4.9] For any n-dimensional random vectors $\mathcal{X}$ and $\mathcal{R}$ ($\mathcal{R}$ has mean vector zero) and $\mathcal{Y} = \mathcal{X} + \mathcal{R}$, it is the case that: for $1 \le j \le n$, $\lambda^j_{\mathcal{Y}} \in [\lambda^j_{\mathcal{X}} + \lambda^n_{\mathcal{R}}, \lambda^j_{\mathcal{X}} + \lambda^1_{\mathcal{R}}]$.*

Therefore, if $\lambda^j_{\mathcal{Y}} \in [\lambda^n_{\mathcal{R}}, \lambda^1_{\mathcal{R}}]$, then this eigenvalue is largely affected by noise ($\mathcal{R}$). Hence, it is not regarded by Kargupta *et al.* as large and, therefore, not regarded as a good approximation of $\lambda^j_{\mathcal{X}}$. On the other hand, $\lambda^j_{\mathcal{Y}} > \lambda^1_{\mathcal{R}}$ is regarded as large and, therefore, is regarded as a good approximation of $\lambda^j_{\mathcal{X}}$. So how can the attacker use this threshold criterion given only Y?

Let $\hat{\Sigma}_Y$ and $\hat{\Sigma}_R$ be the standard sample covariance matrices computed from Y and R; let $\hat{\lambda}^1_Y \ge \ldots \ge \hat{\lambda}^n_Y$ and $\hat{\lambda}^1_R \ge \ldots \ge \hat{\lambda}^n_R$ be the associated eigenvalues, respectively. The above criterion can be modified to consider $\hat{\lambda}^j_Y > \hat{\lambda}^1_R$ as large. But how should the attacker estimate an upper-bound on $\hat{\lambda}^1_R$? This question is answered using a result from large random matrix theory alluded to in the opening paragraph of this subsection. Intuitively, as R grows large, the eigenvalues computed from R can be bounded by the attacker. And when m is large relative to n, these bounds are quite good. Formally stated [21, 39], as $m, n \to \infty$ and $\frac{m}{n} \to Q \ge 1$,

$$\hat{\lambda}^{max}_R = \sigma^2(1 + 1/\sqrt{Q})^2 \ge \hat{\lambda}^1_R \ge \hat{\lambda}^n_R \ge \hat{\lambda}^{min}_R = \sigma^2(1 - 1/\sqrt{Q})^2.$$

As such, $\hat{\lambda}^{max}_R$ serves as the estimate of an upper-bound on $\hat{\lambda}^1_R$. Moreover, for Q large relative to σ^2, this bound will be quite good as all eigenvalues of $\hat{\Sigma}_R$ will be concentrated in a small band. Since the attacker is assumed to know σ^2, then she can compute $\hat{\lambda}^{max}_R$ and will deem any $\hat{\lambda}^j_Y > \hat{\lambda}^{max}_R$ as large.

The spectral filtering algorithm is given in Algorithm 3. The empirical results show that when the variance of the noise is low and the original data does not contain many inherent random components, the recovered data can be reasonably close to the original data. However, two important questions remain to be answered. 1) What are the theoretical bounds on the estimation accuracy? 2) What are the fundamental factors that determine the quality of the data estimation? The first is touched on in Section 15.3.3 and the second in Section 15.3.4.

15.3.3 SVD Filtering

Guo *et al.* [17] revisited spectral filtering to address the issue of an optimal choice of k and to develop bounds on the estimation accuracy. They showed that when $k = \min\{1 \le j \le n | \hat{\lambda}^j_Y < 2\sigma^2\} - 1$, the estimated data is approximately optimal, *i.e.*, the benefits due to the inclusion of the kth eigenvector is greater than the information loss due to the noise projected along the kth eigenvector. They further proposed a singular value decomposition-based data reconstruction approach, and proved the equivalence of this approach to spectral filtering. A lower bound and upper bound of the estimation error in terms

Protocol 3 Spectral Filtering

Require: Y, the perturbed data matrix and σ^2, the variance of the random noise.

Ensure: $\hat{X}$, an estimate of the original data matrix X.

1: Compute the sample mean of Y and subtract it from every column of Y.

2: Compute the standard sample covariance $\hat{\Sigma}_Y$ of Y, its eigenvalues $\hat{\lambda}_Y^1 \geq \ldots \geq \hat{\lambda}_Y^n$, and their associated normalized eigenvectors $\hat{v}_Y^1, \ldots, \hat{v}_Y^n$.

3: Compute $k = \max\{1 \leq j \leq n | \hat{\lambda}_Y^j > \hat{\lambda}_R^{max}\}$. Let $\tilde{\hat{V}}_Y$ denote the matrix $[\hat{v}_Y^1 \cdots \hat{v}_Y^k]$.

4: Set $\hat{X}$ to $\tilde{\hat{V}}_Y \tilde{\hat{V}}_Y^T Y$.

of Frobenius matrix norm were also derived. We refer readers to [14, 17] for more details.

15.3.4 PCA Filtering

Huang *et al.* [18] observe that a key factor in determining the accuracy of spectral filtering is the degree of correlation that exists among the attributes of $\mathcal{X}$ relative to σ^2. The higher the degree, the greater the accuracy in estimating the original data. Indeed, for small k, the higher the degree of correlation, the more variance will be captured by the first k PCs. The addition of $\mathcal{R}$ does not change this property. The attributes of $\mathcal{R}$ are uncorrelated and thus, the amount of variance captured by *any* direction is the same. Therefore, removing the last $n-k$ PCs of $\mathcal{X}$ does not cause much variance loss but will cause $100\frac{n-k}{n}$ percent of the variance in $\mathcal{R}$ to be lost.

Based on this observation, Huang *et al.* [18] proposed a filtering technique based on PCA. A major difference with spectral filtering, is that PCA filtering does not use matrix perturbation theory and spectral analysis to estimate the dominant PCs of X. Instead PCA filtering takes a more direct approach based on the fact that

$$\Sigma_{\mathcal{Y}} = \Sigma_{\mathcal{X}} + \Sigma_{\mathcal{R}} = \Sigma_{\mathcal{X}} + \sigma^2 I. \tag{15.2}$$

The first equality is due to the independence of $\mathcal{X}$ and $\mathcal{R}$ and the second by assumption. Therefore, the attacker can directly estimate $\Sigma_{\mathcal{X}}$ as $\hat{\Sigma}_Y - \sigma^2 I$, then compute the top k PCs of this. The PCA filtering procedure is given in Algorithm 4.

The original dataset estimate can be written as the sum of two parts: $\hat{X} = \tilde{\hat{V}}_X \tilde{\hat{V}}_X^T Y = \tilde{\hat{V}}_X \tilde{\hat{V}}_X^T X + \tilde{\hat{V}}_X \tilde{\hat{V}}_X^T R$. Therefore, the recovery error [1] is determined

[1] assuming the estimated sample covariance $\hat{\Sigma}_X$ is very close to $\Sigma_{\mathcal{X}}$

Protocol 4 PCA Filtering

Require: Y, the perturbed data matrix; σ^2, the variance of the random noise; and $1 \leq k \leq n$, the number of PCs to keep.

Ensure: $\hat{X}$, an estimate of the original data matrix X.

1: Compute the sample mean of Y and subtract it from every column of Y.

2: Compute the standard sample covariance $\hat{\Sigma}_Y$ of Y, and produce $\hat{\Sigma}_X = \hat{\Sigma}_Y - \sigma^2 I$ an estimate of $\Sigma_{\mathcal{X}}$.

3: Compute the eigenvalues of $\hat{\Sigma}_X$, $\hat{\lambda}_X^1 \geq \ldots \geq \hat{\lambda}_X^n$. Compute their their associated normalized eigenvectors, $\hat{v}_X^1, \ldots, \hat{v}_X^n$. Let $\tilde{\hat{V}}_X$ denote the matrix $[\hat{v}_X^1 \cdots \hat{v}_X^k]$.

4: Set $\hat{X}$ to $\tilde{\hat{V}}_X \tilde{\hat{V}}_X^T Y$.

by the the percentage of variance captured by the first k PCs of $\mathcal{X}$ and the noise. It can be shown that the mean squared recovery error caused by the noise part is $\sigma^2 \frac{k}{n}$. These results echo the empirical results observed in spectral filtering and suggests an approach for choosing k.

15.3.5 MAP Estimation Attack

Different from eigen-analysis, MAP estimation considers both prior and posterior knowledge via Bayes' theorem to estimate original dataset. For each $1 \leq i \leq m$, the attacker will produce $\hat{x}_i$ an estimate of x_i using[2] y_i. Let $f_{\mathcal{X}}$ and $f_{\mathcal{R}}$ denote the *p.d.f* of $\mathcal{X}$ and $\mathcal{R}$, respectively. Given $x \in \mathbb{R}^n$ and $y \in \mathbb{R}^{n'}$, let $f_{\mathcal{X}|\mathcal{Y}=y}$ and $f_{\mathcal{Y}|\mathcal{X}=x}$ denote the *p.d.f* of $\mathcal{X}$ conditioned on $\mathcal{Y} = y$ and the *p.d.f* of $\mathcal{Y}$ conditioned on $\mathcal{X} = x$, respectively. The MAP estimate of x_i is[3]

$$\begin{aligned}
\hat{x}_i &= argsup\{f_{\mathcal{X}|\mathcal{Y}=y_i}(x) : x \in \mathbb{R}^n\} \\
&= argsup\{f_{\mathcal{Y}|\mathcal{X}=x}(y_i) f_{\mathcal{X}}(x) : x \in \mathbb{R}^n\} \\
&= argsup\{f_{\mathcal{R}}(y_i - x) f_{\mathcal{X}}(x) : x \in \mathbb{R}^n\}. \qquad (15.3)
\end{aligned}$$

The second equality is due to Bayes' theorem and the third due to the fact that $\mathcal{Y} = \mathcal{X} + \mathcal{R}$ and $\mathcal{R}$ is independent of $\mathcal{X}$.

Huang *et al.* [18] considered the case where both $f_{\mathcal{X}}$ and $f_{\mathcal{R}}$ are multivariate normal (and the attacker knows this). The following closed form expression can then be derived with $\mu_{\mathcal{X}}$ denoting the mean vector of $\mathcal{X}$.

$$\hat{x}_i = (\Sigma_{\mathcal{X}}^{-1} + (1/\sigma^2)I)^{-1}(\Sigma_{\mathcal{X}}^{-1}\mu_{\mathcal{X}} + y_i/\sigma^2).$$

[2]Due to independence, the attacker will gain nothing more if using all of Y.

[3]Here $argsup\{\}$ is based on $supA$ which denotes the smallest upper bound on a set A (if A is upper-bounded, $supA$ always exists.

The assumption that $f_{\mathcal{X}}$ is multi-variate normal and known to the attacker is quite strong. Other cases are worth comment (in each, $f_{\mathcal{R}}$ is multi-variate normal and known to the attacker). When $f_{\mathcal{X}}$ is known but not multivariate normal, it may be difficult to derive a closed-form expression for $\hat{x}_i$. In this case, the attacker can use numerical methods such as Newton's gradient descent methods. When $f_{\mathcal{X}}$ is not known, the MAP estimate reduces to the maximum likelihood estimate (MLE) by assuming $f_{\mathcal{X}}$ is uniform over some interval. Therefore, $f_{\mathcal{X}}$ can be dropped from (15.3) and $\hat{x}_i = y_i$. However, this estimate may suffer from accuracy problems due to dropping $f_{\mathcal{X}}$.

It is worth noting that the MAP approach has been widely studied in statistical disclosure control. For example, Trottini *et al.* [44] used this approach to study the linkage privacy breaches in the scenario where microdata is masked by both additive and multiplicative noise. In their settings, the attacker tries to identify the identity (of a person) linked to a specific record, which is different from the primary focus of this chapter - data record recovery.

15.3.6 Distribution Analysis Attack

Recall that techniques exist for estimating $f_{\mathcal{X}}$ from Y. This is quite useful as $f_{\mathcal{X}}$ represents a useful data mining pattern. However, in some cases, this reconstructed distribution can be used by the attacker to gain extra knowledge about the private data. For example, assume the each entry of $\mathcal{R}$ is uniformly distributed over $[-1, 1]$ and the observed perturbed data $\mathcal{Y} = 1$. If there is no additional information, the attacker can determine $\mathcal{X} \in [0, 2]$. However, if a large amount of data is available, the reconstructed distribution will have a high degree of accuracy. Assume the attacker can perfectly recover $f_{\mathcal{X}}$ which is:

$$f_{\mathcal{X}}(x) = \begin{cases} 0.5, & 0 \leq x \leq 1; \\ 0.5, & 5 \leq x \leq 6; \\ 0, & \text{otherwise.} \end{cases}$$

Then, the estimate of $\mathcal{X}$ given $\mathcal{Y} = 1$ is localized to a smaller interval $[0, 1]$ instead of $[0, 2]$. When data has a multi-variate distribution, the attacker can determine intervals $I_1, I_2, \ldots, I_n$, which are narrow in one or more dimensions, and for which the number of data records that fall in the interval is very small. Such intervals make outliers/minorities more identifiable than they would seem when merely looking at the perturbed data set. This kind of disclosure leads to a bigger open problem - *when do data mining results cause privacy breach?* Further discussions can be found in [4, 9, 31, 16, 12].

15.3.7 Summary

This section surveyed recent research that investigated the vulnerability additive data perturbation. The research showed, in many cases, the private information can be reasonably well derived from the perturbed data. The primary attack techniques presented are summarized in Table 15.1.

Table 15.1. Summarization of Attacks on Additive Perturbation

Categories	*Related Work*	*General Assumptions*
Eigen-Analysis	[14, 17, 18, 22]	the degree of correlation between the original data attributes is high relative to σ^2
MAP Estimation	[18]	data and noise arose from a multi-variate normal distribution
Distribution Analysis	[4, 9, 16]	reconstructed distribution describes the original data with sufficient accuracy

One possible improvement on additive perturbation is to use colored noise with similar correlation structure to the original data [23, 43], *i.e.*, $\mathcal{R} \sim (0, \Sigma_{\mathcal{R}})$, where $\Sigma_{\mathcal{R}} = \beta\Sigma_{\mathcal{X}}$ for $\beta > 0$. With this method, the covariance of the perturbed data is

$$\Sigma_{\mathcal{Y}} = \Sigma_{\mathcal{X}} + \beta\Sigma_{\mathcal{X}} = (1+\beta)\Sigma_{\mathcal{X}}.$$

The correlation coefficients of the perturbed attributes are the same as that of the original attributes:

$$\rho_{\mathcal{Y}_i,\mathcal{Y}_j} = \frac{1+\beta}{1+\beta}\frac{Cov(\mathcal{X}_i,\mathcal{X}_j)}{\sqrt{Var(\mathcal{X}_i)Var(\mathcal{X}_j)}} = \rho_{\mathcal{X}_i,\mathcal{X}_j}.$$

This kind of perturbation puts noise on the principal components of the original data, therefore, separating noise from the data using eigen analysis becomes difficult. However, this approach is not free from problem either. Domingo-Ferrer *et al.* [9] pointed out that the reconstructed distribution (using their p-dimensional reconstruction algorithm, a multivariate generalization of the approach describe in [5] for the univariate case) may still lead to disclosure in some cases. The higher the dimensionality, the more likely is the disclosure.

In summary, additive perturbation has its roots in statistical disclosure control. It offers a simply way to mask private data while allowing aggregate statistics to be queried; and making more sophisticated privacy preserving data mining possible. However, recent work from PPDM community has shown this technique vulnerable to attack in many cases (*e.g.*, high correlations between many attributes). Therefore, careful attention must be paid when applying this technique in practice.

Before closing this section, we note that several researchers have proposed privacy metrics *e.g.*, interval-based [5], entropy-based [4], mixture models [49]. However, the relationship between these and the recovery accuracy of the attack techniques is not clear.

15.4 Attacking Matrix Multiplicative Data Perturbation

The data owner replaces the original data X with

$$Y = MX, \tag{15.4}$$

where M is an $n' \times n$ matrix chosen to have certain useful properties. If M is orthogonal ($n' = n$ and $M^T M = I$) [7, 36, 37], then the perturbation exactly preserves Euclidean distances, *i.e.*, for any columns x_1, x_2 in X, their corresponding columns y_1, y_2 in Y satisfy $||x_1 - x_2|| = ||y_1 - y_2||$.[4] If each entry of M is generated independently from the same distribution with mean zero and variance σ^2 (n' not necessarily equal to n) [28, 30], then the perturbation approximately preserves Euclidean distances on expectation up to constant factor $\sigma^2 n'$. If M is the product of a discrete cosine transformation matrix and a truncated perturbation matrix [33], then the perturbation approximately preserves Euclidean distances.

Because matrix multiplicative perturbation preserves Euclidean distance with either small or no error, it allows many important data mining algorithms to be applied to the perturbed data and produce results very similar to, or exactly the same as those produced by the original algorithm applied to the original data, *e.g.*, hierarchical clustering, k-means clustering. However, the issue of how well X is hidden is not clear and deserves careful study. Without any prior knowledge, an attacker can do very little (if anything) to accurately recover X. However, no prior knowledge seems an unreasonable assumption in many situations. Motivated by this line of reasoning, several researchers have investigated the vulnerabilities of matrix multiplicative perturbation using various forms of prior knowledge [8, 15, 28–30]. In the bulk of this section (15.4.1 and 15.4.2), we discuss attack techniques based on two types of prior knowledge.

1 **Known input-output (I/O):** The attacker knows some small collection of original data records and the attacker knows the mapping between these known original data records and their perturbed counterparts in Y. In other words, the attacker has a set of input-output pairs.

2 **Known sample:** The attacker has a collection of independent samples (columns of S) from $\mathcal{X}$ (S may or may not overlap with X).

The first two attacks are based on the known I/O prior knowledge assumption. The first one [29] assumes an orthogonal perturbation matrix while the

[4] Conversely, any function $T : \mathbb{R}^n \rightarrow \mathbb{R}^n$ which preserves Euclidean distance (for all $x, y \in \mathbb{R}^n$, $||x - y|| = ||T(x) - T(y)||$) and fixes the origin is equivalent to left-multiplication by an $n \times n$ orthogonal matrix.

second [28] assumes a randomly generated perturbation matrix. The third attack is based on the known sample prior knowledge assumption and assumes an orthogonal perturbation matrix. It works by examining certain features of the original and perturbed data distributions (*i.e.*, the *p.d.f.* of $\mathcal{X}$ and $\mathcal{Y}$), namely the eigenvectors of $\Sigma_\mathcal{X}$ and $\Sigma_\mathcal{Y}$. These features have two important properties: (i) they are related to each other in a natural way allowing M to be estimated, and (ii) they can be accurately extracted from S and Y.

Before moving on, we emphasize the fact that the perturbation technique considered here, matrix multiplicative, is completely different than multiplicative data perturbation mentioned in the introduction. There each element of X is *separately multiplied* by a randomly generated number.

15.4.1 Known I/O Attacks

Without loss of generality, the attacker is assumed to know X_p $(1 \leq p < m)$, the first p columns of X (of course, the attacker also knows Y_p, the first p columns of Y). In other words, the attacker knows a set of input/output pairs (x_1, y_1), ..., (x_p, y_p) where $y_j = Mx_j$.

Orthogonal Perturbation Matrix. Liu *et al.* [29] assumed M is orthogonal. Unlike all other attacks in this chapter, they *do not assume* that the original data records arose as independent samples from $\mathcal{X}$. Their attacker uses Y_p and X_p to produce, $\hat{M}$, an estimation of M. Then, for any $p \leq i \leq m$, the attacker will produce $\hat{x}_i$, an estimation of x_i as

$$\hat{x}_i = \hat{M}^T y_i. \tag{15.5}$$

The rationale for (15.5) is: if $\hat{M} \approx M$, then $\hat{x}_i \approx M^T y_i = M^T(Mx_i) = x_i$. In choosing $\hat{M}$, the attacker knows that M must be in $\mathbb{M}(X_p, Y_p)$, the set of all $n \times n$, orthogonal matrices, O, such that $OX_p = Y_p$. However, with no additional information for further narrowing down this space of the possibilities, the attacker will assume each is equally likely to be M. Therefore, she will choose $\hat{M}$ uniformly from $\mathbb{M}(X_p, Y_p)$.

Given an error tolerance $\epsilon > 0$, the attacker's success probability, $\rho(x_i, \epsilon)$, is defined as the probability that the relative Euclidean distance between x_i and $\hat{x}_i$ is no larger than ϵ, *i.e.*, $Pr(||\hat{x}_i - x_i|| \leq ||x_i||\epsilon)$. Liu *et al.* developed closed form expression

$$\rho(x_i, \epsilon) = \begin{cases} \left(\frac{1}{\pi}\right) 2arcsin\left(\frac{||x_i||\epsilon}{2d(x_i, X_p)}\right) & \text{if } ||x_i||\epsilon < 2d(x_i, X_p); \\ 1 & \text{otherwise,} \end{cases} \tag{15.6}$$

where $d(x_i, X_p)$ denotes the Euclidean distance of x_i to the space of vectors spanned by the columns of X_p, *i.e.*, $inf\{||x - x_i||{:}x$ is in the column space of $X_p\}$. Equation (15.6) illustrates that the sensitivity of a tuple, x_i, to breach depends upon its length relative to its distance to the column space of X_p, *i.e.*, $\frac{||x_i||}{2d(x_i,X_p)}$. Tuples whose relative length is large are particularly sensitive to breach. In particular when x_i is in the column space of X_p, the attacker's success probability equals one. Liu *et al.* also described how the attacker can compute $||x_i||$ and $d(x_i, X_p)$ for any $p \leq i \leq m$, and therefore, determine which tuple is most sensitive to breach.

Chen *et al.* [8] also discussed a known I/O attack technique. They however consider a combination of matrix multiplicative and additive perturbation: $Y = MX + R$. They considered the case when the number of linearly independent data tuples (columns in X_p) is no smaller than the data dimensionality, n (rows in X_p). They pointed out that $\hat{M}$, an estimate of M, can be produced using linear regression, then x_i estimated as $\hat{M}^{-1}y_i$.

Random Perturbation Matrix. Liu [28] developed a MAP-based known I/O attack which works under the assumption that M is an $n' \times n$ matrix whose entries were generated independently from a normal distribution with mean zero and variance σ^2 (n' may be $\leq n$ or $> n$).[5] The larger n' is, the more closely preserved are Euclidean distances between data tuples (up to constant factor $\sigma^2 n'$), but, the better the known I/O attack will work at breaching privacy. Therefore, a trade-off must be balanced in setting n'.

For simplicity, we assume that the columns of Y_p are linearly independent.[6] For any $p \leq i \leq m$, the attacker will produce $\hat{x}_i$ an estimate of x_i. If x_i is linearly dependent on the columns of X_p, the attacker can discover this as y_i will be linearly dependent on the columns of Y_p. In this case, the attacker will set $\hat{x}_i = X_p(Y_p^T Y_p)^{-1} Y_p^T y_i$ which equals x_i (perfect recovery).[7] Henceforth, we assume x_i is linearly independent of the columns of X_p. Therefore, the attacker will only consider estimates, $\hat{x} \in \mathbb{R}^n$, which are also linearly independent of the columns of X_p (for brevity, we write "l.i. $\hat{x}$" to mean that $\hat{x}$ is linearly independent of the columns of X_p). Finally, since the columns of Y_p are assumed to be linearly independent, then it follows that the columns of X_p are too.

Let $\mathcal{M}$ be an $n' \times n$ matrix of random variables each independently and identically distributed as normal with mean zero and variance σ^2. The columns of Y arose as independent samples from random vector $\mathcal{Y} = \mathcal{M}\mathcal{X}$. Using the

---

[5] They do assume that the original data records arose as independent samples from $\mathcal{X}$.

[6] This assumption is not essential. It can be eliminated at the cost of a more complicated attack algorithm. However, the fundamental idea remains the same.

[7] There exists $z_i \in \mathbb{R}^p$ such that $X_p z_i = x_i$ and $Y_p z_i = y_i$. Since the columns of Y_p are assumed to be linearly independent, then by [13, pg. 96], the matrix $(Y_p^T Y_p)^{-1} Y_p^T$ exists. Thus, $X_p(Y_p^T Y_p)^{-1} Y_p^T y_i = X_p(Y_p^T Y_p)^{-1}(Y_p^T Y_p) z_i = X_p z_i = x_i$.

MAP approach, the attacker will choose l.i. $\hat{x}$ so as to maximize the likelihood that $\mathcal{X}$ equals $\hat{x}$ given that $\mathcal{Y}$ equals y_i and $\mathcal{M}X_p$ equals Y_p. This analysis is based on the following key observation (whose proof follows directly from manipulating moment-generating functions). For any matrix B, let $\overline{B}$ denote the column vector which results from stacking the columns of B.

THEOREM 15.2 *For any $n \times q$ matrix A with linearly independent columns, $\overline{\mathcal{M}A}$ is distributed as an (qn')-variate Gaussian with mean vector zero and covariance matrix*

$$\Sigma_{\overline{\mathcal{M}A}} = \sigma^2 \begin{bmatrix} A^TA & 0 & 0 & \cdots & 0 \\ 0 & A^TA & 0 & \cdots & 0 \\ 0 & 0 & A^TA & \cdots & 0 \\ \vdots & \vdots & \vdots & \ddots & \vdots \\ 0 & 0 & 0 & \cdots & A^TA \end{bmatrix}$$

Let $[X_p, \hat{x}]$ and $[Y_p, y_i]$ denote matrices which result from attaching $\hat{x}$ and y_i as an additional right-most column onto X_p and Y_p. Observe that $[X_p, \hat{x}]$ has linearly independent columns. Let $f_{\mathcal{X}|\mathcal{Y}=y_i, \overline{\mathcal{M}X_p}=\overline{Y_p}}$ denote the *p.d.f.* of $\mathcal{X}$ conditioned on $\mathcal{Y} = y_i$ and $\overline{\mathcal{M}X_p} = \overline{Y_p}$; let $f_{\overline{\mathcal{M}[X_p,\hat{x}]}}$ denote the *p.d.f.* of $\overline{\mathcal{M}[X_p, \hat{x}]}$. Using the MAP approach, the attacker will choose

$$\hat{x}_i = argsup\{f_{\mathcal{X}|\mathcal{Y}=y_i, \overline{\mathcal{M}X_p}=\overline{Y_p}}(\hat{x}) : \text{l.i. } \hat{x} \in \mathbb{R}^n\}.$$

Using Bayes' rule, it can be shown that

$$\hat{x}_i = argsup\{f_{\overline{\mathcal{M}[X_p,\hat{x}]}}(\overline{[Y_p, y_i]}) f_{\mathcal{X}}(\hat{x}) : \text{l.i. } \hat{x} \in \mathbb{R}^n\},$$

thus, Theorem 15.2 implies

$$\hat{x}_i = argsup\{\phi(\overline{[Y_p, y_i]}) f_{\mathcal{X}}(\hat{x}) : \text{l.i. } \hat{x} \in \mathbb{R}^n\}, \tag{15.7}$$

where ϕ is the $((p+1)n')$-variate Gaussian distribution with mean vector zero and covariance matrix $\Sigma_{\overline{\mathcal{M}[X_p,\hat{x}]}}$. For simplicity we assume that the attacker knows nothing about $f_{\mathcal{X}}$ and, following a common practice, uses a uniform distribution over some interval in place of $f_{\mathcal{X}}$ in (15.7).[8] Thus,

$$\hat{x}_i = argsup\{\phi(\overline{[Y_p, y_i]}) : \text{l.i. } \hat{x} \in \mathbb{R}^n\}. \tag{15.8}$$

Producing a closed-form expression for $\hat{x}_i$ in (15.8) is desirable, but quite difficult. Instead, the attacker can turn to numerical approaches. Experiments

[8] A more complicated approach could have the attacker using the fact that the columns of X_p arose as independent samples from $\mathcal{X}$, and use X_p to inform a better substitution for $f_{\mathcal{X}}$ in (15.7).

were reported in [28] where the attacker used the Matlab implementation[9] of the Nelder-Mead simplex algorithm [35] to solve this optimization problem. The results show that the accuracy of the attack technique increases with n' or the number of known input-output pairs.

15.4.2 Known Sample Attack

The attacker is assumed to know a collection of independent samples (columns of S) from $\mathcal{X}$ (S may or may not overlap with X). Furthermore, the attacker assumes M is orthogonal.

The approach is based on the observation that the eigenvectors of $\mathcal{Y}$ are equal to those of $\mathcal{X}$ *left-multiplied by* M (up to a factor of ± 1). Therefore by estimating $\Sigma_{\mathcal{Y}}$ and $\Sigma_{\mathcal{X}}$ and matching their eigenvectors, the attacker can produce, $\hat{M}$, an estimation of M. Using this, data record x_i ($1 \leq i \leq m$) is estimated as $\hat{x}_i = \hat{M}^T y_i$.

The following results (proved in [29]) establishes the key match between the normalized eigenspaces.

THEOREM 15.3 *The eigenvalues of $\Sigma_{\mathcal{X}}$ and $\Sigma_{\mathcal{Y}}$ are the same and for all $1 \leq j \leq n$, $M\mathbb{V}^j_{\mathcal{X}} = \mathbb{V}^j_{\mathcal{Y}}$, where $M\mathbb{V}^j_{\mathcal{X}}$ equals $\{Mv : v \in \mathbb{V}^j_{\mathcal{X}}\}$.*

COROLLARY 15.4 *Let $\mathbb{I}_n$ be the space of all $n \times n$, matrices with each diagonal entry ± 1 and each off-diagonal entry 0 (2^n matrices in total). There exists $D_0 \in \mathbb{I}_n$ such that $M = V_{\mathcal{Y}} D_0 V^T_{\mathcal{X}}$.*

First assume that the attacker knows the covariance matrices $\Sigma_{\mathcal{X}}$ and $\Sigma_{\mathcal{Y}}$ and, thus, computes $V_{\mathcal{X}}$ and $V_{\mathcal{Y}}$. By Corollary 15.4, the attacker can perfectly recover M if she can choose the right D from $\mathbb{I}_n$. To do so, the attacker utilizes S and Y, in particular, the fact that these arose as independent samples from $\mathcal{X}$ and $\mathcal{Y} = M\mathcal{X}$. For any $D \in \mathbb{I}_n$, if $D = D_0$, then $V_{\mathcal{Y}} D V^T_{\mathcal{X}} S$ and Y have both arisen as independent samples from $\mathcal{Y}$. The attacker will estimate M as $\hat{M} = V_{\mathcal{Y}} D V^T_{\mathcal{X}}$, where D was chosen from $\mathbb{I}_n$ so as to maximize the likelihood that $V_{\mathcal{Y}} D V^T_{\mathcal{X}} S$ and Y arose from the same random vector. To make this choice, the attacker can use a multi-variate two-sample hypothesis test for equal distributions [42]. The smaller the p-value, the more convincingly the null hypothesis (that $V_{\mathcal{Y}} D V^T_{\mathcal{X}} S$ and Y have both arisen as independent samples from $\mathcal{Y}$) can be rejected. Therefore, $D \in \mathbb{I}_n$ is chosen to maximize the p-value.

Finally, the attacker can eliminate the assumption at the start of the previous paragraph by replacing $\Sigma_{\mathcal{X}}$ and $\Sigma_{\mathcal{Y}}$ with estimates computed from S and Y. Using the standard sample covariance matrices, the pseudo-code for the attack technique is shown in algorithm 5. A weakness lies in its computation cost, $O(2^n(m+p)^2)$. For high-dimensional data, the technique is infeasible.

[9] http://www.mathworks.com/access/helpdesk/help/techdoc/ref/fminsearch.html

Protocol 5 Eigen-Analysis Attack

Require: Y, the perturbed data matrix and S, the sample data matrix.
Ensure: $\hat{X}$, an estimate of the original data matrix X.

1: Compute standard, sample covariance matrices of S and Y and $\hat{V}_{\mathcal{X}}$ and $\hat{V}_{\mathcal{Y}}$ their normalized eigenvector matrices.
2: Choose $D \in \mathbb{I}_n$ so as to maximize the p-value of two-sample hypothesis test for equal distributions on $\hat{V}_{\mathcal{Y}} D \hat{V}_{\mathcal{X}}^T S$ and Y.
3: Set $\hat{M}$ to $\hat{V}_{\mathcal{Y}} D \hat{V}_{\mathcal{X}}^T$ and $\hat{X}$ to $\hat{M}^T Y$.

It should be noted the eigen-analysis attack does not work if each entry of M were generated independently from some distribution with mean zero and variance σ^2. In that case, $\Sigma_{\mathcal{Y}}$ will equal γI for some constant $\gamma > 0$, thereby killing any useful matching like that in Theorem 15.3.

15.4.3 Other Attacks Based on ICA

Before finishing the section, we briefly describe some attacks based on independent component analysis (ICA) [19].

ICA Overview. Given an n'-variate random vector $\mathcal{V}$, one common ICA model posits that this random vector was generated by a linear combination of independent random variables, *i.e.*, $\mathcal{V} = A\mathcal{S}$ with $\mathcal{S}$ an n-variate random vector with independent components. Typically, $\mathcal{S}$ is further assumed to satisfy the following additional assumptions: (i) at most one component is distributed as a Gaussian; (ii) $n' \geq n$; and (iii) A has rank n.

One common scenario in practice: there is a set of unobserved samples (the columns of $n \times q$ matrix S) that arose from $\mathcal{S}$ which satisfies (i) - (iii) and whose components are independent. But observed is $n' \times q$ matrix V whose columns arose as linear combination of the rows of S. The columns of V can be thought of as samples that arose from a random vector $\mathcal{V}$ which satisfies the above generative model. There are ICA algorithms whose goal is to recover S and A up to a row permutation and constant multiple. This ambiguity is inevitable due to the fact that for any diagonal matrix (with all non-zeros on the diagonal) D, and permutation matrix P, if A, S is a solution, then so is $(ADP), (P^{-1}D^{-1}S)$.

Other Attacks. Liu *et al.* [30] considered matrix multiplicative data perturbation where M is an $n' \times n$ matrix with each entry generated independently from the some distribution with mean zero and variance σ^2. They discussed the application of the above ICA approach to estimate X directly from Y: $\mathcal{S} = \mathcal{X}, \mathcal{V} = \mathcal{Y}, S = X, V = Y$, and $A = M$. They argued the approach to be problematic because the ICA generative model imposes assumptions not likely to hold in many practical situations: the components of $\mathcal{X}$ are independent with at most one such being Gaussian distributed. Moreover, they pointed out that the row permutation and constant multiple ambiguity further hampers accurate recovery of X. A similar observation is made later by Chen *et al.* [8].

Guo and Wu [15] considered matrix multiplicative perturbation assuming only that M is an $n \times n$ matrix (orthogonal or otherwise). Further they assumed a weaker variant of the known I/O holds: the attacker knows, $\widetilde{X}$, a collection of original data columns from X but does not know to which of the columns in Y these correspond. They develop an ICA-based attack technique for estimating the remaining columns in X. To avoid the ICA problems described in the previous paragraph, they instead applied ICA *separately* to $\widetilde{X}$ and Y producing representations $(A_{\widetilde{X}}, S_{\widetilde{X}})$ and (A_Y, S_Y). They argued that these representations are related in a natural way allowing X to be estimated. Their approach is similar in spirit to the known sample attack described earlier which related S and Y through representations derived through eigen-analysis.

15.4.4 Summary

This section discussed the vulnerabilities of matrix multiplicative data perturbation to certain attacks based on prior knowledge. The primary attack techniques discussed are summarized in Table 15.2.[10]

Table 15.2. Summarization of Attacks on Matrix Multiplicative Perturbation

Categories	*Related Work*	*General Assumptions*
Linear algebra/measure theory	[29]	known I/O, M is orthogonal
MAP Estimation	[28]	known I/O, M is $n' \times n$ with entries generated independently from $\mathcal{N}(0, \sigma^2)$,
Eigen-Analysis	[29]	known sample, M is orthogonal,
ICA	[8, 30]	M has rank n, the data attributes are largely independent and at most one is Gaussian
ICA	[15]	M is $n \times n$, weak known I/O

Chen *et al.* [8] discussed a modification of matrix multiplicative data perturbation to improve its resilience to attack. They examine the combination of matrix multiplicative and additive data perturbation. They argue that this approach offers additional privacy protection, but the utility of the perturbed data

[10] All the attack techniques, except known I/O with orthogonal M, implicitly assume that the original data records arose independently from $\mathcal{X}$.

is negatively affected since additive noise does not preserve Euclidean distance well.

15.5 Attacking *k*-Anonymization

Before concluding this chapter, we briefly survey a very recent body of research aimed at analyzing the vulnerabilities of the popular k-anonymity model [38, 41]. Here, the private data X is perturbed such that each of the resulting records is identical to at least $k-1$ others with respect to a pre-defined set of attributes called *quasi-identifiers*. All of the other attributes are called *sensitive attributes* and these are not modified by the perturbation. This perturbation can be carried out by judicious *value generalization* (*e.g.*, zip 95120 $\rightarrow$ 951**) or *tuple suppression*, and it is aimed at preventing linkage attacks through the quasi-identifiers.

Recently, Machanavajjhala *et al.* [32] developed a background knowledge attack on k-anonymity which we call a *homogeneity attack*. They showed how a lack of diversity among the sensitive attribute values can be used to establish a linkage between individuals and sensitive values. To remedy this problem, they proposed a new privacy definition called l-diversity such that in each equivalence class there are at least l "well-represented" sensitive values. Along the same line, Wong *et al.* [48] proposed an (α, k)-anonymization model such that the relative frequency of the sensitive value in every equivalence class is less than or equal to α. Li *et al.* [25] later developed attacks on l-diversity (*skewness attack* and *similarity attack*), and argued that l-diversity is neither necessary nor sufficient to prevent attribute disclosure. To cope with these problems, they proposed an improved framework called t-closeness, which requires the distribution of a sensitive attribute in any equivalence class to be close to the distribution of the attribute in the original data set.

Wang *et al.* [46] considered the privacy breach caused by the attacker's data mining capabilities. They presented an approach (that combines association rule hiding and k-anonymity) to limit the confidence of inferring sensitive properties about the existing individuals.

Aggarwal [2] also argued the original k-anonymity model to be problematic. He considered the case of high dimensional data and pointed out that the exponential number of quasi-identifier combinations can allow precise inference attacks unless an unacceptably high amount of information loss is suffered.

15.6 Conclusion

This chapter provides a detailed survey of attack techniques on additive and matrix multiplicative perturbation. It also presents a brief overview of attacks on k-anonymization. These attacks offer insights into vulnerabilities data perturbation techniques under certain circumstances. In summary, the following information could lead to disclosure of private information from the perturbed data.

1. Attribute Correlation: Many real world data has strong correlated attributes, and this correlation can be used to filter off additive white noise. See, *e.g.*, [14, 17, 18, 22].

2. Known Sample: Sometimes, the attacker has certain background knowledge about the data such as the *p.d.f.* or a collection of independent samples which may or may not overlap with the original data. See, *e.g.*, [28, 29, 18].

3. Known Inputs/Outputs: Sometimes, the attacker knows a small set of private data and their perturbed counterparts. This correspondence can help the attacker to estimate other private data. See, *e.g.*, [28, 15, 29].

4. Data Mining Results: The underlying pattern discovered by data mining also provides a certain level of knowledge which can be used to guess the private data to a higher level of accuracy. See, *e.g.*, [4, 9, 31, 16, 12, 46].

5. Sample Dependency: Most of the attacks (except the known I/O developed by [29]) discussed in this chapter assume the data as independent samples from some unknown distribution. This assumption may not hold true for all real applications. For certain types of data, such as the time series data, there exists auto correlation/dependency among the samples. How this dependency can help the attacker to estimate the original data is still an open problem.

Notes

The contributions of C. Giannella and K. Liu were equal.

Acknowledgements

The authors wish to thank the U.S. National Science Foundation for their support through awards IIS-0329143 and IIS-0093353. The authors also wish to thank Kamalika Das, Souptik Datta, and Ran Wolff for their assistance.

References

[1] N. R. Adam and J. C. Worthmann. Security-control methods for statistical databases: a comparative study. *ACM Computing Surveys (CSUR)*, 21(4):515–556, 1989.

[2] Charu C. Aggarwal. On k-anonymity and the curse of dimensionality. In *Proceedings of the 31st VLDB Conference*, pages 901–909, Trondheim, Norway, 2005.

[3] Charu C. Aggarwal and Philip S. Yu. A condensation based approach to privacy preserving data mining. In *Proceedings of the 9th International*

Conference on Extending Database Technology (EDBT'04), pages 183–199, Heraklion, Crete, Greece, March 2004.

[4] D. Agrawal and C. C. Aggarwal. On the design and quantification of privacy preserving data mining algorithms. In *Proceedings of the 20th ACM SIGMOD-SIGACT-SIGART Symposium on Principles of Database Systems*, pages 247–255, Santa Barbara, CA, 2001.

[5] R. Agrawal and R. Srikant. Privacy-preserving data mining. In *Proceedings of the ACM SIGMOD Conference on Management of Data*, pages 439–450, Dallas, TX, May 2000.

[6] R. Brand. Microdata protection through noise addition. *Lecture Notes in Computer Science - Inference Control in Statistical Databases*, 2316:97–116, 2002.

[7] K. Chen and L. Liu. Privacy preserving data classification with rotation perturbation. In *Proceedings of the 5th IEEE International Conference on Data Mining (ICDM'05)*, pages 589–592, Houston, TX, November 2005.

[8] K. Chen, G. Sun, and L. Liu. Towards attack-resilient geometric data perturbation. In *Proceedings of the 2007 SIAM International Conference on Data Mining (SDM'07)*, Minneapolis, MN, April 2007.

[9] J. Domingo-Ferrer, F. Sebé, and J. Castellà-Roca. On the security of noise addition for privacy in statistical databases. *Privacy in Statistical Databases*, LNCS3050:149–161, 2004.

[10] A. Evfimevski, J. Gehrke, and R. Srikant. Limiting privacy breaches in privacy preserving data mining. In *Proceedings of the ACM SIGMOD/PODS Conference*, San Diego, CA, June 2003.

[11] S. E. Fienberg and J. McIntyre. Data swapping: Variations on a theme by dalenius and reiss. Technical report, National Institute of Statistical Sciences, Research Triangle Park, NC, 2003.

[12] A. Friedman, R. Wolff, and A. Schuster. Providing k-anonymity in data mining. *Journal of VLDB*, 2006 (to be published).

[13] G. Strang. *Linear Algebra and Its Applications (3rd Ed.)*. Harcourt Brace Jovanovich College Publishers, New York, 1986.

[14] S. Guo and X. Wu. On the use of spectral filtering for privacy preserving data mining. In *Proceedings of the 21st ACM Symposium on Applied Computing*, pages 622–626, Dijon, France, April 2006.

[15] S. Guo and X. Wu. Deriving private information from arbitrarily projected data. In *Proceedings of the 11th Pacific-Asia Conference on Knowledge Discovery and Data Mining (PAKDD'07)*, Nanjing, China, May 2007.

[16] S. Guo, X. Wu, and Y. Li. Deriving private information from perturbed data using iqr based approach. In *Proceedings of the Second International Workshop on Privacy Data Management (PDM'06)*, Atlanta, GA, April 2006.

[17] S. Guo, X. Wu, and Y. Li. On the lower bound of reconstruction error for spectral filtering based privacy preserving data mining. In *Proceedings of the 10th European Conference on Principles and Practice of Knowledge Discovery in Databases (PKDD'06)*, pages 520–527, Berlin, Germany, September 2006.

[18] Z. Huang, W. Du, and B. Chen. Deriving private information from randomized data. In *Proceedings of the 2005 ACM SIGMOD Conference*, pages 37–48, Baltimroe, MD, June 2005.

[19] A. Hyvärinen and E. Oja. Independent component analysis: Algorithms and applications. *Neural Networks*, 13(4):411–430, June 2000.

[20] I. T. Jolliffe. *Principal Component Analysis.* Springer Series in Statistics. Springer, second edition, 2002.

[21] D. Jonsson. Some limit theorems for the eigenvalues of a sample covariance matrix. *Journal of Multivariate Analysis*, 12:1–38, 1982.

[22] H. Kargupta, S. Datta, Q. Wang, and K. Sivakumar. On the privacy preserving properties of random data perturbation techniques. In *Proceedings of the IEEE International Conference on Data Mining (ICDM'03)*, pages 99–106, Melbourne, FL, November 2003.

[23] J. Kim. A method for limiting disclosure in microdata based on random noise and transformation. In *Proceedings of the American Statistical Association on Survey Research Methods*, pages 370–374, Washington, DC, 1986.

[24] J. J. Kim and W. E. Winkler. Multiplicative noise for masking continuous data. Technical Report Statistics #2003-01, Statistical Research Division, U.S. Bureau of the Census, Washington D.C., April 2003.

[25] N. Li, T. Li, and S. Venkatasubramanian. t-closeness: Privacy beyond k-anonymity and l-diversity. In *Proceedings of the 23rd International Conference on Data Engineering (ICDE'07)*, pages 106–115, Istanbul, Turkey, April 2007.

[26] X.-B. Li and S. Sarkar. A tree-based data perturbation approach for privacy-preserving data mining. *IEEE Transactions on Knowledge and Data Engineering (TKDE)*, 18(9):1278–1283, 2006.

[27] C. K. Liew, U. J. Choi, and C. J. Liew. A data distortion by probability distribution. *ACM Transactions on Database Systems (TODS)*, 10(3):395–411, 1985.

[28] K. Liu. *Multiplicative Data Perturbation for Privacy Preserving Data Mining*. PhD thesis, University of Maryland, Baltimore County, Baltimore, MD, January 2007.

[29] K. Liu, C. Giannella, and H. Kargupta. An attacker's view of distance preserving maps for privacy preserving data mining. In *Proceedings of the 10th European Conference on Principles and Practice of Knowledge Discovery in Databases (PKDD'06)*, pages 297–308, Berlin, Germany, September 2006.

[30] K. Liu, H. Kargupta, and J. Ryan. Random projection-based multiplicative data perturbation for privacy preserving distributed data mining. *IEEE Transactions on Knowledge and Data Engineering (TKDE)*, 18(1):92–106, January 2006.

[31] M. Kantarcioğlu, J. Jin, and C. Clifton. When do data mining results violate privacy? In *Proceedings of the 10th ACM SIGKDD Conference (KDD'04)*, pages 599–604, Seattle, WA, August 2004.

[32] A. Machanavajjhala, J. Gehrke, D. Kifer, and M. Venkitasubramaniam. l-diversity: Privacy beyond k-anonymity. *ACM Transactions on Knowledge Discovery from Data*, 1(1), 2006.

[33] S. Mukherjee, Z. Chen, and A. Gangopadhyay. A privacy preserving technique for euclidean distance-based mining algorithms using fourier-related transforms. *The VLDB Journal*, 15(4):293–315, 2006.

[34] K. Muralidhar and R. Sarathy. Data shuffling - a new masking approach for numerical data. *Management Science*, 52(5):658–670, May 2006.

[35] J. A. Nelder and R. Mead. A simplex method for function minimization. *Computer Journal*, 7:308–313, 1965.

[36] S. R. M. Oliveira and O. R. Zaïane. Privacy preserving clustering by data transformation. In *Proceedings of the 18th Brazilian Symposium on Databases*, pages 304–318, Manaus, Amazonas, Brazil, October 2003.

[37] S. R. M. Oliveira and O. R. Zaïane. Privacy preservation when sharing data for clustering. In *Proceedings of the International Workshop on Secure Data Management in a Connected World*, pages 67–82, Toronto, Canada, August 2004.

[38] P. Samarati. Protecting respondents identities in microdata release. *IEEE Transactions on Knowledge and Data Engineering*, 13(6):1010–1027, November/December 2001.

[39] J. W. Silverstein and P. L. Combettes. Signal detection via spectral theory of large dimensional random matrices. *IEEE Transactions on Signal Processing*, 40(8):2100–2105, 1992.

[40] G. W. Stewart and Ji-Guang Sun. *Matrix Perturbation Theory*. Academic Press, 1990.

[41] L. Sweeney. k-anonymity: a model for protecting privacy. *International Journal on Uncertainty, Fuzziness and Knowledge-based Systems*, 10(5):557–570, 2002.

[42] G. J. Szekély and M. L. Rizzo. Testing for equal distributions in high dimensions. *InterStat*, November(5), 2004.

[43] P. Tendick. Optimal noise addition for preserving confidentiality in multivariate data. *Journal of Statistical Planning and Inference*, 27(2):341–353, 1991.

[44] M. Trottini, S. E. Fienberg, U. E. Makov, and M. M. Meyer. Additive noise and multiplicative bias as disclosure limitation techniques for continuous microdata: A simulation study. *Journal of Computational Methods in Sciences and Engineering*, 4:5–16, 2004.

[45] V. S. Verykios, A. K. Elmagarmid, E. Bertino, Y. Saygin, and E. Dasseni. Association rule hiding. In *IEEE Transactions on Knowledge and Data Engineering*, volume 16, pages 434–447, 2004.

[46] K. Wang, Benjamin C. M. Fung, and Philip S. Yu. Handicapping attacker's confidence: an alternative to k-anonymization. *Knowledge and Information Systems*, 11(3):345–368, 2007.

[47] E. P. Wigner. On the statistical distribution of the widths and spacings of nuclear resonance levels. *Proceedings of the Cambridge Philosophical Society*, 47:790–798, 1952.

[48] R. Chi-Wing Wong, J. Li, A. Wai-Chee Fu, and K. Wang. (α, k)-anonymity: an enhanced k-anonymity model for privacy preserving data publishing. In *Proceedings of the 12th ACM SIGKDD Conference (KDD'06)*, pages 754–759, Philadelphia, PA, August 2006.

[49] Y. Zhu and L. Liu. Optimal randomization for privacy preserving data mining. In *Proceedings of the 10th ACM SIGKDD Conference (KDD'04)*, pages 761–766, Seattle, WA, August 2004.

Chapter 16

Private Data Analysis via Output Perturbation

A Rigorous Approach to Constructing Sanitizers and Privacy Preserving Algorithms

Kobbi Nissim
Department of Computer Science,
Ben-Gurion University of the Negev,
Be'er Sheva, Israel.
kobbi@cs.bgu.ac.il

To R., Y., and N.

Abstract We describe output perturbation techniques that allow for a provable, rigorous sense of individual privacy. Examples where the techniques are effective span from basic statistical computations to sophisticated machine learning algorithms.

Keywords: Private query processing, output perturbation.

16.1 Introduction

Rapidly increasing volumes of sensitive individual information are maintained by governments, statistical agencies and private enterprises, the latter making them increasingly ubiquitous as electronic collection and archiving evolves. The potential social benefits from analyzing these databases are enormous. A challenge, however, is to compute and release useful information about the data while protecting the privacy of individual data contributors. Our focus is on such analyses.

Applying to intuition, one may claim that statistical analysis and datamining procedures already answer this challenge. After all, these analyses are aimed at finding *large scale* phenomena, hence, applying them to data collections should not result in a significant leakage of private individual information. This intuition, however, seems hard to substantiate, as it may so happen that the results of several such "harmless looking" analyses may be combined in a way that would cause privacy breaches.

In this chapter we describe a formal approach, analogous to that taken in the theoretical research of cryptography. We first present a simple and rather intuitive privacy definition that allows us to argue about its implications (so as to hopefully understand what kind of privacy is provided), and then construct analyses that preserve privacy.

Before we continue, we differentiate our goal from another goal pursued in the privacy literature, namely the construction of efficient secure multiparty protocols for datamining tasks[1]. This is the problem of applying the cryptographic tool of *secure multiparty computation* to collections of sensitive individual information that are distributed among several parties, each of which is not willing to explicitly share its information with the other parties. Extremely rich theory exists, starting from the foundational work of [34, 21, 7, 5], showing that essentially every analysis may be performed such that the parties collaboratively compute it over their joint data without any of them learning more than what is implied by the intended outcome of the analysis. These strong results follow by generic transformations of insecure computations to secure computations, that result with only a polynomial overhead.

When applied to large datasets, the generic techniques of creating secure multiparty protocols are inefficient in practice. Hence, more efficient secure protocols for specific functionalities are sought after, e.g. protocols where the total communication is sublinear in the dataset size. A breakthrough result in this direction is due to Lindel and Pinkas [24]. They showed how to securely compute an ID3 decision tree when the dataset is vertically split between two parties. The approach taken in [24] and much of the following research is to choose one of the existing algorithms/heuristics for a datamining problem, and implement an efficient secure protocol for it, avoiding using the generic techniques when they do not yield efficient protocols. The privacy guarantee is that the participating parties would not learn any information beyond what is implied by the outcome of the chosen algorithm/heuristic. This is a very different notion of privacy from what we seek herein – in general there is no guarantee that the outcome of the datamining analysis procedure itself preserves *individual privacy*. It may leak some information pertaining to individuals, or small groups, and no matter how secure the implementation is, it would also leak this information.

In the rest of this chapter we consider a simple formal model – *statistical databases* – that serves as an underlying model for our discussion. In this model we present a privacy definition, capturing the intuition that an

[1]The term *privacy preserving datamining* is often used in the literature in connection with both goals.

individual's privacy is preserved if the inclusion of her data in the analysis has a minor effect on the outcome. At first sight it may seem that this definition is so restrictive that it would prohibit any useful computation, however, this is not the case – it is possible to construct analyses that yield useful outcome, and yet preserve privacy in a rigorous sense. Our focus is on the basic techniques for constructing such statistical and machine learning analyses. We start in Section 16.4 with the basic technique of adding noise of magnitude proportional to a property of the query function called *sensitivity*, and show the effectiveness of this idea for simple functions. These simple functions are used in Section 16.5 as the building blocks for more complex functionalities. Section 16.6 includes a brief overview of more recent techniques that have emerged from the basic techniques. We conclude with related work and bibliographic notes.

16.2 The Abstract Model – Statistical Databases, Queries, and Sanitizers

As the underlying model for our discussion of privacy we will consider a simple abstract model that we will refer to as a *statistical database*. Roughly speaking, a statistical database is a centralized database, controlled by a single trusted party that interacts with users who wish to issue queries to the database. We note, however, that our definitions and results carry also to many other settings. In particular, settings where the data is distributed among several parties and settings where the collection of individual data does not physically or formally consist a database.

DEFINITION 16.1 (STATISTICAL DATABASE) *A statistical database* $\mathbf{x}$ *of size* n *over domain* $\mathcal{D}$ *is an ordered collection of* n *entries*

$$\mathbf{x} = (x_1, \dots, x_n) ,$$

where each entry is taken from the domain $\mathcal{D}$.

The definition of statistical databases is very general. In particular, the domain $\mathcal{D}$ can be points in $\mathbb{R}^d$, text, images, or any other (arbitrarily complex) set of possible entries. Furthermore, we do not make any assumption regarding to how the entries x_i of the database are selected (i.e. whether the database entries are sampled from some underlying distribution, whether the entries are independent of each other, etc.).

As a means of accessing the information stored in a statistical database $\mathbf{x}$ we will assume the existence of an algorithmic mechanism that has access to the statistical database $\mathbf{x}$. We call this mechanism a *sanitizer*, emphasizing its goal of preserving the privacy of the underlying data by only releasing answers from which the dependency on individual information was "cleared". Users access the statistical database by issuing queries to the sanitizer, where a query to a database **x** is any function

$$f : \mathcal{D}^n \rightarrow \mathcal{R} .$$

For simplicity of exposition, we will only consider real valued functions f, i.e.,

$$f : \mathcal{D}^n \rightarrow \mathbb{R}^d .$$

We note, however, that the techniques we present in the sequel do generalize to other metric spaces.

EXAMPLE 16.2 (SUM QUERIES) *A family of queries, that turns to be extremely useful, is that of sum queries. These are queries of the form*

$$\mathtt{sum}_g(\mathbf{x}) = \sum_{i=1}^{n} g(i, x_i) , \tag{16.1}$$

where $g : \mathbb{N} \times \mathcal{D} \rightarrow [0, 1]$. *Sum queries allow expressing basic statistical functions (such as counts, averages etc.) as well as more complicated computations.*

EXAMPLE 16.3 *Consider a database held and maintained by a hospital, containing patient information as depicted below.*

#	SSN	Sex	Age	Disease	Smoking
1	631-35-1210	M	41	Heart	Yes
2	051-34-1430	F	32	Cancer	No
⋮					
n	615-84-1924	M	37	Viral	No

We can view this database as a statistical database, where the domain $\mathcal{D}$ *corresponds to the possible values for a record of format*

$$(\mathit{SSN}, \mathit{Sex}, \mathit{Age}, \mathit{Disease}, \mathit{Smoking}) .$$

Many different queries may be generated by specifically setting the function g *in a sum query. Such queries may e.g. used for comparing the odds of having cancer of smokers and non-smokers. Let*

$$\begin{aligned} g_1(x) &\triangleq x.\mathit{Smoking} = \mathit{Yes} \\ g_2(x) &\triangleq (x.\mathit{Smoking} = \mathit{Yes}) \wedge (x.\mathit{Disease} = \mathit{Cancer}) \\ g_3(x) &\triangleq (x.\mathit{Smoking} = \mathit{No}) \wedge (x.\mathit{Disease} = \mathit{Cancer}) . \end{aligned}$$

The odds are $\mathtt{sum}_{g_2}(\mathbf{x})/\mathtt{sum}_{g_1}(\mathbf{x})$ *and* $\mathtt{sum}_{g_3}(\mathbf{x})/(n - \mathtt{sum}_{g_1}(\mathbf{x}))$ *for smokers and non-smokers respectively.*

Note that unlike a common practice in the privacy literature (see e.g. [31, 32]), we do not assume a classification of record fields into *identifying* and *sensitive*. An implication of this choice is that *all* parts of individual information is treated as sensitive. This may seem as an over-conservative choice. However, it saves the need to decide which information is sensitive, and protects against the risk that harmless looking pieces of information, or the relationships between them, would eventually be linked to sensitive information (e.g. using datamining techniques), and hence become sensitive themselves.

Furthermore, we allow queries to directly address individuals in the database. For example, in the definition of sum queries (Equation 16.1) the function g is explicitly given the record 'identity' i, and hence distinct functions $g_i(\cdot) = g(i, \cdot)$ may be applied to distinct individual records. One implication of this choice is that in our model privacy is not a derivative of anonymity – privacy has to be maintained when the attacker is able to separately address in his query function each of the individual contributors to the database, i.e. even if anonymity is breached.

A large collection of techniques for constructing sanitizers appear in the literature, and we refer the reader to the survey in [1] for a classification of sanitization techniques. Roughly speaking, sanitizers may decide not to answer some queries, and to modify query results. We will restrict our attention to sanitizers that preserve privacy by adding noise to query answers, so as to mask out the effects of individual records, but still leave global trends visible. This intuitively appealing technique is commonly referred to as *output perturbation*.

The answer given by the output perturbation sanitizer on a query f is distributed according to:

$$\mathbf{San}(\mathbf{x}, f) = f(\mathbf{x}) + Y ,$$

where Y – often refereed to as *noise* – is a random variable taken from a probability distribution $\mathcal{N}$. In general, the noise distribution $\mathcal{N}$ may depend on the query and on the actual values stored in the statistical database, i.e. $\mathcal{N} = \mathcal{N}(f, \mathbf{x})$. In most of our discussion, however, we will consider probability distributions that depend on the query type, but *do not depend* on the actual values stored in the statistical database[2].

We will not touch upon questions of how statistical databases and their sanitizers are actually implemented, but rather on their functionality. A typical example where the model of statistical database directly applies is the database of information collected by statistical agencies such as the U.S. Census Bureau. Similarly, collections of individual data records collected and maintained by health care organizations, financial organizations, search engines, etc. may

[2] Note that when $\mathcal{N}$ is a function of $\mathbf{x}$ special care has to be taken as the noise itself may become an unexpected source of information leakage. See [25] and Section 16.6.1.

be viewed as statistical databases. As noted above, our results also apply to distributed setups e.g. by reducing a distributed setup to a centralized setup using standard cryptographic techniques of Secure Multiparty Computation [34, 21, 7, 5]. Efficient secure multiparty computation protocols may be designed for specific sanitizers, as in [15].

16.3 Privacy

In an attack on the privacy of a statistical database, an adversarial attacker that has complete knowledge of the sanitizer algorithm and privacy parameters communicates with the sanitizer issuing queries $f_1, f_2, \ldots$ and receiving answers $a_1, a_2, \ldots$ where a_i is distributed according to $\mathbf{San}(\mathbf{x}, f_i)$. The issue of attack detection, if at all possible, is beyond the scope of our discussion, and we assume the sanitizers answers all queries as if it communicates with a legitimate user. The attacker may choose the queries *adaptively*, i.e. the choice of query f_{i+1} may depend on the answers $a_1, \ldots, a_i$ to the previous queries. The definition we present in this section captures the requirement that individual privacy is preserved even in presence of any such attacker.

DEFINITION 16.4 (HAMMING DISTANCE, NEIGHBOR DATABASES) *The Hamming Distance between two databases of the same size is defined as the number of entries on which they differ:*

$$\mathbf{dist}_H(\mathbf{x}, \mathbf{x}') = \left|\{i : x_i \neq x_i'\}\right| .$$

Two databases that differ on a single individual entry, i.e. $\mathbf{x}, \mathbf{x}'$ *such that* $\mathbf{dist}_H(\mathbf{x}, \mathbf{x}') = 1$ *are called* neighbor databases.

We can now state our privacy definition. It is reminiscent of (and was inspired by) the notion of *indistinguishability of ciphertexts* introduced by Goldwasser and Micali [20] in the context of probabilistic encryption. Informally, a sanitizer is private if no adversary $\mathcal{A}$ gains significant knowledge about an individual entry of the statistical database beyond what $\mathcal{A}$ could have learned by interacting with a similar (neighbor) database where that individual entry is arbitrarily modified, or removed. This is formalized as a requirement that for all pairs of neighbor databases $\mathbf{x}, \mathbf{x}'$, and all possible sanitizer answers, the probability that an adversary obtains a specific answer when interacting with the sanitizer on the database $\mathbf{x}$ is within an e^{ε} multiplicative factor from the probability the same answer is obtained on $\mathbf{x}'$, where $\varepsilon > 0$ – the privacy parameter – is chosen by the privacy policy.

DEFINITION 16.5 (ε-PRIVACY (ε-DIFFERENTIAL PRIVACY) [16]) *A sanitizer* **San** *is* ε*-private if for all neighbor statistical databases* $\mathbf{x}, \mathbf{x}' \in \mathcal{D}^n$*, and for all subsets of possible answers* T *(i.e. subsets of the support of*

$\mathbf{San}(\cdot)$*):*

$$\frac{\Pr[\mathbf{San}(\mathbf{x}) \in \mathcal{T}]}{\Pr[\mathbf{San}(\mathbf{x}') \in \mathcal{T}]} \leq e^{\varepsilon} . \tag{16.2}$$

The probability is taken over the coin tosses of the sanitizer.

In similarity to the security parameter of cryptographic primitives, the parameter ε controls the *leakage* of information about individual entries of the statistical database. When ε is small, $e^{\varepsilon} \approx 1 + \varepsilon$, and hence the requirement is roughly that for all sets of possible transcripts $\mathcal{T}$ the probability of $\mathbf{San}(\mathbf{x}) \in \mathcal{T}$ is about the same as that of $\mathbf{San}(\mathbf{x}') \in \mathcal{T}$.

An immediate consequence of Definition 16.5 is that the sanitizer **San** cannot be deterministic (unless it computes a constant function). Otherwise, there would exist neighbor databases $\mathbf{x}, \mathbf{x}'$ and an answer t in the support of $\mathbf{San}()$ such that $\mathbf{San}(\mathbf{x}') = t$ but $\mathbf{San}(\mathbf{x}') \neq t$ and hence the ratio $\Pr[\mathbf{San}(\mathbf{x}) = t]/\Pr[\mathbf{San}(\mathbf{x}') = t]$ is unbounded.

Notation. For simplicity of exposition, we will only consider sanitizers where $\mathbf{San}(\mathbf{x})$ is sampled from a continuous distribution. We will use the notation $h_{\mathbf{x}}^{\mathbf{San}}(t)$ for the probability density function of this distribution and $h_{\mathbf{x}}^{\mathbf{San}}(t|A)$ for this probability density function conditioned on the event A. We will usually abuse notation and write $h_{\mathbf{x}}$ for $h_{\mathbf{x}}^{\mathbf{San}}$.

When $\mathbf{San}(\mathbf{x})$ is sampled from a continuous distribution $h_{\mathbf{x}}^{\mathbf{San}}$ we can state an equivalent requirement to Equation 16.2:

$$\frac{h_{\mathbf{x}}^{\mathbf{San}}(t)}{h_{\mathbf{x}'}^{\mathbf{San}}(t)} \leq e^{\varepsilon} \quad \text{for all possible sanitizer answers } t. \tag{16.3}$$

NOTE 16.6 *Readers familiar with the cryptographic notion of indistinguishability of ciphertexts might have expected the requirement in Equation 16.2 to be that the distributions* $\mathbf{San}(\mathbf{x})$ *and* $\mathbf{San}(\mathbf{x}')$ *would be* statistically close. *In our setting, however, the difference between these distributions should not be negligible, as a negligible difference would disallow any utility*[3]. *When the difference ε is not negligible, the requirement of statistical difference ε is insufficient – it is possible to have two distributions* $\mathbf{San}(\mathbf{x})$ *and* $\mathbf{San}(\mathbf{x}')$ *where with probability $\Theta(\varepsilon)$ an attacker receives an answer $a \in_R$* $\mathbf{San}(\mathbf{x})$ *that is not in the support of* $\mathbf{San}(\mathbf{x}')$ *(or vice versa), and hence is able to tell these cases apart. For small ε the multiplicative requirement of Equation 16.2 is more stringent*

[3]This follows by a standard hybrid argument noting that for any two statistical databases $\mathbf{x}, \mathbf{x}'$ there exist $m \leq n + 1$ statistical databases $\mathbf{x} = \mathbf{x}_1, \mathbf{x}_2, \ldots, \mathbf{x}_m = \mathbf{x}'$ such that $\mathbf{x}_i, \mathbf{x}_{i+1}$ are neighbor databases for all $1 \leq i < m$.

than what we could get with statistical difference ε, in particular, Equation 16.2 trivially implies that the statistical difference between **San**(**x**) *and* **San**(**x**$'$) *is at most $e^{\varepsilon} - 1 \approx \varepsilon$, as*

$$\begin{aligned}\Pr[\mathbf{San}(\mathbf{x}) \in \mathcal{T}] - \Pr[\mathbf{San}(\mathbf{x}') \in \mathcal{T}] &\leq (e^{\varepsilon} - 1)\Pr[\mathbf{San}(\mathbf{x}') \in \mathcal{T}] \\ &\leq (e^{\varepsilon} - 1).\end{aligned}$$

There exist equivalent semantic security flavored *versions of Definition 16.5, that are somewhat less intuitive than Definition 16.5. The reader is referred to [19, 18, 6, 16] for further discussion of these definitions.*

16.3.1 Interpreting the Privacy Definition

Before diving into the techniques of constructing output perturbation sanitizers that respect Definition 16.5, we highlight some of its properties (the reader is referred to [18, 6, 16, 13] for further details).

Many of the privacy definitions that appear in the literature are strongly tied with specific sanitization techniques, and specify particular properties of the sanitization output designed to subvert specific attacks. An example is the definition of k-anonymity [31, 32] where the information is released in a tabular form so that the identifying information for each individual in the release equals that of least k-1 other individuals in the release.

In contrast, definition of ε-privacy is not tied with a specific technique or output format. This is important, as privacy does not rely on the assumption that a specific sanitization technique is 'good'. Separating privacy and sanitization techniques allows meaningful comparison of sanitizers even when they employ different techniques and use different output formats. Moreover, ε-privacy is *not* a property of a specific outcome of the sanitization algorithm, but of the sanitization algorithm itself (intuitively, that no attacker 'wins' the sanitizer in a distinguishing game. No assumptions are made regarding the attacker behavior, except that it accesses the information in **x** via **San**$(\cdot)$).

The Case of Independent Entries. We start with the simple case where the entries of the statistical database are chosen i.i.d. from some distribution over the domain[4]. If this is the case, an ε-private sanitizer preserves the privacy of an individual entry (wlog x_1) in a very strong sense – we can bound the change in beliefs about x_1 for an attacker that is given $x_2, \ldots, x_n$ for free.

More formally, let $b : \mathcal{D} \to \{0, 1\}$ be a predicate (i.e. $b(x) = 1$ if x satisfies some property, and $b(x) = 0$ otherwise). The attacker's a priori belief that

[4]An assumption of total independence is not realistic, however, it helps to illustrate the kind of privacy guaranteed by Definition 16.5. We emphasize that Definition 16.5 guarantees meaningful privacy even when the entries of the statistical database are not chosen independently.

$b(x_1)$ holds is $\Pr[b(x_1) = 1]$, where the probability is taken over the choice of x from the domain $\mathcal{D}$. As x_1 is independent of $x_2, \ldots, x_n$, the attacker's knowledge of these entries does not affect his beliefs about x_1, hence

$$\Pr[b(x_1) = 1] = \Pr[b(x_1) = 1 | x_2, \ldots, x_n] .$$

The attacker's a posteriori belief that $b(x_1)$ holds, given the sanitizer answer a and the actual values of $x_2, \ldots, x_n$ is

$$\Pr[b(x_1) = 1 | a, x_2, \ldots, x_n] .$$

We now show that the attacker's relative change in $\Pr[b(x_1) = 1]$ is bounded by e^{ε}. In our calculation we use two substitute values $z, z' \in \mathcal{D}$ for x_1 that we now choose: given an answer a (i) z maximizes $h_{(z,x_2,\ldots,x_n)}(a)$ subject to $b(z) = 1$, and (ii) z' minimizes $h_{(z',x_2,\ldots,x_n)}(a)$.

$$\begin{aligned}\frac{\Pr[b(x_1) = 1 | a, x_2, \ldots, x_n]}{\Pr[b(x_1)]} &= \frac{\Pr[b(x_1) = 1 | a, x_2, \ldots, x_n]}{\Pr[b(x_1) | x_2, \ldots, x_n]} \\ &= \frac{h_{\mathbf{x}}(a | b(x_1) = 1, x_2, \ldots, x_n)}{h_{\mathbf{x}}(a | x_2, \ldots, x_n)} \\ &\leq \frac{h_{(z,x_2,\ldots,x_n)}(a)}{h_{(z',x_2,\ldots,x_n)}(a)} \\ &\leq e^{\varepsilon} .\end{aligned}$$

In this calculation we first used Bayes rule, then our choice of z, z' and, finally, the ε-privacy of the sanitizer.

Differential Privacy. Ideally, we would like sanitizers not to reveal any information about individuals, as is captured by the following citation from Dalenius: *"Access to a statistical database should not enable one to learn anything about n individual that could not be learned without access"* [11]. Dwork and Naor [13] have shown that this ideal cannot be achieved when the database has utility. I.e. for any sanitizer, and any definition of compromise, there exists an auxiliary knowledge (information available to the attacker other than the access to the sanitizer) such that the sanitizer enables the compromise. That is, without communicating with the sanitizer, the auxiliary knowledge is useless for the attacker whereas combining it with the information learned from the sanitizer results in a compromise[5]. The conclusion is that an *absolute* guarantee of privacy (i.e. that what an attacker learns about an individual with

[5] Interestingly, the argument in [13] exemplifies that even the privacy of an individual whose information is not included in the database may be compromised this way!

access to the database could also be learned without the access) in presence of arbitrary auxiliary information cannot be achieved.

In contrast, Definition 16.5 succeeds in guaranteeing meaningful privacy even in the presence of arbitrary auxiliary knowledge. That is because instead of comparing the attacker's knowledge with and without access to the sanitizer, it compares the attacker's knowledge when it accesses (via the sanitizer) databases (i) with the individual's information and (ii) with the individual's information arbitrarily modified, or even removed. This is a fair comparison as it does not require the sanitizer to protect against those 'breaches' that occur, say, even when the individual's information is excluded.

A consequence of Definition 16.5 is that even if a participant removed her data from the database $\mathbf{x}$, no consequence of the computation $\mathbf{San}(\mathbf{x})$ would become significantly more or less likely. For example, Suppose an individual debates whether she should or should not contribute her true data to the medical database of Example 16.3. Assuming the database is accessed by insurance providers via an ε-private sanitizer, her decision may affect the probability of receiving coverage by a factor of at most e^{ε}. Similarly, her decision might affect the expectancy of her insurance premium by a factor of e^{ε} at most.

We note that, by the analysis below of the privacy of small groups (Lemma 16.8), a similar phenomenon holds for any group of c contributors (with a degradation in the privacy parameter from ε to $c\varepsilon$).

Composition – Single vs. Multiple Queries. In the beginning of this section we have described an adaptive adversarial attacker that communicates with the statistical database by issuing a multitude of adaptively chosen queries to the sanitizer. In contrast, Definition 16.5 deals with only a single query. The following simple lemma bridges this gap.

Informally, we get that each interaction with an ε-private sanitizer results in an ε additive decrease in privacy. Hence, to maintain ε'-privacy when q queries are made it is sufficient that each of the queries would be answered with an ε-private sanitizer with $\varepsilon = \varepsilon'/q$. For simplicity, we prove the following lemma for non-adaptive queries. A similar argument holds for adaptive queries (see [16]).

LEMMA 16.7 *Let* $\mathbf{San}_i$ *be* ε_i*-private for* $i = 1, \ldots, q$. *The sanitizer that answers according to* $\mathbf{San}_1, \ldots, \mathbf{San}_q$ *(where the randomness of each sanitizer is chosen independently of the other sanitizers) is* ε'*-private for* $\varepsilon' = \sum_i \varepsilon_i$.

Proof: Denote by $h_{\mathbf{x}}^{(i)}$ the probability density function corresponding to the distribution on $\mathbf{San}_i(\mathbf{x})$. The probability density function corresponding to the distribution on $(\mathbf{San}_1(\mathbf{x}), \ldots, \mathbf{San}_q(\mathbf{x}))$ is

$$\bar{h}_{\mathbf{x}}(t_1, \ldots, t_q) = \prod_{i=1}^{q} h_{\mathbf{x}}^{(i)}(t_i) .$$

We hence get the desired bound:

$$\frac{\bar{h}_{\mathbf{x}}(t_1, \ldots, t_q)}{\bar{h}_{\mathbf{x}'}(t_1, \ldots, t_q)} = \frac{\prod_{i=1}^{q} h_{\mathbf{x}}^{(i)}(t_i)}{\prod_{i=1}^{q} h_{\mathbf{x}'}^{(i)}(t_i)} = \prod_{i=1}^{q} \frac{h_{\mathbf{x}}^{(i)}(t_i)}{h_{\mathbf{x}'}^{(i)}(t_i)} \leq e^{\sum_i \varepsilon_i} = e^{\varepsilon'} .$$

∎

Privacy of Small Groups. The requirement of Definition 16.5 is made to neighbor databases, and hence it directly deals with the privacy of individuals. In some cases, individuals guarantees are insufficient — such as when the attacker is aware of a group of individuals that have identical or similar information. A sanitizer **San** that withstands Definition 16.5 provides privacy also for small groups, where the degradation in the privacy parameter depends linearly on the group size:

LEMMA 16.8 *Let* **San** *be ε-private and consider databases* $\mathbf{x}, \mathbf{x}'$ *that differ on c entries. Then*

$$\frac{h_{\mathbf{x}}(t)}{h_{\mathbf{x}'}(t)} \leq e^{c\varepsilon} .$$

Proof: Assume $\mathbf{x}, \mathbf{x}'$ differ on entries $i_1, \ldots, i_c$. There exists databases

$$\mathbf{x} = \mathbf{x}_0, \mathbf{x}_1, \ldots, \mathbf{x}_c = \mathbf{x}'$$

such that $\mathbf{x}_i, \mathbf{x}_{i+1}$ are neighbor databases for all $0 \leq i < c$. By Equation 16.2, it follows that

$$\frac{h_{\mathbf{x}}(t)}{h_{\mathbf{x}'}(t)} = \prod_{i=0}^{c-1} \frac{h_{\mathbf{x}_i}(t)}{h_{\mathbf{x}_{i+1}}(t)} \leq e^{c\varepsilon} .$$

∎

Hence, reasonable privacy is guaranteed for small groups — for $c \ll 1/\varepsilon$ the degradation in privacy is approximately linear in the group size as $e^{c\varepsilon} \approx 1 + c\varepsilon$ — and privacy disintegrates as the group size c grows.

ε-Privacy as a Lipschitz Condition. A mapping $f : M \rightarrow M'$ between two spaces M, M' a function is said to satisfy the *Lipschitz condition* (also called *Lipschitz continuous*) if there exists a constant $\mathcal{L}$ such that for all $x, x' \in M$

$$\frac{\mathbf{dist}_{M'}(f(x), f(x'))}{\mathbf{dist}_M(x, x')} \leq \mathcal{L} .$$

The smallest $\mathcal{L}$ for which the above inequality holds is called the *Lipschitz constant* of f.

If we view **San** as a function mapping statistical databases to distributions over possible answers, then Equation 16.2 can be viewed as a Lipschitz condition on **San** as follows. Define the following metric over distributions of possible answers:

$$\mathbf{dist}_{\div}(\mathcal{S}_1, \mathcal{S}_2) \triangleq \max_{\mathcal{T}} \left| \ln \frac{Pr_{\mathcal{S}_1}[\mathcal{T}]}{Pr_{\mathcal{S}_2}[\mathcal{T}]} \right| ,$$

where $\mathcal{T}$ runs over all subsets of possible samples from $\mathcal{S}_1, \mathcal{S}_2$. It is easy to see that $\mathbf{dist}_{\div}$ is indeed a metric[6]. We can now rewrite Equation 16.2 as a Lipschitz condition with Lipschitz constant $\mathcal{L} = \varepsilon$:

$$\frac{\mathbf{dist}_{\div}(\mathbf{San}(\mathbf{x}), \mathbf{San}(\mathbf{x}'))}{\mathbf{dist}_H(\mathbf{x}, \mathbf{x}')} \leq \varepsilon .$$

16.4 The Basic Technique: Calibrating Noise to Sensitivity

We will now see that it is possible to release some *global* information about a statistical database while preserving privacy as in Definition 16.5. As our first example, we will consider sum queries (Equation 16.1).

Intuitively, in order to satisfy Definition 16.5 using the technique of output perturbation, a large enough noise is needed to be added to the result of computing $f(\mathbf{x})$, so as to mask out the potential difference between $f(\mathbf{x})$ and $f(\mathbf{x}')$ for all its neighbor databases $\mathbf{x}'$. In the case of sum queries, the difference between $f(\mathbf{x})$ and $f(\mathbf{x}')$ is bounded by 1. Following this intuition, it should be possible to guarantee Definition 16.5 while releasing *noisy* answers to sum queries, where the noise magnitude is constant (i.e. independent of the database size). We show that this intuition is correct. The consequence is that it is possible to answer sum queries quite accurately while preserving privacy.

We will use the (one dimensional) Laplace distribution $\mathbf{Lap}(\lambda)$ with zero mean and variance $2\lambda^2$, that has density function

$$\mathbf{Lap}(\lambda): \quad h(t) = \frac{1}{2\lambda} e^{-\frac{|t|}{\lambda}} .$$

A property of the Laplace distribution that we will use extensively is that for all t, t':

$$\frac{h(t)}{h(t')} = \frac{e^{\frac{-|t|}{\lambda}}}{e^{\frac{-|t'|}{\lambda}}} = e^{\frac{|t'|-|t|}{\lambda}} \leq e^{\frac{|t-t'|}{\lambda}} , \tag{16.4}$$

where the inequality follows by the triangle inequality.

[6] $\mathbf{dist}_{\div}(\mathcal{S}_1, \mathcal{S}_3) = \max_{\mathcal{T}} \left| \ln \frac{Pr_{\mathcal{S}_1}[\mathcal{T}]}{Pr_{\mathcal{S}_2}[\mathcal{T}]} + \ln \frac{Pr_{\mathcal{S}_2}[\mathcal{T}]}{Pr_{\mathcal{S}_3}[\mathcal{T}]} \right| \leq \max_{\mathcal{T}} \left(\left| \ln \frac{Pr_{\mathcal{S}_1}[\mathcal{T}]}{Pr_{\mathcal{S}_2}[\mathcal{T}]} \right| + \left| \ln \frac{Pr_{\mathcal{S}_2}[\mathcal{T}]}{Pr_{\mathcal{S}_3}[\mathcal{T}]} \right| \right) \leq \max_{\mathcal{T}} \left| \ln \frac{Pr_{\mathcal{S}_1}[\mathcal{T}]}{Pr_{\mathcal{S}_2}[\mathcal{T}]} \right| + \max_{\mathcal{T}} \left| \ln \frac{Pr_{\mathcal{S}_2}[\mathcal{T}]}{Pr_{\mathcal{S}_3}[\mathcal{T}]} \right| = \mathbf{dist}_{\div}(\mathcal{S}_1, \mathcal{S}_2) + \mathbf{dist}_{\div}(\mathcal{S}_2, \mathcal{S}_3)$.

Our sanitizer answers a sum query $f_g = \sum_i g(i, x_i)$ where $g : \mathbb{N} \times \mathcal{D} \to [0,1]$ by computing the exact answer and adding Laplace noise with $\lambda = 1/\varepsilon$:

$$\mathbf{San}(\mathbf{x}, f_g) = \sum_i g(i, x_i) + Y \quad \text{where} \quad Y \sim \mathbf{Lap}(1/\varepsilon) .$$

We now show that this sanitizer is indeed ε-private, i.e. that Equation 16.3 is satisfied:

$$\frac{h_{\mathbf{x}}(t)}{h_{\mathbf{x}'}(t)} = \frac{h(t - f(\mathbf{x}))}{h(t - f(\mathbf{x}'))} \le e^{\varepsilon |f(\mathbf{x}) - f(\mathbf{x}')|} \le e^{\varepsilon} .$$

The first inequality follows by Equation 16.4 and the second by noting that for every two statistical databases $\mathbf{x}, \mathbf{x}'$ that differ on a single entry (indexed i),

$$|f(\mathbf{x}) - f(\mathbf{x}')| = |g(i, x_i) - g(i, x_i')| \le 1 .$$

The technique presented above for the simple case of sum queries may be generalized to many other queries f, where changing a single entry in $\mathbf{x}$ has limited influence on the value $f(\mathbf{x})$. For simplicity, we will consider query functions f that map the database to vectors of reals, and will use the ℓ_1 norm on $\mathbb{R}^d$ as our distance metric (denoted $\|\cdot\|_1$). We note, however, that all the results of this section generalize to other metric spaces, with appropriate modifications.

The main result of this section is that privacy can be preserved by calibrating the noise magnitude to a combinatorial property of the query f that we call *global sensitivity*. This is a measure how an entry x_i of a statistical database $\mathbf{x}$ may influence the outcome of the query f:

DEFINITION 16.9 ([16]) *For $f : \mathcal{D}^n \to \mathbb{R}^d$, the global ℓ_1 sensitivity of f is*

$$\mathbf{GS}_f = \max_{\mathbf{x}, \mathbf{x}' : \, \mathbf{dist}_H(\mathbf{x}, \mathbf{x}') = 1} \left\| f(\mathbf{x}) - f(\mathbf{x}') \right\|_1 . \tag{16.5}$$

Note that $\mathbf{GS}_f$ is a property inherent in the query function f, and is in particular independent of the actual content of the statistical database. The following theorem states that adding Laplace noise with magnitude proportional to $\mathbf{GS}_f$ and inversely proportional to ε ensures ε-privacy:

THEOREM 16.10 ([16]) *For all $f : \mathcal{D}^n \to \mathbb{R}^d$ such that $\mathbf{GS}_f < \infty$ the following sanitizer is ε-private:*

$$\mathbf{San}(\mathbf{x}, f) = f(\mathbf{x}) + (Y_1, \ldots, Y_d) ,$$

where $Y_1, \ldots, Y_d$ are random variables drawn i.i.d. from $\mathbf{Lap}(\mathbf{GS}_f/\varepsilon)$.

Proof: The proof closely follows our reasoning for the case of sum queries. Let $\bar{h}() : \mathbb{R}^d \to \mathbb{R}^+$ be the joint probability density function of the random variables $Y = Y_1, \ldots, Y_d$. By independence, we have that:

$$\begin{aligned}\bar{h}(t) = \bar{h}(t_1, \ldots, t_d) &= \prod_{i=1}^{d} h(t_i) = \left(\frac{1}{2\lambda}\right)^d \cdot \prod_{i=1}^{d} e^{-|t_i|/\lambda} \\ &= \left(\frac{1}{2\lambda}\right)^d \cdot e^{-\|t\|_1/\lambda},\end{aligned}$$

and hence we get that (similarly to Equation 16.4):

$$\frac{\bar{h}(t)}{\bar{h}(t')} = \frac{e^{\frac{-\|t\|_1}{\lambda}}}{e^{\frac{-\|t'\|_1}{\lambda}}} = e^{\frac{\|t'\|_1 - \|t\|_1}{\lambda}} \leq e^{\|t-t'\|_1/\lambda}. \tag{16.6}$$

To see that Equation 16.3 is satisfied, note that for all $\mathbf{x}, \mathbf{x}'$ such that $\mathbf{dist}_H(\mathbf{x}, \mathbf{x}) = 1$ and for all $t = (t_1, \ldots, t_d)$:

$$\frac{h_{\mathbf{x}}(t)}{h_{\mathbf{x}'}(t)} = \frac{\bar{h}(t - f(\mathbf{x}))}{\bar{h}(t - f(\mathbf{x}'))} \leq e^{\|f(\mathbf{x}) - f(\mathbf{x}')\|_1/\lambda} \leq e^{\mathbf{GS}_f/\lambda} = e^{\varepsilon},$$

where the first inequality follows from Equation 16.6, the second inequality from Definition 16.9 and the last equality by substituting $\lambda = \mathbf{GS}_f/\varepsilon$. ∎

For an alternative proof, note that the construction in Theorem 16.10 is actually a composition of two Lipschitz continuous functions:

1 The query function $f : \mathcal{D}^n \to \mathbb{R}^d$ is Lipschitz continuous with Lipschitz constant $\mathbf{GS}_f$, with respect to the metrics $\mathbf{dist}_H$ (over $\mathcal{D}^n$) and $|\cdot\|_1$ (over $\mathbb{R}^d$).

2 The perturbation function $\mathcal{P}(z)$ mapping $z \in \mathbb{R}^d$ to the probability distribution $z + \mathbf{Lap}^d(\mathbf{GS}_f/\varepsilon)$ is Lipschitz continuous with Lipschitz constant $\varepsilon/\mathbf{GS}_f$, with respect to the metrics $|\cdot\|_1$ (over $\mathbb{R}^d$) and $\mathbf{dist}_{\div}$ (over distributions over $\mathbb{R}^d$).

The resulting function $\mathbf{San}(\mathbf{x}) = \mathcal{P}(f(\mathbf{x}))$ is Lipschitz continuous with Lipschitz constant bounded by $\mathbf{GS}_f \cdot \frac{\varepsilon}{\mathbf{GS}_f} = \varepsilon$.

16.4.1 Applications: Functions with Low Global Sensitivity

Theorem 16.10 gives a simple but extremely powerful recipe for constructing output perturbation sanitizers. It implies that if a query function f has low global sensitivity, then it can be released by the sanitizer relatively accurately — with noise magnitude $\mathbf{GS}_f/\varepsilon$. We start by showing that the global sensitivity of many useful functions is indeed low, and hence, for these functions the technique of Theorem 16.10 may be directly applied. In Section 16.5 we extend the applicability of Theorem 16.10 by using these simple functionalities in constructing more complex ones.

Sum Queries. For sum queries $\mathtt{sum}_g(\mathbf{x}) = \sum_{i=1}^{n} g(i, x_i)$ where $g : \mathbb{N} \times \mathcal{D} \to [0, 1]$ we have

$$\begin{aligned} \mathbf{GS}_{\mathtt{sum}} &= \max_{\mathbf{x},\mathbf{x}' : \, \mathbf{dist}_H(\mathbf{x},\mathbf{x}')=1} \left(\sum_{i=1}^{n} \left(g(i, x_i) - g(i, x'_i) \right) \right) \\ &= \max_{x,x' \in \mathcal{D}, i \in [n]} \left(g(i, x) - g(i, x') \right) \leq 1 \, . \end{aligned}$$

As we have already seen above, this amounts to answering the query $\mathtt{sum}_g(\mathbf{x})$ with noise magnitude $1/\varepsilon$, independent of n. This is valuable as sum queries can be used for computing basic statistics like counts and means.

Mean and Covariance. Assume $v : \mathcal{D} \to \mathbb{R}^d$ is some function mapping elements of $\mathcal{D}$ into column vectors in $\mathbb{R}^d$. Applying v on the entries of a database $\mathbf{x} \in \mathcal{D}^n$ results in a collection of n vectors

$$V(\mathbf{x}) = \{v(x_i)\}_{i \in [n]} \, .$$

The mean and covariance of V are defined as

$$\begin{aligned} \mu_V &= \mathop{\mathbf{avg}}_{u \in V} u = \mathop{\mathbf{avg}}_{i} v(x_i) \, , \\ C_V &= \mathop{\mathbf{avg}}_{u \in V} u u^T - \mu_V \mu_V^T = \mathop{\mathbf{avg}}_{i} v(x_i) v(x_i)^T - \mu_V \mu_V^T \, . \end{aligned}$$

Define $\mathtt{mean}_v : \mathcal{D}^n \to \mathbb{R}^d$ and $\mathtt{cov}_v : \mathcal{D}^n \to \mathbb{R}^{d \times d}$ to be the functions that on $\mathbf{x} \in \mathcal{D}^n$ return μ_V and C_V as above.

We consider the case where $\|v(x)\|_1 \leq \gamma$ for all $x \in \mathcal{D}$ (without bounding $v(x)$ a single change to the database can inflict unbounded change on $\mathtt{mean}_v(\mathbf{x})$ and $\mathtt{cov}_v(\mathbf{x})$), and will incorporate this bound into our sensitivity analysis.

Note that $\mathtt{mean}_v$, $\mathtt{cov}_v$ are simply sums, and hence may be expressed in terms of sum queries. Hence, employing Lemma 16.7, we can compute $\mathtt{mean}_v$ by invoking d sum queries with noise parameter $\varepsilon' = \varepsilon/d$, resulting in noise magnitude proportional to $1/\varepsilon' = d/\varepsilon$. A similar analysis for $\mathtt{cov}_v$ yields noise magnitude proportional to d^2/ε. However, taking a closer look at these functionalities allows us to answer these with noise magnitude that is independent of d, as we now show.

In our analysis we change a single entry x_j of the database and bound its effect on μ_V and C_V. Denote by δ the difference $v(x_j) - v(x'_j)$. By our assumption on $v()$, we get that

$$\|\delta\|_1 \leq \max_{x,x' \in \mathcal{D}} \|v(x) - v(x')\|_1 \leq 2\gamma .$$

We get that $\|\texttt{mean}(\mathbf{x}) - \texttt{mean}(\mathbf{x}')\|_1 = \frac{1}{n}\|\delta\|_1 \leq 2\gamma/n$, and hence

$$\mathbf{GS}_{\texttt{mean}} \leq 2\gamma/n .$$

To analyze $\mathbf{GS}_{\texttt{cov}}$ we treat C_V as a vector of dimension d^2 and apply the ℓ_1 norm to this vector. We get that a change to a single entry x_j can change the $\mu_V \mu_V^T$ term by at most

$$\begin{aligned}
\|(\mu_V + \frac{1}{n}\delta)(\mu_V + \frac{1}{n}\delta)^T - \mu_V \mu_V^T\|_1 &= \|\frac{1}{n}\mu_V \delta^T + \frac{1}{n}\delta \mu_V^T + \frac{1}{n^2}\delta\delta^T\|_1 \\
&= \|\frac{1}{n}\mu_V \delta^T + \frac{1}{n}\delta(\mu_V + \frac{1}{n}\delta)^T\|_1 \\
&\leq 4\gamma^2/n .
\end{aligned}$$

The last inequality follows as $\|\delta\|_1 \leq 2\gamma$, and both $\|\mu_V\|_1$ and $\|\mu_V + \frac{1}{n}\delta\|_1$ are bounded by γ.

A similar analysis yields a bound on the change in $\mathbf{avg}_i\, v(x_i)v(x_i)^T$:

$$\begin{aligned}
\|\operatorname*{\mathbf{avg}}_i v(x'_i)v(x'_i)^T - \operatorname*{\mathbf{avg}}_i v(x_i)v(x_i)^T\|_1 &= \frac{1}{n}\|(x_j + \delta)(x_j + \delta)^T - x_j x_j^T\|_1 \\
&= \frac{1}{n}\|x_j \delta^T + \delta(x_j + \delta)^T\|_1 \\
&\leq 4\gamma^2/n .
\end{aligned}$$

Where the last inequality follows from $\|\delta\|_1 \leq 2\gamma$ and $\|v(x_j)\|_1, \|v(x'_j)\|_1 \leq \gamma$. Combining these two inequalities we get that $\|\texttt{cov}(\mathbf{x}) - \texttt{cov}(\mathbf{x}')\|_1 \leq 4\gamma^2 + 4\gamma^2 = 8\gamma^2$, and hence,

$$\mathbf{GS}_{\texttt{cov}} \leq 8\gamma^2/n .$$

Histograms. A histogram partitions the domain $\mathcal{D}$ into k disjoint bins, and counts the number of database elements that fall within each bin. The partitioning is according some partitioning function $q : \mathcal{D} \to [k]$ where $q(x) = j$ is interpreted as "x belongs to the jth bin".

$$\texttt{hist}_q(\mathbf{x}) = (|\{i : q(x_i) = 1\}|, \ldots, |\{i : q(x_i) = k\}|) .$$

As with $\texttt{mean}_v$ and $\texttt{cov}_v$ above, it is possible to compute $\texttt{hist}_q(\mathbf{x})$ by issuing k sum queries and applying Lemma 16.7. This would result in noise magnitude $\frac{k}{\varepsilon} \cdot \mathbf{GS}_{\texttt{sum}} = \frac{k}{\varepsilon}$. A closer look at the histogram function reveals that the dependency of the noise magnitude on k may be eliminated, as it may be reduced to $\frac{2}{\varepsilon}$. To see that, note that changing a database entry x_i may result in changing at most two of the counts (i.e. that of the original bin for x_i and that of the new bin for x_i), each by a quantity of one. We hence get that

$$\mathbf{GS}_{\texttt{hist}} = 2 .$$

Subset Sum. This is a simple extension of the histogram function, allowing summation over disjoint subsets of the database entries. Here, again, $q : \mathcal{D} \rightarrow [k]$ is a partitioning function, but instead of counting the number of elements in each bin, we sum over a function $g : \mathcal{D} \rightarrow \mathbb{R}^d$ where $\|g(x)\|_1 \leq \gamma$ for all $x \in \mathcal{D}$.

$$\texttt{subsets}_{q,g}(\mathbf{x}) = \left(\sum_{q(x_i)=1} g(x_i), \ldots, \sum_{q(x_i)=k} g(x_i) \right) .$$

Combining the arguments we used for bounding $\mathbf{GS}_{\texttt{mean}}$ and for $\mathbf{GS}_{\texttt{hist}}$ we get

$$\mathbf{GS}_{\texttt{subsets}} = 4\gamma .$$

Our last two examples for this section are query families that are not naturally expressed as sum queries:

Distance to Property. Given a property $P \subseteq \mathcal{D}^n$, the distance of a specific statistical database $\mathbf{x} \in \mathcal{D}^n$ from P is the Hamming distance between $\mathbf{x}$ and the nearest point in P, i.e.

$$\texttt{distance}_P(\mathbf{x}) = \min_{\mathbf{x}' \in P} \mathbf{dist}_H(\mathbf{x}, \mathbf{x}')$$

Note that $\texttt{distance}_P(\mathbf{x})$ is the minimal number of entries of $\mathbf{x}$ that need to be changed so that the property P holds. changing a single entry of $\mathbf{x}$ results in a change in $\texttt{distance}_P(\mathbf{x})$ of at most 1, and hence

$$\mathbf{GS}_{\texttt{distance}} = 1 .$$

Caveat: This result ignores the question whether $\texttt{distance}_P$ is computationally tractable. An approximation to $\texttt{distance}_P$ may exhibit global sensitivity that is greater than 1.

Functions with Low Query Complexity. Let $f : \mathcal{D}^n \to \mathbb{R}^d$ be a function that can be accurately computed by an algorithm that peeks at a small fraction of the database entries. I.e., there exists a randomized algorithm $\mathcal{A}$ such that for all inputs $\mathbf{x}$:

$$\Pr[\|\mathcal{A}(\mathbf{x}) - f(\mathbf{x})\|_1 \leq \sigma] > \beta = \frac{1+\alpha}{2}$$
$$\text{and} \quad \Pr[\mathcal{A} \text{ reads } x_j] \leq \alpha \quad \text{for all } 1 \leq j \leq n$$

Assume $\mathbf{x}, \mathbf{x}'$ differ on the jth entry. Denote by $\mathcal{A}^{-j}(\mathbf{x})$ the distribution on $\mathcal{A}(\mathbf{x})$ conditioned on not reading x_j. We get that

$$\Pr[\|\mathcal{A}^{-j}(\mathbf{x}) - f(\mathbf{x})\|_1 \leq \sigma] > \frac{\beta - \alpha}{1-\alpha} \geq \frac{1}{2} \, .$$

The same argument holds for $\mathbf{x}'$. As $\mathcal{A}^{-j}(\mathbf{x})$ and $\mathcal{A}^{-j}(\mathbf{x}')$ are equally distributed, we get, using the union bound, that

$$\Pr[\|\mathcal{A}^{-j}(\mathbf{x}) - f(\mathbf{x})\|_1 > \sigma \text{ or } \|\mathcal{A}^{-j}(\mathbf{x}) - f(\mathbf{x}')\|_1 > \sigma] < \frac{1}{2} + \frac{1}{2} = 1 \, .$$

Hence, there exists a point $p \in \mathbb{R}^d$ in the support of $\mathcal{A}$ satisfying

$$\|p - f(\mathbf{x})\|_1 \leq \sigma \text{ and } \|p - f(\mathbf{x}')\|_1 \leq \sigma \, ,$$

implying $\|f(\mathbf{x}) - f(\mathbf{x}')\|_1 \leq 2\sigma$. As the above argument holds for every two databases that differ on a single entry we get that

$$\mathbf{GS}_f \leq 2\sigma \, .$$

16.5 Constructing Sanitizers for Complex Functionalities

As we have seen above, Theorem 16.10 directly yields output perturbation sanitizers for functions whose global sensitivity can be analyzed, and turns to be low. For many functions, however, a direct calculation of global sensitivity is complicated (sometimes computationally intractable), or yields high global sensitivity, even when the function is expected to be insensitive for typical inputs.

Lemma 16.7 suggests a partial remedy to these problems (we discuss other techniques in Section 16.6). It implies that simple functions, that exhibit low global sensitivity, may be combined in algorithms computing more complex functions. Suppose algorithm $\mathcal{A}$ is constructed so that it behaves as if its input is stored in a statistical database, and accesses it at most q times by simulating ε'-private sanitizers $\mathbf{San}_1, \ldots, \mathbf{San}_q$ where $\varepsilon' = \varepsilon/q$, then the outcome of algorithm $\mathcal{A}$ is assured to preserve ε-privacy.

We demonstrate this idea by presenting two types of results. In sections 16.5.1 and 16.5.2 we modify well known machine learning algorithms — k-means, Singular Value Decomposition and Principle Component Analysis — so that the resulting algorithms preserve ε-privacy. The input to these algorithms is a collection of n points $p_1, \ldots, p_n \in \mathbb{R}^d$, where each point corresponds to an individual's information. While the original algorithms may access their input in a point by point manner, the modified algorithms access their input via a small number q of insensitive queries The exact answers to these queries are replaced with noisy answers so that each answer preserves ε'-privacy. For that, we view the collection of n points as a statistical database, where each database entry consists a point. (See [6] for the private version of other machine learning algorithms — the Perceptron Algorithm, and constructing ID3 classification trees.)

The last result of this section is a more general result, translating a large family of algorithms into their ε-private version, while retaining their accuracy. A little more specifically, the result in Section 16.5.3 shows a strong connection between learning and privacy — any learning task that can be performed in the statistical queries learning model of Kearns [22] can also be performed while preserving ε-privacy.

Note. In our analysis, we will need to bound the location of the input points, and will assume they satisfy $\|p_i\|_1 \leq \gamma$ for all $1 \leq i \leq n$.

16.5.1 k-Means Clustering

Clustering is the task of partitioning n data points $p_1, \ldots, p_n$ into k disjoint sets of 'similar' points. One approach to solving this problem is known as Lloyd's Algorithm. This algorithm iteratively updates k cluster centers $c_1, \ldots, c_k$ by moving each center to the mean of the points that are closer to it than to the other centers.

k-Means Iteration:

Input: points $p_1, \ldots, p_n \in \mathbb{R}^d$, and centers $c_1, \ldots, c_k \in \mathbb{R}^d$.

1 [Partition the points into k sets]

$$S_j \leftarrow \{p_i : c_j \text{ is the closest center to } p_i\}, \text{ let } s_j \leftarrow |S_j| .$$

2 [Move each center to the mean of its associated points]
for $1 \leq j \leq k$:

$$\text{Let } m_j \leftarrow \sum_{i \in S_j} p_i, \text{ and set } c'_j \leftarrow \frac{m_j}{s_j} .$$

This rule is repeated either for a fixed number of iterations, or until a convergence criteria is satisfied.

At first sight it may seem that the k-Means Iteration cannot be implemented privately. In particular, unless noise renders it useless, a partitioning of the points according to their nearest centers would breach privacy. However, an equivalent computation can be performed without revealing the partitioning. Using our hist and subsets queries and setting $g : \mathbb{R}^d \to \mathbb{R}^d$ to be the identity function $g(p) \triangleq p$, and $q : \mathbb{R}^d \to [k]$ to be the function associating points to their centers, i.e.,

$$q(p) \triangleq \underset{j\in[k]}{\mathbf{argmin}} \left(\mathbf{dist}(p, c_j) \leq \mathbf{dist}(p, c_i) \text{ for all } i \in [k]\right) ,$$

we can rewrite an equivalent algorithm as:

Modified k-Means Iteration:

Input: points $p_1, \ldots, p_n \in \mathbb{R}^d$, and centers $c_1, \ldots, c_k \in \mathbb{R}^d$.

1 [Compute the number of points in each of the sets S_j]

$$(\overline{s}_1, \ldots, \overline{s}_k) \leftarrow \texttt{hist}_q(p_1, \ldots, p_n) .$$

2 [Compute the sum of points in each of the sets S_j]

$$(\overline{m}_1, \ldots, \overline{m}_k) \leftarrow \texttt{subsets}_{q,g}(p_1, \ldots, p_n) .$$

3 [Update each mean]
for $1 \leq j \leq k$:

$$\overline{c}'_j \leftarrow \frac{\overline{m}_j}{\overline{s}_j} .$$

As our last step, we replace hist and subsets with their noisy version, adding Laplace noise to each to each coordinate according to our analysis in the previous section:

$$\begin{aligned} \overline{s}_j &= s_j + \hat{s}_j, \text{ where } \hat{s}_j \sim \left(\mathbf{Lap}(2/\varepsilon')\right)^d, \text{ and} \\ \overline{m}_j &= m_j + \hat{m}_j, \text{ where } \hat{m}_j \sim \left(\mathbf{Lap}(4\gamma/\varepsilon')\right)^d . \end{aligned}$$

where $\varepsilon' = \varepsilon/2$. Appealing to Lemma 16.7, the outcome of the modified algorithm preserves ε privacy.

We get that, as long as the number of points in each cluster is large, $\overline{s}_j$ is a good estimate of s_j, and hence $\overline{c}'_j$ is very close to c'_j of the non private computation. A little more formally:

LEMMA 16.11 *For each $1 \le j \le k$, if $s_j \gg 1/\varepsilon$ then with high probability*

$$\|\overline{c}'_j - c'_j\|_1 = O\left(\frac{\|c_j\|_1 + \gamma d}{\varepsilon s_j}\right) .$$

Proof:

$$\begin{aligned} \|\overline{c}'_j - c_j\|_1 &= \|\frac{\overline{m}_j}{\overline{s}_j} - \frac{m_j}{s_j}\|_1 \\ &= \|\frac{m_j + \hat{m}_j}{\overline{s}_j} - \frac{m_j}{s_j}\|_1 \\ &\le \|\frac{m_j}{s_j}\|_1 \cdot \left|\frac{s_j - \overline{s}_j}{\overline{s}_j}\right| + \|\hat{m}_j\|_1 \cdot \left|\frac{1}{\overline{s}_j}\right| \\ &= \|c_j\|_1 \cdot \left|\frac{s_j - \overline{s}_j}{\overline{s}_j}\right| + \|\hat{m}_j\|_1 \cdot \left|\frac{1}{\overline{s}_j}\right| . \end{aligned}$$

From our assumption that $s_j \gg 1/\varepsilon$, we get that with high probability $|(s_j - \overline{s}_j)/\overline{s}_j| = O(1/\varepsilon s_j)$ and $|\hat{m}_j\|_1/|\overline{s}_j| = O(\gamma d/\varepsilon s_j)$, The lemma follows. ∎

16.5.2 SVD and PCA

Many datamining algorithms treat their data points $p_1, \ldots, p_n \in \mathbb{R}^d$ as an $d \times n$ matrix A (whose columns correspond to the points), and analyze the top k eigenvectors of the matrix AA^T. This analysis can be performed while preserving ε-privacy.

Notice that

$$AA^T = \sum_{i=1}^{n} p_i p_i^T .$$

Hence, an analysis similar to that of cov leads to the following natural algorithm:

SVD:

Input: The matrix $A \in \mathbb{R}^{d\times n}$ and a parameter $0 < k \le n$.

1 [Approximate AA^T]

$$B \leftarrow \sum_i p_i p_i^T + Y \text{ where } Y \sim \left(\mathbf{Lap}(4\gamma^2)\right)^{d\times d} .$$

2 Compute the top k eigenvectors of B.

We omit the noise analysis, but note that eigenvectors are quite robust in the presence of independent zero-mean noise as is added in the procedure above (moreover, the noise magnitude does not depend on n). Hence, although B is not exactly $A^T A$, one expects the eigenvectors computed to be close to those of $A^T A$.

Principle Component Analysis (PCA) is a related technique [30] where the top k eigenvectors of the covariance matrix are computed. Again, our analysis of the noise that should be added to the covariance matrix in order to preserve ε-privacy yields a natural algorithm for a privacy preserving version of PCA.

16.5.3 Learning in the Statistical Queries Model

Our last example for this section is a generic transformation of algorithms in the statistical queries learning model [22] to algorithms that access their data via noisy sum queries, and hence their outcome preserves privacy. In the statistical query model, a latent probability distribution over the domain $\mathcal{D}$ is assumed, and, instead of accessing samples of this distribution, learning algorithms repeatedly invoke the following computational primitive:

Statistical Query:

Input: A predicate $p : \mathcal{D} \rightarrow \{0, 1\}$ and an additive accuracy parameter τ.

Output: The expected fraction of samples satisfying $p()$, to within additive error τ.

Conceptually, the framework models drawing a sufficient number of samples so that the observed count of samples satisfying p is a good estimate of the actual expectation.

A learning algorithm learns a concept — predicate c out of a concept class C — if it produces a predicate such that the probability of misclassification under the latent distribution is bounded by some parameter δ. The transformation from learning algorithms in the statistical learning model to private learning algorithms is rather straightforward. If the number of samples is large enough, then we can use noisy `sum` queries to estimate the probability of $p()$ within the required accuracy.

Statistical Query Emulation:

Input: p, τ, error probability δ', allotted privacy parameter ε.

1 [Check if query can be computed accurately enough without breaching privacy]

$$\text{Set } \varepsilon' \leftarrow \frac{\ln(1/\delta')}{\tau n} \text{; If } \varepsilon < \varepsilon' \text{ then halt.}$$

2 Answer $\frac{\texttt{sum}_p + Y}{n}$ where $Y \sim \mathbf{Lap}(1/\varepsilon')$.

3 [Update allotted privacy parameter]

$$\text{Set } \varepsilon \leftarrow \varepsilon - \varepsilon' \,.$$

THEOREM 16.12 *Let $\mathcal{A}$ be an algorithm that δ-learns a concept class C using at most q statistical queries of accuracy $\{\tau_1, \dots, \tau_q\}$. If*

$$n \geq \frac{1}{\varepsilon} \ln\left(\frac{q}{\delta}\right) \sum_{i=1}^{q} \frac{1}{\tau_i} \,,$$

then $\mathcal{A}$ that accesses **Private Statistical Query** *with error parameter $\delta' = \delta/q$ can 2δ-learn C on n elements while preserving ε-privacy.*

Proof: Note in the ith call to **Private Statistical Query** $\varepsilon'_i = \frac{\ln(q/\delta)}{\tau_i n}$, hence the procedure never halts as

$$\frac{1}{n} \ln\left(\frac{q}{\delta}\right) \sum_{i=1}^{q} \frac{1}{\tau_i} \leq \varepsilon \,.$$

To see that the misclassification probability of $\mathcal{A}$ grows by at most δ note that the value of $|Y|$ is distributed according to the exponential distribution, and satisfies $\Pr[|Y| > z] = e^{-z/\lambda}$. Hence, the probability that in the ith iteration $|Y| > n\tau_i$ is bounded by $e^{-n\tau_i\varepsilon'} = \delta/q$. Using the union bound, the probability that in any of the iterations $|Y| > n\tau_i$ is bounded by δ.

Finally, using Lemma 16.7 we get that the outcome of $\mathcal{A}$ preserves ε-privacy. ∎

The importance of Theorem 16.12 is in its generality. Although it would probably not yield the most efficient algorithm for specific learning tasks (e.g. in terms of the number of samples needed), it shows that an important collection of learning problems can be solved while preserving ε-privacy.

16.6 Beyond the Basics

As we have seen in Section 16.4.1, Theorem 16.10 directly yields simple output perturbation sanitizers for a variety of functions — those which exhibit low global sensitivity. However, in some cases Theorem 16.10 cannot be directly used. E.g. when one is interested in a query f that does not exhibit low global sensitivity (when compared with the magnitude of $f(\mathbf{x})$), or the global sensitivity of f is hard to analyze (or intractable), or when the range of f does not lend itself to a natural metric. In Section 16.5 we have seen one technique to get around these shortcomings, by expressing complex functionalities in terms of simple, insensitive ones, that are easy to analyze.

We review some of the more recent techniques for creating algorithms that preserve ε-privacy. The presentation of this section is not self contained as we only attempt to present the main ideas.

16.6.1 Instance Based Noise and Smooth Sensitivity

The framework of Theorem 16.10 considers the global, i.e. *worst-case*, sensitivity of the query function f. However, for many interesting functions, the worst-case sensitivity is high due to instances that do not typically occur in practice. As an example, consider the median function:

EXAMPLE 16.13 (MEDIAN) *Let $x_1, \ldots, x_n$ be real numbers taken from a bounded interval $[0,1]$. The median of $\mathbf{x} = x_1, \ldots, x_n$ is its middle ranked element. Assuming (for simplicity) that n is odd, and that $x_1 \leq x_2 \leq \cdots \leq x_n$, we can write:* $\mathtt{med}(\mathbf{x}) = x_{\frac{n+1}{2}}$. *Although* $\mathtt{med}$ *is usually considered insensitive, it exhibits high global sensitivity. To see that, consider the case where*

$$x_1 = \cdots = x_{\frac{n+1}{2}} = 0 \text{ and } x_{\frac{n+1}{2}+1} = \cdots = x_n \,.$$

Note that $\mathtt{med}(x_1, \ldots, x_n) = 0$*, and that by setting* $x_{\frac{n+1}{2}} = 1$*, we get* $\mathtt{med}(x_1, \ldots, x_n) = 1$*. Hence,* $\mathbf{GS}_{\mathtt{med}} = 1$*. Applying Theorem 16.10 hence results with noise magnitude* $\mathbf{GS}_{\mathtt{med}}/\varepsilon$ *that, for small* ε*, completely destroys the information.*

A first natural attempt at fixing this problem is to consider a local variant of Equation 16.5, and perturb the query function result with noise poroportional to it:

$$\mathbf{LS}_f(\mathbf{x}) = \max_{\mathbf{x}': \mathbf{dist}_H(\mathbf{x},\mathbf{x}')=1} \|f(\mathbf{x}) - f(\mathbf{x}')\|_1 \,.$$

(Observe that $\mathbf{GS}_f = \max_{\mathbf{x}} \mathbf{LS}(\mathbf{x})$.) This attempt fails, as we show now for the median.

EXAMPLE 16.13 (MEDIAN (CONT.)) *It is easy to see that given an instance* $\mathbf{x}$*, the maximum change in* $\mathtt{med}(\mathbf{x})$ *occurs when* x_1 *is set to 1 or when* x_n *is set to 0. This observation yields an expression for the local sensitivity in terms of the values next to the median:*

$$\mathbf{LS}_{\mathtt{med}}(\mathbf{x}) = \max\left(x_{\frac{n+1}{2}} - x_{\frac{n+1}{2}-1}, x_{\frac{n+1}{2}+1} - x_{\frac{n+1}{2}}\right) \,.$$

For inputs where a constant fraction of the population is uniformly concentrated around the median we get $\mathbf{LS}_{\mathtt{med}}(\mathbf{x}) \propto \frac{1}{n} \ll \mathbf{GS}_{\mathtt{med}}$.

Releasing $\mathtt{med}(\mathbf{x})$ *with noise sampled from* $\mathbf{Lap}(\mathbf{LS}_{\mathtt{med}}(\mathbf{x})/\varepsilon)$ *fails to satisfy Definition 16.5. For instance, the probability of receiving a non-zero answer when* $x_1 = \cdots = x_{\frac{n+1}{2}+1} = 0$ *and* $x_{\frac{n+1}{2}+2}, \ldots, x_n > 0$ *is zero, whereas the probability of a non-zero answer on its neighbor database where* $x_{\frac{n+1}{2}+1} > 0$ *is one.*

This example illustrates that special care has to be taken when adding instance based noise. As the noise is correlated with the instance $\mathbf{x}$, it may itself be the cause of information leakage. To prevent this kind of leakage, a variant of local sensitivity — smooth sensitivity — was defined in [25], such that adding noise proportional to the smooth sensitivity at $\mathbf{x}$ is safe. Unlike local sensitivity, smooth sensitivity does not change abruptly as $\mathbf{x}$ changes, and hence an adversary cannot distinguish well the noise distributions on neighbor databases, as was in the example above.

We will only preset the definition of smooth sensitivity, without getting into the details of constructing output perturbation sanitizers with instance based noise[7].

DEFINITION 16.14 (SMOOTH SENSITIVITY) *An ε-smooth upperbound on $\mathbf{LS}_f$ is a function satisfying*

$$\mathbf{S}_f(\mathbf{x}) \geq \mathbf{LS}_f(\mathbf{x}) \qquad \textit{for all databases } \mathbf{x}\text{ ; } \textit{and}$$
$$\mathbf{S}_f(\mathbf{x}) \leq e^{\varepsilon}\mathbf{S}_f(\mathbf{x}') \qquad \textit{for all neighbor databases } \mathbf{x}, \mathbf{x}' \text{ .}$$

Clearly, $\mathbf{S}_f(\mathbf{x}) = \mathbf{GS}_f$ is an ε-smooth upperbound on $\mathbf{LS}_f$, but the definition allows for cases where $\mathbf{S}_f(\mathbf{x}) \ll \mathbf{GS}_f$, and hence a gain with respect to Theorem 16.10.

It turns out that a minimal ε-smooth upperbound on $\mathbf{LS}_f$ exists. This function is called the *ε-smooth sensitivity* of f and satisfies for every smooth upperbound $\mathbf{S}_f$ on $\mathbf{LS}_f$:

$$\mathbf{S}^*_f(\mathbf{x}) \leq \mathbf{S}_f(\mathbf{x}) \text{ for all } \mathbf{x} \in \mathcal{D}^n \text{ .}$$

It can be shown that

$$\mathbf{S}^*_f(\mathbf{x}) = \max_{\mathbf{x}' \in \mathcal{D}^n} \left(\mathbf{LS}_f(\mathbf{x}') \cdot e^{\varepsilon \cdot \mathbf{dist}_H(\mathbf{x},\mathbf{x}')} \right) \text{ .} \tag{16.7}$$

Equation 16.7 implies that low noise may be added at the instance $\mathbf{x}$ if the local sensitivity at its 'neighborhood' is low (i.e. $\mathbf{LS}_f(\mathbf{x}')$ is low for those instances $\mathbf{x}'$ where $\mathbf{dist}_H(\mathbf{x}, \mathbf{x}')$ is small), as the influence of far instances decays exponentially with $\mathbf{dist}_H(\mathbf{x}, \mathbf{x}')$.

Computing $\mathbf{S}^*_f(\mathbf{x})$ may prove to be tricky, and if an approximation to $\mathbf{S}^*_f(\mathbf{x})$ is used for the noise magnitude, it has to be a smooth upperbound on $\mathbf{LS}_f$ by itself. We omit these details, and refer the reader to [25] where it is shown how to compute $\mathbf{S}^*_f(\mathbf{x})$ for queries like median, minimum, and graph problems such as MST cost and the number of triangles in a graph.

[7] The technicalities include (i) a relaxation of Definition 16.5 where breaches may occur with negligible probability, and (ii) conditions on the noise process (in analogy to Equation 16.6).

16.6.2 The Sample-Aggregate Framework

The sample-aggregate framework of [25] is a generic technique for creating a 'smoothed' version $\bar{f}$ of a query function f. Assume $f(\mathbf{x})$ is a function that can be well approximated on random samples taken from $x_1, \ldots, x_n$. We abuse notation and write $f(S)$ for the approximation of $f(\mathbf{x})$ where $S \subset \{x_1, \ldots, x_n\}$ although f is formally defined to take an n-tuple as input.

The function f is evaluated on several random samples, and the results of these evaluations $f(S_1), \ldots, f(S_t)$ are combined using an *aggregation function* g:

$$\bar{f} = g(f(S_1), \ldots, f(S_t)) \ .$$

The main observation is that to preserve privacy it is sufficient to add noise whose magnitude depends on the smooth sensitivity[8] of the aggregation function g. To illustrate why privacy would be preserved, assume (for simplicity) that each entry from $x_1, \ldots, x_n$ appears in exactly one of the samples $S_1, \ldots, S_t$. As Definition 16.5 is only concerned with neighbor databases, we only need to care about a change in a single entry $x_i \in S_j$. however, a change in x_i may only affect a single of the inputs to g (i.e. $f(S_j)$). Even if the change in $f(S_j)$ is significant, it is enough to mask it by adding noise proportional to the smooth sensitivity of g. (The complete argument is a little more involved. In particular, x_i may appear in several of the subsets.)

The crux of this technique is finding good aggregation functions, i.e. functions g whose outcome (plus the required noise) would faithfully represent $f(S_1), \ldots, f(S_t)$. In particular, when $f(S_1), \ldots, f(S_t)$ are well concentrated (or 'clustered'), the aggregation $g(f(S_1), \ldots, f(S_t))$ should return a point that is close to the cluster center, *and* the noise level should be low. Furthermore, we would like g, and its smooth sensitivity to be efficiently computable. An aggregation function satisfying these requirements — the *center of attention* — was proposed in [25].

The sample-aggregate technique was applied to Lloyd's algorithm, and to the problem of learning the parameters of a mixture of k spherical Gaussian distributions when the data x consists of polynomially-many (in the dimension and k) i.i.d. samples from the distribution.

As with the result of Section 16.5.3, an application of sample-aggregate need not always result in the optimal sanitizer. It serves, however, as a strong feasibility result that is appealing to our intuition, showing that all functions that are well approximated on random samples can be computed privately with

[8] In principle, adding noise proportional to global sensitivity would also work.

relatively low noise[9]. Furthermore, aggregation may serve as a strong algorithmic tool in the construction of private data analysis algorithms.

16.6.3 A General Sanitization Mechanism

We conclude this short section with another generic technique for constructing ε-private sanitizers, recently put forward by McSherry and Talwar [28]. In this technique the query function $f : \mathcal{D}^n \to \mathcal{R}$ is replaced with a score (or output quality) function $q : \mathcal{D}^n \times \mathcal{R} \to \mathbb{R}$. Intuitively, $q(\mathbf{x}, r)$ may represent a query function $f(\mathbf{x})$ by giving high scores to pairs $(\mathbf{x}, r)$ such that $f(\mathbf{x}) = r$, e.g., for real valued functions we may define $q(\mathbf{x}, r) = -\|f(\mathbf{x}) - r\|_1$.

We can define a property of the function q that is analogous to our definition of sensitivity:

$$\Delta q = \max_r \max_{\mathbf{x},\mathbf{x}' :\, \mathbf{dist}_H(\mathbf{x},\mathbf{x}')=1} q(\mathbf{x}, r) - q(\mathbf{x}', r) \; .$$

The ε-private sanitizer then picks an answer $r \in \mathcal{R}$ with probability proportional to $e^{\frac{\varepsilon}{2\Delta q} q(\mathbf{x},r)}$.

This sanitization mechanism improve on the results of Section 16.4 as it does not require the range $\mathcal{R}$ of f to be a metric space. Note, however, that this generality may come with a price, as sampling the answer r may be a non-trivial.

16.7 Related Work and Bibliographic Notes

There is a vast body of work on private data analysis, pertaining to research disciplines as statistics, security, databases and cryptography. Our goal in this short section is not to review this rich work, but rather to reference that part of recent work on private data analysis which is most relevant to our presentation.

The recent interest in perturbation techniques for privacy was rekindled in part due to a work by R. Agrawal and Srikant [6]. They considered a non-interactive *input perturbation* model, akin to [33], where individual information is sanitized once by adding noise. D. Agrawal and Aggarwal [4], and later Evfimievski, Gehrke, and Srikant [19] improved on the privacy definition of [6]. In particular, [19] defined privacy in terms of the change between the a priori and the a posteriori probability of arbitrary predicates applied to individual records, and identified a sufficient criteria of the randomization operator for guaranteeing privacy.

Dinur, Dwork, and Nissim [12, 18] initiated a formal study of privacy in data analysis, in light of modern cryptographic research. [12] studied the required

[9] In fact, the feasibility result is stronger: for all instances $\mathbf{x}$ where f can be well approximated on random samples, it is possible to learn $f(\mathbf{x})$ with low noise.

noise magnitude for output perturbation when an attacker issues subset-sum queries. Stating the problem as a decoding with noise, they showed that an attacker that makes an exponential number of queries, can reconstruct almost the entire database, unless noise magnitude is $\Omega(n)$; and, similarly, that a polynomial time attacker that makes $\approx n$ random queries may accurately reconstruct almost the entire database, unless noise magnitude is $\approx \sqrt{n}$. These results were strengthened by the recent work of Dwork, McSherry, and Talwar [17]. The negative results of [12] guided some of the developments towards the techniques described in this chapter.

The first formal definitions of privacy [19, 12, 18, 8, 6] were well understood when the database entries were sampled i.i.d. from some distribution, and without attacker access to auxiliary information. Trying to capture the more general case, led to the notion of 'informed adversary' in [6]. The definition presented herein (Definition 16.5 from [16]) is more intuitive and easier to understand. It was influenced by the impossibility result of Dwork and Naor [13]. The notion of differential privacy is due to Dwork and McSherry [13].

The first positive results in the line that led to the techniques presented here are by Dinur, Dwork, and Nissim [12, 18]. These results dealt with sum queries, and capitalized on limiting the number queries made to the database to be sublinear in its size. Blum et al. [6] have further built on these results, showing that for many algorithms there exists a version that accesses their inputs via a limited number of noisy sum queries, and hence preserves privacy. These techniques were generalized from sum queries to arbitrary queries in the framework of calibrating noise to global sensitivity [16]. This framework resulted in simplified proofs of privacy as well as noise magnitudes that were lower than what was possible by previous work. The work in [25] introduced the concept of smooth sensitivity and presented the sample-aggregate framework. The recent work of [28] presented the general sanitization mechanism.

The area of private data analysis has been very dynamic in the last few years, yielding many interesting results that are not included in this chapter, some of which we reference briefly below. Chawla et al. [8, 9] have initiated a study of non-interactive sanitization under a formal definition of privacy requiring the inability of an adversary to *isolate* an individual in the database. Another approach to non-interactive sanitization, based on techniques similar to those in this writeup was put forward by Dwork and Nissim [18]. Kenthapadi et al. [23] studied auditing — a technique where the sanitizer may refuse to answer 'dangerous' queries — and have put forward a notion of *simulatable auditing* where the decision whether to allow a query does not leak information. Dwork et al. [15] presented the first distributed noise generation protocols. Mishra and Sandler [27] presented *pseudorandom sketches*. It turns out that in terms of utility there is a separation between interactive and non-interactive sanitization. A weak result is implied by the impossibility result of [12] and stronger results were presented in [16, 17]. Chaudhuri and Mishra [10] discuss the privacy of random samples. Barak et al. [4] show how to construct consistent contingency tables. McSherry and Talwar [28] initiate an exploration of the relationship between privacy and game theory and show that privacy has implications to mechanism design.

Acknowledgements

I am greatly indebted to Cynthia Dwork, Frank McSherry and Adam Smith for generously sharing with me many of their insights about privacy.

References

[1] Nabil R. Adam and John C. Wortmann. Security-control methods for statistical databases: a comparative study. In *ACM Computing surveys*, Vol. 21, No. 4, pages 515–556, 1989.

[2] Dakshi Agrawal and Charu C. Aggarwal. On the design and quantification of privacy preserving data mining algorithms. In *Proceedings of the 20th Symposium on Principles of Database Systems (PODS)*, pages 247–255, 2001.

[3] Rakesh Agrawal and Ramakrishnan Srikant. Privacy-preserving data mining. In *Proceedings of the 2000 SIGMOD International Conference on Management of Data*, Vol. 29, No. 2, pages 439–450, 2000.

[4] Boaz Barak, Kamalika Chaudhuri, Cynthia Dwork, Satyen Kale, Frank McSherry, and Kunal Talwar. Privacy, accuracy, and consistency too: a holistic solution to contingency table release. In *Proceedings of the 26th Symposium on Principles of Database Systems (PODS)*, pages 273–282, 2007.

[5] Michael Ben-Or, Shafi Goldwasser, and Avi Wigderson. Completeness theorems for noncryptographic fault-tolerant distributed computations. In *Proceedings of the 20th Symposium on the Theory of Computing (STOC)*, pages 1–10, 1988.

[6] Avrim Blum, Cynthia Dwork, Frank McSherry, and Kobbi Nissim. Practical privacy: The SULQ framework. In *Proceedings of the 24th Symposium on Principles of Database Systems (PODS)*, Pages 128–138, 2005.

[7] David Chaum, Claude Crépeau, and Ivan Damgård. Multiparty unconditionally secure protocols. In *Proceedings of the 20th Symposium on the Theory of Computing (STOC)*, pages 11–19, 1988.

[8] Shuchi Chawla, Cynthia Dwork, Frank McSherry, Adam Smith, and Hoeteck Wee. Toward privacy in public databases. In *Theory of Cryptography Conference (TCC)*, pages 363–385, 2005.

[9] Shuchi Chawla, Cynthia Dwork, Frank McSherry, and Kunal Talwar. On the utility of privacy-preserving histograms. In *21st Conference on Uncertainty in Artificial Intelligence (UAI)*, 2005.

[10] Kamalika Chaudhuri and Nina Mishra When Random Sampling Preserves Privacy. In *Proceedings of the 26th Annual International Cryptology Conference (CRYPTO)*, LNCS 4117, Springer, pages 198–213, 2006.

[11] Tore Dalenius. Towards a methodology for statistical disclusure control. In *statistik Tidskrift*, Vol. 15, pages 429–444, 1997.

[12] Irit Dinur and Kobbi Nissim. Revealing information while preserving privacy. In *Proceedings of the 22nd Symposium on Principles of Database Systems (PODS)*, pages 202–210, 2003.

[13] Cynthia Dwork. Differential Privacy. In *Proceedings of the 33rd International Colloquium on Automata, Languages and Programming (ICALP)*, LNCS 4052, pages 1–12, 2006.

[14] Cynthia Dwork. Ask a Better Question, Get a Better Answer. A New Approach to Private Data Analysis. In *Proceedings of the 11th International Conference on Database Theory (ICDT)*, LNCS 4353, pages 18–27, 2007.

[15] Cynthia Dwork, Krishnaram Kenthapadi, Frank McSherry, Ilya Mironov, and Moni Naor. Our data, ourselves: Privacy via distributed noise generation. In *25th Annual International Conference on the Theory and Applications of Cryptographic Techniques (EUROCRYPT)*, LNCS 4004, pages 486–503, 2006.

[16] Cynthia Dwork, Frank McSherry, Kobbi Nissim, and Adam Smith. Calibrating Noise to Sensitivity in Private Data Analysis. In *Theory of Cryptography Conference (TCC)*, pages 265–284, 2006.

[17] Cynthia Dwork, Frank McSherry, and Kunal Talwar, The price of privacy and the limits of LP decoding. In *Proceedings of the 39th Annual ACM Symposium on Theory of Computing (STOC)*, pages 85–94, 2007.

[18] Cynthia Dwork and Kobbi Nissim. Privacy-preserving datamining on vertically partitioned databases. In *Advances in Cryptology - CRYPTO 2004, 24th Annual International Cryptology Conference (CRYPTO)* LNCS 3152, pages 528–544, 2004.

[19] Alexandre V. Evfimievski, Johannes Gehrke, and Ramakrishnan Srikant. Limiting privacy breaches in privacy preserving data mining. In *Proceedings of the 22nd Symposium on Principles of Database Systems (PODS)*, pages 211–222, 2003.

[20] Shafi Goldwasser and Silvio Micali. Probabilistic encryption. In *Journal of Computer and System Sciences*, Vol. 28, No. 2, pages 270–299, April 1984.

[21] Oded Goldreich, Silvio Micali, and Avi Wigderson. How to play any mental game. A Completeness Theorem for Protocols with Honest Majority. In *Proceedings of the 19th Annual ACM Symposium on Theory of Computing (STOC)*, pages 218–229, 1987.

[22] Michael Kearns, Efficient Noise-Tolerant Learning from Statistical Queries, In *Journal of the ACM* Vol. 45, No. 6, pages 983 – 1006, 1998. See also *Proceedings of the Twenty-Fifth Annual ACM Symposium on Theory of Computing (STOC)*, pages 392–401, 1993.

[23] Krishnaram Kenthapadi, Nina Mishra, and Kobbi Nissim. Simulatable auditing In *Proceedings of the Twenty-fourth ACM SIGACT-SIGMOD-SIGART Symposium on Principles of Database Systems (PODS)*, Pages 118–127, 2005.

[24] Yehuda Lindell and Benny Pinkas. Privacy preserving data mining. In *Journal of Cryptology*, Vol. 15, No. 3, pages 177–206, 2002.

[25] Kobbi Nissim, Sofya Raskhodnikova, and Adam Smith. Smooth Sensitivity and Sampling in Private Data Analysis. In *Proceedings of the 39th Annual ACM Symposium on Theory of Computing (STOC)*, pages 7584, 2007.

[26] Ashwin Machanavajjhala, Johannes Gehrke, Daniel Kifer, and Muthuramakrishnan Venkitasubramaniam. l-Diversity: Privacy Beyond k-Anonymity. In *Proceedings of the 22nd International Conference on Data Engineering, (ICDE)*, page 24, 2006.

[27] Nina Mishra and Mark Sandler. Privacy via pseudorandom sketches. In *Proceedings of the Twenty-Fifth ACM SIGACT-SIGMOD-SIGART Symposium on Principles of Database Systems (PODS)*, pages 143–152.

[28] Frank McSherry and Kunal Talwar. Mechanism Design via Differential Privacy. To appear, FOCS 2007.

[29] Shubha U. Nabar, Bhaskara Marthi, Krishnaram Kenthapadi, Nina Mishra, and Rajeev Motwani. Towards Robustness in Query Auditing. In *Proceedings of the 32nd International Conference on Very Large Data Bases (VLDB)*, pages 151-162, 2006.

[30] M. J. O'Connell, Search Program for Significant Variables, In *Computer Physics Communications*, Vol. 8, No. 1, Pages 49-55, 1974.

[31] Latanya Sweeney. k-anonymity: a model for protecting privacy. In *International Journal of Uncertainty, Fuzziness, and Knowledge-Based Systems*, 10(5):557–570, 2002.

[32] Latanya Sweeney. Achieving k-anonymity privacy protection using generalization and Suppression. In *International Journal of Uncertainty, Fuzziness, and Knowledge-Based Systems*, Vol. 10, No. 5, pages 571–588, 2002.

[33] Stanley L. Warner. Randomized response: A survey technique for eliminating evasive answer bias. In *Journal of the American Statistical Association*, Vol. 60, No. 309, pages 63–69, 1965.

[34] Andrew C. Yao. Protocols for secure computations. In *Proceedings of the 23th IEEE Symposium on Foundations of Computer Science (FOCS)*, pages 160–164, 1982.

Chapter 17

A Survey of Query Auditing Techniques for Data Privacy

Shubha U. Nabar
Stanford University
sunabar@cs.stanford.edu

Krishnaram Kenthapadi
Microsoft Search Labs
krishnaram.kenthapadi@microsoft.com

Nina Mishra
University of Virginia
nmishra@cs.virginia.edu

Rajeev Motwani
Stanford University
rajeev@cs.stanford.edu

Keywords: Query auditing, offline auditing, online auditing.

17.1 Introduction

This chapter is a survey of query auditing techniques for detecting and preventing disclosures in a database containing private data. Informally, auditing is the process of examining past actions to check whether they were in conformance with official policies. In the context of database systems with specific data disclosure policies, auditing is the process of examining queries that were answered in the past to determine whether answers to these queries could have been used by an individual to ascertain confidential information forbidden by the disclosure policies. Techniques used for *detecting* disclosures could potentially also be used or extended to *prevent* disclosures, and so in addition to the retroactive auditing mentioned above, researchers have also studied an online variant of the auditing problem wherein the task of an online auditor is to *deny* queries that could potentially cause a breach of privacy.

Other common approaches to tackling the disclosure prevention problem include adding noise to the the data or otherwise perturbing the query results supplied to the user. However statisticians are generally averse to potential biases introduced by adding noise. One commonly stated reason is that the data collection process is already prone to biases and imperfections due to factors such as too few respondents, the cost of gathering data, and inaccurate answers provided by respondents. Since important decisions are made based on this data, they prefer to receive answers without additional noise. It is in this context that query restriction techniques become relevant in disclosure prevention. The work on offline (or retroactive) auditing has also similarly focused on the case where answers supplied to users are exact.

The main focus of this chapter will be on statistical databases with a single private attribute that only permit aggregate queries such as `sum`, `max`, `min` or `median` over this private attribute. An instructive example is a company database with *employee salary* as a private attribute. Or a set of medical records with a boolean private attribute indicating whether or not a patient was HIV-positive. We will first review the most commonly used notion of disclosure in the statistical database literature called *full disclosure* and then review algorithms and hardness results for offline auditing that have been developed for different classes of queries under this definition.

A natural question to ask is whether offline auditors could directly be used as online auditors as well. The answer to the question, as we shall see, is *no* due to the fact that query denials can leak information. Researchers have proposed the paradigm of *simulatability* to surmount this problem, and developed simulatable auditors for different classes of queries to prevent full disclosure. We will review some of them.

The notion of full disclosure is not entirely satisfactory as a measure of disclosure, so we will next present a recently proposed measure called *partial disclosure* as well as simulatable online auditors that have been proposed for different classes of queries under this definition. We will conclude the chapter with a brief survey of results in another auditing scenario where the information to be protected is an arbitrary view of the database; and finally end with a discussion of the limitations of present day auditing techniques.

17.2 Auditing Aggregate Queries

Most work on aggregate queries has focused on the case of a single numerical private attribute that is either real valued (from a bounded or unbounded range) or boolean. Additionally, most auditing algorithms developed are for queries of only one kind, with hardness results for auditing combinations of queries. Before proceeding further, we will formalize some of the terminology used in the remainder of this section.

Let $X = \{x_1, \ldots, x_n\}$ be the set of private attribute values of n individuals in a database. An aggregate query $q = (Q, f)$ specifies a subset of the records $Q \subseteq \{1, \ldots, n\}$ and a function f such as `sum`, `max`, `min` or `median`. The result, $f(Q)$, is f applied to the subset $\{x_i \mid i \in Q\}$. We call Q the *query set* of q.

17.2.1 Offline Auditing

We now survey some of the results in the offline auditing literature.

Full Disclosure. Given the set of private values X and a set of aggregate queries $\mathcal{Q} = \{q_1, \ldots, q_t\}$ posed over this data set that were correspondingly answered $\{a_1, \ldots, a_t\}$, the goal of an offline auditor is to determine if an individual's private value can be deduced. Traditionally, the definition of disclosure that has been used is the notion of full disclosure defined below.

DEFINITION 17.1 (FULL DISCLOSURE) *An element $x_i \in X$ is fully disclosed by a query set $\mathcal{Q}$ if it can be uniquely determined, i.e., in all possible data sets X consistent with the answers $a_1, \ldots, a_t$, to queries $q_1, \ldots, q_t$, x_i is the same.*

As a simple example, if the query set consisted of a single query asking for the sum of the salaries of all the female employees in the company, and Alice was the only female employee in the company, then the answer to this query uniquely determines Alice's salary.

In general the answers to many different queries can be stitched together by a user to uniquely determine an individual's private value. The goal of the auditor then is to prevent such a full disclosure.

Examples of Offline Auditors. As one example of such an auditor, consider a set of `sum` queries posed over X, the elements of which are real-valued from an unbounded range. To determine if the answers to these queries can be used to uniquely deduce some private value, the auditor essentially needs to solve a system of linear equations. It maintains a matrix where the rows correspond to queries and the columns to private values. Each query is represented by a vector of 1s and 0s, indexing the private elements that were in the `sum` query. The matrix of query vectors is diagonalized via a series of elementary row operations and column interchanges. If the resulting matrix has a row with only one 1 and $n-1$ 0s, then some element is uniquely determined. Since only a linearly independent set of query vectors need to be examined, the matrix is of size at most $n \times n$, and the diagonalization can be carried out in time $O(n^3)$.

Since finding a maximal set of linearly independent query vectors requires $O(n^2|\mathcal{Q}|)$ time, sum queries can be audited in polynomial time.

THEOREM 17.2 *Let $X \in \mathbb{R}^n$ be a data set of private values. There is an algorithm to determine if an $x_i \in X$ is fully disclosed by a set of* `sum` *queries $\mathcal{Q}$ and corresponding answers $\mathcal{A}$ that runs in time $O(n^3 + n^2|\mathcal{Q}|)$.*

Besides `sum` queries, offline auditors for exact determination of full disclosure also exist for combinations of `max` and `min` queries, `median` queries and `average` queries over real-valued data. Unfortunately, no significant progress has been made in auditing arbitrary combinations of aggregate queries. For example, the following hardness result has been proved via a reduction from set partition.

THEOREM 17.3 *There is no polynomial time full-disclosure auditing algorithm for* `sum` *and* `max` *queries unless P=NP.*

The auditing problem has also been examined when the private attribute is boolean. Surprisingly, full-disclosure auditing of `sum` queries over boolean data is coNP-hard. There exists an efficient polynomial time algorithm, however, in the special case where the queries are 1-dimensional, i.e., for some ordering of the elements in X, the query set for each query involves a consecutive sequence of x_i's. Considering such restrictions of the general auditing problem is useful in practice, since in reality, users would rarely be able to pose queries over arbitrary subsets of the data. Rather, they would use conditions over some attribute or combinations of attributes to select specific records in the data set to aggregate. For example, a realistic query would ask for the total number of HIV-positive people in a particular age group. The set of queries asking for the total number of HIV-positive people in various age groups would form a set of 1-dimensional `sum` queries over a boolean private attribute. Such assumptions about the structure of queries can yield even more efficient auditors. For example, the `sum` auditor over real-valued data can be made to run in linear time over 1-dimensional `sum` queries.

17.2.2 Online Auditing

In recent years, researchers have also become interested in the online auditing problem as a means of preventing data disclosure. Given a sequence of queries, $q_1, \ldots, q_{t-1}$ that have already been posed, corresponding answers $a_1, \ldots, a_{t-1}$ that have already been supplied, and a new query q_t, the task of an online auditor is to determine if the new query should be answered as such, or denied in order to prevent a privacy breach. Here each of the previous answers a_i, is itself either the true answer $f_i(Q_i)$ to query q_i, or a "denial".

The earliest online auditors prevented disclosures by restricting the size and overlap of queries that could be answered. For the case of `sum` queries, for instance, it was shown that for queries with query sets of exactly k elements, each pair of query sets overlapping in at most r elements, any data set can be compromised in $(2k - (l + 1))/r$ queries by an attacker who knows l values a priori. For fixed k, r and l, if the auditor denies answers to query $(2k - (l + 1))/r$ and on, then the data set is definitely not compromised, i.e., no private value can be uniquely determined. Such an auditing scheme is rather limited: if $k = n/c$ for some constant c and $r = 1$, then after only a constant number of distinct queries, the auditor would have to deny all further queries since there are only about c queries where no two overlap in more than one element. This motivated a search for auditors that could provide greater utility.

The next natural question is whether offline auditors can directly solve the online auditing problem. Whenever a new query is posed, the online auditor checks to see if the answer to this query in combination with all previous query responses can be used to uniquely determine a private value. If so, the query is denied, else it is answered exactly. While it would seem that such an approach should work, in actuality it does not as we demonstrate next.

Example where Denials Leak: Suppose that the underlying data set is real-valued and that a query is denied only if some value is fully disclosed. Suppose that the attacker poses the first query $\mathtt{sum}(x_1, x_2, x_3)$ and the auditor answers 15. Suppose also that the attacker then poses a second query $\mathtt{max}(x_1, x_2, x_3)$ and the auditor denies the answer. The denial tells the attacker that if the true answer to the second query were given then some value could be uniquely determined. Note that $\mathtt{max}(x_1, x_2, x_3) \not< 5$ since then the sum could not be 15. Further, if $\mathtt{max}(x_1, x_2, x_3) > 5$ then the query would not have been denied since no value could be uniquely determined. Consequently, $\mathtt{max}(x_1, x_2, x_3) = 5$ and the attacker learns that $x_1 = x_2 = x_3 = 5$ — a privacy breach of all three entries. The issue here is that denials reduce the space of possible consistent solutions, and we have not explicitly accounted for this.

In this example only a few values were compromised. However, it is possible to construct examples where a large fraction of private values can be uniquely determined. Intuitively, denials that depend on the answer to the current query leak information because users can ask why a query was denied, and the reason is in the data. If the decision to answer or deny a query depends on the actual data, it reduces the set of possible consistent solutions for the underlying data.

Another naive solution to the leakage problem is to deny whenever the offline algorithm does, and to also randomly deny queries that would normally be answered. While this solution seems appealing, it has its own problems. Most importantly, although it may be that denials leak less information, leakage is not generally prevented. Furthermore, the auditing algorithm would need to remember which queries were randomly denied, since otherwise an attacker could repeatedly pose the same query until it was answered. A difficulty then arises in determining whether two queries are equivalent. The computational hardness of this problem depends on the query language, and may be intractable, or even undecidable. As a work around to this problem, the *simulation* paradigm (used vastly in cryptography) was proposed and is described next.

Simulatable Auditing. The idea for simulatable auditing came from the following observation: Query denials have the potential to leak information if in choosing to deny, the auditor uses information that is unavailable to the attacker (the answer to the newly posed query). A successful attacker capitalizes on this leakage to infer private values. The requirement of a simulatable auditor then, is that the attacker should be able to simulate or mimic the auditors decisions to answer or deny a query. In such a scenario, because the attacker can equivalently determine for himself when his queries will be denied, denials provably do not leak information. More formally, let $\mathcal{Q} = \{q_1, \ldots, q_t\}$ be any sequence of queries and $\mathcal{A} = \{a_1, \ldots, a_t\}$ be their corresponding answers. Here each a_i is either the exact answer $f_i(Q_i)$ to query q_i on the data set X, or a denial.

DEFINITION 17.4 (ONLINE AUDITOR) *An online auditor B is a function of $\mathcal{Q}$, $\mathcal{A}$ and X that returns as output either an exact answer to q_t or a denial.*

DEFINITION 17.5 (SIMULATABLE AUDITOR) *An online auditor B is simulatable, if there exists another auditor B' that is a function of only $\mathcal{Q}$ and $\mathcal{A} \setminus a_t$ and whose output on q_t is always equal to that of B.*

An attractive property of simulatable auditors is that the auditor's response to denied queries does not convey any new information to the attacker (beyond what is already known given the answers to the previous queries). Hence denied queries need not be taken into account in future decisions that the auditor makes.

Note that the auditor that restricted the size and overlap of queries was simulatable since it never actually looked at the answers to queries in choosing to deny. As another example of a simulatable auditor, the `sum` auditor over real-valued data from Section 17.2.1.0 is also simulatable since all that is examined in making the decision to deny or answer is the matrix of query vectors and never the actual answers to any of the queries, let alone the answer to the current query. In contrast to the query-size-and-overlap-restricting auditor, this auditor has also been shown to provide fairly high utility for large data sets — in a sequence of random sum queries over a data set, the first denial can be expected to occur only after a linear number of queries.

A more general sufficient condition for ensuring simulatability is that in making its decision, with each new query, the auditor should determine if there is any possible data set, consistent with all past responses, in which the answer to the current query would cause some element to be fully disclosed. If so, the query should be denied, else it can be answered. Since this is a condition that an attacker could check for himself and predict denials, denials leak no information. Using this idea, simulatable online auditors have been constructed for `max` and `min` queries.

In the example from the previous section, the query $q_1 = \texttt{sum}\{x_1, x_2, x_3\}$ would be answered, since no matter the answer, no element from the data set could be uniquely pinned down. The second query $q_2 = \texttt{max}\{x_1, x_2, x_3\}$ would always be denied, since there is a possible answer to this query, consistent with the answer to q_1 that would cause a private value to be uniquely determined. Note that if the actual answer to q_2 had been greater than $\frac{1}{3}f_1(Q_1)$, q_2 would in reality have been safe to answer, and thus we lose some utility due to the requirement of simulatability.

Partial Disclosure. The notion of full disclosure as a measure of privacy breach has certain shortcomings. Even if a private value cannot be uniquely determined, it might still be determined to lie in a tiny interval, or even in a large interval with a heavily skewed distribution — and some might consider this to be sufficient disclosure. Researchers proposed a new definition of privacy to mitigate this issue by modeling the change in an attacker's confidence about the values of private data points. In this definition, it is assumed that the data is drawn from some distribution $\mathcal{D}$ on $(-\infty, \infty)^n$ that is known to both the attacker and the auditor. See Section 17.2.2.0 for some discussion about this assumption.

Let $\mathcal{Q} = \{q_1, \ldots, q_t\}$ be a sequence of queries on the data set X and let $\mathcal{A} = \{a_1, \ldots, a_t\}$ be the corresponding answers. Here each a_i is either the true answer to query q_i on X or a denial. We allow the auditor to be randomized, i.e., it's decision to answer or deny a query need not be deterministic.

DEFINITION 17.6 (RANDOMIZED AUDITOR) *A randomized auditor is a randomized function of $\mathcal{Q}$, $\mathcal{A}$, X and $\mathcal{D}$ that returns as output either an exact answer to q_t on X or a denial.*

We say that the sequence of queries and corresponding answers is λ-safe for an element x_i and an interval $I \subseteq (-\infty, \infty)$ if the attacker's confidence that $x_i \in I$ does not change significantly upon seeing the queries and answers. Consider for example a private value such as salary: if a sequence of queries and answers does not change an attacker's confidence about a private individual's salary, then the sequence is safe.

DEFINITION 17.7 (λ-SAFE) *The sequence of queries and answers, $q_1, \ldots, q_t$, $a_1, \ldots, a_t$ is said to be λ-safe with respect to a data element x_i and* *an interval $I \subseteq (-\infty, \infty)$ if the following Boolean predicate evaluates to 1:*

$$\text{Safe}_{\lambda,i,I}(q_1,\ldots,q_t,a_1,\ldots,a_t) = \begin{cases} 1 & \textit{if } 1/(1+\lambda) \leq \frac{\Pr_{\mathcal{D}}(x_i \in I | q_1,\ldots,q_t,a_1,\ldots,a_t)}{\Pr_{\mathcal{D}}(x_i \in I)} \leq (1+\lambda) \\ 0 & \textit{otherwise} \end{cases}$$

Partial disclosure is defined in terms of the following predicate that evaluates to 1 if and only if $q_1,\ldots,q_t$, $a_1,\ldots,a_t$ is λ-safe for all entries and all intervals[1]:

$$\text{AllSafe}_{\lambda}(q_1,\ldots,q_t,a_1,\ldots,a_t) = \begin{cases} 1 & \text{if } \text{Safe}_{\lambda,i,I}(q_1,\ldots,q_t,a_1,\ldots,a_t) = 1, \text{ for every } i \in [n] \text{ and} \\ & \text{every interval } I \\ 0 & \text{otherwise} \end{cases} \tag{17.1}$$

We now turn to the privacy definition. Consider the following (λ, T)-*privacy game* between an attacker and an auditor, where in each round t (for up to T rounds):

1 The attacker (adaptively) poses a query $q_t = (Q_t, f_t)$.

2 The auditor determines whether q_t should be answered. The auditor responds with $a_t = f_t(Q_t)$ if q_t is allowed and with $a_t =$ "denied" otherwise.

3 The attacker wins if $\text{AllSafe}_{\lambda}(q_1,\ldots,q_t,a_1,\ldots,a_t) = 0$.[2]

DEFINITION 17.8 (PRIVATE RANDOMIZED AUDITOR) *An auditor is* (λ, δ, T)-private *if for any attacker A*

$$\Pr[A \textit{ wins the } (\lambda, T)\textit{-privacy game}] \leq \delta\,.$$

Here the probability is taken over the distribution $\mathcal{D}$ that the data comes from and the coin tosses of the auditor and the attacker.

Since here too, one would like to ensure that denials leak no information, the condition of simulatability is imposed on auditors that are designed. Consider $\mathcal{Q}$, $\mathcal{A}$ and X as before. Then,

DEFINITION 17.9 (SIMULATABLE RANDOMIZED AUDITOR) *A randomized auditor B is simulatable, if there exists another auditor B' that is a randomized function of only $\mathcal{Q}$, $\mathcal{A} \setminus a_t$ and $\mathcal{D}$ such that the output of B' on q_t is computationally indistinguishable from that of B.*

---

[1] In reality, the privacy definition only considers all intervals that have a significant prior probability mass.

[2] Hereafter, we will refer to the predicates without mentioning the queries and answers for the sake of clarity.

Discussion on Privacy Definition. Note that the above definition of privacy makes the assumption that the distribution from which the data is drawn is known to the attacker. In reality it need not be. In this scenario, the predicate AllSafe needs to be evaluated with respect to the attacker's prior distribution, since compromise occurs only if there is a substantial change in his beliefs. However, if the attacker's distribution can be arbitrarily far from the true data distribution, there is not much that the auditor can release without causing partial disclosure of some private value, since it is required to release exact answers if at all. For example, consider a database that contains *height* as a private attribute, and consider an attacker whose prior belief is that all men are less than a foot tall. If by querying the data, the attacker suddenly learns that this is not true and there is substantial change in his posterior distribution, the privacy breach would be massive. In reality, his prior beliefs are so far off the mark, that there is no aggregate query about the heights that the auditor can truthfully answer without compromising privacy, not even the average height of all people in the database.

Instead the data distribution that we assume the auditor and the attacker share is supposed to represent such common sense facts and it allows for more useful information to be released. There are many circumstances where such an assumption is realistic. For example, distributions of attributes such as *age* or *salary* may be known from previous data releases or even published by the auditor itself.

A General Approach for Constructing Private Randomized Auditors. A query is thus safe to answer if doing so is not likely to cause a significant change in the attacker's confidence that an x_i lies in any interval. Also, the decision to deny must be simulatable. We now describe a general approach that could be used to construct such simulatable randomized auditors. Figure 17.1 gives a high level picture.

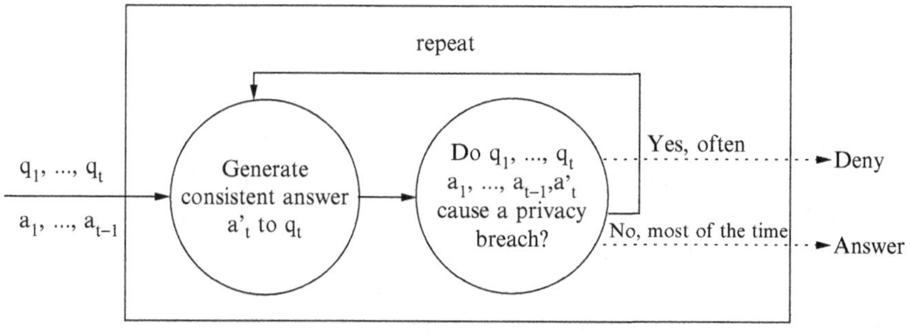

Figure 17.1. Skeleton of a simulatable private randomized auditor

The basic idea is to have the auditor generate random data sets (of n private values) consistent with answers to past queries. The data sets are generated according to the distribution $\mathcal{D}$ conditioned on the past answers. The auditor then checks to see if answering the new query on these random data sets causes a significant change in the attacker's confidence about any x_i. If the answer is 'no' for a sizable fraction of the generated data sets, the query is safe to answer. Since the true answer to the query is never looked at in this process, the auditors are simulatable and denials provably do not leak information.

The left circle in Figure 1.1 thus represents the process of generating a possible answer a'_t to the new query according to $\mathcal{D}$ conditioned on past answers, and the right circle represents the evaluation of the predicate AllSafe (Equation 1.1) that checks to see whether privacy is violated for any x_i and any interval I if a'_t were revealed in conjunction with all previous answers. For each new query this procedure is repeated many times, and the decision to deny is based on the fraction of sampled consistent answers that cause a privacy breach. By repeating often enough and choosing an appropriate cut-off for denials, it can be shown using Chernoff bounds, that the above procedure gives us a (λ, δ, T)-private auditor.

One technicality arises from the fact that the AllSafe predicate needs to be evaluated with respect to an infinite number of intervals. It can be shown that requiring the a-priori and posteriori probabilities of a private value to be close on arbitrarily small intervals would cause no queries to be answered at all, and therefore existing literature focuses on protecting the privacy of only intervals that have a significant a-priori probability mass. While there may also be an infinite number of such intervals, it can be shown that if each x_i is drawn independently according to some distribution $\mathcal{H}$ on $(-\infty, \infty)$, then we only need to check for privacy with respect to a finite number of non-overlapping intervals $\mathcal{I}$. Thus far randomized auditors have only been designed for data sets where the private values are drawn iid from such an underlying distribution.

Randomized Auditor for Sum Queries. We will now briefly describe how the above generic approach can be tuned to obtain a private randomized auditor for `sum` queries (where each query is of the form $\texttt{sum}(Q_j)$ for some query set, Q_j).

Prior to describing the solution, we give some intuition. Assume for simplicity that each private value is drawn uniformly at random from the range $[0, 1]$. Then the data set $X = \{x_1, \ldots, x_n\}$ can be any point in the unit cube $[0, 1]^n$ with equal probability. A sum query and its corresponding answer induce a hyperplane. The data sets consistent with one sum query and its answer are then those points in $[0, 1]^n$ that fall on this hyperplane. Each successive query and answer reduces the space of possible consistent data sets to those points in $[0, 1]^n$ that fall in the intersection of the induced hyperplanes, i.e., the consistent data sets lie in a convex polytope. Because the prior distribution is uniform, the posterior distribution (given the queries and answers) inside the convex polytope is also uniform. Thus it would suffice to sample data sets uniformly at random from this convex polytope to generate the consistent answers required in the left circle of Figure 1.1. Further we can determine if the answer to the query in a sampled data set would cause a privacy breach (in the right circle of Figure 1.1): Suppose that $\mathcal{P}$ is the current convex polytope. To determine if a partial disclosure has occurred for a particular individual x_i and a particular interval $I \in \mathcal{I}$, consider the definition of privacy breach:

$$\frac{Pr_{\mathcal{D}}\{x_i \in I | q_1, \ldots q_t, a_1, \ldots a_t\}}{Pr_{\mathcal{D}}\{x_i \in I\}} = \frac{\Pr_{\mathcal{D}}\{x_i \in I | \vec{x} \in \mathcal{P}\}}{|I|}$$

The probability in the numerator can be estimated by sampling from the convex polytope $\mathcal{P}$ and counting the fraction of the sampled points for which x_i lies inside I. If the fraction above is greater than $(1 + \lambda)$ or less than $\frac{1}{1+\lambda}$ then the query is unsafe for this sampled data set.

Rather than a uniform prior distribution, we can assume an even more general log-concave distribution, since algorithms exist for sampling from it. The class of log-concave distributions forms a common generalization of uniform distributions on convex sets and Gaussian distributions. A distribution over a domain T is said to be log-concave if it has a density function g such that the logarithm of g is concave on its support. That is, the density function $g : T \to \mathbb{R}_+$ is *log-concave* if it satisfies $g(\alpha x + (1-\alpha)y) \geq g(x)^\alpha g(y)^{1-\alpha}$ for every $x, y \in T$ and $0 \leq \alpha \leq 1$. These distributions constitute a broad class and play an important role in stochastic optimization.

Assume that each element x_i is independently drawn according to the same log-concave distribution H over $\mathbb{R}$. Let $\mathcal{D} = H^n$ denote the joint distribution. Using the properties of log-concave functions, it can be shown that the joint distribution $\mathcal{D}$ is also log-concave and further, the posterior distribution, $\mathcal{D}$ conditioned on $\wedge_{j=1}^{t}(\mathtt{sum}(Q_j) = a_j)$ is also log-concave. In addition, there exist randomized, polynomial-time algorithms for sampling (with a small error) from a log-concave distribution.

Without going into the technical details, we will sketch how one can adapt the generic randomized auditor from Section 17.2.2.0 for the problem of auditing `sum` queries. An algorithm for sampling from a log-concave distribution can be used to estimate the posterior probabilities required for evaluating the AllSafe predicate in the right circle of Figure 1.1. This algorithm can also be used in the left circle of Figure 1.1 for sampling data sets and hence consistent answers from the posterior distribution $\mathcal{D}$ conditioned on previous answers. The AllSafe predicate is evaluated for a λ' smaller than λ to accommodate the sampling algorithm's inability to sample exactly from the underlying log-concave distribution.

Besides `sum` queries, randomized auditors have also been developed for `max` queries where the sampling procedure for uniform priors is much more efficient, and for combinations of `max` and `min` queries. We do not go in to the details in this chapter, instead we next very briefly discuss auditing in another scenario when the queries are not aggregate queries.

17.3 Auditing Select-Project-Join Queries

Other than aggregate queries, auditing has also been studied in the context of select-project-join queries when the information to be kept confidential is a forbidden view of the database. The secret view itself is also specified via a select-project-join query. For example the database may consist of a single relation, Employee(*name, department, phone*), and the forbidden view may be of the form $\pi_{name,phone}$(Employee). Here π represents the projection of the table on to the $name$ and $phone$ attributes. The forbidden view thus represents that the $name$ and $phone$ attributes of the Employee relation, or perhaps some combination of them, are sensitive and should not be revealed. The task of an offline auditor then is to determine whether a set of select-project-join queries answered in the past disclosed any information about the forbidden view, and the task of an online auditor is to deny queries when their answers could disclose information about the forbidden view.

The precise semantics of what the forbidden view represents in terms of what should be kept private could vary from system to system. For example, the above forbidden view could represent the requirement that not a single phone number or name in the database should be disclosed. Alternatively, it could represent the requirement that it is only the association between the name and phone number of any individual in the database that should be kept private and so on. The first ever formal notion of forbidden view privacy suggested in the literature was the notion of perfect privacy defined below. It assumes an underlying distribution $\mathcal{D}$ that the tuples of the database are drawn from.

DEFINITION 17.10 (PERFECT PRIVACY) *Let $\mathcal{D}$ be the underlying distribution according to which tuples of the database are drawn. A set of queries, $\mathcal{Q}$, are said to respect* perfect privacy *of a forbidden view $\mathcal{V}$ if for any set of answers to the queries, $\vec{a}$, and any instantiation of the forbidden view, v,*

$$Pr_{\mathcal{D}}\{\mathcal{V} = v | \mathcal{Q} = \vec{a}\} = Pr_{\mathcal{D}}\{\mathcal{V} = v\}$$

If the distribution $\mathcal{D}$ is such that each tuple t_i from the (finite) domain of possible tuples is included in the database with some probability p_i, independently of other tuples, the condition of checking for perfect privacy of a set of queries reduces to a purely logical statement. We will introduce some definitions before stating the result.

DEFINITION 17.11 (CRITICAL TUPLE) *A tuple t from the finite domain of possible tuples is critical for a query Q, if there exists a possible instance of the database, I, where the presence or absence of t makes a difference to the result of Q, i.e.,* $Q(I - \{t\}) \neq Q(I)$.

We then get the following characterization of query-view privacy which applies for queries that follow a set semantics.

THEOREM 17.12 *A set of queries,* $\mathcal{Q}$, *violates perfect privacy of a forbidden view,* $\mathcal{V}$, *if and only if there exists a tuple in the domain of possible tuples that is critical to both* $\mathcal{V}$ *and some query in* $\mathcal{Q}$.

This useful result implies that for a set of queries to violate perfect privacy of the forbidden view, some query in the set must violate it. Thus an offline auditor auditing a set of queries to check for violations of perfect privacy needs to audit each query in turn, and an online auditor interested in maintaining perfect privacy of the forbidden view can make its decisions to answer or deny each new query independently of past queries. Collusion between users is not a problem. In addition, since tuple criticality and therefore query denials are independent of the actual database instance, such an online auditor is simulatable and denials do not leak information.

Unfortunately, checking the condition in Theorem 17.12 is Π_2^P-complete, even when the forbidden view and the queries are conjunctive. Auditors have been developed, however, for particular subclasses of conjunctive queries. Even so an online auditor that maintains perfect privacy of its forbidden view would result in a very strict denial policy. For instance, going back to the example of the Employee relation, suppose the forbidden view is π_{phone}(Employee), then even just the query π_{name}(Employee) asking for the names of all employees would be denied, even though it does not access a single phone number. This is because every single tuple in the domain of possible tuples would be critical to both the forbidden view and the query. The idea is that just by revealing information about the size of the relation, the query reveals some information about the forbidden view and should be denied. The notion of perfect privacy of the forbidden view may thus be a little too strong.

Ongoing research aims to relax the notion of privacy of a forbidden view, thereby permitting auditors that would provide more utility to a user. These new notions of privacy also permit more efficient auditors that can run in polynomial time for large classes of queries. See Section 17.5 for recommended reading on this topic.

17.4 Challenges in Auditing

We describe challenges and future directions in auditing where further research is warranted.

Privacy Definition: There has been a steady evolution of privacy definitions and notions of compromise over the years starting from full disclosure (Definition 17.1) to more recent notions of partial disclosure (Definition 17.8) and perfect privacy (Definition 17.10). But there is certainly room for further improvement. One assumption made by the more recent definitions is that there is one probability distribution $\mathcal{D}$ from which the data is generated and which is known to both the attacker and the auditor. In reality, there are two other distributions, the attacker's prior and the auditor's prior. While it may be reasonable to assume that these three distributions are close, current definitions and auditors all assume that these three distributions are the same. In the case of aggregate queries, another problem is that current definitions only consider the privacy of a single individual to be important, whereas in reality, it may be important to protect the privacy of groups of individuals such as families. In the case of select-project-join queries, the notion of perfect privacy is far too strong causing many seemingly innocuous queries to be deemed suspicious.

Algorithmic Limitations: Online simulatable algorithms for auditing aggregate queries following the general framework suggested in this chapter have several limitations. They require sampling a data set consistent with a given set of queries and answers. In practice, this procedure may be computationally prohibitive given the massive size of data sets, although such sampling algorithms have been steadily improving over the years. In addition, as already mentioned, it is assumed that both the attacker and auditor know the distribution $\mathcal{D}$ from which the data is generated. Algorithms that could overcome these sampling requirements would make great improvements.

Section 17.2 largely focused on auditing one kind of query: `sum`. In reality, a large variety of queries are posed to data sets. While there has been some investigation into auditing `max`, `min`, `median` queries, intermingling these queries has proven to be a greater challenge. For example, under full disclosure, it is NP-hard to audit intermingled `sum` and `max` queries, while polynomial time algorithms are known for auditing exclusively `sum` queries and exclusively `max` queries. While there are situations in which only one kind of query need be considered (e.g. when releasing contingency tables `sum` queries are the only kind of queries that are answered), ultimately, in order for auditing to be truly useful, we will need to allow richer queries of varied types, such as those posed in data mining applications such as clustering or decision tree classification.

As mentioned in Section 17.3 checking for perfect privacy violations of the forbidden view for a very simple kind of probability distribution is Π_2^P-complete even just for conjunctive queries and views. While, auditors have been developed for various subclasses of conjunctive queries, weakening the requirement of "perfect privacy" may go a long way in enabling the design of efficient auditors for larger classes of queries. There has already been some effort in this direction, where assumptions are imposed on the distribution from which the data is drawn.

Collusion: Collusion is a largely unaddressed issue in most interactive data sharing mechanisms today. In the absence of any obstacles to collusion, the online auditors from Section 17.2 would need to pool together aggregate queries posed by all users in the past in order to determine potential privacy breaches. This could result in a user receiving more than his fair share of denials. On a related note, online auditors might need to maintain a large audit trail of queries posed in the past. While the auditors we saw in this chapter were able to maintain a query history of bounded size, or even no query history at all, this need not be true in general, and with the possibility of collusion, larger query histories may need to be stored for longer periods of time. The notion of perfect privacy (17.10) is so strong that past queries need not even be considered in determining privacy breaches — a set of queries leak information about a forbidden view only if some query in the set leaks some (potentially negligible) amount of information about the view. However, strengthening the privacy definition in this way, results in only more denials, and is not a satisfactory solution to the collusion problem.

Utility: While there have been some initial analyses on the utility of online auditors, utility is a dimension that is not well understood. How should we even define utility? One line of work attempts to study the expected number of denials in a random sequence of aggregate queries. However, it is unlikely that users would be able to pose aggregate queries over arbitrary subsets of the data and queries are likely to come from a non-uniform distribution. Furthermore, there might be some important, fairly generic queries, that should always be answered, such as the total number of HIV-positive people in the country. An auditor that would deny such a query could be construed as providing weak utility.

In general, we would like to ensure that a database will not be rendered useless with too many denials. To this end, it might well be worthwhile to sacrifice some privacy for greater utility.

17.5 Reading

A general, though somewhat dated, overview of disclosure control methods for statistical databases can be found in [1]. Some of the representative work in offline auditors for aggregate queries and full disclosure can be found in [4, 8, 106, 12]. [4] describes offline auditors for `sum` and `max` queries over real-valued data. [8] considers auditing subcube queries, [19] considers the case of auditing `average` and `median` queries, while [12] considers the case when the private attribute is boolean. Most of the above work treats online and offline auditing interchangeably — the difference is not made explicit — and in [11] the issue of denials leaking information is uncovered. [11] proposes the simulatable auditing paradigm as a solution, and [10] and [18] construct online simulatable auditors for different kinds of queries and different kinds of data distributions. Chapter 2 in [10] is an extended and refined version of [11]. Algorithms for uniform sampling from convex polytopes and from log-concave distributions can be found in [13, 3, 7, 9, 14]. [18] also contains an initial analysis of the utility of online auditors. The earliest examples of online auditors that restrict the size and overlap of queries can be found at [6].

The work on auditing select-project-join queries presented in this chapter can be found in [16]. [15] contains algorithms for auditing specific classes of conjunctive queries to check for perfect privacy violations of the forbidden view. [5] considers a data distribution that is a variant of that considered in [16] where tuples are drawn independently of one another, but the expected size of the database is a constant. The authors show that privacy violations for conjunctive queries and views can be determined algorithmically in this situation as the size of the domain of the tuples grows to infinity. [2] builds a practical system for detecting "suspicious" select-project-join queries, however the privacy guarantees of their definition of suspiciousness are not made explicit. [17] suggests other notions of suspiciousness that lie in between those of [16] and [2] both in terms of their disclosure detection guarantees and the ease of auditing under them.

References

[1] N. Adam and J. Wortmann. Security-control methods for statistical databases: a comparative study. *ACM Computing Surveys*, 21(4):515–556, 1989.

[2] R. Agrawal, R. Bayardo, C. Faloutsos, J. Kieman, R. Rantzau, and R. Srikant. Auditing Compliance with a Hippocratic Database. In *Proceedings of the International Conference on Very Large Databases (VLDB)*, 2004.

[3] D. Applegate and R. Kannan. Sampling and integration of near log-concave functions. In *Proceedings of the ACM Symposium on Theory of Computing (STOC)*, pages 156–163, 1991.

[4] F. Chin. Security Problems on Inference Control for SUM, MAX, and MIN Queries. *J. ACM*, 33(3):451–464, 1986.

[5] N. Dalvi, G. Miklau, and D. Suciu. Asymptotic Conditional Probabilities for Conjunctive Queries. In *Proceedings of the International Conference on Database Theory (ICDT)*, 2007.

[6] D. Dobkin, A. Jones, and R. Lipton. Secure Databases: Protection against User Influence. *ACM Transactions on Database Systems (TODS)*, 4(1):97–106, 1979.

[7] A. Frieze and R. Kannan. Log-sobolev inequalities and sampling from log-concave distributions. *Annals of Applied Probability*, 9(1):14–26, February 1999.

[8] J. Kam and J. Ullman. A model of statistical databases and their security. *ACM Transactions on Database Systems (TODS)*, 2(1):1–10, 1977.

[9] R. Kannan, L. Lovasz, and M. Simonovits. Random walks and an $O^*(n^5)$ volume algorithm for convex bodies. *Random Structures and Algorithms*, 11, 1997.

[10] K. Kenthapadi. Models and Algorithms for Data Privacy. *Ph.D. Thesis, Computer Science Department, Stanford University*, 2006.

[11] K. Kenthapadi, N. Mishra, and K. Nissim. Simulatable Auditing. In *Proceedings of the ACM Symposium on Principles of Database Systems (PODS)*, pages 118–127, 2005.

[12] J. Kleinberg, C. Papadimitriou, and P. Raghavan. Auditing Boolean Attributes. *Journal of Computer and System Sciences*, 6:244–253, 2003.

[13] L. Lovasz and S. Vempala. Logconcave functions: Geometry and efficient sampling algorithms. In *Proceedings of the IEEE Symposium on Foundations of Computer Science (FOCS)*, 2003.

[14] L. Lovasz and S. Vempala. Simulated annealing in convex bodies and an $O^*(n^4)$ volume algorithm. In *Proceedings of the IEEE Symposium on Foundations of Computer Science (FOCS)*, pages 650–659, 2003.

[15] A. Machanavajjhala and J. Gehrke. On the Efficiency of Checking Perfect Privacy. In *Proceedings of the ACM Symposium on Principles of Database Systems (PODS)*, 2006.

[16] G. Miklau and D. Suciu. A Formal Analysis of Information Disclosure in Data Exchange. *Journal of Computer and System Sciences*, 2006.

[17] R. Motwani, S. U. Nabar, and D. Thomas. Auditing SQL Queries. In *Proceedings of the International Conference on Data Engineering (ICDE)*, 2008.

[18] S. U. Nabar, B. Marthi, K. Kenthapadi, N. Mishra, and R. Motwani. Towards Robustness in Query Auditing. In *Proceedings of the International Conference on Very Large Databases (VLDB)*, 2006.

[19] S. Reiss. Security in Databases: A Combinatorial Study. *J. ACM*, 26(1):45–57, 1979.

Chapter 18

Privacy and the Dimensionality Curse

Charu C. Aggarwal
IBM T. J. Watson Research Center
Hawthorne, NY 10532
charu@us.ibm.com

Abstract Most privacy-transformation methods such as k-anonymity or randomization use some kind of transformation on the data for privacy-preservation purposes. In many cases, the data can be indirectly identified with the use of a combination of attributes. Such attributes may be available from public records and they may be used to link the sensitive records to the target of interest. Thus, the sensitive attributes of the record may be inferred as well with the use of publicly available attributes. In many cases, the target of interest may be known to the adversary, which results in a large number of combinations of attributes being known to the adversary. This is a reasonable assumption, since privacy attacks will often be mounted by an adversary with some knowledge of the target. As a result, the number of attributes for identification increases, and results in almost unique identification of the target. In this paper, we will examine a number of privacy-preservation methods and show that in each case the privacy-preservation approach becomes either ineffective or infeasible.

Keywords: High dimensional privacy, dimensionality curse for privacy.

18.1 Introduction

With increasing ability to collect data, the dimensionality and size of the available data has increased considerably in recent years. Numerous data transformation methods have been proposed which try to preserve the privacy of the data from adversarial attacks. One of the key adversarial attacks is with the use of public information. We note that the straightforward approach of simply removing the identifier fields is not sufficient from a privacy point of view. This is because many attributes such as age, sex or zip-code are available from public records. Consider a data set which contains both public and sensitive fields. For example, a medical data set will typically contain information about the individual such as zip code or sex (which is public), and will also contain information about diagnosis, which is sensitive or private. In this case, one can use the public fields in order to identify individual, even if unique identifier keys such as the name or social-security number have been removed. Therefore, the public fields can be used in order to make an identification of the sensitive fields. Such public fields are also referred to as *pseudo-identifier* fields. Many privacy-transformation methods such as k-anonymity or l-diversity have been specifically designed to preserve the privacy of identification in the presence of public information.

We note that the definition of a pseudo-identifier field has been treated rather conservatively in the privacy literature. Typically, only attributes which are available from public records are referred to as pseudo-identifiers. A key problem in privacy is that of *background knowledge*. Typically, an adversary is familiar with the target of interest, and may know far more about an individual than is available from public information. In such cases, the boundary between a pseudo-identifier and a sensitive attribute is blurred. For example, consider a database which contains salary information about an individual. While this field is clearly sensitive, it can also be considered a pseudo-identifier with respect to the organization where the individual is employed.

The problem of high dimensionality arises in the context of records in which the adversary may have partial or background knowledge [12] about various attributes of the record. *Since we do not know in advance which attributes the adversary may know, we may not have a choice but to include all fields in the anonymization process.* In such cases, the number of fields available for anonymization increases greatly. In such cases, we will show that the privacy-preservation problem becomes increasingly difficult for the different definitions of privacy. This difficulty may show up in different ways; in some cases such as condensation, it may show as a huge degradation in the quality of the transformed data, in cases such as randomization, it may show up as a privacy breach, and in cases such as l-diversity it may become difficult to design any feasible solution at all. Nevertheless, in all cases, the problem encountered in high dimensionality seems to be quite fundamental, and it cannot be easily resolved with the use use of better algorithms or techniques.

This paper is organized as follows. In the next section, we will discuss the dimensionality issues for k-anonymization. In section 3, we will discuss the dimensionality issues in the context of the condensation based approach. In section 4, we will explore the randomization technique in the context of dimensionality. In section 5, we will explore the l-diversity technique in the context of dimensionality. Section 6 contains the conclusions and summary.

18.2 The Dimensionality Curse and the k-anonymity Method

It is well known that increasing dimensionality makes the k-anonymization problem more difficult from a computational perspective [7, 13]. For ease in exposition, we will assume that any dimension in the database is a potentially identifying quasi-identifier. This assumption can be made without loss of generality, since we can restrict our analysis only to such identifying attributes. We will further assume the use of quantitative attributes. This assumption can also be made without loss of generality. The results can be easily extended to categorical data, since both the quantitative and categorical data domains can be represented in binary form.

We note that all anonymization techniques depend upon some notion of spatial locality in order to perform the generalization. However distance functions begin to show loss of intra-record distinctiveness in high dimensional space. It has been argued in [8], that under certain reasonable assumptions on the data distribution, the distances of the nearest and farthest neighbors to a given target in high dimensional space is almost the same for a variety of data distributions and distance functions. In such a case, the concept of spatial locality becomes ill defined, since the contrasts between the distances to different data points do not exist. Generalization based approaches to privacy preserving data mining are deeply dependent upon spatial locality, since they use the ambiguity of different data points within a given spatial locality in order to preserve privacy. We will see that privacy preservation by anonymization becomes impractical in very high dimensional cases, since it leads to an unacceptable level of information loss.

In Figure 18.1, we have illustrated two cases of generalization of data points into a range along each dimension. In Figure 18.1(a), 2-anonymization is achieved by simple discretization without much optimization. In Figure 18.1(b), more careful clustering methods are utilized to achieve 2-anonymity, so that the sizes of the bounding rectangles are reduced. The latter is also an example

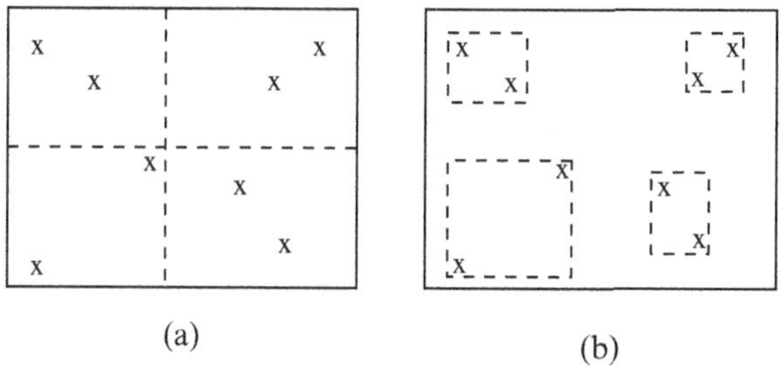

Figure 18.1. Some Examples of Generalization for 2-Anonymity

of optimized axis-parallel generalizations. It is not necessary to generalize using axis-parallel ranges only. In general, the problem of finding the optimal k-anonymous representation is known to be NP-hard [13].

In order to analyze the behavior of anonymization approaches with increasing dimensionality, we consider the case of data in which individual dimensions are independent and identically distributed. The resulting bounds provide insight into the behavior of the anonymization process with increasing *implicit* dimensionality. We construct a bounding box around a target point $\overline{X_d}$ in order to generalize it. The value of the data point $\overline{X_d}$ in this grid cube is generalized to the corresponding partially specified range of this bounding box. For data point $\overline{X_d}$ to maintain k-anonymity, this bounding box must contain *at least* $(k-1)$ other points. First, we will consider the case when the generalization of each point uses a maximum fraction f of the data points along each of the d partially specified dimensions. Thus, data points which do not satisfy this condition may need to be *suppressed* [14]. It has been suggested [14], that suppression of a larger percentage of the data leads to an unacceptable aggregate change in the statistical characteristics of the data for mining purposes. In the following analysis, we will show the difficulty of preserving k-anonymity using the approach of partial range masking.

LEMMA 18.1 *Let $\mathcal{D}$ be a set of N points drawn from the d-dimensional distribution $\mathcal{F}^d$ in which individual dimensions are independently distributed. Consider a randomly chosen grid cell, such that each partially masked dimension contains a fraction f of the total data points in the specified range. Then, the probability P^q of exactly q points in the cell is given by* $\frac{N!}{q! \cdot (N-q)!} \cdot f^{q \cdot d} \cdot (1-f^d)^{(N-q)}$.

Proof: We note that the probability of a data point in a grid cell with range specificity of f along each of the d dimensions is given by $x = f^d$. Then, the probability that a given grid cube contains exactly q points is given by the binomial distribution with parameters N and x. Therefore, we can use the binomial distribution formula to define the corresponding probability P^q:

$$P^q = \frac{N!}{q! \cdot (N-q)!} \cdot x^q \cdot (1-x)^{(N-q)} \tag{18.1}$$

A direct corollary of the above result is the following:

COROLLARY 18.2 *Let B_k be the event that the grid cell corresponding to the partially specified dimensions contains k or more data points. The corresponding probability $P(B_k)$ is given by:*

$$P(B_k) = \sum_{q=k}^{N} \frac{N!}{q! \cdot (N-q)!} \cdot f^{q \cdot d} \cdot (1-f^d)^{(N-q)} \tag{18.2}$$

Proof: We note that $P(B_k) = \sum_{q=k}^{N} P^q$. By substituting $x = f^d$ from Equation 18.1, we get the corresponding result.

We note that a set of partially specified dimensions violates the conditions of k-anonymity, when the corresponding set of partially specified ranges contain at least one data point, but less than k data points. Therefore, we need to find the conditional probability denoted by $P(B_k|B_1)$. The value of this conditional probability is defined by the Lemma below.

LEMMA 18.3 *Let B_k be the event that the set of partially masked ranges contains at least k data points. Then the following result for the conditional probability $P(B_k|B_1)$ holds true:*

$$P(B_k|B_1) = \frac{\sum_{q=k}^{N} \frac{N!}{q!\cdot(N-q)!} \cdot f^{q\cdot d} \cdot (1 - f^d)^{(N-q)}}{\sum_{q=1}^{N} \frac{N!}{q!\cdot(N-q)!} \cdot f^{q\cdot d} \cdot (1 - f^d)^{(N-q)}} \tag{18.3}$$

Proof: We know from elementary probability theory that:

$$P(B_k|B_1) = P(B_k \cap B_1)/P(B_1) \tag{18.4}$$

However, the event B_k is a special case of B_1. This is because if a set of masked ranges contain at least k points, the corresponding set of ranges must also be non-empty. Therefore, we have:

$$P(B_k \cap B_1) = P(B_k) \tag{18.5}$$

Therefore, we have:

$$P(B_k|B_1) = P(B_k)/P(B_1) \tag{18.6}$$

By substituting for the value of $P(B_k)$ and $P(B_1)$ in Equation 18.2, we get the desired result.

We note the following simple observation:

OBSERVATION 18.2.1 *For all $k > 2$, we have $P(B_k|B_1) \leq P(B_2|B_1)$.*

The above observation is true because the event B_k is subsumed by the event B_2 for any value of k larger than 2. Therefore, by finding an upper bound on $P(B_2|B_1)$, we can also find a upper bound on the probability that k-anonymity is achieved on a randomly chosen (non-empty) set of non-empty grid changes. Next, we observe the following:

$$P(B_2|B_1) = \frac{1 - N \cdot f^d \cdot (1 - f^d)^{(N-1)} - (1 - f^d)^N}{1 - (1 - f^d)^N} \tag{18.7}$$

The above observation can be easily verified by substituting the values of $k = 2$, $P(B_k)$ and $P(B_1)$ in Equation 18.3 of Lemma 18.3. We are simply expressing the events $P(B_2)$ and $P(B_1)$ in the complementary form[1] of the binomial expression. Next, we will show that the probability of achieving 2-anonymity in a non-empty grid cell is zero for the limiting case of high dimensionality. We formalize this result as follows:

LEMMA 18.4 *The limiting probability for achieving 2-anonymity in a set of partially specified ranges, each containing a fraction $f < 1$ of the data points is zero. In other words, we have:*

$$lim_{d \to \infty} P(B_2|B_1) = 0 \tag{18.8}$$

Proof: By substituting $x = f^d$ in Equation 18.7, we get:

$$P(B_2|B_1) = 1 - \frac{N \cdot x \cdot (1-x)^{N-1}}{1-(1-x)^N} \tag{18.9}$$

We note that as $d \to \infty$, we have $x \to 0$. This is because $f < 1$. Consequently, we get:

$$\lim_{d \to \infty} P(B_2|B_1) = 1 - \lim_{x \to 0} \frac{N \cdot x \cdot (1-x)^{N-1}}{1-(1-x)^N} \tag{18.10}$$

Since both the numerator and denominator tend to zero in the limiting case, we can use L'Hopital's rule to differentiate the numerator and denominator. Therefore, we have:

$$P(B_2|B_1) =$$
$$1 - \lim_{x \to 0} \frac{N \cdot (1-x)^{(N-1)} - N \cdot x \cdot (1-x)^{(N-2)}}{N \cdot (1-x)^{(N-1)}}$$

It is easy to verify that this expression evaluates to zero.

The following result follows directly:

COROLLARY 18.5 *The limiting probability for achieving k-anonymity in a non-empty set of masked ranges containing a fraction $f < 1$ of the data points is zero. In other words, we have:*

$$lim_{d \to \infty} P(B_k|B_1) = 0 \tag{18.11}$$

[1]Another way of deriving this would be to simply use the fact that the event of k or more data points occurring in the unit cube is the complementary event to that of less than k points in the unit cube.

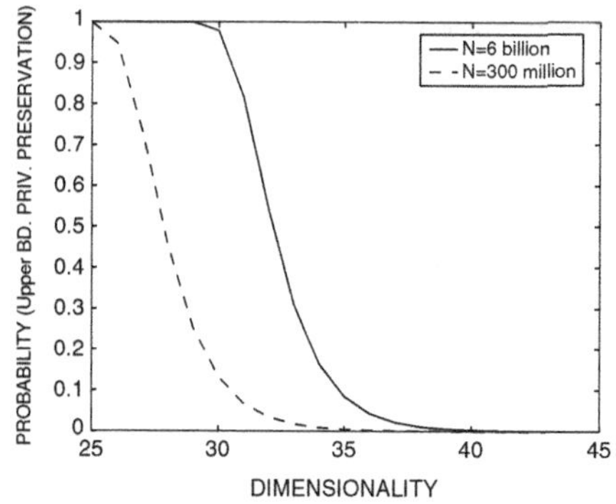

Figure 18.2. Upper Bound of 2-anonymity Probability in an Non-Empty Grid Cell

This result follows because of our earlier observation that $P(B_k|B_1) \leq P(B_2|B_1)$. In order to derive a further practical understanding of this bound, let us consider some practical values of f. While it is clear that larger values of the population size (denoted by N) and f result in increased privacy, it is interesting to analyze some practical limits on these numbers. Therefore, we will set f and N to the largest practical values possible and calculate the variation of privacy probability with increasing dimensionality. Therefore we will set $f = 0.5$, and N to the values of $3 * 10^8$ and $6 * 10^9$. The latter two values represent the populations of the United States and the earth respectively. In Figure 18.2, we have plotted the 2-anonymity bound with increasing value of the dimensionality d. It is clear that even for modest values of the dimensionality between 25 and 35, the probability of achieving 2-anonymity within a non-empty grid cell fall off rapidly. Furthermore, we note that these are *upper bounds* for very liberally set values, and represent the probability of 2-anonymity preservation in *each non-empty cell*. In order for privacy to be preserved over the entire data set, the privacy of each non-empty cell must be preserved. Consequently, the overall probability for 2-anonymity preservation would be much lower than that predicted by Figure 18.2. We note that while these results are derived for uniformly distributed data, they conceptually represent the behavior of the privacy preservation process with increasing *implicit dimensionality* of the data set. In the empirical section, we will also illustrate the cases when correlations are present in the data and show that a very large fraction of the records would continue to violate the privacy requirements. This would require a large level of suppression.

We will illustratean analysis of the technique on a family of synthetic data sets. The synthetic data sets were generated as Gaussian clusters with randomly distributed centers in the unit cube. The radius along each dimension of each of the clusters was a random variable with a mean of 0.075 and standard

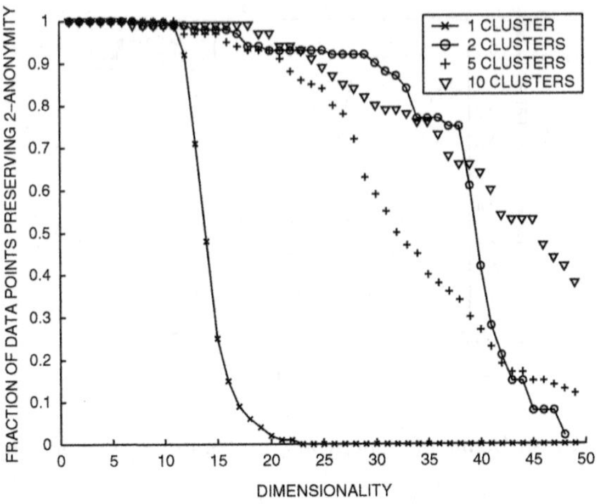

Figure 18.3. Fraction of Data Points Preserving 2-Anonymity with Data Dimensionality (Gaussian Clusters)

deviation of 0.025. Thus, a given cluster could be elongated differently along different dimensions by varying the corresponding standard deviation. Each data set was generated with $N = 10000$ data points in a total of 50 dimensions. Finally, the data set was normalized such that the variance along each dimension was 1 unit. We generated the data sets with different numbers (1, 2, 5 and 10) of clusters in order to test the effectiveness of the method with data skew. A larger number of clusters lead to a greater amount of skew in the data. We tested the two measures on the bounds for the privacy preservation process using projections of different dimensionality from the generated data set. Since the original data set was 50-dimensional, projections up to 50 dimensions could be generated. In Figure 18.3, we have illustrated the behavior of a generalization approach in which each attribute is divided into only two ranges. The number of dimensions on the X-axis represents those which are partially specified using these two ranges, whereas all other dimensions are *fully suppressed.* On the Y-axis, we have illustrated the percentage of data points which maintain 2-anonymity using this generalization. We note that all other data points (which violate the 2-anonymity condition) would need to be suppressed. A high percentage of suppression is never acceptable from a data mining point of view [14]. It is interesting to see that while a greater number of clusters (and corresponding skew) in the underlying data helps the anonymization, the percentage of data points which continue to preserve privacy falls off rapidly with increasing data dimensionality. When the data sets contained more than 45 dimensions, *almost all the data points* violated the 2-anonymity condition. Another interesting characteristic of the results of Figure 18.3 is that for the case of 1 cluster, the shape of the corresponding curve resembles that of

Figure 18.1. The main difference is that in this case, the rate of privacy preservation falls off much more rapidly. This is because the results in Figure 18.1 only represent upper bounds on the true probability of privacy preservation.

18.3 The Dimensionality Curse and Condensation

In the previous discussion, we analyzed the privacy requirements for the case of randomly chosen masked attributes. Next, we will analyze the case where the masking can be performed in a more effective way with optimization techniques such as clustering. An example is the anonymization approach of [4] which uses the technique of multi-group cluster formation without the use of bounding rectangles. In the following discussion, we will try to find a lower bound on the information loss for achieving 2-anonymity using any kind of optimized group formation. We will show that in this case, the privacy preservation process requires an unacceptably high loss of information in order to satisfy the anonymity requirements. In order to facilitate further discussion, we will establish certain notations and definitions. We assume that all points are distributed in the unit cube. In Table 18.1, we have introduced some notations and definitions, which we will use throughout this paper.

We assume that a set S of k data points are merged together in one group for the purpose of condensation. Let $M(S)$ be the maximum euclidian distance between any pair of data points in this group. We note that larger values of

Table 18.1. Notations and Definitions

Notation	Definition
d	Dimensionality of the data space
N	Number of data points
$\mathcal{F}$	1-dimensional data distribution in $(0, 1)$
X_d	Data point from $\mathcal{F}^d$ with each coord. drawn from $\mathcal{F}$
$dist_d^k(x, y)$	Distance between $(x^1, \ldots x^d)$ and $(y^1, \ldots y^d)$ using L_k metric $= \sum_{i=1}^{d} [(x_1^i - x_2^i)^k]^{1/k}$
$\lVert \cdot \rVert_k$	Distance of a vector to the origin $(0, \ldots, 0)$ using the function $dist_d^k(\cdot, \cdot)$
$E[X]$, $var[X]$	Expected value and variance of a random variable X
$Y_d \rightarrow_p c$	A sequence of vectors $Y_1, \ldots, Y_d$ converges in probability to a constant vector c if: $\forall \epsilon > 0 \; lim_{d \rightarrow \infty} P[dist_d(Y_d, c) \leq \epsilon] = 1$

$M(S)$ represent a greater loss of information, since the points within a group cannot be distinguished for the purposes of data mining. Similarly, let $M(\mathcal{D})$ represent the corresponding measure for the global database $\mathcal{D}$. This provides us a global base for the overall contrast between different data points. Then, we define the *relative condensation loss* $\mathcal{L}(S)$ for that group of k entities as follows:

DEFINITION 18.6 *The relative condensation loss $\mathcal{L}(S)$ for the group S is defined as the following ratio:*

$$\mathcal{L}(S) = M(S)/M(\mathcal{D}) \tag{18.12}$$

Intuitively speaking, the above definition measures how much of the *relative contrast* between the data points (in a group) is lost with respect to the base contrast of the remaining data set. A value of $\mathcal{L}(S)$ which is close to one implies that most of the distinguishing information is lost as a result of the privacy preservation process. We further note that $\mathcal{L}(S)$ represents the *very minimum* level of information loss that any anonymization or condensation technique is likely to be achieve. This is because a particular algorithm for condensation or anonymization may use domain specific considerations [14], which are not always optimal from the information preservation perspective. In the following analysis, we will show how the value of $\mathcal{L}(S)$ is affected by the dimensionality d.

In order to provide a better understanding of the results, we will first analyze the behavior of a uniform distribution of $N = 3$ data points, and deal with the particular case of 2-anonymity. For ease in analysis, we will assume that one of these 3 points is the origin O_d, and the remaining two points are A_d and B_d which are uniformly distributed in the data cube. We also assume that the closest of the two points A_d and B_d need to be merged with O_d in order to preserve 2-anonymity of O_d. Later, we will generalize the results to the case of $N = n$ data points. Since the information loss $\mathcal{L}(\cdot)$ depends upon relative distances among data points, we will start by establishing some convergence results about the distances between A_d, B_d, and O_d in high dimensionality.

LEMMA 18.7 *Let $\mathcal{F}^d$ be uniform distribution of $N = 2$ points. Let us assume that the closest of the 2 points to O_d is merged with O_d to preserve 2-anonymity of the underlying data. Let q_d be the Euclidean distance of O_d to the merged point, and let r_d be the distance of O_d to the remaining point. Then, we have: $lim_{d\to\infty} E\left[r_d - q_d\right] = C$, where C is some constant.*

Proof: Let A_d and B_d be the two points in a d dimensional data distribution such that each coordinate is independently drawn from the data distribution $\mathcal{F}$. Specifically $A_d = (P_1 \ldots P_d)$ and $B_d = (Q_1 \ldots Q_d)$ with P_i and Q_i being drawn from $\mathcal{F}$. Let $PA_d = \{\sum_{i=1}^{d}(P_i)^2\}^{1/2}$ be the distance of A_d to the origin

O_d, and $PB_d = \{\sum_{i=1}^{d}(Q_i)^2\}^{1/2}$ the distance of B_d from O_d. The difference of distances is $PA_d - PB_d = \{\sum_{i=1}^{d}(P_i)^2\}^{1/2} - \{\sum_{i=1}^{d}(Q_i)^2\}^{1/2}$.

It can be shown[2] that the random variable $(P_i)^2$ has mean $\frac{1}{3}$ and standard deviation $(\frac{2}{3})\sqrt{(\frac{1}{5})}$. This means that $(PA_d)^2/d \rightarrow_p 1/3, \quad (PB_d)^k/d \rightarrow_p 1/3$ and therefore we have:

$$PA_d/d^{1/2} \rightarrow_p (1/3)^{1/2}, \quad PB_d/d^{1/2} \rightarrow_p (1/3)^{1/2} \qquad (18.13)$$

We intend to show that $|PA_d - PB_d| \rightarrow_p C'''$ for some constant C'''. We can express $|PA_d - PB_d|$ in the following numerator/denominator form:

$$|PA_d - PB_d| = \frac{|(PA_d)^2 - (PB_d)^2|}{(PA_d) + (PB_d)} \qquad (18.14)$$

Now, we will analyze the convergence behavior of the numerator and denominator individually. By dividing numerator and denominator on RHS by the same value, we get:

$$|PA_d - PB_d| = \frac{|((PA_d)^2 - (PB_d)^2)|/\sqrt{d}}{\frac{PA_d}{d^{1/2}} + \frac{PB_d}{d^{1/2}}} \qquad (18.15)$$

Consequently, using Slutsky's theorem[3] and the results of Equation 18.13 we obtain

$$\left(\frac{PA_d}{d^{1/2}}\right) + \left(\frac{PB_d}{d^{1/2}}\right) \rightarrow_p 2/\sqrt{3} \qquad (18.16)$$

Having characterized the convergence behavior of the denominator of the right hand side of Equation 18.15, let us now examine the behavior of the numerator: $|(PA_d)^2 - (PB_d)^2|/\sqrt{d} = |\sum_{i=1}^{d}((P_i)^2 - (Q_i)^2)|/\sqrt{d} = |\sum_{i=1}^{d} R_i|/\sqrt{d}$. Here R_i is the new random variable defined by $((P_i)^2 - (Q_i)^2)\ \forall i \in \{1, \ldots d\}$. This random variable has zero mean and standard deviation which is $\sqrt{2} \cdot \sigma$ where σ is the standard deviation of $(P_i)^2$. The sum of different values of R_i over d dimensions will converge to a normal distribution with mean 0 and standard deviation $\sqrt{2} \cdot \sigma \cdot \sqrt{d}$ because of the central limit theorem. Consequently, the mean average deviation of this distribution will be $C \cdot \sigma$ for some constant C. Therefore, we have:

$$\lim_{d \rightarrow \infty} E\left[\frac{|(PA_d)^2 - (PB_d)^2|}{\sqrt{d}}\right] \leq = C'' \qquad (18.17)$$

---

[2]This is because $E[P_i^2] = 1/3$ and $E[P_i^4] = 1/5$.

[3]**Slutsky's Theorem:** Let $Y_1 \ldots Y_d \ldots$ be a sequence of random vectors and $h(\cdot)$ be a continuous function. If $Y_d \rightarrow_p c$ then $h(Y_d) \rightarrow_p h(c)$.

Here C'' is a new constant defined by a product of the above mentioned constants. Since the denominator of Equation 18.15 shows probabilistic convergence to $2/\sqrt{3}$, we can combine the results of Equations 18.15, 18.16 and 18.17 to obtain the following result for some constant $C'''=C'' \cdot \sqrt{3}/2$.

$$\lim_{d\to\infty} E\left[\|PA_d - PB_d\|\right] = C''' \tag{18.18}$$

We can easily generalize the result for a database of $N = n$ uniformly distributed points. The following corollary provides the result.

COROLLARY 18.8 *Let $\mathcal{F}^d$ be uniform distribution of $N = n$ points. Let us assume that the closest of the n points is merged with O_d to preserve 2-anonymity. Let q_d be the Euclidean distance of O_d to the merged point, and let r_d be the distance of the furthest point from O_d. Then, we have: $C''' \leq lim_{d\to\infty} E\left[r_d - q_d\right] \leq (n-1) \cdot C'''$, where C''' is some constant.*

Proof: This is because if L is the expected difference between the maximum and minimum of two randomly drawn points, then the same value for n points drawn from the same distribution must be in the range $(L, (n-1) \cdot L)$.

A further corollary of the above results is as follows:

COROLLARY 18.9 *Let $\mathcal{F}^d$ be uniform distribution of $N = n$ points. Let us assume that the closest of the n points is merged with O_d to preserve 2-anonymity. Let q_d be the Euclidean distance of O_d to the merged point, and let r_d be the distance of the furthest point from O_d. Then, we have: $lim_{d\to\infty} E\left[\frac{r_d - q_d}{r_d}\right] = 0$, where C''' is some constant.*

Proof: This result can be proved by showing that $r_d \to_p \sqrt{d}$. Note that the distance of each point to the origin in d-dimensional space increases at this rate. Combining the result with Corollary 18.8, we see that both the lower and upper bounds on the expression converge to 0.

Let S be the two point set represented by O_d and the closest point to O_d. We note that the information loss $M(S)/M(\mathcal{D})$ for 2-anonymity can be expressed[4] as $1 - E\left[\frac{r_d - q_d}{r_d}\right]$. It is easy to see that the value of the information loss converges to 1 in the limiting case in order to achieve 2-anonymity. We also note that the bounds for 2-anonymity also provide lower bounds for the general case of k-anonymity. Therefore, the following result holds:

THEOREM 18.10 *For any set S of data points to achieve k-anonymity, the information loss on the set of points S must satisfy:*

$$lim_{d\to\infty} E[M(S)/M(\mathcal{D})] = 1 \tag{18.19}$$

[4]Here we are approximating $M(\mathcal{D})$ to r_d since the origin of the cube is probabilistically expected to be one of extreme corners among the maximum distance pair in the database.

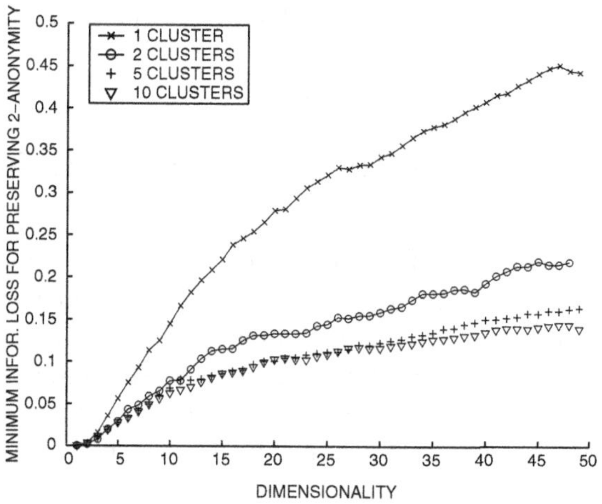

Figure 18.4. Minimum Information Loss for 2-Anonymity (Gaussian Clusters)

Thus, these results show that with increasing dimensionality, all the discriminatory information in the data is lost in order to achieve k-anonymity. In the next section, we will experimentally examine the behavior of these privacy metrics over a variety of data domains and distributions.

As in the case of the k-anonymity method, we used the same family of synthetic data sets in order to illustrate the effects of dimensionality on the condensation technique. In Figure 18.4, we have illustrated the minimum information loss for data sets of different dimensionalities. This corresponds to the loss $\mathcal{L}(\cdot)$ as defined earlier in this paper. It is easy to see from Figure 18.4 that the level of information loss increases rapidly with increasing dimensionality. As in the previous case, the data sets with a smaller number of clusters were more difficult cases. Therefore, the information loss is higher in these cases as well. This is because the presence of greater number of closely clustered regions in the data helps in creating masked groups of anonymized data with lower information loss. However, the overall trends show that even the clustered behavior of the data cannot compensate for the sparsity effects in high dimensionality. This means that either a large portion of the attributes have to be *completely masked* in such cases, or the effectiveness of the anonymization process needs to be compromised. On the other hand, the complete suppression of a large number of attributes reduces the effectiveness of data mining algorithms on the anonymized data.

Thus, the results show that the anonymity model is open to inference attacks when a large number of concepts exist in the data. This also corresponds to a high *implicit dimensionality* of the underlying data. The behavior of privacy preserving data mining algorithms with increasing dimensionality is similar to that of other data mining algorithms, which fail to perform effectively in the high dimensional case because of data sparsity. This sparsity also makes the data more susceptible to inference attacks.

18.4 The Dimensionality Curse and the Randomization Method

In this section, we will explore the effects of the curse of dimensionality on the randomization method. In randomization [1, 2], we add a perturbing distribution to the original data. Even though individual record values are distorted, it is possible to accurately reconstruct aggregate distributions and design data mining algorithms which work with these distributions. One nice characteristic of the k-anonymity model is that it is specifically designed to guarantee privacy in the presence of public information. This is not true of randomization, since the added noise is drawn from a fixed distribution. This paper is designed to introduce the analytical effects of public information into the analysis of randomization. Earlier work on randomization [9, 10] uses spectral analysis to approximately reconstruct attribute *values* without the use of public information. However, attribute value approximation is a subtly different goal from *personal* identification with the use of linkage to public databases. We explore the following issues:

(1) We use a public-information sensitive methodology to analyze the randomization approach.
(2) As in the case of k-anonymity [3], the effectiveness of randomization degrades rapidly with increasing dimensionality. We quantify the required perturbation to achieve a given privacy level as a function of dimensionality.
(3) The use of public information makes the choice of perturbing distribution more critical than previously thought. We analyze two widely used perturbing distributions (gaussian and uniform) and show that gaussian perturbations have overwhelming advantages in high dimensional cases.
(4) The use of public information in the analysis exposes the susceptibility of the randomization method to many natural properties of real data sets such as clusters or outliers.
(5) We demonstrate that the inclusion of public information makes the randomization method vulnerable in unexpected ways. Thus, the goal of privacy preservation may be more elusive than previously thought for the randomization method.

18.4.1 Effects of Public Information

In this section, we will introduce the concepts of likelihood fit and k-randomization which quantify the ability to re-identify the data in the presence of public information. This creates an analogous randomization framework to that of k-anonymity. We assume that the database $\mathcal{D}$ contains N records and d dimensions. The random perturbations for the different dimensions have distributions denoted by $f_{Y_1}(y) \ldots f_{Y_d}(y)$. The corresponding standard deviations of these distributions are denoted by $\sigma_1 \ldots \sigma_d$. Without loss of generality, we may assume that each of the perturbing distributions has zero mean. Let us consider a record $X = (x_1 \ldots x_d)$ to which the perturbation $Y = (y_1 \ldots y_d)$ is added. Then the perturbed data is denoted by $Z = (z_1 \ldots z_d) = (x_1 + y_1, \ldots x_d + y_d)$. Now let us consider an adversary who has access to the publicly available database $\mathcal{D}_p$. Since the perturbing distribution is publicly known, the adversary can calculate the *potential* perturbation of the record Z with respect to each record in the public database $\mathcal{D}_p$. This can be used to calculate the probability that these set of d-dimensional perturbations fit the set of distributions denoted by $f_{Y_1}(y) \ldots f_{Y_d}(y)$. The natural way of calculating the fit of a set of models to a set of observations is the *log-likelihood* fit. In the event that one of the records in the public database has an unusually high degree of fit, this allows the adversary the ability to guess whether the current record truly corresponds to any particular record in the public database.

Let us consider the public record $X = (x_1 \ldots x_d)$. We would like to calculate the likelihood that the perturbed record $Z = (z_1 \ldots z_d)$ corresponds to this publicly available record. In order to do so, the adversary can compute the *potential fit* of the perturbed record to the public database record X. Next, we define the *potential perturbation* of a given record Z to the public database record X.

DEFINITION 18.11 *The potential perturbation $Q(Z, X)$ of a perturbed record $Z = (z_1 \ldots z_d)$ with respect to the public database record $X = (w_1 \ldots w_d)$ is denoted by $Q(Z, X) = (q_1(Z, X) \ldots q_d(Z, X)) = Z - X = (z_1 - x_1 \ldots z_d - x_d)$. The ith component of $Q(Z, X)$ is denoted by $q_i(Z, W) = z_i - x_i$.*

The above definition simply states that in order for the public database record X to correspond to the perturbed record Z, the perturbation for the ith dimension would need to be $q_i(Z, X) = z_i - x_i$. What is the likelihood that the publicly known perturbing distributions $f_{Y_i}(y)$ generate these potential perturbations over the d different dimensions? We note that the log-likelihood that the model $f_{Y_i}(y)$ fits the potential perturbation $q_i(Z, X)$ is given by $\log(f_{Y_i}(q_i(Z, X))) = \log(f_{Y_i}(z_i - x_i))$. We define the corresponding *potential fit* of the dimensions in $Q(Z, X)$ to the distributions denoted by $f_{Y_1}(y) \ldots f_{Y_d}(y)$ as the sum of the log-likelihood fits over the different dimensions.

DEFINITION 18.12 *The potential fit $\mathcal{F}(Z, X)$ of the perturbed data Z to the record X is given by $\sum_{i=1}^{d} \log(f_{Y_i}(q_i(Z, X)))$.*

The higher the value of the log-likelihood fit, the greater the probability that the public database record X corresponds to the perturbed data Z. For a given public database $\mathcal{D}_p$, an adversary can try to match the record in $\mathcal{D}_p$ which has the highest level of fit to the perturbed record Z. We observe that the log likelihood fit is an indirect representation of the Bayes a-posteriori probability that the perturbed data record fits a particular record X:

OBSERVATION 18.4.1 *Consider a database $\mathcal{D}_p$ which is known to contain the true representation of the perturbed record Z with equal a-priori probability. Then the posterior probability $\mathcal{B}(Z, X, \mathcal{D}_p)$ of a particular record $X \in \mathcal{D}_p$ to correspond to Z is given by:*

$$\mathcal{B}(Z, X, \mathcal{D}_p) = \frac{e^{\mathcal{F}(Z,X)}}{\sum_{V \in \mathcal{D}_p} e^{\mathcal{F}(Z,V)}} \tag{18.20}$$

The above observation is easy to verify, since the perturbations over different dimensions are independent and the value of $e^{\mathcal{F}(Z,X)}$ is simply equal to the product of the corresponding probability densities. By applying the Bayes formula in conjunction with equal a-priori probability, we get the desired result. Thus, the log likelihood is an indirect representation of the Bayes probability, and the use of this particular representation is chosen for the sake of numerical and algebraic convenience.

In many cases, the log likelihood fit can provide considerable insights to an adversary in including or excluding particular database records. For example, the log likelihood fit may be a significantly better fit to one record in the public database compared to any other record. In such a case, the corresponding Bayes probability $\mathcal{B}(Z, X, \mathcal{D}_p)$ may approach 1, and the said record can be identified to a high degree of probability. Therefore, anonymity is lost. Another extreme case is one in which the perturbing distribution has a finite range (such as the uniform distribution), and the value of $f_{Y_i}(q_i(Z, X))$ to be zero. In such a case, the corresponding log likelihood fit is $-\infty$, and it is possible to exclude the record X as a fit with Z.

In general, we would like the perturbation to be sufficient, so that at least some other spurious records in the data set have a higher fit to the correct public database record than the true record. Larger perturbations reduce the log-likelihood fit of the true record $X \in \mathcal{D}$ corresponding to Z, and increase the probability that another spurious record in $\mathcal{D}$ may have a higher log-likelihood fit than X by chance. This is desirable from the point of view of privacy preservation. When there are at least k records in $\mathcal{D}$ which have higher (or equal) log likelihood fit than X, then the record X is said to be k-randomized. In such a case, no public database can be used to distinguish X from the k other records within $\mathcal{D}$ which are a better fit to the randomized representation of X. Now, we will define the concept of k-randomization formally.

DEFINITION 18.13 *A (randomized) record $Z \in \mathcal{D}$ with original representation X is said to be k-randomized when there are at least k records $\{X_1 \ldots X_k\} \in \mathcal{D}$ for which the following is true:*

$$\mathcal{F}(Z, X) \leq \mathcal{F}(Z, X_i) \tag{18.21}$$

This means that the randomized record Z cannot be used to distinguish its true representation X from the k records $X_1 \ldots X_k$ in $\mathcal{D}$. By performing k-randomization of every record in the database $\mathcal{D}$, it is possible to achieve an equivalent level of k-anonymity for the randomization approach. However, since the randomization approach does not use a trusted server and can be performed at *data collection time* (without knowledge of other records), the exact level of randomization may not be known or precisely controlled a-priori. This is different from the k-anonymity model which performs the privacy transformation in a controlled way so as to explicitly *engineer* k-anonymity. Here, our aim in defining the randomization level of a record is to use it as an *analytical tool* for judging the effectiveness of a given level of perturbation. The only a-priori control parameter is the perturbation standard deviation, and the randomization level is computed a-posteriori. Thus, the *calculated* randomization level of a point X is denoted by $kr(X)$ and is equal to the number of randomized points in the database which fit the randomized version of X at least as well as (the randomized representation of) X itself. We make the following observation about the expected value of $kr(X)$:

OBSERVATION 18.4.2 *Let $X = (x_1 \ldots x_d)$ be a d-dimensional point from the database $\mathcal{D}$. Let $Z = (z_1 \ldots z_d)$ represent the randomization of X. Then, the expected randomization level $E[kr(X)]$ is as follows:*

$$E[kr(X)] = \sum_{X' \in \mathcal{D}} P(\mathcal{F}(Z, X') \geq \mathcal{F}(Z, X)) \tag{18.22}$$

As in the case of k-anonymity, this value is at least 1 to account for the case when $X' = X$. Next, we generalize the point specific randomization level to the entire database.

DEFINITION 18.14 *The average randomization level of the* database *$\mathcal{D}$ is defined as the average value of $kr(X)$ over all points in $\mathcal{D}$.*

Since the calculated randomization level $kr(X)$ may vary with data point X, we also define a worst-case quantification. In this context, we define the randomization level at *quantile* q.

DEFINITION 18.15 *The randomization level of database $\mathcal{D}$ at quantile q is computed as the lowest quantile q of the randomization level array $kr(\cdot)$.*

18.4.2 Effects of High Dimensionality

In this section, we will analyze the effect of different perturbing distributions on the effectiveness of randomization. We will also analyze the effects of dimensionality on the effectiveness of randomization. The two most common distributions used for perturbation are the uniform and the gaussian distribution [1]. In this section, we will analyze the effects of both.

18.4.3 Gaussian Perturbing Distribution

The gaussian perturbation with standard deviation σ_i on the ith dimension is defined as follows:

$$f_Y(y) = \frac{1}{\sqrt{2 \cdot \pi}\sigma_i} e^{-\frac{y^2}{2 \cdot \sigma_i^2}} \tag{18.23}$$

Let us consider the record $X = (x_1 \ldots x_d)$ which is perturbed to the randomized record denoted by $Z = (z_1 \ldots z_d)$. Then, the log likelihood fit $\mathcal{F}(Z, X)$ is given by $\mathcal{F}(Z, X) = \sum_{i=1}^d \log(f_{Y_i}(q_i(Z, X))) = \sum_{i=1}^d \log(f_{Y_i}(z_i - x_i))$. By substituting the value of $f_{Y_i}(y)$ according to Equation 18.23, we get:

$$\mathcal{F}(Z, X) = -(d/2) \cdot \log(2 \cdot \pi) - \sum_{i=1}^{d} \log(\sigma_i) - \sum_{i=1}^{d} \frac{(z_i - x_i)^2}{2 \cdot \sigma_i^2} \tag{18.24}$$

Let us now consider another record $X' = (x'_1 \ldots x'_d) \in \mathcal{D}$ which is in the neighborhood of X. We would like to calculate the probability that the likelihood fit $\mathcal{F}(Z, X')$ is at least equal to that of $\mathcal{F}(Z, X)$. As evident from Observation 18.4.2, this probability $P(\mathcal{F}(Z, X') \geq \mathcal{F}(Z, X))$ plays a key role in defining the expected randomization level $E[kr(X)]$. Therefore, our future analysis will quantify the value of $P(\mathcal{F}(Z, X') \geq \mathcal{F}(Z, X))$. We will show the following result about this probability:

LEMMA 18.16 *Let $X = (x_1 \ldots x_d)$ and $X' = (x'_1 \ldots x'_d)$ be the two d-dimensional points from the database $\mathcal{D}$, such that $\Delta = (\delta_1 \ldots \delta_d) = X - X'$. Let $Z = (z_1 \ldots z_d)$ represents the randomization of X and σ_i^2 be the variance of the gaussian perturbation along the ith dimension. Then, we have:*

$$P(\mathcal{F}(Z, X') \geq \mathcal{F}(Z, X)) = P\left(\sum_{i=1}^{d} \delta_i^2/(2\sigma_i^2) \leq \sum_{i=1}^{d} -\delta_i \cdot y_i/\sigma_i^2\right) \tag{18.25}$$

Here y_i is the random variable representing the gaussian perturbation along the ith dimension.

Proof: By substituting the values of $\mathcal{F}(Z, X)$ and $\mathcal{F}(Z, X')$ from Equation 18.24, and canceling the common terms, we get:

$P(\mathcal{F}(Z, X') \geq \mathcal{F}(Z, X)) =$

$$= P\left(\sum_{i=1}^{d} -(z_i - x'_i)^2/\sigma_i^2 \geq \sum_{i=1}^{d} -(z_i - x_i)^2/\sigma_i^2\right)$$
$$= P\left(\sum_{i=1}^{d} (z_i - x_i + \delta_i)^2/\sigma_i^2 \leq \sum_{i=1}^{d} (z_i - x_i)^2/\sigma_i^2\right)$$

The last relationship is obtained by replacing $X' = X - \Delta$, and reversing the sign of the inequality by negating both sides. Now, we note that $z_i - x_i$ is simply the value of the random perturbation y_i which is derived from a gaussian distribution. Therefore, let us replace $z_i - x_i$ by y_i for algebraic convenience. Therefore, we have:

$$P(\mathcal{F}(Z, X') \geq \mathcal{F}(Z, X)) = P\left(\sum_{i=1}^{d} (y_i + \delta_i)^2/\sigma_i^2 \leq \sum_{i=1}^{2} y_i^2/\sigma_i^2\right) \quad (18.26)$$

$$= P\left(\sum_{i=1}^{d} \delta_i^2/(2 \cdot \sigma_i^2) \leq -\sum_{i=1}^{d} \delta_i \cdot y_i/\sigma_i^2\right)$$

The last relationship is obtained by simple algebraic expansion of $(y_i + \delta_i)^2$ and subsequent simplification.

While the above lemma provides an algebraic expression for this bound, a more intuitive interpretation with respect to dimensionality and distribution needs to be constructed. In order to do so, we will make use of the well known Chebychev inequality. First, we will prove a simple lemma which we will need in a later section.

LEMMA 18.17 *Let y_i be the gaussian perturbation along the ith dimension with variance σ_i^2. Let $V = -\sum_{i=1}^{d} y_i \cdot \delta_i/\sigma_i^2$. Then, we have:*

$$E[V^2] = \sum_{i=1}^{d} \delta_i^2/\sigma_i^2 \quad (18.27)$$

Proof: We note that $y_1 \ldots y_d$ are independent perturbations along the d dimensions. Therefore, by expanding the expression for V^2, and using independence to simplify expectation of products of random variables, we get:

$$E[V^2] = \sum_{i=1}^{d} \delta_i^2 \cdot E[y_i^2]/\sigma_i^4 + 2 \cdot \sum_{i=1}^{d} \sum_{j=i+1}^{d} \delta_i \cdot \delta_j \cdot E[y_i] \cdot E[y_j]/(\sigma_i^2 \cdot \sigma_j^2) \quad (18.28)$$

Since y_i is a gaussian with variance σ_i^2 about a mean of zero, we have $E[y_i] = 0$ and $E[y_i^2] = \sigma_i^2$. By substituting this in Equation 18.27, we get the desired result.

THEOREM 18.18 *Let $X = (x_1 \ldots x_d)$ and $X' = (x'_1 \ldots x'_d)$ be two d-dimensional points from the database $\mathcal{D}$, such that $\Delta = (\delta_1 \ldots \delta_d) = X - X'$.*

Let Z represent the randomization of X and σ_i^2 be the variance of the gaussian perturbation along the ith dimension. Then, we have:

$$P(\mathcal{F}(Z, X') \geq \mathcal{F}(Z, X)) \leq 4/(\sum_{i=1}^{d} \delta_i^2/\sigma_i^2) \tag{18.29}$$

Proof: As in Lemma 18.17, let us define $V = -\sum y_i \cdot \delta_i/\sigma_i^2$. From Lemma 18.16, we get:

$$P(\mathcal{F}(Z, X') \geq \mathcal{F}(Z, X)) = P\left(\sum_{i=1}^{d} \delta_i^2/(2 \cdot \sigma_i^2) \leq V\right) \tag{18.30}$$

$\leq P\left(V^2 \geq (\sum_{i=1}^{d} \delta_i^2/(2 \cdot \sigma_i^2))^2\right)$ (squaring both sides and recognizing that $\delta_i^2/(2 \cdot \sigma_i^2)$ is always positive)

$\leq E[V^2]/(\sum_{i=1}^{d} \delta_i^2/(2 \cdot \sigma_i^2))^2$ (Chebychev Inequality)

By substituting the expression for $E[V^2]$ from Lemma 18.17, we get the desired result.

We note that the variance of the perturbing distribution along each dimension is typically chosen proportional to the corresponding variance of the original data. This is a natural choice in order to provide a similar level of perturbation over the different dimensions.

ASSUMPTION 18.4.1 **Proportionality Assumption:** *If the variance of the original data along the ith dimension is denoted by σ_i^o, then the perturbing variance σ_i is chosen such that $C_1 \cdot \sigma_i \leq \sigma_i^o \leq C_2 \cdot \sigma_i$ for some constants C_1 and C_2.*

The proportionality assumption automatically helps us reword the results of Theorem 18.18 as follows:

THEOREM 18.19 *Let $X = (x_1 \ldots x_d)$ and $X' = (x'_1 \ldots x'_d)$ be two d-dimensional points from the database $\mathcal{D}$, such that $\Delta = (\delta_1 \ldots \delta_d) = X - X'$. Let $Z = (z_1 \ldots z_d)$ represents the randomization of X. Let σ_i^2 be the variance of the gaussian perturbation along the ith dimension, and $(\sigma_i^o)^2$ be the variance of the original data along dimension i. Then, under the proportionality assumption, for some constant C_3, we have:*

$$P(\mathcal{F}(Z, X') \geq \mathcal{F}(Z, X)) \leq C_3/(\sum_{i=1}^{d} \delta_i^2/(\sigma_i^o)^2) \tag{18.31}$$

We note that denominator of the right hand side of the relationship of Theorem 18.19 contains the term $\sum_{i=1}^{d} \delta_i^2/(\sigma_i^o)^2$. This is simply the distance between X and X', when the original data is normalized by the variance along each dimension. Therefore, it is intuitively clear that a data point X' which is spatially close to X has a higher chance of satisfying the requirement $\mathcal{F}(Z, X') \geq \mathcal{F}(Z, X)$ which increases the randomization level of X. However, with increasing dimensionality, the concept of spatial locality becomes more problematic. According to [8], the sparsity of high dimensional data ensures that the distance to other points in the data $\sum_{i=1}^{d} \delta_i^2/(\sigma_i^o)^2$ grows with d^* in high dimensional space, where d^* is the implicit dimensionality of the data. Therefore, even if X' is chosen to be the nearest neighbor of X, the value of $P(\mathcal{F}(Z, X') \geq \mathcal{F}(Z, X))$ tends to zero with increasing value of d. From Observation 6.1, the expected randomization level $E[kr(X)]$ is critically dependent upon this probability, and therefore, the randomization level of X also reduces with increasing dimensionality. We summarize this result as follows:

CONCLUSION 18.4.1 *The expected randomization level reduces with increasing dimensionality for a fixed level of perturbation.*

How strong is this revealing effect of high dimensionality? We note that the Chebychev inequality is extremely weak in practice. Therefore, the above results represent a fairly weak bound. In practice, it is possible to get much tighter bounds with the use of a few approximations on Lemma 18.16. We note that the right hand side of Lemma 18.16 contains $V = -\sum_{i=1}^{d} y_i \cdot \delta_i/\sigma_i^2$. Since each y_i is independent, the variance of V is equal to the sum of the individual variances. This works out to $\sigma^2(V) = \sum_{i=1}^{d} \delta_i^2/\sigma_i^2$. We further note that $E[V] = 0$. Now, we make the approximation that V is normally distributed. This may be fairly close to the truth for large values of d, since each component of V (which is $-y_i \cdot \delta_i/\sigma_i^2$) is a unit normal distribution scaled by δ_i/σ_i.

The right hand side of Lemma 18.16 can be expressed as $P(V \geq \sum_{i=1}^{d} \delta_i^2/(2 \cdot \sigma_i^2)) = 1 - \Phi((\sum_{i=1}^{d} \delta_i^2/(2 \cdot \sigma_i^2))/\sigma(V))$. Here $\Phi(\cdot)$ is the cumulative normal distribution. Since $\sigma(V) = \sqrt{\sum_{i=1}^{d} \delta_i^2/\sigma_i^2}$, we can summarize as follows:

APPROXIMATION 18.4.1 *Let $X = (x_1 \ldots x_d)$ and $X' = (x_1' \ldots x_d')$ be two d-dimensional points from the database $\mathcal{D}$, such that $\Delta = (\delta_1 \ldots \delta_d) = X - X'$. Let $Z = (z_1 \ldots z_d)$ represents the randomization of X. Let σ_i^2 be the variance of the gaussian perturbation along the ith dimension. Then, we have:*

$$P(\mathcal{F}(Z, X') \geq \mathcal{F}(Z, X)) = 1 - \Phi((\sqrt{\sum_{i=1}^{d} \delta_i^2/\sigma_i^2})/2) \qquad (18.32)$$

Here $\Phi(\cdot)$ is the cumulative normal distribution. The corresponding expected randomization level of the data point X is obtained by summing $P(\mathcal{F}(Z, X') \geq \mathcal{F}(Z, X))$ over all points $X' \neq X$ in the database $\mathcal{D}$.

We note that the cumulative normal distribution $\Phi(\cdot)$ is approximately equal to 1 for an argument value greater than 3. Therefore, the expression $\sum_{i=1}^{d} \delta_i^2/\sigma_i^2$ needs to be at most 36 in order for the probability $P(\mathcal{F}(Z, X') \geq \mathcal{F}(Z, X))$ to not be (nearly) zero. Consider the case of a uniformly distributed data set in which we pick $\sigma_i = C \cdot \sigma^o$. In such a case, we can show [8] that the distance value $\sum_{i=1}^{d} \delta_i^2/\sigma_i^2$ grows as d/C^2. This means that C must grow with $\sqrt{d}$ in order for the probability $P(\mathcal{F}(Z, X') \geq \mathcal{F}(Z, X))$ to be significantly larger than zero. Since Observation 6.1 ties the probability $P(\mathcal{F}(Z, X') \geq \mathcal{F}(Z, X))$ to the expected randomization level $E(kr(X)]$, this indicates that the value of C should grow with $\sqrt{d}$ for the randomization level to be constant with increasing dimensionality. While the result of [8] is true for the case of uniform distribution of the original data, it provides the intuition that the perturbing standard deviation along each dimension should grow as the square root of the *implicit dimensionality* of the data. We summarize this result as follows:

CONCLUSION 18.4.2 *Under the proportionality assumption, the perturbing gaussian distribution along each dimension should have a standard deviation which grows with the square root of the implicit dimensionality of the underlying data in order to retain the same level of randomization.*

In practice, only a small number of data points X' (which lie in the locality of X) are likely to have dominant values for $P(\mathcal{F}(Z, X') \geq \mathcal{F}(Z, X))$ in the right hand side of Observation 6.1. The value of each of these terms depend inversely upon the normalized distance $\sum_{i=1}^{d} \delta_i^2/(\sigma_i^o)^2$ between X and X'. Thus, for data sets with the same global variance, the expected randomization level $E[kr(X)]$ is likely to be higher when non-empty localities of the data are dense and highly clustered. This provides the following result:

CONCLUSION 18.4.3 *The presence of clusters is helpful in increasing the randomization level for data sets with similar global variance.*

This is a nice property of the randomization method, since most real data sets exhibit clustered behavior. We further note that while Approximation 18.4.1 provides an understanding of the randomization level of each data point, it may often be more desirable to examine the worst-case randomization behavior of the entire data set. As discussed earlier, the *local* magnitudes of the normalized distances $\sum_{i=1}^{d} (\delta_i/\sigma_i^o)^2$ have a strong inverse relationship with the expected randomization level $E[kr(X)]$. Therefore, for data sets with the same global variance, a variation in the local density distribution can affect the worst-case randomization more sharply.

CONCLUSION 18.4.4 *A data set with varying density distribution is likely to have a significantly lower worst-case randomization level than the average randomization level.*

The presence of outliers is the extreme case, since the density within the locality of an outlier is significantly lower than the average case density.

CONCLUSION 18.4.5 *The presence of outliers may reduce the worst-case randomization level without significantly affecting the average-case randomization behavior of the data.*

These results show that the randomization approach is susceptible to the presence of the density variations and outliers. The intuition for this is that unlike methods such as k-anonymity, the current methods for randomization of individual data points are applied without assumption of knowledge about the rest of the data. This is an issue which needs to be addressed in future research on randomization.

18.4.4 Uniform Perturbing Distribution

We assume that the perturbation along the ith dimension is uniformly distributed with range $[0, a_i]$, and the corresponding standard deviation σ_i is equal to $a_i/\sqrt{12}$. For simplicity, we assume that the range of the perturbation a_i is larger than the range of the non-perturbed data along dimension i. This is not really restrictive, since it is needed to preserve a minimum level of privacy along the ith dimension. Therefore, if $\Delta = (\delta_1 \ldots \delta_d) = X - X'$, we must have $|\delta_i| \leq a_i$.

THEOREM 18.20 *Let* $X = (x_1 \ldots x_d)$ *and* $X' = (x'_1 \ldots x'_d)$ *be two d-dimensional points from the randomized database* $\mathcal{D}$, *such that* $\Delta = (\delta_1 \ldots \delta_d) = X - X'$ *and* $Z = (z_1 \ldots z_d)$ *represents the randomization of* X. *Let* $[0, a_i]$ *be the range of the uniform perturbation along the* i*th dimension. Then, we have:*

$$P(\mathcal{F}(Z, X') \geq \mathcal{F}(Z, X)) = \pi_{i=1}^{d}(1 - |\delta_i|/a_i) \tag{18.33}$$

Proof: Since the distribution is uniform with density $1/a_i$, the value of $\mathcal{F}(Z, X)$ is simply $d \cdot \log(1/a_i)$. Now we note that the value of $\mathcal{F}(Z, X')$ is defined as follows:

$$\mathcal{F}(Z, X') = \sum_{i=1}^{d} \log(f_Y(z_i - x'_i)) = \tag{18.34}$$

$$= \sum_{i=1}^{d} \log(f_Y(z_i - x_i + \delta_i)) = \sum_{i=1}^{d} \log(f_Y(y_i + \delta_i))$$

Here y_i is the uniformly distributed perturbation in the range $[0, a_i]$. We note that each of the d terms on the right hand side is either $\log(1/a_i)$ or $-\infty$ depending upon whether or not $(y_i + \delta_i)$ lies in the range $[0, a_i]$. Therefore $\mathcal{F}(Z, X')$ can *never* be larger than $\mathcal{F}(Z, X)$. The value of $\mathcal{F}(Z, X')$ can at most be equal to $\mathcal{F}(Z, X)$, if and only if for each and every dimension i, $y_i + \delta_i$ lies in the range $[0, a_i]$. Since y_i is uniformly distributed in the range $[0, a_i]$, it is easy to verify that the probability of $(y_i + \delta_i)$ lying in the range $[0, a_i]$ is $(1 - |\delta_i|/a_i)$. By using the independence of the different values of y_i, the result follows.

A simple corollary of the above result is as follows;

COROLLARY 18.21 *Let $X = (x_1 \dots x_d)$ and $X' = (x'_1 \dots x'_d)$ be two d-dimensional points from the randomized database $\mathcal{D}$, such that $\Delta = (\delta_1 \dots \delta_d) = X - X'$ and $Z = (z_1 \dots z_d)$ represents the randomization of X. Let $[0, a_i]$ be the range of the uniform perturbation along the ith dimension. Then, we have:*

$$P(\mathcal{F}(Z, X') \geq \mathcal{F}(Z, X)) \leq (1 - \sum_{i=1}^{d} (|\delta_i|/(d \cdot a_i)))^d \qquad (18.35)$$

Proof: This corollary simply follows from Theorem 18.20 and the fact that the geometric mean of a set of non-negative values is at most equal to the arithmetic mean.

As in the previous case, let us examine what happens in a uniformly distributed data set, when the range a_i is chosen to be $C \cdot \sigma_i^o$ for some constant C using the proportionality assumption. In this case, the results of [8] indicate that in the high dimensional case, $\sum_{i=1}^{d} |\delta_i|/\sigma_i^o$ is expected to increase as $B \cdot d$ for some constant B. Then, we can use the result of Corollary 18.21 to derive the following;

$$P(\mathcal{F}(Z, X') \geq \mathcal{F}(Z, X)) \leq (1 - B/C)^d \qquad (18.36)$$

Note that when the value of C is chosen to be $B \cdot d$, the value of the above expression is $(1 - 1/d)^d$. This is bounded above by $1/e$, where e is the base of the natural logarithm. By choosing C smaller than $B \cdot d$, it is possible for this probability $P(\mathcal{F}(Z, X') \geq \mathcal{F}(Z, X))$ to fall off rapidly to zero. This would result in lower randomization levels. We summarize as follows:

CONCLUSION 18.4.6 *Under the proportionality assumption, the perturbing uniform distribution along each dimension should have a range (or standard deviation) which grows* **at least** *linearly with the implicit dimensionality of the underlying data.*

Recall that the in the case of the gaussian distribution, the required standard deviation grows proportionally only with the *square-root* of the dimensionality. Therefore, a greater level of randomization (and hence information loss) may be sustained when the uniform distribution is used. We also emphasize that unlike the case of the gaussian distribution, the above intuition on the choice

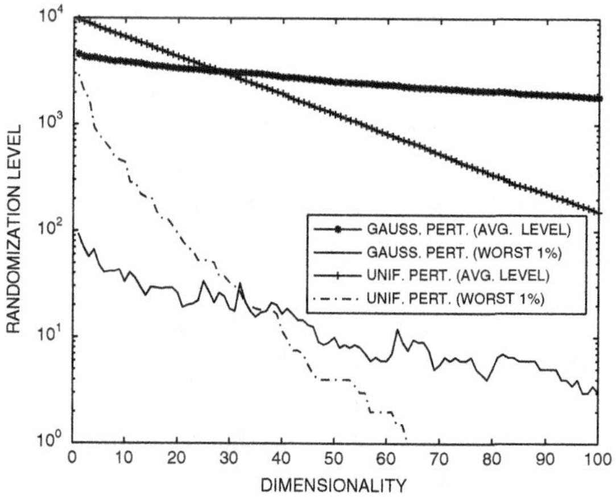

Figure 18.5. Randomization Level with Increasing Dimensionality, Perturbation level $= 8 \cdot \sigma^o$ $(UniDis)$

of the perturbing distribution is only a lower bound. This is because of the use of the inequality between the geometric and arithmetic mean. This inequality can be extremely loose in practice, when the values of $|\delta_i|/a_i$ are very different from one another over the different dimensions. Therefore, even the lower bound on the standard deviation of the perturbing distribution for the uniformly distributed case is significantly higher than the required standard deviation for the gaussian distribution. We summarize as follows:

CONCLUSION 18.4.7 *In the high dimensional case, gaussian perturbations provide higher randomization than the uniform perturbation.*

In Figure 18.5, we have illustrated the effect of increasing dimensionality on the $UniDis$ data set. These results were obtained by applying the technique to projections of the data of different dimensionality. The X-axis on each chart illustrates the data dimensionality, whereas the Y-axis illustrates both the average and worst case randomization levels $\mathcal{AR}(\mathcal{D})$ and $\mathcal{WR}(\mathcal{D})$ respectively for different perturbing distributions. Since the randomization level varied widely for different data sets, distributions, and dimensionalities, we made it a point to use a logarithmic scale on the Y-axis. As pointed out earlier, the base data sets were normalized so that the variance along each dimension was $\sigma^o = 1$. The corresponding perturbation variance was set of $8 \cdot \sigma^o$ in each case. Since the variance of the original data set was always the same, this set of charts helps us compare the relative behavior of different data sets and perturbing distributions with varying dimensionality.

One immediate observation from each of the (logarithmically scaled) charts in Figures 18.5 was that both the average and worst case randomization levels reduced rapidly with increasing dimensionality for different data sets. For example, in Figure 18.5, the average randomization (with uniform perturbations) for the 1-dimensional data set was 9646.1, whereas the average randomization level for the 100-dimensional case was 151.7. This means that for the 1-dimensional case, 96.46% of the original 10,000 points fit a given data point as well as the true point. On the other hand, this number reduced to only 1.51% in the 100-dimensional case. Even more interesting behavior was observed by examining the lowest 1% quantile of the data. This corresponds to $\mathcal{WR}(\mathcal{D})$. In this case, the randomization level was 2907 for the 1-dimensional case. *However, for any instantiation of the data set beyond 64 dimensions, the randomization level was only 1 for the entire lower* 1% *quantile of the data*, when uniform perturbations were used. We note that a randomization level of 1 denotes no privacy, since the data point itself contributes to a randomization level of 1. This behavior was specific to the uniform perturbing distribution, and happened in spite of a high perturbation level of $8 \cdot \sigma^o$ for each dimension.

As evident from Figure 18.5, the behavior of the gaussian perturbing distribution was much more robust with increasing dimensionality, even though the uniform perturbation turned out to be superior for the lower dimensional cases. For example, for the 1-dimensional case in Figure 18.5, the average randomization level for the gaussian perturbing distribution 4552.2 which was less than half the randomization level of the uniform distribution. However, when the dimensionality increases to 100, the average randomization level was 1824.4, which was more than an order of magnitude higher than the randomization level 151.7 for the uniform perturbations. An even more interesting case was the behavior of the worst 1%-quantile of the data. While the uniform perturbation had no privacy of the worst 1% quantile for dimensionalities beyond 64, the gaussian perturbation had a randomization level of between 5 and 10 for dimensionalities higher then 64. Thus, the results show that while the curse of dimensionality results in a reduction of privacy with increasing dimensionality, the effect was more pronounced in the uniformly distributed case. Since a better choice of perturbing distribution seems to moderate the effects of the dimensionality curse, this underlines the importance of judiciously choosing the perturbing distribution in the randomization method. A more detailed analysis of the randomization technique with different kinds of data sets may be found in [5].

18.5 The Dimensionality Curse and l-diversity

We will also briefly study the effect of dimensionality on the l-diversity method. Clearly, while k-anonymity is effective in preventing *identification* of a record, it may not always be effective in preventing inference of the sensitive values of the attributes of that record. For example, if the sensitive values of an attribute take on the same value within an anonymized group, one can infer the value of the corresponding sensitive value. Therefore, the technique of l-diversity was proposed which not only maintains the minimum group size of k, but also focusses on maintaining the diversity of the sensitive attributes. Therefore, the l-diversity model [11] for privacy is defined as follows:

DEFINITION 18.22 *Let a q^*-block be a set of tuples such that its non-sensitive values generalize to q^*. A q^*-block is l-diverse if it contains l "well represented" values for the sensitive attribute S. A table is l-diverse, if every q^*-block in it is l-diverse.*

When there are multiple sensitive attributes, the concept of l-diversity needs to be maintained separately for each sensitive attribute. With increasing number of sensitive attributes it becomes increasingly infeasible to construct a feasible solution. This is because different sensitive attributes may behave independently of one another and large group sizes may be required in order to maintain l-diversity across every attribute. The practical difficulty of maintaining l-diversity across a large number of sensitive attributes has been discussed in [11] in the experimental section of the paper. We make the following observation.

OBSERVATION 18.5.1 *In order to maintain l-diversity across r sensitive attributes, a block of records with minimum size of $O(l \cdot r)$ may be required in the worst case.*

This is easy to verify, since all the diverse values for one attribute may be homogeneous in the next attribute and so on. Such group sizes can be very large with increasing diversity and dimensionality. For example, in order to maintain 5-diversity over 100 sensitive attributes, this may require group sizes as large of 500. It may often be difficult to preserve locality in such large groups, as a result of which the generalizations across pseudo-identifiers will be very large.

18.6 Conclusions and Research Directions

In this paper, we examined the effects of the dimensionality curse on a variety of privacy-preservation methods such as l-anonymity, condensation, randomization and l-diversity. The results seem to suggest that the curse of dimensionality may be a fundamental one from the point of view of privacy,and cannot be easily solved using more effective algorithms and techniques. The true situation may not be quite as bleak in many cases when the data sets have spacial structure which can be exploited in order to design effective privacy-preservation methods. For example, in [6], the special structure of text and market basket data sets was used in order to design privacy-preservation methods for two data domains which are inherently high dimensional.

References

[1] Agrawal R., Srikant R. Privacy-Preserving Data Mining. *Proceedings of the ACM SIGMOD Conference*, 2000.

[2] Agrawal D. Aggarwal C. C. On the Design and Quantification of Privacy-Preserving Data Mining Algorithms. *ACM PODS Conference*, 2002.

[3] Aggarwal C. C. On k-anonymity and the curse of dimensionality. *VLDB Conference*, 2005.

[4] Aggarwal C. C., Yu P. S.: A Condensation approach to privacy preserving data mining. *EDBT Conference*, 2004.

[5] Aggarwal C. C.: On Randomization, Public Information and the Curse of Dimensionality. *ICDE Conference*, 2007.

[6] Aggarwal C. C., Yu P. S.: On Privacy-Preservation of Text and Sparse Binary Data with Sketches. *SIAM Conference on Data Mining*, 2007.

[7] Bayardo R.J., Agrawal R.: Data Privacy through Optimal k-Anonymization. *Proceedings of the ICDE Conference*, pp. 217–228, 2005.

[8] Hinneburg A., Aggarwal C.. Keim D.: What is the nearest neighbor in high dimensional spaces? *VLDB Conference*, 2000.

[9] Huang Z., Du W., Chen B.: Deriving Private Information from Randomized Data. pp. 37–48, *ACM SIGMOD Conference*, 2005.

[10] Kargupta H., Datta S., Wang Q., Sivakumar K.: On the Privacy Preserving Properties of Random Data Perturbation Techniques. *ICDM Conference*, pp. 99–106, 2003.

[11] Machanavajjhala A., Gehrke J., Kifer D., and Venkitasubramaniam M.: l-Diversity: Privacy Beyond k-Anonymity. *ICDE*, 2006.

[12] Martin D., Kifer D., Machanavajjhala A., Gehrke J., Halpern J.: Worst-Case Background Knowledeg. *ICDE Conference*, 2007.

[13] Meyerson A., Williams R. On the complexity of optimal k-anonymity. *ACM PODS Conference*, 2004.

[14] Samarati P.: Protecting Respondents' Identities in Microdata Release. IEEE Trans. Knowl. Data Eng. 13(6): 1010–1027, 2001.

Chapter 19

Personalized Privacy Preservation

Yufei Tao
Department of Computer Science and Engineering
Chinese University of Hong Kong
Sha Tin, New Territories, Hong Kong
taoyf@cse.cuhk.edu.hk

Xiaokui Xiao
Department of Computer Science and Engineering
Chinese University of Hong Kong
Sha Tin, New Territories, Hong Kong
xkxiao@cse.cuhk.edu.hk

Abstract Unlike conventional methods that exert the same amount of privacy control over all the tuples in the microdata, *personalized privacy preservation* applies various degrees of protection to different tuples, subject to the preferences of the data owners. This chapter explains the formulation of personal preferences, and the computation of anonymized tables that fulfill the privacy requirement of everybody. Several theoretical results regarding privacy guarantees will also be discussed. Finally, we point out the open research issues that may be explored in the future.

Keywords: Personalized, k-anonymity, l-diversity.

19.1 Introduction

In earlier chapters, we have seen several principles, such as k-anonymity [12] and l-diversity [9], for determining the degree of privacy preservation in data publication. A common feature of those principles is that, they impose an *identical* amount of privacy protection on all the tuples in the microdata. In other words, they cannot be deployed to achieve various degrees of privacy preservation on different tuples.

Ideally, a data owner should be allowed to specify how much protection is necessary for her/his tuple. This is reasonable because, for example, a flu patient would typically demand weaker protection than a person that contracted HIV. In fact, *personalization* is an inherent notion in privacy preservation, whose objective is to guard the privacy of *individuals* anyway. By applying a "universal" privacy standard, the data publisher would be offering insufficient protection to some tuples while exerting excessive privacy control on other tuples. Note that although "excessive control" poses no harm on privacy, it reduces the utility of the resulting dataset.

k-anonymity and l-diversity can be extended to accommodate personal preferences in a straightforward manner. Take k-anonymity for example (the rationale extends to l-diversity in a straightforward manner). We can permit every person to pick a tailored k to indicate the smallest size that the QI-group containing her/his tuple should have. As a result, if the microdata has cardinality n, there are totally n personalized k-values: $k_1, k_2, ..., k_n$, each associated with a tuple. We will discuss this approach in more detail later in Section 19.6.

In this chapter, we focus on another, more sophisticated, form of personalization, which allows a data owner to formulate semantically-richer privacy preferences. To clarify the motivation, let us first see a defect of k-anonymity and l-diversity. Figure 19.1a demonstrates a microdata table, where *Age*, *Sex*, *Zipcode* are the QI-attributes, and *Disease* is the sensitive attribute. Column *row-#* is not part of the microdata, but is included for row referencing. The column *guarding-node* will be explained later. Figure 19.1b shows a 2-anonymous version of Figure 19.1a. Since Andy's record appears in a QI-group containing *gastric-ulcer* and *dyspepsia*, an adversary is able to learn that Andy has contracted either of these two diseases. While this is acceptable according to 2-anonymity, Andy may not want anyone to think (with high confidence) "Andy must have some stomach problem". Such a requirement cannot be guaranteed in Figure 19.1b, since both *gastric-ulcer* and *dyspepsia* are stomach diseases. On the other hand, it is possible that Linda regards *flu* as a common disease,

row #	**Age**	**Sex**	**Zipcode**	**Disease**	guarding node
1 (Andy)	5	M	12000	gastric ulcer	*stomach disease*
2 (Bill)	9	M	14000	dyspepsia	*dyspepsia*
3 (Ken)	6	M	18000	pneumonia	*respiratory infection*
4 (Nash)	8	M	19000	bronchitis	*bronchitis*
5 (Joe)	12	M	22000	pneumonia	*pneumonia*
6 (Sam)	19	M	24000	pneumonia	*pneumonia*
7 (Linda)	21	F	58000	flu	∅
8 (Jame)	26	F	36000	gastritis	*gastritis*
9 (Sarah)	28	F	37000	pneumonia	*respiratory infection*
10 (Mary)	56	F	33000	flu	*flu*

(a) Microdata

row #	**Age**	**Sex**	**Zipcode**	**Disease**
1	[1, 10]	M	[10001, 15000]	gastric ulcer
2	[1, 10]	M	[10001, 15000]	dyspepsia
3	[1, 10]	M	[15001, 20000]	pneumonia
4	[1, 10]	M	[15001, 20000]	bronchitis
5	[11, 20]	M	[20001, 25000]	pneumonia
6	[11, 20]	M	[20001, 25000]	pneumonia
7	[21, 60]	F	[30000, 60000]	flu
8	[21, 60]	F	[30000, 60000]	gastritis
9	[21, 60]	F	[30000, 60000]	pneumonia
10	[21, 60]	F	[30000, 60000]	flu

(b) A 2-anonymous table

Figure 19.1. Microdata and generalization

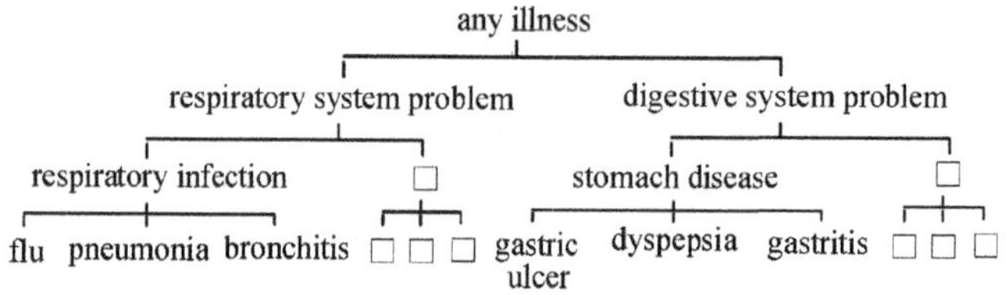

Figure 19.2. The taxonomy of attribute *Disease*

and agrees to release her true diagnosis result (to enhance the effectiveness of research). In this case, it is not necessary to apply any generalization on tuple 7. Such preference variations cannot be captured, even if data owners are allowed to select their own "k-values". Although k-anonymity is used in the example, the same defect applies to l-diversity as well.

The above defect can be remedied through a "guarding node" approach [15], which requires a taxonomy on the sensitive attribute. Figure 19.2 demonstrates a simple taxonomy on attribute *Disease*, which is accessible by the public, and organizes all diseases as leaves of a tree. An intermediate node carries a name summarizing the diseases in its subtree. Some part of the tree is omitted since it is not relevant to our discussion. A personal preference is formulated through a node in the taxonomy. As an example, for tuple 1 in Figure 19.1a, Andy may specify node *stomach-disease*, i.e., the *guarding node* for his privacy (which will be formalized in the next section). Thus, nobody should be able to infer, with significant confidence, that he suffered from any disease (i.e., *gastric-ulcer*, *dyspepsia*, or *gastritis*) in the subtree of the node. In other words, in Andy's opinion, allowing the public to associate him with *dyspepsia* or *gastritis* is as serious as revealing his true disease. On the other hand, for tuple 7 in Figure 19.1a, Linda may specify $\emptyset$, which is an implicit node underneath all the leaves of the taxonomy. The empty-set preference implies that she is willing to release her actual diagnosis result *flu*; therefore, tuple 7 can be published directly.

Next, we will discuss the guarding-node technique in more detail. Specially, Section 19.2 formalizes the underlying concepts. Section 19.3 elaborates the process of a privacy attack carried out by an adversary. Section 19.4 clarifies the derivation of privacy breach probabilities. Section 19.5 explains an algorithm for computing a generalized table which fulfills the requirements of all guarding nodes. Section 19.6 reviews two other forms of personalized privacy protection in the literature. Finally, Section 19.7 concludes this chapter with a discussion of future work.

19.2 Formalization of Personalized Anonymity

Let T be a relation storing private information about a set of individuals. The attributes in T are classified in 4 categories: (i) an *identifier attribute* A^i which uniquely identifies a person, and must be removed when T is released to the public, (ii) a *sensitive attribute* $\mathcal{A}^s$ (e.g., *Disease* in Figure 19.1a), whose values may be confidential for an individual (subject to her/his preferences), (iii) d *quasi-identifier* (QI) *attributes* $\mathcal{A}_1^{qi}$, ..., $\mathcal{A}_d^{qi}$, whose values can be published, but may reveal a personal identify with the aid of external information (*Age*, *Sex*, *Zipcode* in Figure 19.1a), and (iv) other attributes that are not relevant to our discussion.

We require that $\mathcal{A}^s$ should be categorical, whereas the other attributes can be either numerical or categorical. All the attributes have finite domains. Following a common assumption in the literature [3, 4, 7–9, 11, 12, 14], we assume that each categorical attribute A is accompanied by a *taxonomy* (as in Figure 19.2 for *Disease*), which indicates the publicly-known hierarchy among the possible values of A.

Our objective is to compute a generalized table T^* such that (i) it contains all the attributes of T except $\mathcal{A}^i$, (ii) it has a generalized tuple for *every* tuple in T, (iii) it preserves as much information of T as possible, and (iv) its publication does not cause any privacy breach, as formulated in the next section.

19.2.1 Personal Privacy Requirements

We start by defining a subtree in the taxonomy of $\mathcal{A}^s$.

DEFINITION 19.1 ($\mathcal{A}^s$ SUBTREE) *For any node x in the taxonomy of $\mathcal{A}^s$, we represent its* **subtree** *as* SUBTR(x), *which includes x itself, and the part of the taxonomy under it.*

A tuple $t \in T$ defines an *association* between an individual o (identified by $t.\mathcal{A}^i$) and a sensitive value $v = t.\mathcal{A}^s$. We denote the association as $\{o, v\}$. To formulate her/his privacy preference, o specifies a *guarding node* as follows:

DEFINITION 19.2 (GUARDING NODE) *For a tuple $t \in T$, its* **guarding node $t.\mathcal{GN}$** *is a node on the path from the root to $t.\mathcal{A}^s$ in the taxonomy of $\mathcal{A}^s$.*

Through $t.\mathcal{GN}$, o indicates that s/he does not want the public to associate her/him with any leaf $\mathcal{A}^s$ value in SUBTR($t.\mathcal{GN}$). Specifically, assume that SUBTR($t.\mathcal{GN}$) contains x leaf values v_1, v_2, ..., v_x. The privacy requirement of $t.\mathcal{GN}$ is *breached* if an adversary thinks that any of the associations $\{o, v_1\}$, ..., $\{o, v_x\}$ exists in T.

DEFINITION 19.3 (BREACH PROBABILITY) *For a tuple $t \in T$, its* **breach probability $P_{breach}(t)$** *equals the probability that an adversary can infer from T^* that any of the associations $\{o, v_1\}$, ..., $\{o, v_x\}$ exists in T, where v_1, ..., v_x are the leaf values in* SUBTR($t.\mathcal{GN}$).

The published table T^* should guarantee that, for all $t \in T$, $\mathbf{P}_{breach}(t)$ is at most p_{breach}, which is a system parameter specifying the amount of confidentiality control.

Figure 19.1a demonstrates the guarding nodes selected by the individuals involved in the microdata. For example, let t be tuple 3 ($t.\mathcal{A}^i$ = *Ken* and $t.\mathcal{A}^s$ = *pneumonia*). The guarding node *respiratory-infection* of t indicates that nobody can infer, with high confidence, that Ken suffered from a disease under *respiratory-infection* in the taxonomy of Figure 19.2. $\mathbf{P}_{breach}(t)$ is the probability that an adversary can infer that *any* of the following 3 associations exists in T: $\{$Ken, *flu*$\}$, $\{$Ken, *pneumonia*$\}$, $\{$Ken, *bronchitis*$\}$.

On the other hand, Ken does not care if somebody conjectures, with any probability, that he contracted *gastric-ulcer* (not in SUBTR($t.\mathcal{GN}$)), since it is very different from his true diagnosis result. In general, the higher $t.\mathcal{GN}$ is in the taxonomy, the stronger privacy must be guaranteed.

Guarding nodes depend entirely on personal preferences, and are not determined by the sensitive values. For instance, Joe and Sam (who, as with Ken, contracted *pneumonia*) set their guarding nodes simply to *pneumonia* (tuples 5, 6 in Figure 19.1a), implying that they do not mind being associated with *flu* or *bronchitis*. Specially, if a patient believes that disclosing $t.\mathcal{A}^s$ to the public does not violate her/his privacy, s/he may simply set $t.\mathcal{GN}$ to $\emptyset$.

19.2.2 Generalization

We first clarify two fundamental concepts.

DEFINITION 19.4 (PARTITION / GENERAL DOMAIN) *If attribute $\mathcal{A}$ is numeric, a* **partition** *is a continuous interval in the domain of $\mathcal{A}$. Otherwise, a* **partition** *consists of all the leaves in the subtree of a node in the taxonomy of $\mathcal{A}$. In any case, a* **general domain** *of $\mathcal{A}$ is a set of disjoint* partitions *whose union forms the original domain of $\mathcal{A}$.*

By a simple transformation, we can use the interval representation for the general domains of both numeric and categorical attributes. Notice that, when $\mathcal{A}$ is categorical, a general domain is determined by a set of nodes in the taxonomy of $\mathcal{A}$, whose subtrees do not overlap, but cover all the leaves. (For instance, in Figure 19.2, nodes *respiratory-system-problem* and *digestion-system-problem* decide a general domain of *Disease*.) Clearly, $\mathcal{A}$ can be converted to a numeric attribute by imposing a 1D ordering on the leaves of its taxonomy: the left-most leaf is mapped to value 1, its neighbor to 2, and so on. Thus, a partition of $\mathcal{A}$ can be denoted as an interval. For example, the partition corresponding to *respiratory-system-problem* in Figure 19.2 is an interval of $[1, 6]$.

DEFINITION 19.5 (GENERALIZATION) *A general domain of an attribute $\mathcal{A}$ uniquely decides a* **generalization function**. *Given a value v in the original domain of $\mathcal{A}$, the function returns the only partition in the general domain that contains v. The partition is the* **generalized value** *of v.*

Clearly, $\mathcal{A}$ can have many generalization functions, since its values can be partitioned into numerous general domains.

For each tuple $t \in T$, we use t^* to represent its generalized tuple in T^*. The generalization is performed in two steps. The first step, the *QI-generalization*, is identical to conventional generalization in [3, 4, 7, 14]. Specifically, we choose a generalization function for every QI attribute $\mathcal{A}_i^{qi}$ ($1 \leq i \leq d$), and obtain the generalized value $t^*.\mathcal{A}_i^{qi}$ for all tuples $t \in T$ (t^* retains the sensitive value of t at this step). Then, the generalized tuples are divided into *QI-groups*, defined as follows.

DEFINITION 19.6 (QI-GROUP) *After* QI-generalization, *a* **QI-group** *consists of the tuples with identical values on all the QI attributes.* **The i-th QI-value** *($1 \leq i \leq d$) of the QI-group equals $t.A_i^{qi}$, where t is an arbitrary tuple in the QI-group.*

In the second step, *SA-generalization* (SA stands for "sensitive attribute"), we consider each QI-group in turn, and select a *tailored* generalization function on $\mathcal{A}^s$. Note that, unlike the previous step where all tuples are processed with identical generalization functions, SA-generalization uses a *different* function for each group. This strategy achieves less information loss, by allowing each group to decide the amount of necessary generalization.

Figure 19.3 shows a possible result of our entire generalization scheme for Figure 19.1a. The table contains 5 QI-groups: the first one includes tuples 1-4, the second involves tuples 5-6, the third only tuple 7, the fourth tuples 8-9, and the fifth group consists of the last tuple. Note that the sensitive value *flu* of tuple 7 is retained directly, while the same disease of tuple 10 is gen-

row #	Age	Sex	Zipcode	Disease
1 (Andy)	[1, 10]	M	[10001, 20000]	gastric ulcer
2 (Bill)	[1, 10]	M	[10001, 20000]	dyspepsia
3 (Ken)	[1, 10]	M	[10001, 20000]	respiratory infection
4 (Nash)	[1, 10]	M	[10001, 20000]	respiratory infection
5 (Joe)	[11, 20]	M	[20001, 25000]	respiratory infection
6 (Sam)	[11, 20]	M	[20001, 25000]	respiratory infection
7 (Linda)	21	F	58000	flu
8 (Jane)	[26, 30]	F	[35001, 40000]	gastritis
9 (Sarah)	[26, 30]	F	[35001, 40000]	pneumonia
10 (Mary)	56	F	33000	respiratory infection

Figure 19.3. A possible result of our generalization scheme

eralized to *respiratory-infection*. This is legal because, as mentioned earlier, SA-generalization may choose a different generalization function for each QI-group.

None of the existing methods permits SA-generalization. In fact, SA-generalization may produce a table that allows more accurate analysis about the correlation between the sensitive attribute $\mathcal{A}^s$ and other attributes. The reason is that, although SA-generalization results in less precise values on $\mathcal{A}^s$, it retains more information on the QI attributes.

19.3 Combinatorial Process of Privacy Attack

Consider an adversary who attempts to infer the sensitive data of an individual o from T^*. In the worst case, s/he has all the QI values $o.\mathcal{A}_1^{qi}$, ..., $o.\mathcal{A}_d^{qi}$ of o. Therefore, s/he inspects only those tuples $t^* \in T^*$ whose QI value $t^*.\mathcal{A}_i^{qi}$ covers $o.\mathcal{A}_i^{qi}$, for all $i \in [1, d]$.

These tuples must form a QI-group. That is, if t^* and t'^* are two such tuples, then $t^*.\mathcal{A}_i^{qi} = t'^*.\mathcal{A}_i^{qi}$ for all $i \in [1, d]$. Actually, if, for instance, $t^*.\mathcal{A}_1^{qi} \neq t'^*.\mathcal{A}_1^{qi}$, the two values are different partitions in the general domain of $\mathcal{A}_1^{qi}$ that both contain $o.\mathcal{A}_1^{qi}$, violating the requirement that all partitions are disjoint.

DEFINITION 19.7 (ESSENTIAL QI-GROUP / $\mathcal{S}_{real}$). *Given an individual o, the* **essential QI-group $\mathcal{EG}(o)$** *is the only QI-group in T^* whose i-th QI-value covers $o.\mathcal{A}_i^{qi}$, for all $i \in [1, d]$. We use* $\boldsymbol{\mathcal{S}_{real}(o)}$ *to refer to the set of individuals, who have tuples in T generalized to $\mathcal{EG}(o)$.*

Note that $\mathcal{S}_{real}(o)$ is unknown to an adversary. To derive $\mathcal{S}_{real}(o)$, the adversary must resort to an external dataset, and retrieve a set $\mathcal{S}_{ext}(o)$ of persons that may be concerned in $\mathcal{EG}(o)$. $\mathcal{S}_{ext}(o)$ is defined as follows.

DEFINITION 19.8 (EXTERNAL INDIVIDUAL SET $\mathcal{S}_{ext}$) *Given an essential QI-group $\mathcal{EG}(o)$, and an external database DB_{ext},* $\boldsymbol{\mathcal{S}_{ext}(o)}$ *consists of the people $o' \in DB_{ext}$, such that $o'.\mathcal{A}_i^{qi}$ ($1 \leq i \leq d$) is covered by the i-th QI-value of $\mathcal{EG}(o)$.*

To illustrate the above concepts, assume that an adversary tries to infer the disease of Ken from Figure 19.3, having his age 6, sex, and zipcode 18000. The essential QI-group $\mathcal{EG}$(Ken) consists of tuples 1-4, i.e., $\mathcal{S}_{real}$(Ken) equals {Andy, Bill, Ken, Nash}. $\mathcal{S}_{real}$(Ken) is unknown to the adversary. Attempting to derive it, s/he accesses an external database, which is the voter registration list in Figure 19.4, and contains all the QI attributes in Figure 19.3. From the external database, the adversary obtains $\mathcal{S}_{ext}$(Ken) = {Andy, Bill, Ken, Nash, Mike}.

In general

$$\mathcal{S}_{real}(o) \subseteq \mathcal{S}_{ext}(o) \tag{19.1}$$

Name	Age	Sex	Zipcode
Andy	5	M	12000
Bill	9	M	14000
Ken	6	M	18000
Nash	8	M	19000
Mike	*7*	*M*	*17000*
Joe	12	M	22000
Sam	19	M	24000
Linda	21	F	58000
Jane	26	F	36000
Sarah	28	F	37000
Mary	56	F	33000

Figure 19.4. The voter registration list

This is a reasonable condition underlying all the previous work. For instance, if Ken does not appear in the voter registration list, his privacy is trivially preserved. In fact, under the circumstances where an arbitrary number of individuals in T may be missing in the external source, the adversary can infer little information, because all tuples of the essential QI-group may actually correspond to the missing individuals.

Next, the adversary adopts a combinatorial approach to infer the A^s value of individual o. We elaborate the approach by distinguishing two cases in Sections 19.3.1 and 19.3.2, respectively. The subsequent discussion uses m, n to represent the sizes of $\mathcal{EG}(o)$ and $\mathcal{S}_{ext}(o)$, respectively. Also, we denote the tuples in $\mathcal{EG}(o)$ as t_1^*, ..., t_m^*, whose original versions in the microdata are t_1, ..., t_m, respectively.

19.3.1 Primary Case

We first consider the case where $T.\mathcal{A}^i$ is the primary key of T, i.e., each individual has at most one tuple in T.

DEFINITION 19.9 (PRIMARY POSSIBLE RECONSTRUCTION). *In the Primary Case, given an individual o, a* **possible reconstruction** *of the essential QI-group $\mathcal{EG}(o)$ includes*

- *m distinct persons o_1, ..., o_m, who constitute a subset of $\mathcal{S}_{ext}(o)$, i.e., o_j ($1 \le j \le m$) is taken as the owner of t_j;*
- *m leaf sensitive values v_1, ..., v_m, such that v_j ($1 \le j \le m$) is in* SUBTR$(t_j^*.\mathcal{A}^s)$*, i.e., v_j is taken as the real sensitive value of t_j.*

EXAMPLE 19.10 We explain the definition by continuing our example, where the adversary has derived $\mathcal{S}_{ext}$(Ken) = {Andy, Bill, Ken, Nash, Mike}. As mentioned earlier, m = 4, n = 5, and t_1^*, ..., t_4^* are tuples 1-4 in Figure 19.3, respectively.

To obtain a possible reconstruction, the adversary first assigns o_1, ..., o_4 to 4 different persons in $\mathcal{S}_{ext}(\text{Ken})$. As a possible assignment, o_1 = Mike, o_2 = Nash, o_3 = Andy, and o_4 = Ken. Then, the adversary sets v_1 to *gastric-ulcer*, which is the only potential value of v_1, because $t_1^*.\mathcal{A}^s$ = *gastric-ulcer* is a leaf node in the *Disease*-taxonomy. For the same reason, v_2 must be *dyspepsia*. On the other hand, v_3 (v_4) can be any of the 3 leaf diseases under $t_3^*.\mathcal{A}^s$ ($t_4^*.\mathcal{A}^s$) = *respiratory-infection*. The possible reconstruction is completed by assuming, for instance, v_3 = *flu* and v_4 = *bronchitis*.

According to the reconstruction, the adversary thinks that Mike, Nash, Andy, Ken contracted *gastric-ulcer*, *dyspepsia*, *flu*, and *bronchitis*, respectively. Note that a reconstruction most likely is not equivalent to the microdata (where Mike does not even exist); instead, it is only a conjecture by the adversary. Nevertheless, the previous reconstruction violates the privacy requirement enforced by the guarding node of tuple 3 in Figure 19.1a (i.e., Ken does not want people to think that he had any respiratory infection). Interestingly, the breach happens when Ken is associated with tuple 4, *instead of his original tuple 3* in the microdata.

It is important to understand the probabilistic nature of possible reconstructions. In fact, o_1, ..., o_4 can be decided in $\text{Permu}(5, 4) = 120$ ways[1]. For each decision, by the reasoning explained earlier, v_1 and v_2 are fixed, but $3^2 = 9$ choices exist for setting v_3 and v_4. Hence, there exist totally $120 \times 9 = 1080$ possible reconstructions.

432 reconstructions breach the privacy requirement of tuple 3 in Figure 19.1. Specifically, a reconstruction is breaching if and only if either o_3 or o_4 equals Ken. If o_3 = Ken, then there are $\text{Permu}(4, 3) = 24$ choices to formulate o_1, o_2, o_4, and 9 possibilities to determine v_1, ..., v_4, leading to $24 \times 9 = 216$ reconstructions. Symmetrically, if o_4 = Ken, there exist another 216 breaching reconstructions.

Without further information, the adversary assumes that each reconstruction happens with identical likelihood. Hence, the breach probability of tuple 3 in the microdata equals 432/1080 = 2/5.

19.3.2 Non-primary Case

We proceed to analyze the case where $T.\mathcal{A}^i$ is not the primary key of T, namely, each individual can appear an arbitrary number of times in T.

DEFINITION 19.11 (NONPRIMARY POSSIBLE RECONSTRUCTION). *In the Non-primary Case, given an individual o, a* **possible reconstruction** *of the essential QI-group $\mathcal{EG}(o)$ includes*

[1] $\text{Permu}(x, y)$ equals the number of permutations by taking y objects out of a set of x objects.

- *a multi-set of individuals* $\{o_1, ..., o_m\}$ *(perhaps with duplicates), where the distinct elements constitute a subset of* $\mathcal{S}_{ext}(o)$*;*
- m *leaf sensitive values* $v_1, ..., v_m$*, such that* v_j $(1 \leq j \leq m)$ *is in* SUBTR($t_j^*.\mathcal{A}^s$).

EXAMPLE 19.12 Let us revisit the situation where the adversary has obtained $\mathcal{S}_{ext}$(Ken) = {Andy, Bill, Ken, Nash, Mike}. The values of m, n, t_1^*, ..., and t_4^* are the same as in Example 19.10.

In a possible reconstruction, the adversary may set all of o_1, ..., o_4 to Ken (which is not allowed in the Primary Case). The way that v_1, ..., v_4 are decided is identical to that in Example 19.10; let us again assume v_1 = *gastric-ulcer*, v_2 = *dyspepsia*, v_3 = *flu*, and v_4 = *bronchitis*. By this reconstruction, the adversary thinks that Ken contracted all the 4 diseases. Evidently, the conjecture does not correctly reflect the microdata, but it causes a privacy breach for tuple 3 in Figure 19.1a.

Since each of o_1, ..., o_4 can independently be any of {Andy, Bill, Ken, Nash, Mike}, 5^4 = 625 choices exist for deciding o_1, ..., o_4. Given each decision, due to the reasons presented in Example 19.10, there are 9 ways to formulate v_1, ..., v_4. Therefore, the total number of possible reconstructions equals 625 $\times$ 9 = 5625.

A reconstruction breaches the privacy constraint of tuple 3 in the microdata, if and only if Ken is assigned to o_3 or o_4. If o_3 = Ken, o_1, o_2, o_4 may be any person in $\mathcal{S}_{ext}$(Ken), and hence, can be assigned in 5^3 = 125 manners. Regardless of the assignment, v_1, ..., v_4 may be set in 9 ways, resulting in $125 \times 9 = 1125$ different reconstructions. Similarly, another 1125 exist if o_4 = Ken, but some of them (where o_3 = o_4 = Ken) have been counted twice. Specifically, if o_3 = o_4 = Ken, there are 25 possibilities for determining o_1 and o_2, whereas, for each possibility, 9 choices exist for deciding v_1, ..., v_4. Hence, the number of double-counted reconstructions equals 25 $\times$ 9 = 225.

Therefore, totally $1125 + 1125 - 225 = 2025$ reconstructions breach the privacy of tuple 3 in Figure 19.1a. Thus, the breach probability of the tuple equals 2025/5625 = 9/25. □

Deriving a breach probability through the above procedures is quite cumbersome. In the next section, we present closed formulae that return the probability directly. Then, it will become simple to verify that publishing the table of Figure 19.3 allows no tuple in Figure 19.1a to be breached with a probability more than 50%.

19.4 Theoretical Foundation

In this section, we solve the probability $\mathbf{P}_{breach}(t_{tar})$ formulated in Definition 19.3, where t_{tar} is an arbitrary tuple in T (the subscript means "target").

Obviously, if the guarding node $t_{tar}.\mathcal{GN}$ of t_{tar} is $\emptyset$, $\mathbf{P}_{breach}(t_{tar}) = 0$, i.e., no privacy control is required. Next, we focus on $t_{tar}.\mathcal{GN} \neq \emptyset$.

Section 19.4.1 first clarifies the notations and their properties, which will be used in our derivation. Then, Section 19.4.2 solves $\mathbf{P}_{breach}(t_{tar})$ into closed formulae.

19.4.1 Notations and Basic Properties

Following the notations in Section 19.3, we use o_{tar} to denote the person identified by $t_{tar}.\mathcal{A}^i$, and t^*_{tar} for the generalized tuple of t_{tar}. Furthermore, let m be the size of the corresponding essential QI-group $\mathcal{EG}(o_{tar})$ (Definition 19.7), whose tuples are represented as t^*_1, ..., t^*_m (one of which is t^*_{tar}), respectively. $\mathcal{S}_{real}(o_{tar})$ refers to the set of individuals whose records (in the microdata T) are generalized to $\mathcal{EG}(o_{tar})$. Finally, we deploy n for the cardinality of $\mathcal{S}_{ext}(o_{tar})$ (Definition 19.8).

As a direct corollary of Formula 19.1, we have:

$$n \geq |\mathcal{S}_{real}(o_{tar})| \tag{19.2}$$

In the Primary Case, $|\mathcal{S}_{real}(o_{tar})|$ always equals m, since every tuple in $\mathcal{EG}(o_{tar})$ is owned by a distinct person. In the Non-primary case, however, $|\mathcal{S}_{real}(o_{tar})|$ may be any value in $[1, m]$. Furthermore, regardless of the size of $\mathcal{EG}(o_{tar})$, $|\mathcal{S}_{real}(o_{tar})|$ can take the minimum value 1, which happens if all the tuples in $\mathcal{EG}(o_{tar})$ belong to the same person.

We introduce b as the number of tuples t^*_j $(1 \leq j \leq m)$ in $\mathcal{EG}(o_{tar})$, such that SUBTR($t^*_j.\mathcal{A}^s$) overlaps SUBTR($t_{tar}.\mathcal{GN}$). For example, assume that t_{tar} is tuple 1 of Figure 19.1a, i.e., $t_{tar}.\mathcal{GN}$ = *stomach-disease*. Thus, in Figure 19.3, $\mathcal{EG}(o_{tar})$ involves tuples 1-4, and $m = 4$. Since SUBTR($t_{tar}.\mathcal{GN}$) overlaps the subtrees of the $\mathcal{A}^s$ values of tuples 1 and 2 in $\mathcal{EG}(o_{tar})$, we have $b = 2$.

We define two functions $\mathcal{F}_{subsize}$ and $\mathcal{F}_{percent}$ related to the tuples $t^* \in T^*$. Specifically, $\mathcal{F}_{subsize}(t^*)$ equals the number of leaf values in SUBTR($t^*.\mathcal{A}^s$) (e.g., $\mathcal{F}_{subsize}(t^*) = 3$ if $t^*.\mathcal{A}^s$ = *respiratory-infection*). On the other hand:

- $\mathcal{F}_{percent}(t^*, t_{tar})$ equals the *percentage* of the leaf values in SUBTR($t^*.\mathcal{A}^s$) that are also in SUBTR($t_{tar}.\mathcal{GN}$).

Thus, it follows that:

- $\mathcal{F}_{percent}(t^*, t_{tar}) = 1$, if $t^*.\mathcal{A}^s$ is in SUBTR($t_{tar}.\mathcal{GN}$);
- $\mathcal{F}_{percent}(t^*, t_{tar}) = 0$, if SUBTR($t^*.\mathcal{A}^s$) is disjoint with SUBTR($t_{tar}.\mathcal{GN}$).

We illustrate $\mathcal{F}_{percent}$ assuming $t_{tar}.\mathcal{GN}$ = *respiratory-infection*. If $t^*.\mathcal{A}^s$ = *respiratory-system-problem*, then $\mathcal{F}_{percent}(t^*, t_{tar})$ = 50%, because $t^*.\mathcal{A}^s$ has

6 leaf diseases, and half of them lie in SUBTR($t_{tar}.\mathcal{GN}$). As another example, if $t^*.\mathcal{A}^s$ is *flu*, which is in SUBTR($t_{tar}.\mathcal{GN}$); therefore, $\mathcal{F}_{percent}(t^*, t_{tar})$ = 100%. Finally, given $t^*.\mathcal{A}^s$ = *stomach-disease* (whose subtree is disjoint with SUBTR($t_{tar}.\mathcal{GN}$)), $\mathcal{F}_{percent}(t^*, t_{tar}) = 0$.

LEMMA 19.13 *For all tuples t_j^* ($1 \leq j \leq m$) in $\mathcal{EG}(o_{tar})$, $\mathcal{F}_{percent}(t_j^*, t_{tar})$ equals 0 or a constant.*

Therefore, in the sequel, we avoid the notation of $\mathcal{F}_{percent}$ by using c to represent the non-zero value of $\mathcal{F}_{percent}(t_1^*, t_{tar})$, ..., $\mathcal{F}_{percent}(t_m^*, t_{tar})$.

19.4.2 Derivation of the Breach Probability

As clarified in Section 19.3, to infer the $\mathcal{A}^s$ value of o_{tar}, an adversary reconstructs $\mathcal{EG}(o_{tar})$ according to Definition 19.9 (or 19.11) in the primal (or non-primal) scenario. In any case, we use n_{recon} to capture the total number of possible reconstructions, and n_{breach} for the number of reconstructions violating the privacy constraint enforced by $t_{tar}.\mathcal{GN}$. It follows that

$$\mathbf{P}_{breach}(t_{tar}) = n_{breach}/n_{recon} \tag{19.3}$$

The next two theorems solve $\mathbf{P}_{breach}(t_{tar})$ for the primal and non-primary cases, respectively.

THEOREM 19.14 *In the Primary Case,* $\mathbf{P}_{breach}(t_{tar}) =$

$$\begin{cases} b/n & \textit{if } t_{tar}^*.\mathcal{A}^s \textit{ is in } \text{SUBTR}(t_{tar}.\mathcal{GN}) \\ b \cdot c/n & \textit{otherwise} \end{cases}$$

EXAMPLE 19.15 We illustrate the theorem using Figures 19.1a, 19.1b, and 19.3. Assume t_{tar} (or t_{tar}^*) to be tuple 3 in Figure 19.1a (or Figure 19.3). Thus, $t_{tar}^*.\mathcal{A}^s = t_{tar}.\mathcal{GN}$ = *respiratory-infection*, and $\mathcal{EG}(o_{tar})$ involves the first 4 tuples of Figure 19.3. According to Figure 19.1b, Andy, Bill, Ken, Nash, Mike are potentially involved in $\mathcal{EG}(o_{tar})$, rendering $n = 5$. Furthermore, $b = 2$, because the subtrees of the $\mathcal{A}^s$ values in tuples 3, 4 (Figure 19.3) overlap SUBTR($t_{tar}.\mathcal{GN}$). Since $t_{tar}^*.\mathcal{A}^s$ is in SUBTR($t_{tar}.\mathcal{GN}$), by Theorem 19.14, $\mathbf{P}_{breach}(t_{tar}) = b/n$ = 2/5, confirming the analysis in Example 19.10.

To demonstrate the second case of the theorem, let t_{tar} (or t_{tar}^*) be tuple 5 in Figure 19.1a (or Figure 19.3). Namely, $t_{tar}^*.\mathcal{A}^s$ = *respiratory-infection*, $t_{tar}.\mathcal{GN}$ = *pneumonia*, and $\mathcal{EG}(o_{tar})$ consists of tuples 5, 6 of Figure 19.3. Only Joe and Sam in Figure 19.1b can be involved in $\mathcal{EG}(o_{tar})$, leading to $n = 2$. Furthermore, $b = 2$, because the $\mathcal{A}^s$ values of both tuples in $\mathcal{EG}(o_{tar})$ have subtrees overlapping SUBTR($t_{tar}.\mathcal{GN}$). In particular, the subtree of the sensitive value in tuple 5 (or 6) of Figure 19.3 has 3 leaf diseases, one of which is in SUBTR($t_{tar}.\mathcal{GN}$). Hence, c equals 1/3. Since $t^*_{tar}.\mathcal{A}^s$ is not in SUBTR($t_{tar}.\mathcal{GN}$), $\mathbf{P}_{breach}(t_{tar}) = b \cdot c/n$ = 1/3. □

THEOREM 19.16 *In the Non-primary Case,* $\mathbf{P}_{breach}(t_{tar}) =$

$$\begin{cases} 1-(1-1/n)^b & \text{if } t^*_{tar}.\mathcal{A}^s \text{ is in SUBTR}(t_{tar}.\mathcal{GN}) \\ 1-(1-c/n)^b & \text{otherwise} \end{cases}$$

EXAMPLE 19.17 Let t_{tar} be tuple 3 of Figure 19.1a. As explained in Example 19.15, n = 5, b = 2, and $t^*_{tar}.\mathcal{A}^s$ is in SUBTR($t_{tar}.\mathcal{GN}$). Theorem 19.16 shows that $\mathbf{P}_{breach}(t_{tar})$ is $1 - (1 - 1/5)^2$ = 9 / 25, which is consistent with the derivation in Example 19.12.

To demonstrate the second case, assume t_{tar} to be tuple 5 in Figure 19.1a. As mentioned in Example 19.15, n = 2, b = 2, c = 1/3, and $t^*_{tar}.\mathcal{A}^s$ is not in SUBTR($t_{tar}.\mathcal{GN}$). Thus, $\mathbf{P}_{breach}(t_{tar})$ is $1 - (1 - 1/(3 \times 2))^2$ = 11/36. □

19.5 Generalization Algorithm

Let v be a value in the domain of attribute $\mathcal{A}$. We use $\mathcal{IL}_{value}(v^*)$ to capture the (amount of) information loss in generalizing v to v^*, which is a partition in the corresponding general domain of $\mathcal{A}$ (Definition 19.5). Formally,

$$\mathcal{IL}_{value}(v^*) = \frac{(\text{the number of values in } v^*) - 1}{\text{the number of values in the domain of } \mathcal{A}} \tag{19.4}$$

For instance, if the domain of *Age* is $[1, 60]$, generalizing age 5 to $[1, 10]$ has information loss $\mathcal{IL}_{value}([1, 10])$ = (10 − 1) / 60. Similarly, since the taxonomy of *Disease* has 12 leaves, generalizing *flu* to *respiratory-infection* results in $\mathcal{IL}_{value}(\textit{respiratory-infection}) = (3-1)/12$, where 3 is the number of leaves under *respiratory-infection*. Obviously, if v is not generalized (i.e., $v = v^*$), $\mathcal{IL}_{value}(v^*)$ equals 0, i.e., no information is lost.

The overall information loss $\mathcal{IL}_{tuple}(t^*)$ of a generalized tuple t^* equals

$$w^s \cdot \mathcal{IL}_{value}(t^*.\mathcal{A}^s) + \sum_{i=1}^{d} w_i^{qi} \cdot \mathcal{IL}_{value}(t^*.\mathcal{A}_i^{qi}) \tag{19.5}$$

where w_1^{qi}, ..., w_d^{qi}, and w^s are positive system parameters, specifying the penalty factor of sacrificing precision on each attribute. Obviously, SA-generalization can be easily disabled by setting $w^s = \infty$, i.e., even the least generalization on $\mathcal{A}^s$ entails infinite information loss.

The total information loss $\mathcal{IL}_{table}(T^*)$ of the entire (generalized) relation T^* is given by

$$\mathcal{IL}_{table}(T^*) = \sum_{\forall t^* \in T^*} \mathcal{IL}_{tuple}(t^*) \tag{19.6}$$

Algorithm *Greedy-Personalized-Generalization*
Input: the microdata T, and the guarding nodes of all tuples
Output: the publishable relation T^*
1. for every QI-attribute $\mathcal{A}_i^{qi}$ $(1 \leq i \leq d)$
2. initialize a generalization function f_i with a single partition covering the entire domain of $\mathcal{A}_i^{qi}$ (see Definitions 19.4 and 19.5)
3. T^* = the relation after applying QI-generalization on T according to $S = \{f_1, ..., f_d\}$
4. G' = the only QI-group in T^*
5. *SA-generalization* (G') //Figure 19.6
/* at this point, T^* becomes publishable */
6. while (true)
7. $T^*_{best} = T^*$; $S_{best} = S$
8. for every possible $S' = \{f'_1, ..., f'_d\}$ obtained from S with a "single split" (see the explanation in Section 19.5.1)
9. T'^* = the relation after applying QI-generalization on T according to S'
10. for each QI-group $G' \in T'^*$
11. *SA-generalization* (G') //Figure 19.6
/* at this point, T'^* becomes publishable */
12. if $\mathcal{IL}_{table}(T'^*) < \mathcal{IL}_{table}(T^*_{best})$
13. $T^*_{best} = T'^*$; $S_{best} = S'$
14. if ($T^*_{best} = T^*$) then go to Line 17 //no next round
15. else
16. $T^* = T^*_{best}$; $S = S_{best}$ //prepare for the next round
17. return T^*_{best}

Figure 19.5. Algorithm for computing personalized generalization

Next, leveraging the findings of the previous section, we propose an algorithm for computing a generalized table T^* with small $\mathcal{IL}_{table}(T^*)$ which guarantees $\mathbf{P}_{breach}(t) \leq p_{breach}$ for each $t \in T$.

19.5.1 The Greedy Framework

As elaborated in Section 19.2.2, our generalization scheme includes two steps. The first phase applies QI-generalization on T, using a set of generalization functions $S = \{f_1, ..., f_d\}$ on the d QI-attributes, respectively. Then, the second step produces the final T^* by performing SA-generalization on the resulting QI-groups, employing a specialized generalization function for each QI-group. Hence, the quality of T^* depends on (i) the choice of S, and (ii) the effectiveness of SA-generalization. We provide a solution for settling the first issue in this subsection, and deal with (ii) in Section 19.5.2.

A generalization function f_i $(1 \leq i \leq d)$ is decided by a general domain of $\mathcal{A}_i^{qi}$ (Definition 19.5), which, in turn, is determined by a set of partitions in the original domain of $\mathcal{A}_i^{qi}$ (Definition 19.4). Therefore, selecting S is equivalent to finding the appropriate partitions of each f_i. Figure 19.5 presents a greedy algorithm for achieving this purpose (the pseudocode also explains the framework of calculating T^*).

At Lines 1-2, we obtain the simplest f_i ($1 \le i \le d$), which contains a single partition, covering the entire domain of $\mathcal{A}_i^{qi}$. Using such f_1, ..., f_d, Line 3 carries out QI-generalization on T, which, apparently, results in a single QI-group. Next, the algorithm invokes *SA-generalization* (elaborated in the next section) on the QI-group (Lines 4-5), which yields a publishable T^*.

The subsequent execution proceeds in *rounds*. Specifically, each round slightly refines *one* of f_1, ..., f_d, and leads to a new T^* with lower information loss. Before explaining the details, we must clarify the refinement of a function, e.g., f_1, without loss of generality.

Refining a generalization function. Refining f_1 means splitting one of its partitions once. For instance, assume that f_1 is on a numeric attribute *Age* with domain $[1, 60]$, and is determined by partitions $[1, 30]$ and $[31, 60]$. Partition $[1, 30]$ may be split into $[1, x]$ and $[x + 1, 30]$, for any $x \in [1, 29]$, i.e., $[1, 30]$ can be split in 29 ways. Similarly, there are also 29 options for splitting $[31, 60]$. Therefore, by a single split, f_1 can be refined into 58 possible generalization functions.

The situation is different, if f_1 concerns a categorical attribute, e.g., *Disease* (strictly speaking, *Disease* is not a QI-attribute in Figure 19.1c; but no confusion should be caused by borrowing it to illustrate the refinement of f_1). For example, suppose that *respiratory-system-problem* is one of the partitions (in the taxonomy of Figure 19.2) deciding f_1. Using the transformation stated in Section 19.2.2, we can represent *respiratory-system-problem* with an interval $[1, 6]$ (by converting the leaf nodes under the partition to values 1-6, respectively). Note that, it is not possible to split the partition into, for instance, $[1, 2]$ and $[3, 6]$. As formulated in Definition 19.4, each partition of a categorical attribute must be a node in the corresponding taxonomy. Here, $[1, 2]$ cannot be mapped to any node in Figure 19.2. In fact, there is only one possible split for *respiratory-system-problem*, i.e., breaking its interval $[1, 6]$ to sub-intervals $[1, 3]$ and $[4, 6]$.

In general, the number of possible refinements for a categorical f_1 equals exactly the number of non-leaf partitions of f_1. For example, assuming that f_1 is determined by *respiratory-system-problem* and *digestive-system-problem*, we can refine it into 2 different generalization functions with a single split.

A round of the greedy algorithm. We are ready to elaborate each round of the algorithm in Figure 19.5. Before a round starts, the algorithm has obtained a publishable table T^*, with a set of QI-generalization functions $S = \{f_1, ..., f_d\}$. At the beginning of the round, we duplicate T^* and S into T^*_{best} and S_{best}, respectively (Line 7).

Next, the algorithm examines (Line 8) all possible sets of refined functions $S' = \{f'_1, ..., f'_d\}$, obtained by performing one split over a single function in S (i.e., S' shares $d-1$ identical functions with S). Given an S', Lines 9-11 perform QI- and SA-generalizations to calculate a publishable T'^*, in the same manner as Lines 3-5, except that multiple QI-groups may be produced after the QI-generalization. If T'^* incurs smaller information loss (computed with Equation 19.6) than our current best solution T^*_{best} (Line 12), T'^* and S' replace T^*_{best} and S_{best} respectively (Line 13).

We provide a heuristic to reduce computation time. Since S' differs from S in only one element, the QI-generalization based on S' can be computed incrementally from that based on S (which is available from the previous round). Furthermore, if the same QI-group G results from both QI-generalizations, its SA-generation does not need to be re-computed. Similarly, in deriving the information loss $\mathcal{IL}_{table}(T'^*)$, the contribution of the tuples in G needs not be re-calculated, either.

The round finishes, after all S' has been considered. Line 14 checks if a better solution (compared to the one discovered prior to this round) has been found. If not, the algorithm terminates by returning T^*_{best}. Otherwise, another round is executed, after setting T^* (or S) to T^*_{best} (or S_{best}) at Line 16.

19.5.2 Optimal SA-generalization

Let G' be an arbitrary QI-group output by performing QI-generalization on T. Without loss of generality, assume that G' contains m tuples t'_1, ..., t'_m. We use G to denote the set of corresponding tuples $\{t_1, ..., t_m\}$ in the microdata T. Specifically, for each $j \in [1, m]$, $t'_j.\mathcal{A}^s = t_j.\mathcal{A}^s$, whereas $t'_j.\mathcal{A}^{qi}_i$ generalizes $t_j.\mathcal{A}^{qi}_i$ $(1 \le i \le d)$.

We aim at applying SA-generation on G' to derive $G^* = \{t^*_1, ..., t^*_m\}$, which achieves two objectives. As discussed in Sections 19.2.2 and 19.4, $\mathbf{P}_{breach}(t_j)$ $(1 \le j \le m)$ depends only on G^* (which is the essential QI-group of the individual that t_j belongs to). Hence, as the first objective, G^* must ensure $\mathbf{P}_{breach}(t_j) \le p_{breach}$.

The second objective is to minimize

$$\sum_{j=1}^{m} \mathcal{IL}_{value}(t^*_j.\mathcal{A}^s) \tag{19.7}$$

where $\mathcal{IL}_{value}$ is given in Equation 19.4. Given the fact that the QI-values of t^*_1, ..., t^*_m have been finalized (before the SA-generalization), fulfilling the second objective essentially minimizes $\sum_{j=1}^{m} \mathcal{IL}_{tuple}(t^*_j)$, where $\mathcal{IL}_{tuple}$ is defined in Equation 19.5. Therefore, after carrying out SA-generalization on all the QI-groups (produced by QI-generalization) in the same manner, the resulting publishable T^* minimizes $\mathcal{IL}_{table}(T^*)$ of Equation 19.6.

LEMMA 19.18 *For any tuples t_x and t_y ($1 \leq x, y \leq m$), if $t_x.\mathcal{GN}$ is in* SUBTR($t_y.\mathcal{GN}$), *then $\mathbf{P}_{breach}(t_x) \leq \mathbf{P}_{breach}(t_y)$ regardless of the SA-generalization applied.*

Therefore, in searching for the optimal SA-generalization, we can avoid checking the breach probabilities of the tuples like t_x in Lemma 19.18, because they must be adequately protected once the privacy information of the other tuples is secured.

LEMMA 19.19 *For any tuple t_j ($1 \leq j \leq m$), if $\mathbf{P}_{breach}(t_j) > p_{breach}$ before SA-generalization, then $t_j^*.\mathcal{A}^s$ must be an ancestor of $t_j.\mathcal{GN}$ after SA-generalization.*

Based on the above properties, Figure 19.6 shows an algorithm that finds the optimal SA-generalization for the given QI-group G'. Line 1 initializes two sets G and G^*. G collects all tuples $t_1, ..., t_m$ in T generalized to G', while $G^* = G'$. Line 2 creates a set S_{prob} as follows. For each tuple $t \in G$, if its guarding node $t.\mathcal{GN}$ is not in SUBTR($t'.\mathcal{GN}$) of any other tuple $t' \in G$, t is added to S_{prob}. By Lemma 19.18, once the privacy requirements of the tuples in S_{prob} are satisfied, the requirements of the other tuples are also fulfilled.

For each tuple $t \in S_{prob}$, the algorithm calculates $\mathbf{P}_{breach}(t)$ according to Theorem 19.14 or 19.16 (based on the current, non-generalized, A^s values in G^*). If $\mathbf{P}_{breach}(t)$ is larger than p_{breach}, t is placed in a set S_{bad} (Line 3), that is, S_{bad} includes the tuples in S_{prob} whose privacy constraints have not been satisfied after QI-generalization.

Next, we consider each tuple $t \in S_{bad}$ in turn (Line 4). Let t^* be its corresponding tuple in G^*. According to Lemma 19.19, we can immediately set $t^*.\mathcal{A}^s$ to the parent of $t.\mathcal{GN}$ (Line 5). After this, $t^*.\mathcal{A}^s$ may become an ancestor of $t'^*.\mathcal{A}^s$ of another tuple $t'^* \in G^*$. This is not allowed because, otherwise, $t^*.\mathcal{A}^s$ and $t'^*.\mathcal{A}^s$ become two overlapping partitions in the general domain of $\mathcal{A}^s$. To remedy this problem, we must also generalize $t'^*.\mathcal{A}^s$ to $t^*.\mathcal{A}^s$ (Lines 6-8).

The algorithm terminates (Line 9) if $\mathbf{P}_{breach}(t)$ does not exceed p_{breach} for any tuple $t \in S_{prob}$. Otherwise ($\mathbf{P}_{breach}(t) > p_{breach}$ for some tuple t), we must decrease $\mathbf{P}_{breach}(t)$ by generalizing $t^*.\mathcal{A}^s$ further (t^* is the tuple in G^* corresponding to t). If $t^*.\mathcal{A}^s$ is already the root of the taxonomy (Line 10), the algorithm returns, reporting that no appropriate SA-generalization can be found (Line 11). In fact, in this case, the $\mathcal{A}^s$ values of all tuples in G^* have been generalized to the root, so that no more generalization is possible.

If $t^*.\mathcal{A}^s$ is not the root, we raise $t^*.\mathcal{A}^s$ "one level up" in the taxonomy, by replacing it with its parent (Line 12). After this, the $\mathcal{A}^s$ values of some other tuples may also need to be raised, due to the reasoning for Lines 6-8. These changes may increase the breach probabilities of some tuples. Hence, the

Algorithm *SA-generalization* (G')
Input: a QI-group G' with tuples t'_1, ..., t'_m after QI-generalization
Output: a set G^* of tuples t^*_1, ..., t^*_m in the final publishable T^*
1. G = the set of tuples t_1, ..., t_m in T generalized to G';
   $G^* = \{t'_1, ..., t'_m\}$
2. S_{prob} = the set of tuples $t \in G$ such that $t.\mathcal{GN}$ is not in the subtree of the guarding node of any other tuple in G
3. S_{bad} = the set of tuples $t \in G$ satisfying $\mathbf{P}_{breach}(t) > p_{breach}$
   /* In the Primary Case, $\mathbf{P}_{breach}(t)$ is computed from Theorem 19.14, replacing n with the size of G. In the Non-primary Case, the computation is based on Theorem 19.16, replacing n with the number of distinct individuals involved in G. */
4. for each tuple $t \in S_{bad}$
5. $t^*.\mathcal{A}^s$ = the parent of $t.\mathcal{GN}$
      //t^* is the tuple in G^* corresponding to t
6.   for each tuple $t'^* \in G^*$ such that $t'^* \neq t^*$
7.   if $t'^*.\mathcal{A}^s$ is in SUBTR$(t^*.\mathcal{A}^s)$
8. $t'^*.\mathcal{A}^s = t^*.\mathcal{A}^s$
9. while there is a tuple $t \in S_{prob}$ satisfying $\mathbf{P}_{breach}(t) > p_{breach}$
10.   if $t^*.\mathcal{A}^s$ is the root of the taxonomy
11.   return NULL //no possible SA-generalization
12. $t^*.\mathcal{A}^s$ = the parent of $t^*.\mathcal{A}^s$

Lines 13-15 are identical to Lines 6-8

Figure 19.6. Algorithm for finding the optimal SA-generalization

algorithm returns to Line 9 to check whether any probability is above p_{breach}. If yes, the above procedures are repeated.

The computation of $\mathbf{P}_{breach}(t)$ deserves further clarification. The value of n in Theorems 19.14 and 19.16 is unavailable when T^* is being computed (i.e., we do not know which external database will be consulted by an adversary). Hence, as a conservative approach, we replace n with its lower bound $|S_{real}(o_{tar})|$ (Inequality 19.2). If the breach probability computed with this lower bound is at most p_{breach}, then the actual breach probability derived by an adversary will definitely be bounded by p_{breach}.

The following theorem proves that Figure 19.6 produces an SA-generalization that minimizes Equation 19.7.

THEOREM 19.20 *Let t^*_1, ..., t^*_m be the tuples returned by the algorithm in Figure 19.6, and t'^*_1, ..., t'^*_m be the tuples obtained by any alternative SA-generalization that prevents privacy breach. For any $j \in [1, m]$, $t^*_j.\mathcal{A}^s$ must be in* SUBTR$(t'^*_j.\mathcal{A}^s)$, *namely, $\mathcal{IL}_{value}(t^*_j.\mathcal{A}^s) \leq \mathcal{IL}_{value}(t'^*_j.\mathcal{A}^s)$.*

19.6 Alternative Forms of Personalized Privacy Preservation

This section reviews several other methods that allow data owners to specify their preferences of privacy protection. Section 19.6.1 analyzes a natural extension of k-anonymity. Then, Section 19.6.2 discusses preservation of location privacy.

19.6.1 Extension of k-anonymity

At the beginning of this chapter, we mentioned an approach to augment k-anonymous generalization with personalization features. Without loss of generality, consider that the microdata T contains n tuples. The owner of the i-th $(1 \leq i \leq n)$ tuple t selects an integer k_i as the *k-value* of t. This value signifies her/his intention that, in the published table T^* (obtained by generalizing T), t must be included in a QI-group with size at least k_i. The objective is to find a T^* that satisfies the requirements of all the owners, and at the same time, minimizes the amount of generalization (according to a certain metric to be discussed shortly). Note that, here generalization is performed only along the QI-attributes of T, namely, every sensitive value in T is published in T^* directly.

For simplicity, assume that all the QI-attributes $\mathcal{A}_1$, ..., $\mathcal{A}_d$ are numeric. As a result, for every tuple t^* in T^*, its value on $\mathcal{A}_i$ $(1 \leq i \leq d)$ is an interval covering a set of consecutive values in the domain of $\mathcal{A}_i$. In other words, $t^*.\mathcal{A}_1$, ..., $t^*.\mathcal{A}_d$ can be regarded as the extents of a rectangle, in the d-dimensional *QI-space* whose axes are $\mathcal{A}_1$, ..., $\mathcal{A}_d$.

Apparently, all the tuples in a QI-group G of T^* have the same rectangle representation in the QI-space, which we define as the *QI-rectangle of G*. Intuitively, the larger the rectangle is, the more information is lost, with respect to the original data of G in T. In the sequel, we measure the "amount of generalization" as the largest perimeter of the QI-rectangles in T^*, i.e., ideally, the optimal T^* should minimize that perimeter. This metric is chosen, as it allows us to visualize the quality of generalization easily. The following discussion, however, is general, and readily extendible to other metrics as well.

A straightforward solution to computing T^* is to first set k to the highest of k_1, ..., k_n, and then invoke any existing algorithm for k-anonymous generalization. Although such T^* fulfills the requirements set forth by all data owners, it entails excessive generalization, and hence, its usefulness for data analysis may be rather limited. As a slightly more complex solution, we may take a partitioning approach. Specifically, T is partitioned into disjoint subsets, such that tuples in the same subset have an identical k-value. Then, for each partition, we independently obtain its k'-anonymous generalization, where k' is the k-value of the tuples in that partition.

The partitioning approach also has two obvious drawbacks. First, a partition may not have enough tuples to make even a single QI-group. For example, suppose that the partition contains tuples whose k-values equal 100, whereas the partition has only 50 tuples. In this case, no generalization from this partition

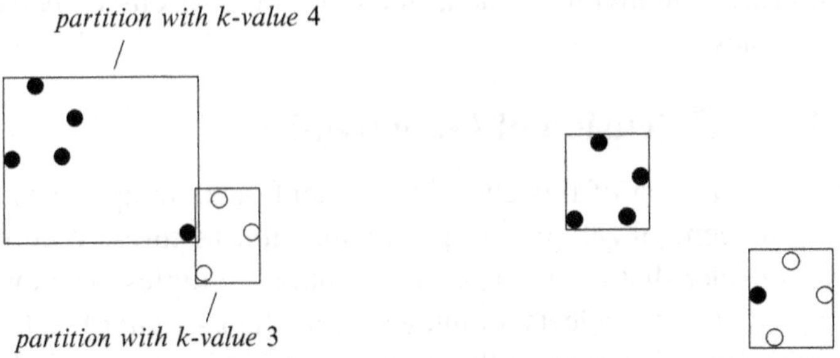

(a) Generalization of the partitioning approach (b) Better generalization

Figure 19.7. Personalized k-anonymous generalization

is possible, and therefore, all the tuples in it must be discarded. When there are many such undersized QI-groups, the number of discarded tuples would be too large for T^* to be useful for data analysis.

Another drawback of the partitioning solution is that, it never leverages the opportunity of placing tuples of different partitions into the same QI-group, in order to reduce the group's perimeter. To illustrate this, consider Figure 19.7a, where the black (white) points have the same k-value 4 (3), and hence, constitute a partition. The rectangles correspond to the QI-rectangles produced by the partitioning approach. Figure 19.7b demonstrates an alternative generalization that satisfies the privacy requirements of all points. Note that a black point is grouped together with the white points, which leads to significant decrease in the perimeter of the QI-rectangle on the other black points.

Based on this idea, Aggarwal and Yu develop an algorithm for finding a good generalized table T^* that achieves small QI-rectangles. We refer the interested readers to [2] for details. It is worth mentioning that this algorithm is originally developed for a condensation approach developed in [1]. Nevertheless, it can be easily adapted to k-anonymity.

19.6.2 Personalization in Location Privacy Protection

In recent years, we have witnessed significant improvements in location-tracking and wireless-communication technologies. These technologies have enabled numerous spatiotemporal applications that require monitoring the locations of moving objects continuously. An example is the intelligent traffic control system, where each vehicle transmits its location periodically (e.g., every 5 minutes) to a server, which provides public services such as congestion warning, travel time estimation (given a source and a destination), etc. Unfortunately, drivers may not necessarily always be happy to disclose their locations precisely. For example, when a driver wishes to hide the place that s/he is visiting, s/he may choose to switch off her/his location-reporting device.

Obviously, such intermittent "incorporation" may cast enormous adverse influences on the quality of the services provided by the underlying application. For instance, the server would no longer be able to ensure the accuracy of the number of vehicles in a district, and hence, be unable to provide reliable congestion analysis.

It is thus clear that the success of many spatiotemporal applications depends on a privacy preserving location reporting mechanism. In particular, the mechanism should convince each individual that, even if s/he sends her/his location continuously, no adversary is able to identify her/his position down to a certain precision. This is the motivation of a growing research area called "location privacy protection".

The purpose of preventing accurate position pinpointing is in conflict with transmitting precise object locations. Fortunately, many spatiotemporal applications do not really demand precise data, but instead, they work fairly well even with "fuzzy" locations. For example, congestion in the Manhattan area can be detected as long as the number of vehicles in that area can be *estimated* with reasonable accuracy. Imagine that, instead of representing each vehicle's position as a point, we adopt a small rectangle that covers its position. This fuzzier representation achieves two objectives. First, it does not bias the estimation of traffic volume considerably (due to the fact that each rectangle is significantly smaller than the Manhattan area). Second, it does not allow an adversary to know exactly where a vehicle is.

The above idea is the rationale of the *cloaking* technique, which is commonly adopted in preserving location privacy. The process (as mentioned earlier) of transforming a point to a rectangle is called *spatial cloaking*. Spatial cloaking, however, hides only geographic information, but does not conceal the time when an individual visits a place. Time concealment leads to *temporal cloaking*, which replaces a timestamp with an interval. Formally, an exact *spatiotemporal location* can be represented as a ternary tuple (x, y, t), where (x, y) denote the 2D coordinates of an object, and t corresponds to the time when this object is located at that position. Spatial cloaking converts (x, y) into a rectangle R and temporal cloaking changes t into an interval I. Note that R and I together define a 3D box, which is called the *cloaking box* of (x, y, t).

Most location-preserving solutions adopt a three-layered architecture: objects, a cloaking server, and an application server. Specifically, each object transmits its spatiotemporal location (x, y, t) to the cloaking server, which is trustable, and responsible for converting the location to a cloaking box. Then, the cloaking server relays the box to the application server, which provides public services based on the cloaking boxes received. The objective is to disallow the application server to recover any exact spatiotemporal location.

As with the privacy-preserving publication scenario (c.f. Section 19.6.1), a principle is required also in location-privacy preservation to measure the degree of privacy protection. k-anonymity can once again be adopted, by simply treating x, y, and t as three QI-values, and a cloaking box as the generalized form. Specifically, the effect of applying k-anonymity is to ensure that the cloaking box of every spatiotemporal location is the same as those of $k-1$ other spatiotemporal locations. Obviously, a higher k promises stronger privacy protection. Unlike the publication problem, however, the data in the location-privacy context are not known in advance, but continuously arrive and are appended to the database along the time dimension. This difference prevents the generalization algorithms in the publication scenario from being applied directly. Several adapted algorithms have been developed in [6, 5, 10].

Gedik and Liu [5] propose a personalized approach to preserve location privacy. Their main idea is to allow each individual to specify, for every spatiotemporal location that s/he reports, (i) a tailored-made k-value, and (ii) a size limit for the extent of the corresponding cloaking box on every geographic or temporal dimension. Gedik and Liu also provide an algorithm to obtain cloaking boxes subject to these personal requirements.

19.7 Summary and Future Work

This chapter introduced several techniques that incorporate personal preferences in privacy preserving data publication. Particular attention was paid to a guarding-node approach, which permits the formulation of complex constraints that require guarding individual sensitive values. We analyzed the privacy guarantees of that approach, and discussed its algorithm for generating anonymized datasets. As a second step, we also explained the rationales of two alternative personalization solutions.

Despite its significant importance in practice, personalized privacy preservation has received only limited attention in the literature. There remain numerous interesting research topics awaiting to be explored. We conclude the chapter by listing some of those topics.

Greater flexibility in formulation. The guarding-node solution has an obvious drawback: a person cannot request guarding against an arbitrary set of sensitive values. For instance, assume that, concerning the hierarchy in Figure 19.2, Ken wishes to prevent the public from associating him with *pneumonia* and *bronchitis*. For this purpose, he must set his guarding node to "respiratory infection", which disallows associating Ken with an extra disease: flu. Although such an additional prevented association causes no harm to Ken's privacy, it is *redundant*, and would most likely trigger unnecessary data distortion in the published relation. This problem is especially serious if Ken would like to protect himself from being suspected of having diseases that scatter in mutually distant parts of the hierarchy. As an example, in additional to his preferences stated before, Ken may not want the public to guess that he contracted *gastritis* (even though his actual disease is drastically different from *gastritis*), which requires setting the guarding node to the lowest ancestor of *pneumonia*, *bronchitis*, and *gastritis*. Due to the vast dissimilarity of the three types of illness, the ancestor would be a node (*any-illness* in Figure 19.2) at a fairly high level in the taxonomy, leading to a huge number of redundant associations.

Non-categorical attributes. The guarding-node approach demands a hierarchy on the sensitive attribute. Hierarchies, however, are typically absent (or difficult to define) on numeric and transactional data, whose protection must be achieved in alternative manners, and is an important problem due to the abundance of such data in practice. Consider the scenario where the government wants to release tax data to social scientists, involving a numeric private attribute (monthly) *Salary*. A tax-payer would be reluctant to have her/his income released, unless s/he knows that her/his amount cannot be "accurately" derived by analysts. Here, the notion of "accuracy" is subject to personalization, e.g., people with higher remuneration may give more stringent constraints. To enable personalization, we could leverage the guarding-node solution mentioned earlier, after transforming *Salary* into a categorical attribute and manually building a hierarchy over the new attribute. Unfortunately, while categorization may be easy (e.g., we could regularly divide the *Salary* domain into a large number of intervals with identical lengths), building the hierarchy is not. To understand why, assume that *Salary* has a domain of [0, \$100k], discretized into 1000 ranges with length \$100. No hierarchy is able to preserve the proximity of all pairs of adjacent ranges. For instance, imagine that every 10 leaf-ranges are grouped as the subtree of a node at the higher level; e.g., the first node at this level corresponds to range [0, \$1k], and is the parent of leaf-ranges [0, \$0.1k), [\$0.1, \$0.2k), ..., [\$0.9k, \$1k). Then, leaf-ranges [\$0.9k, \$1k) and [\$1k, \$1.1k) are placed into two different subtrees, even though they encompass consecutive salary amounts.

Transactional data usually emerges in a database that requires frequent itemset mining. For example, a tuple collected at Amazon.com may have the form "*Age* = 30, *Gender* = female, *Shipping-address* = ..., *Items-purchased* = {Toshiba 42HP66 HDTV, Dolce 5-piece Dining Set, ...}". To enable market research, the company may release such tuples to a third party, after ensuring adequate preservation on customers' purchase history (i.e., *Items-purchased* is sensitive). It is not clear how to extend the guarding-node technique to permit personal preferences in such an environment, since concealing individual items may result in substantial information loss about their combination. For example, suppose that *electronic-product* (*furniture*) has 100 child nodes in the item taxonomy, one of which is *HDTV* (*dining-set*). If a person specifies {*electronic-product*, *furniture*} as the guarding node, the effect is to hide the original itemset {*HDTV*, *dining-set*} into 10000 different (2-sized) itemsets, where 10000 is the cardinality of the cartesian product of the children sets of *electronic-product* and *furniture*. In general, if an item set contains m items $i_1, i_2, ..., i_m$, it may not be feasible to replace them even with their immediate parents $p_1, p_2, ..., p_m$ in the taxonomy — in this case, no accurate mining would be possible, because the itemset $\{i_1, i_2, ..., i_m\}$ would become indistinguishable with any other itemset in the cartesian product of the children sets of $p_1, p_2, ..., p_m$, respectively (the cardinality of the cartesian product increases exponentially with m).

Automatic guarding node formulation. Sometimes expecting every individual to provide an appropriate privacy constraint may not be realistic. For example, the medical records collected in the old days do not bear any personal preference; it would be intractable to contact all the corresponding owners for preference submission. Even worse, in certain applications, asking a customer for privacy constraints may "back-fire", harming the business itself. For example, a customer shopping at Amazon.com would be confused about such a request, and even discouraged from using the service again after being alerted with privacy concerns. As mentioned earlier, if no personal preference is given, we could at least set a guarding node to the sensitive value itself, thus achieving the same extent of protection as in a conventional non-personalized approach. Nevertheless, a good privacy preserving solution should be able to automatically identify the "rare" sensitive values, and offer greater protection to them. For instance, HIV apparently deserves better protection than the common illness flu. This observation suggests that, given a sensitive value that is statistically insignificant, we could manually set its guarding node to one of its ancestors in the taxonomy, such that the joint frequency of all the leaf values underneath this ancestor is sufficiently high. The real situation, unfortunately, is far more complex, due to the correlation among various attributes. In particular, even though a sensitive value itself has a reasonably high frequency, its combination with a QI-value may be rare. In this case, stronger privacy preservation is needed on the combination. How to achieve this with guarding nodes is a challenging issue.

Alternative forms of personalization. So far our discussion has used "guarding nodes" as the primary means for specifying personalized requirements, but certainly this is not the only means. A good formulation should be based on the needs of individuals involved in different applications, for developing new forms of personal constraints. We are particularly interested in constraints that forbid analysts from discovering sensitive data patterns (e.g., association rules, decision trees, clusters, etc.), in contrast to specific values. Examples of these constraints are: "prevent the discovery of the association rule {*Occupation* = Lawyer, *Gender* = Male, *Salary* = \$100k} → *Marital-status* = Divorced/Single", "do not allow an analyst to tell the gender of an individual from her/his other attributes", or "conceal the cluster of wealthy clients". A pioneering attempt towards this goal has been made in [13].

References

[1] C. C. Aggarwal and P. S. Yu. A condensation approach to privacy preserving data mining. In *EDBT*, pages 183–199, 2004.

[2] C. C. Aggarwal and P. S. Yu. On variable constraints in privacy preserving data mining. In *SDM*, 2005.

[3] R. Bayardo and R. Agrawal. Data privacy through optimal k-anonymization. In *ICDE*, pages 217–228, 2005.

[4] B. C. M. Fung, K. Wang, and P. S. Yu. Top-down specialization for information and privacy preservation. In *ICDE*, pages 205–216, 2005.

[5] B. Gedik and L. Liu. Location privacy in mobile systems: A personalized anonymization model. In *ICDCS*, pages 620–629, 2005.

[6] M. Gruteser and D. Grunwald. Anonymous usage of location-based services through spatial and temporal cloaking. In *MobiSys*, 2003.

[7] V. Iyengar. Transforming data to satisfy privacy constraints. In *SIGKDD*, pages 279–288, 2002.

[8] K. LeFevre, D. J. DeWitt, and R. Ramakrishnan. Incognito: Efficient full-domain k-anonymity. In *SIGMOD*, pages 49–60, 2005.

[9] A. Machanavajjhala, J. Gehrke, and D. Kifer. l-diversity: Privacy beyond k-anonymity. In *ICDE*, 2006.

[10] M. F. Mokbel, C.-Y. Chow, and W. G. Aref. The new casper: Query processing for location services without compromising privacy. In *VLDB*, pages 763–774, 2006.

[11] P. Samarati. Protecting respondents' identities in microdata release. *TKDE*, 13(6):1010–1027, 2001.

[12] L. Sweeney. Achieving k-anonymity privacy protection using generalization and suppression. *International Journal on Uncertainty, Fuzziness and Knowledge-based Systems*, 10(5):571–588, 2002.

[13] K. Wang, B. C. M. Fung, and P. S. Yu. Template-based privacy preservation in classification problems. In *ICDM*, pages 466–473, 2005.

[14] K. Wang, P. S. Yu, and S. Chakraborty. Bottom-up generalization: A data mining solution to privacy protection. In *ICDM*, pages 249–256, 2004.

[15] X. Xiao and Y. Tao. Personalized privacy preservation. In *SIGMOD*, pages 229–240, 2006.

Chapter 20

Privacy-Preserving Data Stream Classification

Yabo Xu,[1] Ke Wang,[1] Ada Wai-Chee Fu,[2] Rong She,[1] and Jian Pei[1]

[1] *School of Computing Science*
Simon Fraser University, Burnaby BC, Canada
yxu,wangk,rshe,jpei@cs.sfu.ca

[2] *Department of Computer Science*
The Chinese University of Hong Kong, China
adafu@cse.cuhk.edu.hk

Abstract In a wide range of applications, multiple data streams need to be examined together in order to discover trends or patterns existing across several data streams. One common practice is to redirect all data streams into a central place for joint analysis. This "centralized" practice is challenged by the fact that data streams often are private in that they come from different owners. In this paper, we focus on the problem of building a classifier in this context and assume that classification evolves as the current window of streams slides forward. This problem faces two major challenges. First, the many-to-many join relationship of streams will blow up the already fast arrival rate of data streams. Second, the privacy requirement implies that data exchange among owners should be minimal. These considerations rule out all classification methods that require producing the join in the current window. We show that Naive Bayesian Classification (NBC) presents a unique opportunity to address this problem. Our main contribution is to adopt NBC to solve the classification problem for private data streams.

Keywords: Privacy, data streams, classification, Naive Bayesian classification.

20.1 Introduction

With today's information explosion, data not only are stored in large amount but also grow rapidly over time. Data streams are such examples, including internet traffic streams, stock trading streams, and telephone call streams. Data streams are characterized as being unbounded, continuously arriving at a high rate, and typically being scanned once (6). To benefit from the information and knowledge contained in data streams, often several related data streams need to be examined together to discover trends or patterns that exist across different data streams. For example, stock streams and news streams are related, traffic report streams and car-accident streams are related, sensor readings of different types are related. In this paper, we focus on building classification models from such data. Our insight is that classification patterns may be jointly determined by the co-occurrence of certain conditions in several related streams. We illustrate this point by a simplified example.

20.1.1 Motivating Example

In stock markets, "favorable trading" refers to stock transactions that are favorable to the engaging party, i.e., selling before a stock plunges or buying before a stock goes up. In order to build classification models to identify "favorable trading", the stock trading stream that records all trading transactions must be examined. However, stock transactions are not isolated or independent events; they are related to other data streams, e.g., phone calls between dealers and managers/staffs of public companies. Thus it is necessary to consider related data streams together.

For example, a classification algorithm may need to look at the following related data:

Trading stream: **T** (τ, Dealer, Type, Stock, Class)
Phone call stream: **P** (τ, Caller, Callee)
Company table: **C** (Company, Stock)
Person table: **S** (Name, Org)

where τ is the timestamp, "Type" is either "sell" or "buy", "Class" ("yes"/"no") refers to the class label of being favorable trading or not. To compute the training set, a SQL query can be used to extract information from the above data as follows:

```
SELECT *
FROM   P, S, T, C
WHERE  S.Name=P.Caller AND P.Callee=T.Dealer
       AND P.τ <T.τ AND T.Stock=C.Stock
```

Essentially, this query performs a join on the related data and each joined tuple represents a connection between a phone call and the trading ensued from this call. The join result is then used to train the classifier.

Figure 20.1 shows a snapshot of data. The join relationship is indicated by the arrows connecting the join attributes. Note that the join between P and T is "many-to-many". For example, "Albert" was called twice and traded twice, generating four tuples in the join stream in Figure 20.2 (The timestamps for each join record are ignored because of the space.), the rule "Org=Company →

Name	Org
Adams	AAA
Ray	BBB
Dennis	CCC

S

P

	Caller	Callee
9:30	Adams	Jack
12:01	Adams	Selina
14:19	Dennis	Albert
15:31	Ray	Albert
15:36	Dennis	Peter

	Dealer	Type	Stock	Class
9:38	Jack	Sell	A	Yes
12:30	Selina	Sell	A	Yes
15:40	Albert	Sell	B	No
15:57	Albert	Buy	C	Yes
16:42	Peter	Sell	C	No

T

Stock	Company
A	AAA
B	CCC
C	BBB

C

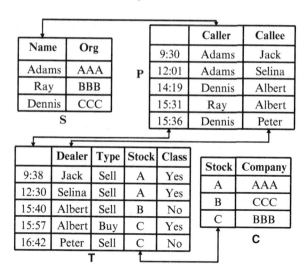

Figure 20.1. Related streams / tables

Class	Caller	Org	Callee	Dealer	Type	Stock	Company
Yes	Adams	AAA	Jack	Jack	Sell	A	AAA
Yes	Adams	AAA	Selina	Selina	Sell	A	AAA
Yes	Ray	BBB	Albert	Albert	Sell	B	CCC
Yes	Ray	BBB	Albert	Albert	Buy	C	BBB
No	Dennis	CCC	Albert	Albert	Sell	B	CCC
No	Dennis	CCC	Albert	Albert	Buy	C	BBB
No	Dennis	CCC	Peter	Peter	Sell	C	BBB

Figure 20.2. The join stream

Class=Yes" holds in 3 out of 4 tuples that have "Org=Company", i.e., with 75% confidence. This suggests that after getting a call, the trading on the caller's company stock tends to be more favorable.

This example illustrates that sometimes classification of certain behaviors (i.e., favorable trading) depends on information contained in several correlated streams and examining such streams together likely produces more accurate classifiers than examining any single input stream alone. The join is a common operation to combine several input streams into a single stream and the training data for classification is defined by this "join stream". Moreover, classification rules evolve as data streams evolve. New favorable trading rules may emerge as a reaction to evade from being identified by existing rules. In order to capture this change, the classifier needs to adapt quickly to the changed data distribution. A solution to this problem faces two major challenges.

- **Privacy preservation.** In the above example, as trade and phone call streams involves trading secrets and individual's privacy, apparently neither the trading company nor the phone service company is willing to disclose their local sensitive data. The common approach of redirecting all streams into a central place immediately violates this privacy constraint. In the literature, privacy-preserving data mining and stream data mining have been studied separately. The traditional privacy-preserving data mining techniques focus on static data and are not applicable to data streams with unbounded data size and continuous arrival of new data. On the other hand, most prior work on stream data mining assumes either a single stream or several streams but no privacy issue, and focuses on the processing speed of stream data. More details in Section 2.

- **Blow-up of the join stream.** As input streams arrive in a fast pace, the classifier must evolve quickly when new structures emerge and old ones are out-of-date. However, the join of multiple input streams is an expensive operation, in fact, much slower than the arrival rate of input streams. Furthermore, the "many-to-many" join relationship, as shown in Table 1, could generate the result join stream that is much larger than input streams. Any method that explicitly generates the join stream will suffer from thid blow-up of data arrival rates and is unlikely to be able to keep pace with the incoming source streams.

20.1.2 Contributions and Paper Outline

We consider several private data streams owned by different sites. One data stream, called *target stream*, contains class labels. The current window of the training data is defined by the join of input streams in their current window. Such joins are called *sliding-window join* (6) and the join result defines a new stream called *join stream*. The specification of window can be either tuple-based or time-based (6). As the window of input streams slides forward, so does the window of the join stream and the classifier must be updated to adapt to the change of window. In practice, only some portion of the data is labeled whereas the remaining is not. For each window, the unlabeled portion will be classified by the classifier built in the previous window, and at the same time, the labeled portion will be used to train the classifier in the current window (1).

Due to the privacy requirement and blow-up of join, however, the join stream cannot be generated explicitly. Hence, the problem we study is to build and update the classifier based on the never-generated join stream, given several private input streams. The construction and update of the classifier must not reveal private information to other sites. This problem is referred to as the *secure join stream classification* (Secure-JSC) hereinafter. Existing classification methods (35)(9)(1) cannot be applied to the Secure-JSC problem because they deal with a single stream and requires the join stream to be explicitly given.

Our insight is that the independence assumption of Naive Bayesian Classifier (NBC) (19) provides a unique opportunity to address the requirements for Secure- JSC: for a given class label, variables are assumed to be independent of each other. Research shows that NBC is reliable even when this assumption is violated (17)(31)(25). The reason for this reliability is that the most likely class label predicted by NBC is typically correct though the estimated probability may be distorted by the independence assumption. In other words, the top ranked class label often is correct though the estimated probability for ranking them was distorted. This reliability has been echoed by the popularity and success of NBC in both research works and practical applications. For most data stream applications, some degree of inaccuracy is tolerable, especially so because the data arrive very fast and there is only the time to scan the data once.

To adopt NBC to Scure-JSC problem, however, we must compute the information required by NBC on the join stream without generating the join and exchanging private information across sites. Our insight is that this can be obtained by computing some "blow-up summary" from examining each input stream and exchanging this summary with other sites. The "blow-up summary" can be computed by examining each input tuple in the current window twice, independent of the number of tuples it joins in other streams. The benefit is twofold: eliminate the need of collecting private data streams in a central place and avoid the expensive join. Though we consider NBC, the idea of "blow-up summary" is applicable to other classification algorithms that require similar statistics such as decision trees.

The rest of this paper is organized as follows. In Section 2, we review related works. In Section 3, we define the problem and discuss core concepts of NBC. In Section 4, we present our algorithm. We evaluate our method in Section 5. Section 6 concludes the paper.

20.2 Related Works

Privacy preserving data mining was first introduced in (3) and (28). These works opened up a rapidly growing area and various privacy preservation techniques have emerged since then. Most works on privacy preserving data mining assume static data. On the other hand, there is a large body of works on stream data management and mining. But this body of works does not deal with privacy issues. The novelty of our work lies in addressing privacy preservation, data streams and the training data defined by a general join of several streams. Below, we focus on related works which address one ore more of these aspects.

In data stream management (6), sliding-window join is proposed to answer queries involving the join of multiple data streams, such as the join size, sum (5)(15), join-distinct (21). Their focus was on how to compute these statistics of the join under resource constraints and techniques such as sampling (11) or load- shedding (10)(32) are used to reduce the cost of join. These works assume either a single stream or multiple streams but no privacy issue. In the Secure- JSC problem, as we explained in the previous section, it is prohibited to first compute the join of multiple streams and then build the classifier. Thus these techniques cannot be applied.

Most stream mining algorithms consider a single stream and simple statistics such as average and standard deviation. Classification on data streams was considered in (16)(20)(35)(9)(1). Other mining problems that involve multiple streams are clustering (24)(7), correlation analysis (37), sequential patterns (13). None of these works consider the privacy issue. Neither do they involve a general join among streams; thus, they do not deal with the blow-up of data arrival rates caused by a many-to-many join.

(18) presents a secure construction of decision tree classifiers from vertically partitioned data, where the join is given by the one-to-one relationship implied by the common key identifier for all partitions. This is not applicable to the general many-to-many join relationship. Recently, (36) proposed a secure construction for decision tree classifiers over distributed tables with the general many-to-many join relationship. Both works consider static data, not stream data.

There are few studies that cover both data streams and privacy preservation. Some prior works (8)(30) focus on the problem of private search over data streams. However, their goal is to protect the privacy of the query over data stream, not the data stream itself. The more related work is (27). It preserves the privacy of data streams by adding randomized noises. No join relationship is involved among streams. This approach cannot be applied to the Secure-JSC problem since the data obfuscation does not preserve the join relationship among streams. In addition, (4) claims its condensation approach can also be applied to the data stream problem because of its support on incremental updating. However, the anonymized data stream appears in a form of aggregate statistics instead of the original records, which makes infeasible to perform join relationships among anonymized streams.

20.3 Problem Statement

20.3.1 Secure Join Stream Classification

Consider n data streams S_1, ..., S_n, distributed among n sites. *Secure Join stream classification* refers to the problem where a classifier needs to be built such that (1) the training instances are defined by a sliding-window join over all data streams; (2) no site learns private information about other data streams. The sliding-window join over S_1, ..., S_n is specified by a join condition, a window specification and window update specification (6)(22).

In this paper, we consider a join condition in the form of a conjunction of equality predicates $S_i.A = S_j.B(i \neq j)$, where each of $S_i.A$ and $S_j.B$, called *join attributes*, represents one or more attributes from S_i and S_j. Since $S_i.A$ and $S_j.B$ are allowed to contain more than one attribute, we need to consider at most one predicate $S_i.A = S_j.B$ between each stream pair S_i and S_j. In the join graph, there is an edge between S_i and S_j if there is a predicate $S_i.A = S_j.B$ in the join condition. We consider join conditions for which the join graph is connected and contains no cycle. Many joins in practice are in fact acyclic, such as chain joins and star joins over the star/snowflake schemas (26).

The window and update specification can be time-based or tuple-based. Our method only depends on the set of tuples in the current window, not on how the window is specified and updated. The term "window" refers to the collection of current windows of all input streams. One of S_1, ..., S_n, called *target stream*, contains the class column. The task is to build a classifier each time the window updates. This means that the classifier must be rebuilt whenever the window on any input stream slides forward. The speed of fastest-sliding window determines the rate of classifier updates.

In the current window, the training set is the set of tuples defined by the sliding-window join. Importantly, the training set is *not* explicitly given, rather, is specified by the input streams and the sliding-window join. Some tuples in the input streams do not contribute any tuple in the join. Such tuples are *dangling*. We do not assume that dangling tuples are removed beforehand; in fact, the removal of dangling tuples is not straightforward due to the privacy requirement.

We assume all sites are assumed to be honest, curious, but not malicious (23). This means that a site may collect intermediate information received from other sites, but will follow the computation correctly. Our privacy model can be described by three types of attributes in each window:

- **Non-private class column**: the class column can be revealed to all sites. This assumption was made previously in (18). When all sites collaborate to build a classifier, we assume these sites are willing to share the information on class labels.

- **Semi-private join attributes**: for a join predicate $S_i.A = S_j.B$, the join attributes $S_i.A$ and $S_j.B$, are semi-private in that the sites of S_i and S_j are willing to share their join values that they both have, i.e., $S_i.A \cap S_j.B$, but not any other join values, i.e., $(S_i.A \cup S_j.B) - (S_i.A \cap S_j.B)$. This model was adopted in the literature for secure join and intersection in (2).

- **Private non-join attributes**: the values of all non-join attributes must not be revealed to any other sites.

In short, any join values known to the joining sites are not private, but everything else is.

20.3.2 Naive Bayesian Classifiers

Consider a single table T (X_1,...,X_n, Class). "Class" denotes the class column whose domain is a collection of class labels C_1, ..., C_m. X_i is a categorical variable. To classify a tuple $x = (x_1, ..., x_n)$, the Naive Bayesian Classifier (NBC) assigns x to the class C_i that maximizes the conditional class probability $P(C_i|x)$ based on the following maximum a posteriori (MAP) hypothesis:

$$argmax_{C_i \in Class} P(C_i|x) = argmax_{C_i \in Class} P(x|C_i)P(C_i) \qquad (3.1)$$

where $P(C_i)$ is the class probability and $P(x|C_i)$ is the conditional probability of x given the class label C_i. Under the independence assumption that variables X_1, ..., X_n are independent given the class label, NBC estimates $P(x|C_i)$ by $P(x|C_i) = \prod_{j=1}^{n} P(x_j|C_i)$. Once $P(x_j|C_i)$ and $P(C_i)$ are collected from the training data, NBC is able to assign a class label to a new tuple x. NBC requires the variables X_i to be categorical (having a small number of distinct values). Continuous attributes can be first discretized (such as equi-width or equi-depth binning) into a small number of intervals before applying NBC.

To compute $P(x_j|C_i)$ and $P(C_i)$, we only need to compute the class count matrix of the form $(x_k, < N_1, ..., N_m >)$ for each distinct value x_k of X_j, where N_j $(1 \leq j \leq m)$ is the number of tuples that has the value x_k and the class label C_j. This data structure has a size proportional to the number of distinct values in X_j.

The above discussion assumes a single table T. For Secure-JSC, T will be the join result of the input streams S_1, ..., S_n in the current window. Generating T would violate the privacy constraint. The challenge is to compute $P(x_j|C_i)$ and $P(C_i)$ on the joined table T without generating T. In the next section, we present such a method.

20.4 Our Approach

We assume that the current window of each input stream can fit in the local memory. The join relationship among streams forms an acyclic join graph, which is a rooted tree. Any stream may be regarded as the root. As our method involves propagation of information along the edges of the tree, we call this tree *propagation tree.*

Let us consider the site for an input stream S_i. Instead of generating the join stream, the site maintains an entry of $(Cls, Count)$ for each tuple t in the current window of S_i. Cls is a class vector in the form of $< N_1, ..., N_m >$ where m is the number of classes and N_i records the number of occurrences of t associated with the class label C_i in the never-generated join stream. $Count$ is the number of occurrences of t in the join stream. Intuitively, the entry $(Cls, Count)$ for t stores all information about t in the current window of the join stream. Thus, instead of keeping every join tuple involving t, we keep t only once and store its number of occurrences and class labels in those occurrences. The size of this data structure is proportional to the window size of S_i. Importantly, having Cls for each tuple t in the current window, the site of S_i is able to compute $P(x_j|C_i)$ and $P(C_i)$ for the all values x_j in the current window. The challenge is computing Cls without performing the join.

To compute the class vectors Cls, we propagate the "blow-up effect" of join. The propagation proceeds in two phases. In the phase of *bottom-up propagation*, Cls and $Count$ are propagated from the leaf nodes to the root. The propagation along an edge blows up Cls and $Count$ according to the join condition on the edge. The detail will be presented shortly. On reaching the root, the Cls for the root reflects the join of all input streams. Next, in the phase of *top-down propagation*, we propagate Cls from the root to all leaf nodes. When reaching all leaf nodes, Cls in each stream have reflected the join effect of all streams. The algorithm is distributed in that each node (site) in the tree performs the propagation as described; there is no central place to collect all data. This approach circumvents the computation of the sliding-window join, thus addresses both the privacy and efficiency requirements.

Now we explain the propagation at each site in details. First, we extend arithmetic operations to class vectors Cls: given an operator $\oplus$ and two Cls's $V_1 =< a_1, ..., a_m >$ and $V_2 =< b_1, ..., b_m >$, $V1 \oplus V2 =< a_1 \oplus b_1, ..., a_m \oplus b_m >$. For example, $< 4, 3 > / < 2, 3 >=< 2, 1 >$.

20.4.1 Initialization

Initially, for each tuple in the target stream, its Cls, $< N_1, ..., N_m >$, is determined as follows: N_i=1 if the class label is C_i or otherwise N_j=0. $Count$ is initialized to 1. For any tuple in any other stream, its Cls is initialized to all zeros $< 0, ..., 0 >$ and $Count$ is 1. This initialization does not require a

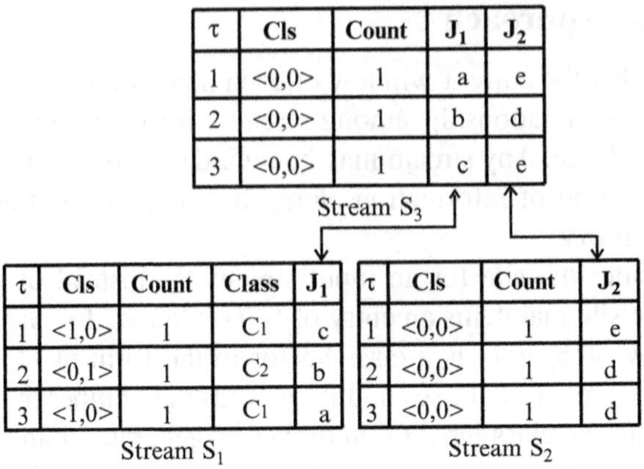
τ | Cls | Count | J1 | J2
1 | <0,0> | 1 | a | e
2 | <0,0> | 1 | b | d
3 | <0,0> | 1 | c | e
Stream S3
τ | Cls | Count | Class | J1
1 | <1,0> | 1 | C1 | c
2 | <0,1> | 1 | C2 | b
3 | <1,0> | 1 | C1 | a
Stream S1
τ | Cls | Count | J2
1 | <0,0> | 1 | e
2 | <0,0> | 1 | d
3 | <0,0> | 1 | d
Stream S2

Figure 20.3. Example with 3 streams at initialization

separate scan of streams and can be combined with the bottom- up propagation discussed in the following subsection.

Example 1. Consider an example with 3 streams with initial Cls and $Count$ shown in Figure 20.3. The join relationships are specified by the arrows: S_1 and S_3 join on J_1, and S_2 and S_3 join on J_2. S_1 is the target stream containing two classes. S_3 is the root of the propagation tree. The root can be arbitrarily selected. We will show later that choosing the input stream with largest window size as the root can optimize the cost of scan of input streams.

20.4.2 Bottom-Up Propagation

This is the phase where the information of Cls and $Count$ are propagated from leaf nodes to the root in a bottom-up order. Consider a parent node S_P and a child node S_C with the join predicate $S_P.J_1 = S_C.J_2$. The propagation from a child to the parent is based on the following observation.

Observation 1: Given a tuple t in S_P, if t joins with k tuples in S_C, t will occur k times in the join between S_P and S_C. These occurrences can be represented by blowing up Cls and $Count$ of t using the aggregated Cls and $Count$ of the k joining tuples in S_C. And if S_P has n child nodes ($n > 1$), the Cls and $Count$ of t in S_P will be blown up by all children to reflect the join with all children streams.

Following *Observation 1* precisely, we define the *blow-up summary* from S_C to S_P as the set $(v, ClsAgg, CountAgg)$. v is a distinct join value in S_C, $ClsAgg = \sum Cls$ and $CountAgg = \sum Count$, where $\sum$ is over all tuples in S_C containing the value v. Since the target stream can be anywhere in the tree, there are two cases in the bottom-up propagation from children to a parent node S_P:

- If the target stream is not in S_P's subtree, we blow up only Count at S_P since $ClsAgg$ is always zero for all child nodes of S_P (recall Cls is initialized to all-zero for a non-target stream);
- If the target stream is in S_P's subtree, exactly one of the child of S_P has non-zero $ClsAgg$ and we blow up both Cls and $Count$ at S_P.

The following lemma gives exactly the computation for blow-up following the above observation and discussion.

Lemma 1. Assume that a parent node S_P has n child nodes. For each tuple t in S_P with the join values $v_1,...,v_n$, where v_i is the join value between S_P and the ith child, let $(v_i, ClsAgg_i, CountAgg_i)$ denote the blow-up summary from ith child. Then

$$t.Count = \prod_{j=1}^{n} CountAgg_j \tag{4.1}$$

If some $ClsAgg_i (1 \leq i \leq n)$ is non-zero,

$$t.Cls = ClsAgg_i \times \prod_{j=1..n, j\neq i} CountAgg_j \tag{4.2}$$

To compute $Count$ and Cls at S_P, each child node S_C propagates its blow-up summary to the parent S_P. After receiving blow-up summaries from all child noes, S_P scans its tuples once and updates $Count$ and Cls of each tuple t as in Lemma 1. S_P also creates the blow-up summary from S_P to its own parent (if any) in the same scan.

Example 2. The bottom-up propagation for Example 1 is shown in Figure 20.4. S_1 and S_2 are scanned (locally) to produce blow-up summaries to propagate to S_3. On receiving the summaries, S_3 blows up Cls and $Count$ of its tuples. For example, consider the tuple t in S_3 as gray scaled in Figure 20.4 (with J_1=b, J_2=d). t has two corresponding summary entries: $(b, < 0, 1 >, 1)$ from S_1 and $(d, < 0, 0 >, 2)$ from S_2. $t.Count = 1 \times 2 = 2$, $t.Cls =< 0, 1 > \times 2 =< 0, 2 >$. These results indicate that t occurs in the join twice, both having the class label C_2, which is exactly the same information as in the join stream.

20.4.3 Top-Down Propagation

At the end of bottom-up propagation, Cls in the root stream reflects the join of all streams. However, Cls in other streams has not reflected the joins performed at their ancestors. Thus we need to propagate in the top-down fashion to push the correct join information to all non-root streams. The propagation is based on the following observation.

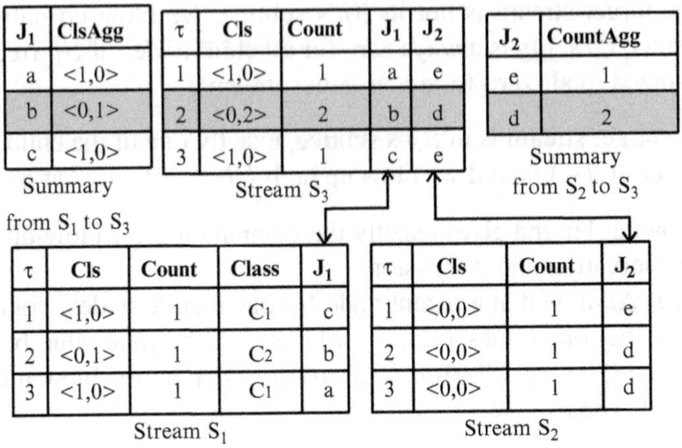

J1 ClsAgg
a <1,0>
b <0,1>
c <1,0>
Summary
from S1 to S3
τ Cls Count J1 J2
1 <1,0> 1 a e
2 <0,2> 2 b d
3 <1,0> 1 c e
Stream S3
J2 CountAgg
e 1
d 2
Summary
from S2 to S3
τ Cls Count Class J1
1 <1,0> 1 C1 c
2 <0,1> 1 C2 b
3 <1,0> 1 C1 a
Stream S1
τ Cls Count J2
1 <0,0> 1 e
2 <0,0> 1 d
3 <0,0> 1 d
Stream S2

Figure 20.4. After bottom-up propagations

Observation 2: For a parent node S_P and a child node S_C, if a tuple t in S_C joins with some tuple in S_P that has the join value v, so do all tuples in S_C that have this join value v. We can view all such tuples as an "equivalence class" on the join value v in S_C, denoted as $S_C[v]$. Similarly, $S_P[v]$ contains all tuples in S_P that have the join value v. Cls of the $S_C[v]$ tuples must be rescaled to reflect all joins not reflected so far at S_C. The rescaling must satisfy the following properties: (1) the relative share of any tuple in $S_C[v]$ remains unchanged because every tuple in $S_C[v]$ will join every tuple in $S_P[v]$, (2) the aggregated $\sum Cls$ in $S_C[v]$ after rescaling is the same as the aggregated $\sum Cls$ in $S_P[v]$.

To perform the top-down propagation, we define the *rescaling summary* from S_P to S_C as the set $(v, ClsAgg)$, where v is a join value in S_P and $ClsAgg$ is the aggregated class vector of all $S_P[v]$ tuples.

Lemma 2. Let t be a tuple in $S_C[v]$ and let $(v, ClsAgg)$ be a rescaling summary entry from S_P. $t.Cls$ is rescaled as follows:

$$t.Cls = ClsAgg \times \frac{t.Count}{S_C[v].CountAgg} \tag{4.3}$$

where $S_C[v].CountAgg$ is the aggregated $\sum Count$ over all $S_C[v]$ tuples.

The ratio $t.Count/S_C[v].CountAgg$ represents t's share in $S_C[v]$. To compute Cls at S_C, the parent node S_P propagates its rescaling summary to S_C. On receiving the rescaling summary from S_P, Cls in S_C are updated as in Lemma 2. In the same scan, the rescaling summary from S_C to its own children (if any) is computed.

Example 3. The top-down propagation is shown in Figure 20.5. At the root S_3, the rescaling summaries to S_1 and S_2 are generated while scanning

S_3 in the bottom- up propagation. On receiving these summaries, S_1 and S_2 rescale their Cls. For example, for the tuple t in S_1 as gray scaled in Figure 20.5, $t.Cls =< 0,1 >$ is rescaled to $< 0,2 > \times(1/1) =< 0,2 >$, where $(b, < 0,2 >)$ is the summary entry corresponding to b, and (1/1) is the share of t in its own equivalence class for J_1=b. The result captures exactly the same information about t as in the join stream: t occurs twice having the class label C_2.

Summary from S_3 to S_1

J_1	ClsAgg
a	<1,0>
b	<0,2>
c	<1,0>

Stream S_3

τ	Cls	Count	J_1	J_2
1	<1,0>	1	a	e
2	<0,2>	2	b	d
3	<1,0>	1	c	e

Summary from S_3 to S_2

J_2	ClsAgg
e	<2,0>
d	<0,2>

Stream S_1

τ	Cls	Count	Class	J_1
1	<1,0>	1	C1	c
2	<0,2>	2	C2	b
3	<1,0>	1	C1	a

Stream S_2

τ	Cls	Count	J_2
1	<2,0>	2	e
2	<0,1>	1	d
3	<0,1>	1	d

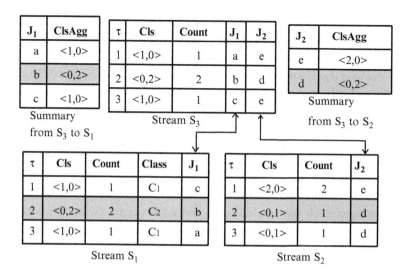

Figure 20.5. After top-down propagations

20.4.4 Using NBC

We now consider classifying a new instance $t =< t_1, ..., t_n >$, where t_j is the sub- record from S_j. At each site j for S_j, let $t_j =< x_1, ..., x_m >$. The site j computes $P(t_j|C_i) = \prod P(x_k|C_i)$ for k=1,...,m, and sends $P(t_j|C_i)$ to a coordinator, which could be any of the participating sites or a third party. After receiving this information from all sites, the coordinator computes $P(t|C_i) = \prod P(t_j|C_i) \times P(C_i)$ for j=1,...,n. The class label C_i that yields the maximum $P(t|C_i)$ is assigned to t. $P(C_i)$ is available to every participating site. No private information, as per our privacy model, is revealed by sending $P(t_j|C_i)$ to the coordinator because $P(t_j|C_i)$ is just a numerical value. If an attribute value x_i in a new instance t is not found in the training data, this value is simply ignored in the posterior computation.

20.4.5 Algorithm Analysis

Privacy. In the bottom-up and top-down propagation, only summaries are passed between parent/child pairs. For non-join attributes, no site transmits their values in any form to other sites. For the join attributes, consider a parent node S_P and a child node S_C with the join predicate $S_P.J_1 = S_C.J_2$. The blow-up summary from S_C to S_P contains entries of the form $(v, ClsAgg, CountAgg)$, where v is a join value in $S_C.J_2$ and $ClsAgg/CountAgg$ contains the class/count information. Since $ClsAgg$ and $CountAgg$ are the aggregate-level information and the class column is non-private, $ClsAgg/CountAgg$ does not pose a problem. $S_P.J_1$ and $S_C.J_2$ are semi-private, thus v can be exchanged between S_P and S_C if $v \in S_P.J_1 \cap S_C.J_2$. This can be ensured by first performing the secure intersection (2) to get $S_P.J_1 \cap S_C.J_2$. Then the blow-up summary from S_C to S_P needs to contain only entries for the join values in the intersection. As for the rescaling summary from S_P to S_C, no secure intersection is needed because all dangling tuples are removed at the end of bottom-up propagation.

Privacy Claim. (1) No private attribute values are transmitted out of its owner site. (2) Semi-private attribute values are transmitted between two joining sites only if they are shared by both sites.

Scalability. In the bottom-up and top-down propagation, one summary is passed between each parent/child pair and each stream (window) is scanned once. At any time, only the summaries for the edges being examined are kept in memory. The size of a summary is proportional to the number of distinct join values, not the number of join tuples. A summary lookup operation takes a constant time in an array or hash table implementation. Therefore, the whole propagation is linear in the window size, thus, independent of the join size. This property is important because the join size can be arbitrarily large compared with the window size, due to the many-to-many join relationships. An additional cost is secure intersection, which is performed once during bottom-up propagation. According to (2), it is loglinear to the number of distinct join values, again, independent of the join size.

Scalability Claim. On sliding each window, the cost of rebuilding NBC is proportional to the window size, not the join size.

The algorithm scans each input stream twice, once at the bottom-up propagation phase and once at the top-down propagation process. The only exception is the root stream, where the bottom-up and top-down propagations meet and two scans can be combined into one. Therefore, choosing the input stream of the largest window size (i.e., the most number of tuples) as the root will minimize the cost of scans, as it saves one scan on the largest stream window.

20.5 Empirical Studies

Our approach aims at two goals, namely, privacy preservation and fast processing of join stream classification. The privacy goal is delivered by limiting the information exchanged among sites, as claimed in Section 4. Therefore, in this section we focus on the performance goal. We would like to answer two questions: (1) whether the formulation of Secure-JSC defines a better training space compared with a single stream alone; (2) whether our algorithm scales up to handle high-speed data streams.

We denote our algorithm as *NB_Join*, as it builds a NBC classifier whose training set is defined on the join of multiple streams. We compared it with following alternatives:

- *NB_Target*: NBC based on the target stream alone. In this case, all non-target streams are ignored.
- *DT_Join*: the decision tree classifier (C4.5) on the join stream. To build the decision tree, the join stream is first computed by actually joining the input streams. Note that his approach does not meet the privacy requirement.
- *DT_Target*: the decision tree classifier on the target stream alone.

For each window, we train the classifier using the first 80% of stream tuples within this window and evaluate the classifier using the remaining 20% of stream tuples in the same window. The testing data are generated by the join of the testing samples from all streams.

We measure performance by "time per input tuple", i.e., time spent on each window divided by the number of input tuples in the window. The "input tuples" refers to the tuples in the input streams, not the join stream. This measure gives an idea about the data arrival rate that an algorithm is able to handle. For *DT_Join*, because it has to generate the join stream before building the classifier, we measure the join time only and ignore the classifier construction time since the join time dominates. Most of sliding-window join algorithms in literature are not suitable for generating the join stream for *DT_Join* because they focus on fast computing special aggregates (15)(21), or producing approximate join results (32) under resource constraints; not the *exact* join result. We implemented the nested loop join algorithm. This choice should not have a major effect because all tuples in the current window are in memory. All programs were coded in C++ and run on a PC with 2GHz CPU, 512M memory and Windows XP.

20.5.1 Real-life Datasets

For experiments on real-life dataset, we obtained UK road accident data from the UK data archive[1]. It contains information about accidents, vehicles and casualties, in order to monitor road safety and determine policies to reduce the road accident casualty toll. There are three tables: "Accident", "Vehicle" and "Casualty". The characteristics of year-2001 data are shown in Figure 20.6 where arrows indicate join relationships: each accident involves one or more vehicles; each vehicle has zero or more casualties. Each table can be regarded as a stream that is timestamped by "date of accident". On average, about 600 "Accident" tuples, 700 "Vehicle" tuples and 850 "Casualty" tuples are added every day. The join stream is specified by the equalities between all common attributes among the three input streams. "Casualty" is the target stream with two casualty classes — class 1: "fatal/serious" (13% of all tuples) and class 2: "slight" (87% of tuples).

Classification Accuracy. Figure 20.7 shows the accuracy of all classifiers being compared. For all methods, the window size is the same and ranges from 10 to 50 days with no window overlapping.

It is apparent that classifiers built on multiple streams are much more accurate. This result confirms that examining correlated streams is advantageous compared with building the classifier on a single stream. In fact, the accuracy obtained by examining the target stream alone is only about 80%, even lower than that obtained by a naive classifier which simply classifies every tuple as belonging to class 2, since 87% of tuples belong to this class.

On the other hand, the results also show that, with the same training set, naive Bayesian classifier has a performance comparable to that of the decision tree. Keep in mind that our method *NB_Join* runs directly on the input streams,

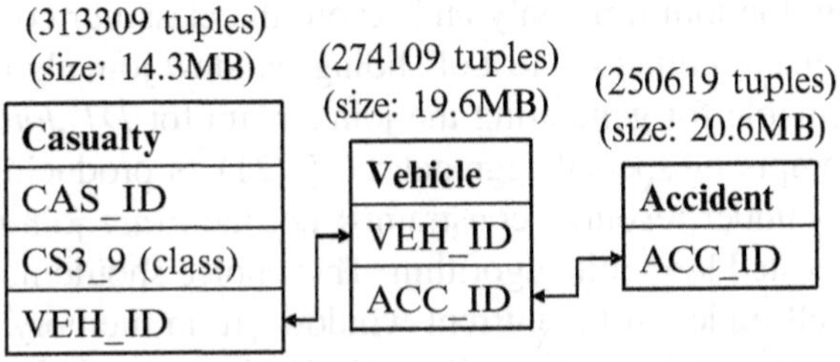

Figure 20.6. UK road accident data (2001)

[1]http://www.data-archive.ac.uk/

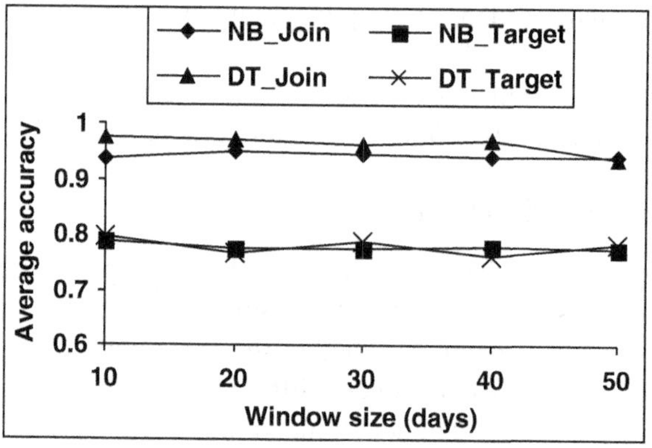

Figure 20.7. Classifier accuracy

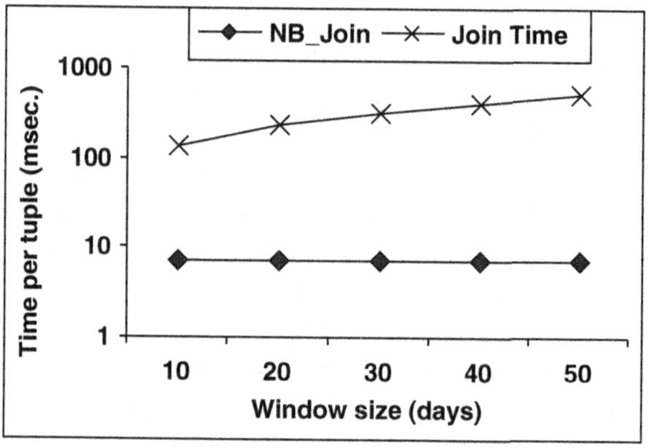

Figure 20.8. Time per input tuple

while the decision tree is built on the join stream. The latter does not meet the privacy requirement and has a high join cost. We will examine the efficiency of these two methods in the next set of experiments.

Time per input tuple. Figure 20.8 compares the time per input tuple. For example, at the window size of 20 days, the join takes about 9.83 seconds whereas *NB_Join* takes only about 0.3 seconds. Therefore, the join time per input tuple is $9.83 \times 106/43{,}900 = 224$ microseconds, where 43,900 is the total number of tuples that arrived in the 20- day window. In contrast, *NB_Join* takes only $0.3 \times 106/43{,}900 = 6.8$ microseconds per input tuple. This means that any method that requires computing the join will be at least 33 times slower than *NB_Join*. As the window size increases, the join time increases quickly due to the increased join cardinality in a larger window; whereas the time per input tuple for *NB_Join* is almost constant. In other words, our approach is linear in the window size, independent of the join stream size. This property makes our approach particularly suitable for multiple correlated streams.

Therefore, though both *NB_Join* and *DT_Join* classifiers exhibit a similar classification accuracy, *NB_Join* is much more efficient than *DT_Join*.

20.5.2 Synthetic Datasets

To further verify our claims, we also conducted experiments on synthetic datasets with various data characteristics. Similar to the experiments on real-life datasets, we want to examine whether the correlation of multiple streams yields benefits for classification under different data characteristics. We also want to evaluate if *NB_Join* can deal with streams with high data arrival rates. As we are not aware of existing data generators to evaluate classification spanning correlated streams, we designed our own data generator.

The data generator. We consider the chain join of k streams S_1, ..., S_k, where S_1 is the target stream. An adjacent pair S_i and S_{i+1} have one join predicate and a non-adjacent pair have no join predicate. All streams have the same number of tuples denoted $|S|$. All join attributes are categorical and have the same domain size D. In addition, all streams have N ranked attributes and N categorical attributes (excluding the join attributes and the class attribute). Categorical values are drawn randomly from a domain of size 20. All ranked attributes have the ranked domain 1,...,10.

Since our goal is to verify that the classifier built on the join stream is more accurate when there are correlations among streams, the dataset must contain certain "structures" for the class label rather than random tuples. We construct the dataset in which the class label in a join tuple is determined by whether at least q percentage of the ranked attributes have a "high" value. A ranked value is "high" if it belongs to the top half of its ranked domain. Since the ranked attributes are distributed among multiple input streams, to ensure the desired property of the class label, the input streams S_1,...,S_k are constructed as follows.

- *Join values.* Each stream S_i consists of D groups: from 1st to Dth group. All tuples in the jth ($1 \leq j \leq D$) group of S_i join with all tuples in the jth group of S_{i+1}, but not any other tuples. The *jth join group* refers to the set of join tuples produced by the jth groups. The size Z_j of the jth group is the same for all streams S_1,...,S_k, and follows Poisson distribution with the mean $\lambda = |S|/D$. The jth join group has the size Z_j^k, with λ^k being the mean. The *blow-up ratio* of the join is defined as $\lambda^k/\lambda = \lambda^{k-1}$, i.e., the ratio between the mean of group size on the join stream and that on input streams.

- *Ranked values.* We generate ranked attributes such that all join tuples in the jth join group have the same class label. In particular, we ensure that all join tuples in the same group have "high" values in the *same* number of ranked attributes, say hj. To this end, we distribute the number hj among S_1,...,S_k randomly, say hj_1,...,hj_k, such that $hj = hj_1 + ... + hj_k$, and all tuples in the jth group for S_i are "high" in hj_i ranked attributes. hj follows uniform distribution in the range $[0, k \times N]$, where $k \times N$ is the total number of ranked attributes.

- *Class labels.* If $hj \geq q \times k \times N$, for some percentage parameter q, we assign the "Yes" class label to every tuple in the jth group of S_1, otherwise, assign the "No" class label.

Finally, to simulate the "concept drifting" in data streams, we change the parameter q every time after generating W tuples. In particular, for every W tuples we randomly determine a q value in the range [0.25, 0.75) following the uniform distribution. W is called the *concept drifting interval.* Usually W is larger than the window size because not every window leads to a change in classification. A dataset generated as above can be characterized by the parameters $(N, |S|, D, \lambda, W)$, where $\lambda = |S|/D$ is the mean of group size and determines the blow-up ratio of join.

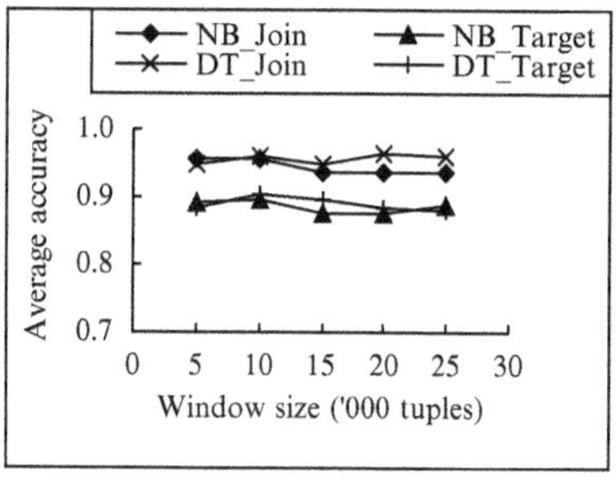

Figure 20.9. Classifier accuracy vs. window size

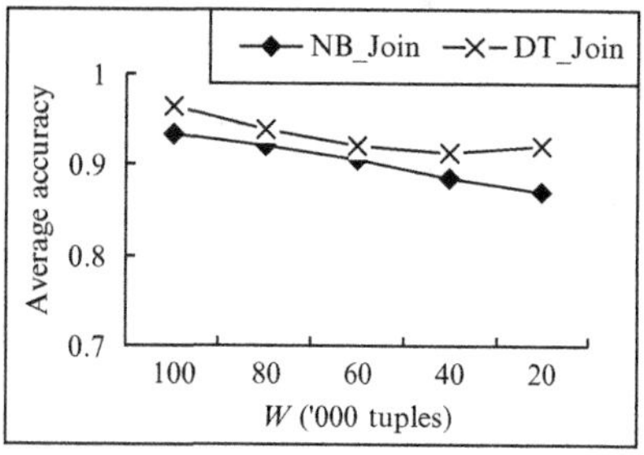

Figure 20.10. Classifier accuracy vs. concept drifting interval

Accuracy. We generated three streams S_1, S_2 and S_3 with the parameter setting N=10, $|S|$=1,000,000, D=200,000, λ=5, W=100,000. Figure 20.9 shows the accuracy vs the window size with 50% window overlapping. *DT_Join* and *NB_Join* are more accurate than their counterparts on the single stream, while both having similar accuracies.

Figure 20.10 shows another experiment, where we fixed the window size w at 20,000 and decreased W from 100,000 to 20,000. Since the previous experiments have confirmed that classifiers built on the join stream have a better accuracy, in this experiment we only show the accuracy of *NB_Join* and *DT_Join*. As expected, the accuracy drops slowly as W decreases, since the structure for the class label changes more frequently.

Time per input tuple. Figure 20.11 shows the time per tuple on the same dataset as in Figure 20.9. The join time is much larger than the time of *NB_Join*. As the window size increases, the join time increases due to the blow-up effect of join, while *NB_Join* spends almost constant time per tuple for any window size.

Figure 20.12 shows time per tuple vs. blow-up ratio of join. The parameters are fixed as N=10, $|S|$=1,000,000, D=200,000, W=100,000. For the join of three streams, the blow-up ratio is λ^2. By varying λ from 2 to 7, the blow-up ratio varies from 4 to 49. The window size is fixed at 20,000. Again, *NB_Join* shows a much better performance and is flat with respect to the blow-up of join. This is because it scans the window exactly twice, independent of the blow-up ratio of the join. On the other hand, the join takes more time per tuple with a larger blow-up ratio because much more tuples are generated.

Figure 20.13 shows time per tuple vs. number of streams. All parameters are still the same as in Figure 20.11. The window size is fixed at 20,000 tuples. We vary the number of steams from 1 to 5. The blow-up ratio for k-stream join is determined by 5^{k-1}. The comparison of the results is similar to Figure 20.12.

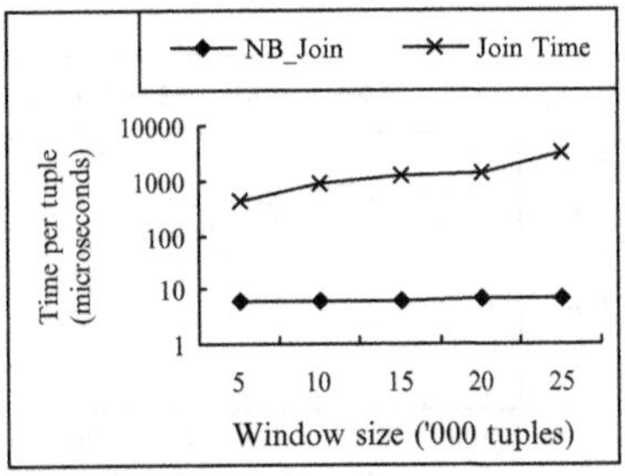

Figure 20.11. Time per input tuple vs. window size

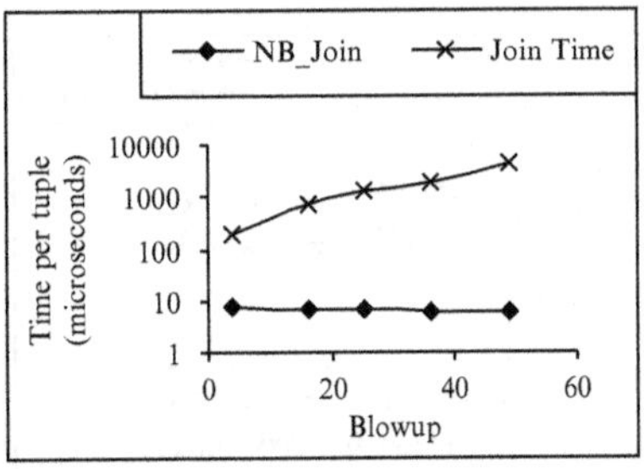

Figure 20.12. Time per input tuple vs. blow-up ratio

20.5.3 Discussion

On both real life and synthetic datasets, our empirical studies showed that when the features for classification are contained in several related streams, the proposed join stream classification has significant accuracy advantage over the conventional method of examining only the target stream.

The main challenge is how such classification can be performed in pace with the high-speed input streams, given that the join stream has an even higher data

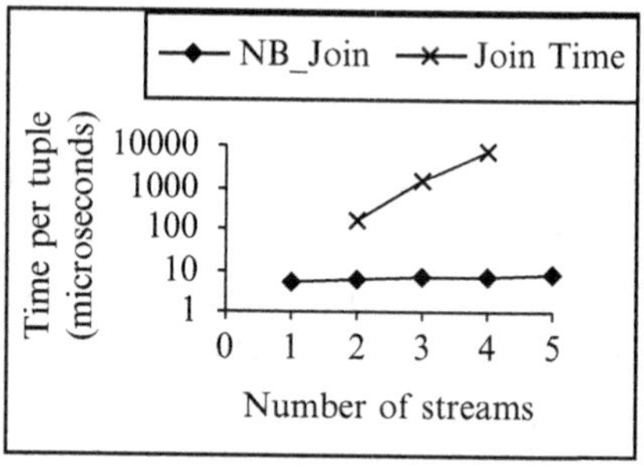

Figure 20.13. Time per input tuple vs. number of streams

arrival rate than that of the input streams. To this end, our experiments showed that our proposed algorithm has a cost linear in the size of input streams, independent of the join size. This feature makes our algorithm superior to other alternative methods.

It is worthy of noting that the classifier must be rebuilt each time the window on any input stream slides forward. This is reasonable when there is no overlap or only small overlaps between windows. However, when windows are significantly overlapped, this strategy tends to repeat the work on the overlapped data. In this case, a more efficient strategy may be incrementally updating the NBC by working only on the difference due to the window sliding. We did not pursue in this direction further because even overlapped tuples still need to be joined with *new* tuples in other streams, which means that the scan of overlapped tuples cannot be avoided. Since our algorithm scans the current window only twice, the benefit of being incremental is limited, especially considering the overhead added.

20.6 Conclusions

Motivated by real life applications, we considered the classification problem where the training data are several private data streams. Joining all streams violates the privacy constraint of such applications and suffers from the blow- up of join. We presented a solution based on Naive Bayesian Classifiers. The main idea is rapidly obtaining the essential join statistics without actually computing the join. With this technique, we can build exactly the same Naive Bayesian Classifier as using the join stream without exchanging private information. The processing cost is linear in the size of input streams and independent of the join size. Empirical studies supported our claim that examining several related streams indeed benefits the quality of classification. Having a much lower processing time per input tuple, the proposed method is able to handle much higher data arrival rate and deal with the general many-to-many join relationships of data streams.

References

[1] C. Aggarwal, J. Han, J. Wang, and P. Yu. (2006). A Framework for On-Demand Classification of Evolving Data Streams. IEEE TKDE, Vol. 18, No. 5, Page:577-589.

[2] R. Agrawal, A. Evfimievski and R. Srikant. (2003). Information sharing across private databases. In Proc. SIGMOD.

[3] R. Agrawal, and R. Srikant. (2000). Privacy-preserving data mining. In Proc. SIGMOD.

[4] C. Agarwal and P. Yu. (2004). A condensation Approach to Privacy Preserving Data Mining. In Proc. EDBT.

[5] Noga Alon, Phillip B. Gibbons, Yossi Matias, and Mario Szegedy. (1999). Tracking Join and Self-Join Sizes in Limited Storage. In ACM PODS.

[6] B. Babcock, S. Babu, M. Datar, R. Motwani, J. Widom. Model and issues in data stream systems. (2002). In ACM PODS, Madison, Wisconsin.

[7] J. Beringer and E. Hullermeier. (2005). Online clustering of parallel data streams. In press for Data & Knowledge Engineering.

[8] J. Bethencourt, D. Song, and B. Waters. (2006). Constructions and Practical Applications for Private Stream Searching. In IEEE Symposium on Security and Privacy.

[9] Y. D. Cai, D. Clutter, G. Pape, J. Han, M. Welge and L. Auvil. (2004). MAIDS: Mining alarming incidents from data streams. In Proc. SIGMOD, demonstration paper.

[10] D. Carney, U. Cetintemel, M. Cherniack, C. Convey, S. Lee, G. Seidman, M. Stonebraker, N. Tatbul, and S. Zdonik. (2002). Monitoring streams - a new class of data management applications. In Proc. VLDB.

[11] S. Chaudhuri, R. Motwani, and V. R. Narasayya. (1999). On random sampling over joins. In Proc. SIGMOD.

[12] K. Chen and L. Liu. (2005). Privacy preserving data classification with rotation perturbation. In ICDM.

[13] G. Chen, X. Wu, X. Zhu. (2005). Sequential pattern mining in multiple streams, In Proc. ICDM.

[14] A. Das, J. Gehrke and M.Riedewald. (2003). Approximate join processing over data streams. In Proc. SIGMOD, Madison, Wisconsin.

[15] A. Dobra, M. Garofalakis, J. Gehrke, and R. Rastogi. (2002). Processing complex aggregate queries over data streams. In Proc. SIGMOD, Madison, Wisconsin.

[16] P. Domingos and G. Hulten. (2000). Mining high-speed data streams. In Proc. SIGKDD.

[17] Pedro Domingos and Michael Pazzani. (1997). On the optimality of the simple Bayesian classifier under zero-one loss. Machine Learning, 29:103-130.

[18] W. Du and Z. Zhan. (2002). Building decision tree classifier on private data. ICDM Workshop on Privacy, Security and Data Mining.

[19] R. O. Duda and P. E. Hart. (1973). Pattern classification and scene analysis. New York: John Wiley & Sons.

[20] J. Gama, R. Racha, P.Medas. (2003). Accurate decision trees for mining high-speed data streams. In Proc. SIGKDD.

[21] S. Ganguly, M. Garofalakis, A. Kumar and R. Rastogj. (2005). Join-distinct aggregate estimation over update streams. In Proc. ACM PODS, Baltimore, Maryland.

[22] L. Golab and M. Tamer Ozsu. (2003) Processing sliding window multi-joins in continuous queries over data streams. In Proc. VLDB.

[23] O. Goldreich. (2001). Secure multi-party computation. Working Draft, Version 1.3.

[24] S. Guha, N. Mishra, R. Motwani, and L. O'Callaghan. (2000). Clustering data streams. In FOCS.

[25] D. J. Hand and K. Yu. (2001). Idiot's Bayes - not so stupid after all? International Statistical Review. 69(3), 385-399.

[26] M. Levene and G. Loizou. (2003). Why is the snowflake schema a good data warehouse design? Information Systems 28(3).

[27] F. Li, J. Sun, S. Papadimitriou, G. Mihala and I. Stanoi. (2007). Hiding in the Crowd: Privacy Preservation on Evolving Streams through Correlation Tracking. In Proc. ICDE.

[28] Y. Lindell and B. Pinkas. (2000). Privacy preserving data mining. In Proc. CRYPTO.

[29] A. Machanavajjhala, J. Gehrke, D. Kifer, and M. Venkitasubramaniam. (2006). l-Diversity: Privacy beyond k-anonymity. ICDE.

[30] R. Ostrovsky and W. Skeith. (2005). Private Searching on Streaming Data. In CRYPTO.

[31] Irina Rish. (2001). An empirical study of the naive Bayes classifier. IJCAI 2001 Workshop on Empirical Methods in Artificial Intelligence.

[32] U. Srivastava, J. Widom. (2004). Memory-limited execution of windowed stream joins. In Proc. VLDB.

[33] L. Sweeney. (2002). k-Anonymity: A Model for Protecting Privacy, International Journal on Uncertainty, Fuzziness and Knowledge-based Systems, 10(5).

[34] J. Vaidya and C. W. Clifton. (2002). Privacy preserving association rule mining in vertically partitioned data. In SIGKDD.

[35] H. Wang, W. Fan, P. Yu and J. Han. (2003). Mining concept-drifting data streams using ensemble classifiers. In Proc. SIGKDD.

[36] K. Wang, Y. Xu, R. She, P. Yu. (2006). Classification Spanning Private Databases. AAAI.

[37] Y. Zhu and D. Shasha. (2002). Statstream: Statistical monitoring of thousands of data streams in real time. In Proc. VLDB.

Index

R-U Confidentiality Maps, 71
μ-Argus Package, 60
k-Anonymity, 20, 71
k-Optimize Algorithm, 21
k-Same, 41
k-anonymity, 105
k-randomization, 448
l-diversity method, 26, 90
m-Invariance, 26
t-closeness Model, 27
t-closeness model, 91
Anatomy, 212
Anonymize-and-Mine, 123
DET-GD, 261
FRAPP Framework, 250, 251
Greedy-Personalized-Generalization Algorithm, 474
IGA, 273
Incognito Algorithm, 21
MASK, 242, 246
MaxFIA, 273
MinFIA, 273
Mine-and-Anonymize, 126
Priority-based Distortion Algorithm, 275
RAN-GD, 261
SA-Generalization Algorithm, 478

Adversarial Attacks on Multiplicative Perturbation, 153
Adversarial Attacks on Randomization, 149
Algebraic Distortion for Associations, 248
Anonymity via Isolation, 97
Anonymized Marginals, 229
Anonymizing Inferences, 92
Application of Privacy, 38
Applications of Randomization, 144
Approximation Algorithms for k-Anonymity, 23, 115
Association Rule Hiding, 33, 249, 267
Association Rule Hiding: Contingency Tables, 291
Association Rule Hiding: Statistical Community Perspective, 291
Association Rule Mining with Vertical Partitioning, 347
Attack on Orthogonal Perturbation Matrix, 370
Attack on Random Perturbation Matrix, 371
Attack-resilient Rotational Perturbations, 177
Attacks on Multiplicative Data Perturbation, 369
Attacks on Perturbation-based Privacy, 360
Auditing Aggregate Queries, 416

Background Knowledge Attack, 26
Bayes Reconstruction Method, 139
Binary Data Anonymization, 27
Bioterrorism Applications of Privacy, 40
Blocking Schemes for Association Rule Hiding, 276
Bootstrapping for Synthetic Data, 66
Border-based Approaches, 277
Bottom Up Generalization for k-Anonymity, 22
Breach Probability for Personalized Anonymity, 472

Calibrating Noise for Query Privacy, 394
Cholesky Decomposition for Generating Private Synthetic Data, 67
Classification from Randomized Data, 144
Classification of Privacy Methods, 84
Classification Rule Hiding, 279
Classifier Downgrading for Privacy, 34
CleanGene, 42
Clustering as k-anonymity, 92
Clustering with Vertical Partitioning, 346
Collaborative Filtering with Randomization, 145
Combinatorial Attack on Privacy, 467
Commutative Encryption, 348
Computational Measures of Anonymity, 94
Condensation, 65
Condensation for Personalized Anonymity, 480
Condensation Methods for k-Anonymity, 22
Constraint Satisfaction Problem, 278
Contingency Tables, 291
Correlated Noise for Multi-dimensional Randomization, 18

Correlated Noise for Randomization, 152
Credential Validation Problem, 40
Cryptographic Methods for Privacy, 28, 85, 313, 337
Curse of Dimensionality and Privacy, 37
Curse of Dimensionality for Privacy, 433

Data Stream Privacy, 487
Data Swapping, 19, 61
Database Inference Control, 269
Datafly System, 39
Decision Trees and k-Anonymity, 130
Differential Entropy as Privacy Measure, 147
Dimensionality Curse for k-anonymity, 435
Dimensionality Curse for l-diversity, 458
Dimensionality Curse for Condensation, 441
Dimensionality Curse for Privacy, 433
Dimensionality Curse for Randomization, 446
Disclosure Limitation: Contingency Tables, 295
Distortion for Association Rule Hiding, 272
Distributed k-Anonymity, 32
Distributed Privacy, 28
Distribution Analysis Attack, 367
Document Indexing on a Network, 31
Dot Product Protocol, 319

EM Reconstruction Method, 141
Entropy for Privacy Quantification, 188
Essential QI-group, 467
External Individual Set, 467

Frequent Itemset Hiding, 267
Full Disclosure in Query Auditing, 417

General Perturbation Schemes, 90
Generalization, 108
Generalization Algorithm for Personalized Anonymity, 473
Generalization Lattices for Genomic Privacy, 42
Genomic Privacy, 42
Global Recoding, 64
Greedy Algorithm for Personalized Anonymity, 474
Greedy Framework for Personalized Anonymity, 474
Group Based Anonymization, 20

Hiding Sensitive Sequences, 281
Histograms for Privacy-Preserving Query Answering, 36
Homeland Security Applications of Privacy, 40
Homogeneity Attack, 26
Homomorphic Encryption, 313, 318
Horizontally Partitioned Data, 30, 313
Hybrid Microdata Approach, 67

ICA-based Attacks, 374
ID3 Decision Tree Mining, 323
Identity Theft, 41
Inference Control, 34, 54
Information Transfer Based Measures of Anonymity, 95
Input-Output Attack on Multiplicative Perturbation, 153

Knowledge Hiding, 267
Known I/O Attacks, 370
Known Sample Attack on Multiplicative Perturbation, 153

Latin Hypercube Sampling, 66
Local Recoding for Utility-Based Anonymization, 214
Local Suppression, 65

Malicious Adversaries, 29
Malicious Adversary, 316
MAP Estimation Attack, 366
Market Basket Data Anonymization with Sketches, 27
Market Basket Data Randomization, 153
Maximum Likelihood Attack on Randomization, 151
Metrics for Privacy Quantification, 184
Microaggregation, 59
Multiple Imputation for Synthetic Data, 65
Multiplicative Perturbation, 158
Multiplicative Perturbations for Randomization, 152
Multiplicity-based Anonymity, 86
Mutual Information as Privacy Measure, 147

Non-perturbative Masking Methods, 63

Oblivious Evaluation of Polynomials, 320
Oblivious Transfer Protocol, 29
Offline Auditor Examples, 417
Offline Query Auditing, 417
Online Query Auditing, 418
Optimal SA-generalization, 476
Optimality of Randomization Models, 143
Outlier Detection over Vertically Partitioned Data, 349

Partial Disclosure in Query Auditing, 421
PCA Attack of Randomization, 149
PCA Filtering-based Privacy Attacks, 365
Personalized Anonymity, 463
Personalized Location Privacy Protection, 480
Personalized Privacy-Preservation, 24, 461
PRAM: Post-Randomization Method, 62
Privacy-Preservation of Application Results, 32
Privacy-Preserving Association Rule Mining, 239
Privacy-Preserving OLAP, 145

Probabilistic Measures of Anonymity, 87
Progressive Disclosure Algorithm, 224
Projection Perturbation, 162
Public Information for Attacking Randomization, 149
Public Information-based Attacks on Randomization, 446

Quantification of Privacy, 15, 146
Quantifying Hidden Failure, 192
Query Auditing, 34, 85, 415
Query Privacy via Output Perturbation, 383

Random Projection for Privacy, 19
Random Projection for Randomization, 152
Randomization, 13, 137
Randomization: Correlated Noise Addition, 58
Randomization: Uncorrelated Noise Addition, 58
Randomized Auditors, 423
Rank Swapping, 61
Reconstruction from Randomization, 139
Retention Replacement Perturbation, 145
Rotational Perturbation, 158

SA-generalization, 466
Samarati's Algorithms for k-Anonymity, 111
Sampling for Privacy, 64
Sanitizers for Complex Functions, 400
Scrub System, 39
Secure Association Rule Mining, 324
Secure Bayes Classification, 324
Secure Comparison Protocol, 319
Secure Computation of $\ln(x)$, 320
Secure Intersection, 321
Secure Join Stream Classification, 493
Secure Multi-party Computation, 29, 85
Secure Set Union, 322
Secure Sum Protocol, 318
Secure SVM Classification, 325
Selection of Privacy Metric, 201
Semi-honest Adversaries, 29
Semi-honest Adversary, 316
Sensitive Attribute Generalization, 466
Sequential Releases for Anonymization, 25
Simulatable Auditing, 420
Sketches for Query Auditing, 37
Sketches for Randomization, 153
Sparse Binary Data Randomization, 153
Spectral Filtering based Privacy Attacks, 363
Statistical Approaches to Association Rule Hiding, 291
Statistical Disclosure Control, 54
Statistical Measures of Anonymity, 85
Stream Classification with Privacy-Preservation, 490
String Anonymization, 27
Suppression, 109
SVD Filtering-based Privacy Attacks, 364
Synthetic Microdata Generation, 65

Text Anonymization with Sketches, 27
Text Data Anonymization, 27
Text Randomization, 153
Threshold Homomorphic Encryption, 327
Time Series Data Stream Randomization, 151
Top and Bottom Coding, 65
Top Down Specialization for k-Anonymity, 22
Tradeoff: Privacy and Information Loss, 69
Trail Re-identification for Genomic Privacy, 42
Transformation Invariant Models, 166

Utility Based Privacy Preservation, 24, 93, 208
Utility Measures, 93, 212

Variance-based Anonymity, 85
Vertically Partitioned Data, 31, 337
Video Surveillance, 41

Watch List Problem, 41
Web Camera Surveillance, 41
Workload Sensitive Anonymization, 25

Yao's Millionaire Problem, 85, 319